17th IUMRS International Conference in Asia (IUMRS-ICA 2016)

IOP Conference Series: Materials Science and Engineering
Volume 182

Qingdao, China
20 – 24 October 2016

Editor:

Yafang Han

ISBN: 978-1-5108-3770-6
ISSN: 1757-8981

Printed from e-media with permission by:

Curran Associates, Inc.
57 Morehouse Lane
Red Hook, NY 12571

Some format issues inherent in the e-media version may also appear in this print version.

Printed by Curran Associates, Inc. (2017)

For permission requests, please contact the Institute of Physics at the address below.

Institute of Physics
Dirac House, Temple Back
Bristol BS1 6BE UK

Phone: 44 1 17 929 7481
Fax: 44 1 17 920 0979

techtracking@iop.org

Additional copies of this publication are available from:

Curran Associates, Inc.
57 Morehouse Lane
Red Hook, NY 12571 USA
Phone: 845-758-0400
Fax: 845-758-2633
Email: curran@proceedings.com
Web: www.proceedings.com

Table of contents

Volume 182

17th IUMRS International Conference in Asia (IUMRS-ICA 2016)
20–24 October 2016, Qingdao, China

Accepted papers received: 28 February 2017
Published online: 10 March 2017

Functional Materials

Biological Materials

IOP Conf. Series: Materials Science and Engineering **182** (2017) 011001 doi:10.1088/1757-899X/182/1/011001

Preface

This is the proceedings of the selected papers presented at IUMRS International Conference in Asia (IUMRS-ICA) held in Qingdao, China, Oct. 20-24, 2016. IUMRS-ICA is one of the most important series conferences of International Union of Materials Research Societies (IUMRS). ICA is held once a year and organized by IUMRS Asia members in turn. Authorized by IUMRS, IUMRS-ICA 2016 was organized by Chinese Materials Research Society (C-MRS).

IUMRS-ICA2016 included 27 symposia covering the fields of: Energy and environment materials, Advanced structural materials, Functional materials, Biological materials, Materials modeling, simulation and characterization. About 2700 participants attended the conference, and the organizers received more than 80 technical papers for this proceedings. By recommendation of symposium organizers and after peer reviewing 54 papers are published in the present volume, which are selected from Symposium A: Solar Energy Materials and Solar Cells, Symposium C: Photocatalytic Materials for Energy Production, Symposium D: Energy Conversion and Storage Materials, Symposium E: Nuclear Materials, Symposium L: Nanomaterials and Energy Devices, Symposium M: Advanced Carbon Materials, Symposium P: The Green and Low-carbon Emission Building Materials, and Symposium AA: Biomaterials.

The editors would like to give the thanks to the symposium chairs and all the paper reviewers of this volume, especially to Prof. Yafang Han, Prof. Xinqing Zhao and Dr. Fenfen Liang for their English polishing of the manuscripts. The thanks should be also given to Mrs. Yiming Jia and Mr. Yunfei Liang for their hard work of compilation and typesetting for each paper in this volume.

Chief Editor

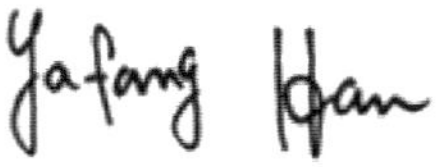

Peer review statement

All papers published in this volume of *IOP Conference Series: Materials Science and Engineering* have been peer reviewed through processes administered by the proceedings Editors. Reviews were conducted by expert referees to the professional and scientific standards expected of a proceedings journal published by IOP Publishing.

Effect of fluorine doped TiO_2 on the property of perovskite solar cell

X Q Zhang[1], Y P Wu[1], Y Huang[1, 2], Z H Zhou[1, 2,*] and S Shen[3]

[1] College of Materials, Xiamen University, Xiamen, Fujian, China 361005
[2] Fujian Key Laboratory of Advanced Materials, Xiamen, Fujian, China 361005
[3] CSIRO Manufacturing, Clayton, VIC 3168, Australia

E-mail: *zzh@xmu.edu.cn

Abstract. Anatase TiO_2 nanoparticles with different amounts of fluorine doping were synthesized by a hydrothermal method using hydrogen titanate nanotubes as a precursor and applied as mesoporous layer for preparing perovskite solar cell.The morphology and structures were characterized by scanning electron microscope (SEM) and X-ray diffraction (XRD), meanwhile, the properties and performances were tested by photoluminescence spectrum (PL) and current density and voltage (J-V) curve. It was found that doping fluorine into TiO_2 made the photoelectric conversion efficiency (PCE) of perovskite solar cell (PSC) to be improved. The best PCE of PSC based on a F-doped TiO_2 was 13.06% and increased by 51% compared to an un-doped TiO_2. The study provided a direction for the exploration of high performance electron transport layer of perovskite solar cell.

Keywords: perovskite solar cell, TiO_2 mesoporous layer, fluorine doping

1. Introduction

Perovskite solar cell (PSC) is one of the most promising next-generation solar cells. Due to the energy band structure matching between TiO_2 and perovskite, photoelectrons can be effectively extracted and transported by the TiO_2 mesoporous layer into PSC. Therefore, modification of TiO_2 mesoporous material is an important approach to control the conversion efficiency of PSC [1-4].

It is well known that element doping is an effective way for TiO_2 modification. Yoshida etc. [5] fabricated Mg-doped anatase TiO_2 nanorods with a microwave hydrothermal method and applied them as the PSC mesoporous layer. They explained that magnesium doping widened the TiO_2 band gap, and improved the degree of the energy band structure matching between TiO_2 and perovskite, which promoted the electron transfer ability of TiO_2, and increased the PSC conversion efficiency by 33%. Other metal element doping, such as Nb-TiO_2 [6], Y-TiO_2 [7], Sn-TiO_2 [8] etc., have also been successively reported. However, the above literatures rarely mentioned the issue of oxygen vacancy, which leads to the existence of "trap state" between the band gaps of TiO_2. It is known that the "trap state" caused by oxygen vacancy in TiO_2 is considered as the disadvantage of charge transfer because of the photoelectron capture for the TiO_2 mesoporous layer in PSC [9-11]. Hence, studying the issue of oxygen vacancy is significant work for the development of PSC.

We have reported the preparation of F-doped TiO_2 and its application for PSC, and found that the oxygen vacancy in TiO_2 could be greatly reduced by the fluorine doping process [12]. It is deduced that one O^{2-} ion needs to be replaced by two F^- ions, and one F^- ion takes O^{2-} position but another F^- ion may be possible to fill into the oxygen vacancy existing in the TiO_2 lattice. As a result, the oxygen vacancy is reduced and the electron transport performance of the TiO_2 is improved. It is the reason that the property of PSC is boosted [12]. For further optimization of F-doping, in this study, we explored TiO_2 with different fluorine doping amounts for the mesoporous layer of PSC and discussed the effect of fluorine doping amounts on the photoelectron conversion efficiency (PCE) of PSC.

2. Experimental

2.1. Materials preparation

All the chemical reagents used are of analytical grade without further purification. The preparation method of F-doped TiO_2 is a modified version reported previously [12], in a typical run: 1 g of P25 TiO_2 powders was dispersed in 75 ml of 10 M NaOH solution, the mixture was injected into a 100 ml autoclave and heated to 130°C for 24 h. After cooled down to room temperature, the as-obtained precipitation was washed and centrifuged with dilute nitric acid and distilled water for several times, then dried at 80°C overnight, and hydrogen titanate precursor powders were obtained. Afterwards, 1 g hydrogen titanate precursor mixed with0, 0.25, 0.5 and 1 g KBF_4, respectively, were added into 75 ml 0.01 M nitric acid solution. The resulted solution was transferred to a 100 ml autoclave and heated to 180°C for 24 h, then washed and centrifuged with distilled water and ethanol for several times and dried overnight. The different F-doped TiO_2 samples were obtained and remarked as S0, S1, S2, S3 and S4, accordingly. Finally, all F-doped TiO_2 slurries were prepared according to the method reported in other work [13].

2.2. Perovskite solar cell fabrication

F-doped SnO_2(FTO) substrates were ultra-sonicated and washed in turn by acetone, isopropyl alcohol and distilled water respectively for 10 minutes. TiO_2 density layer with a thickness of about 60 nm was deposited by spin coating with a rate of 4000 r/s for 30 s using the titanium diisopropoxide bis(acetylacetonate) solution and calcinated at 500°C for 30 min. TiO_2 mesoporous layer with a thickness of about 200 nm was deposited by spin coating with a rate of 4000 r/s for 30 s using the as-prepared TiO_2 slurry and calcinated at 500°C for 30 min. Then, perovskite layer was coated by a standard two-step spin coating process. Firstly, a 460 mg ml^{-1} PbI_2 DMF solution was coated on the mesoporous TiO_2 layer with a speed of 3000 r/s for 30 s. Secondly, 20 mg ml^{-1} MAI solution was coated with the same spin speed and time. The substrate was dried with a plate at 100°C for 10 min, the HTM (Spiro-MeOTAD) layer was coated with a speed of 3000 r/s for 30 s. Finally, Ag electrode with a thickness of 100 nm was evaporated thermally on the top of the device.

2.3. Testing and characterizations

Bruker AXS (D8 advance) X-ray diffraction was used to analyze the phase structure of the samples, the spectra were collected under the following conditions: Irradiated with graphite monochromatic copper λ (Cu, Kα) = 0.154 nm, at 40 kV and 40 mA over a range of 10~90° with a step of 0.0167° for 0.1 s. Scanning electron microscope (SEM, Hitachi SU-70, 10 kV) was used for morphology characterization. Current density and voltage curve (J-V) were measured under a solar intensity of AM 1.5, including a 450 W xenon lamp light source (Newport 6279 ns) and a Keithley 2400 source table. The photoluminescence characterization was conducted by using fluorophotometer (Fluoromax-4c, HORIBA) in the range of 700-830 nm with an excited wavelength of 460 nm.

3. Results and discussion

Figure 1(a) shows the XRD patterns of the different TiO_2 samples. The peaks of all patterns are in accord with the lattice planes of anatase TiO_2 listed in the PDF#73-1764 card, confirming the formation of anatase TiO_2 in all cases. Besides, some peaks, in accord with KBF_4, appear in the S4 pattern. The average grain sizes are obtained by the Scherrer equation [14] and the values of all TiO_2samples are listed in the Table 1. It demonstrates that the grain size of TiO_2 changes slightly by the fluorine doping.

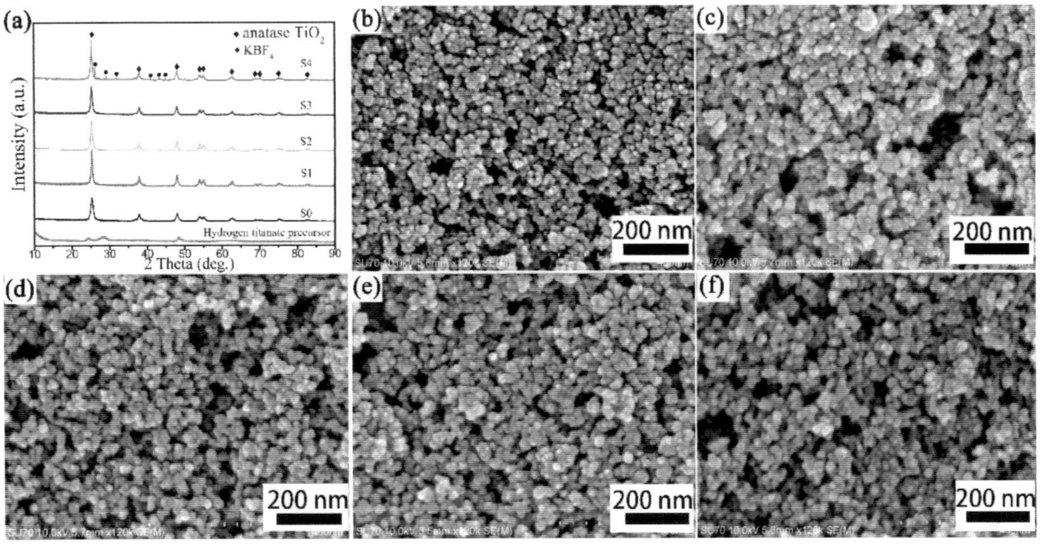

Figure 1. XRD patterns (a) and SEM images of the different TiO_2 samples (b:S0, c:S1, d:S2, e:S3, f:S4).

The SEM images of the different TiO_2 samples were shown in figure 1(b) ~ (f). It is obvious that all samples possess the similar uniform grain size of about 10 ~ 20 nm, and the results agree to the above XRD results. According to XRD and SEM results, it is confirmed that the fluorine doping in TiO_2 does not affect the grain size of the TiO_2 particles.

How does the fluorine doping in TiO_2 affect the property of perovskite solar cell? The PSC with the different TiO_2 samples as mesoporous layer were prepared with the same process [15], the J-V curves (shown in figure 2(a)) of all PSC were tested under AM1.5 light intensities similarly, and the relative parameters are also shown in table 1. It is obvious that the PCE of the PSC based on S0 is 8.63%, by contrast, the PCE of the PSC based on S1 ~ S3 improve dramatically. The PCE of the PSC based on S3 reaches a maximum of 13.06%, which boosted by 51.3% compare to S0. Meanwhile, the PCE of the PSC based on S4 declines distinctly, which can be explained by the above XRD result. XRD peaks of KBF_4have been confirmed in the XRD pattern of S4 (figure 1(a)). So there exists some surplus KBF_4 in sample S4. The surplus KBF_4 may be adsorbed on the surface of the TiO_2 grain, which will hinder the photoelectron transport between TiO_2 grains, resulting the notably decline of PCE of PSC.

Table 1. Grain size of the different TiO_2 samples and the property of PSC.

Samples	crystal size (nm)	J_{sc} (mA/cm^2)	V_{oc} (V)	FF	PCE (%)
S0	9.9	20.97	0.98	0.42	8.63
S1	11.2	21.67	0.98	0.43	9.13
S2	10.9	22.97	1.01	0.52	12.06
S3	10.4	24.40	1.01	0.53	13.06
S4	11.0	18.63	0.98	0.35	6.39

For further confirming that the fluorine doping improves the electron transport of TiO_2 and leads the boost of the PCE, the photoluminescence (PL) spectrum test of different TiO_2/perovskite (PVSK) were carried out and the results are shown in figure 2(b). It is observed that all PL spectra show an evident peak near 760 nm, which is accord with other reports [16-18]. The order of the peak intensity is PVSK> (S4/PVSK) > (S0/PVSK) > (S1/PVSK) > (S2/PVSK) > (S3/PVSK), demonstrating that the order of efficiency of photo-generated electron extraction and transport is S3>S2>S1>S0>S4, which agrees with above J-V result.

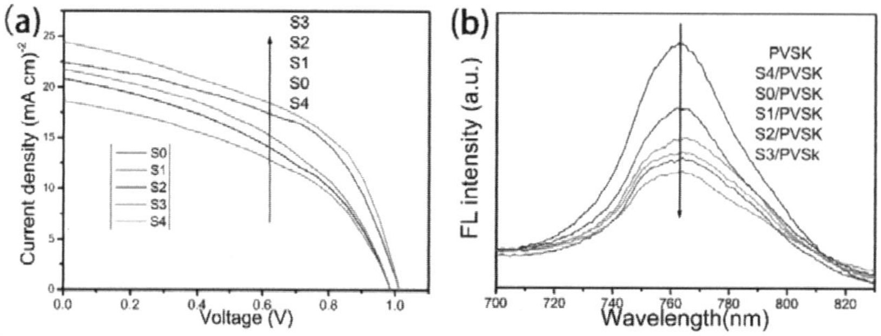

Figure 2. (a) the J-V curve of the PSC with different TiO_2 samples and (b) the photoluminescence spectrum of different TiO_2/PVSK devices.

4. Conclusion

F-doped TiO_2 samples with different fluorine doping amounts were prepared by a two-step hydrothermal method. Compared to the un-doped TiO_2, F-doped TiO_2 can not only maintain the TiO_2 crystal type and grain size, but also improve the electron transport ability of TiO_2. When F-doped TiO_2 samples were applied as the mesoporous layer of PSC, the PSC with sample S3 possesses a maximum PCE of 13.06%, which boosts by 51.6% comparing to that with un-doped TiO_2. It demonstrates that doping fluorine into TiO_2 is an effective means to improve the PCE of PSC. The above study may provide a direction for the exploration of high performance electron transport layer of PSC.

Acknowledgment
This work is supported by the Science and Technology Major Program (2014HZ0005) of Fujian Province, China.

References

[1] Kojima A, Teshima K, Shirai Y and Miyasaka T 2009 *J. Am. Chem. Soc.* **131** 6050
[2] Grätzel M 2014 *Nat. Mater.* **13** 838
[3] Sum T C and Mathews N 2014 *Energy Environ. Sci.* **7** 2518
[4] Saliba M *et al.* 2016 *Energy Environ. Sci.* **9** 1989
[5] Manseki K, Ikeya T, Tamura A, Ban T, Sugiura T and Yoshida T 2014 *RSC Adv.* **4** 9652
[6] Yang M, Guo R, Kakel K, Liu Y, O'Shea K, Bone R, Wang X, He J and Li W 2014 *J. Mater. Chem. A.* **2** 19616
[7] Qin P *et al.* 2014 *Nanoscale* **6** 1508
[8] Zhang X, Bao Z, Tao X, Sun H, Chen W and Zhou X 2014 *RSC Adv.* **4** 64001
[9] Konrr F J, Mercado C C and McHale J J 2008 *J. Phys. Chem. C* **112** 12786
[10] Gopal N O, Lo H-H, Ke T-F, Lee C-H, Chou C-C, Wu J-D, Sheu S-C, and Ke S-C 2012 *J. Phys. Chem. C* **116** 16191
[11] Hurum D C, Agrios A G, Crist S E, Gray K A, Rajh T and Thurnauer M C 2006 *J. Electron. Spectrosc. Relat. Phenom.* **150** 155
[12] Zhang X, Wu Y, Huang Y, Zhou Z and Shen S 2016 *J Alloys Compd.* **681** 191
[13] Lee J, Lee T, Yoo P J, Grätzel M, Mhaisalkar S and Park N 2014 *J. Mater. Chem. A* **2** 9251
[14] Tripathi A K, Mathpal M C, Kumar P, Singh M K, Soler M A G and Agarwal A 2015 *J. Alloys Compd.* **622** 37
[15] Burschka J, Pellet N, Moon S, Baker R H, Gao P, Nazeeruddin M K and Grätzel M 2013 *Nature* **499** 316
[16] Liu D, Yang J and Kelly T L 2014 *J. Am. Chem. Soc.* **136** 17166
[17] Xie Y, Shao F, Wang Y, Xu T, Wang D and Huang F 2015 *ACS Appl. Mater. Interfaces* **7** 12937
[18] Sun S, Salim T, Mathews N, Duchamp M, Boothroyd C, Xing G, Sum T C and Lam Y M 2014 *Energy Environ. Sci.* **7** 399

Preparation of silicon target material by adding Al-B master alloy in directional solidification

Pengting Li[1, 2,*], Kai Wang[1, 2], Shiqiang Ren[1, 2], Dachuan Jiang[1, 2] and Yi Tan[1, 2]

[1] School of Materials Science and Engineering, Dalian University of Technology, Dalian 116024, China

[2] Key Laboratory for Solar Energy Photovoltaic System of Liaoning Province, Dalian 116024, China

E-mail: *ptli@dlut.edu.cn

Abstract. The silicon target material was prepared by adding Al-6B master alloy in directional solidification. The microstructure was characterized and the resistivity was studied in this work. The results showed that the purity of the silicon target material was more than 99.999% (5N). The resistivity was ranges from 0.002 to 0.030 Ω·cm along the ingot height. It was revealed that the particles of AlB_2 in Al-6B master alloy would react spontaneously and generate clusters of [B] and [Al] in molten silicon at 1723 K. After directional solidification, the content of B and Al were increasing gradually with the increase of solidified fraction. The measured values of B were in good agreement with the curve of the Scheil equation below 80% of the ingot height. The mean concentration of B was about 17.20 ppmw and the mean concentration of Al was about 8.07 ppmw after directional solidification. The measured values of Al were fitting well with the curve of values which the effective segregation coefficient was 0.00378. It was observed that B co-doped Al in directional solidification polysiliconcouldregulate resistivity mutually. This work provides the theoretical basis and technical support for industrial production of the silicon target material.

Keywords: B co-doped Al, Segregation, Resistivity, Silicon target material

1. Introduction

Silicon target is widely used in coating field, such as touch screen with a transparent conductive oxide film (TCO), thin film solar cells, LOW-E glass, and so on [1-3]. Silicon target as a semiconductor material, it has high hardness, wear resistance, anti-corrosion properties, and other advantages. Usually silicon target material can be prepared by directional solidification in the influence of temperature gradient [4-7], the purity of the silicon target material is required to be more than 5N, to prevent the dopant of metal impurities during the sputter coating process, and high-purity target is conducive to the uniformity of the coating. The low resistivity value is less than 0.03 Ω·cm, low resistivity silicon target can ensure the conductivity of the coating. The performance of purity and resistivity are the most important factors for the use of silicon target.

During the preparation of silicon target material, group III elements such as B and Ga were used to control the resistivity. B and Ga are well-known p-type dopants. They act as acceptors in a silicon crystal. Huang et al [8] showed that the axial resistivity variation in a Ga and B co-doped Si crystal was smaller than that in simply Ga-doped Si crystal. B has a much larger equilibrium segregation coefficient (k_0=0.8) compared to Ga. Wang et al [9] showed that using a mathematical model and

IOP Conf. Series: Materials Science and Engineering 182017) 012002 doi:10.1088/1757-899X/182/1/0120(

numerical analysis, that the axial resistivity distribution in silicon crystal growth can be controlled by simultaneously doping two different types of impurities, boron (B) and phosphorus (P). At present, silicon target material with high purity and low resistivity was prepared by adding Si-B master alloy in the industrial production which the resistivity could be continuously and precisely controlled. The traditional way is to add high purity Si-B master alloy, the content of B is usually 250 ppmw, and the cost of traditional adding way is higher.

In this study, Al-6 wt. % B (Al-6B for short) master alloy was added to control the resistivity. The segregation behaviour and distribution characteristics of dopant elements were studied during the directional solidification process. This work provides the theoretical basis and technical support for industrial production of silicon target material.

2. Experimental procedures

Industrial directional solidification furnace (GT450) was used in this experiment, which is shown in figure 1. The produce system consists of a quartz crucible, a heat exchange block, heaters and heat insulations cage. The inner dimensions of the crucible were 1040×1040×480 mm with a silicon feedstock capacity of approximately 850 kg. Prior to processing by directional solidification under low vacuum condition, 825.44 kg silicon material was washed sufficiently with water to remove solid residues and extraneous impurities from the surface. The process of directional solidification is as follows: heating→melting→growing→ annealing→cooling.

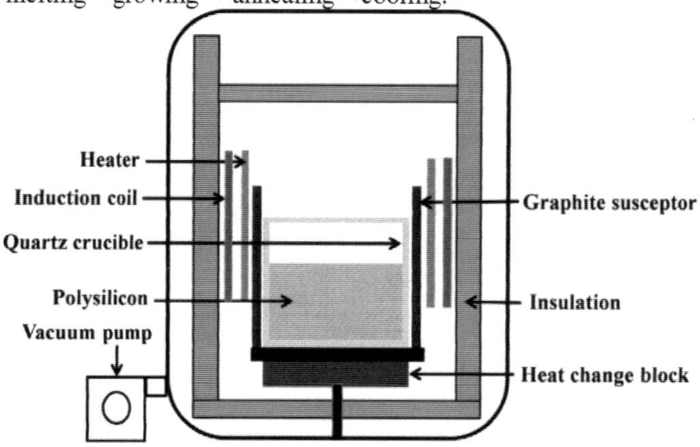

Figure 1. Schematic illustration of the experimental setup.

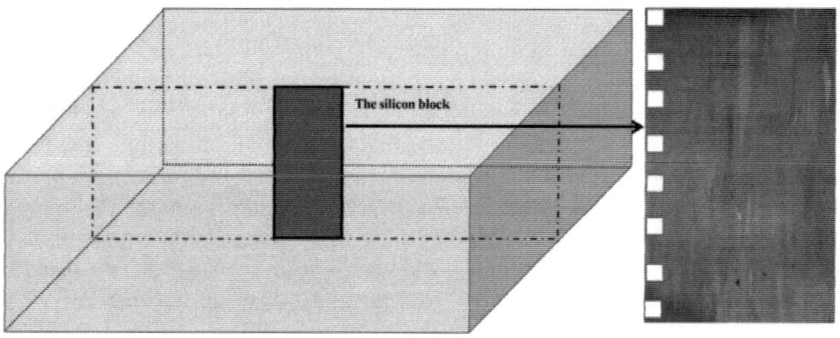

Figure 2. Schematic illustration of the silicon ingot and the samples.

After the casting process finished, analysis of the sample was cut in the ingot as shown in the figure 2. The obtained polysilicon ingot was with a dimension of 1000 ×1000 ×357 mm. A silicon block was cut parallel along the growth direction from the middle of the ingot for observation. Eight samples were cut from the bottom to the top of the block by diamond saw. The impurity concentrations of each sample were determined by Inductively Coupled Plasma Mass Spectrometer (ICP-MS). Four-point probe resistivity was performed to study the electrical performance of the whole ingot. The microstructure and composition were observed by Scanning Electron Microscopy (SEM) and Energy Dispersive Spectrometer (EDS).

3. Experimental results

3.1. The microstructure of the Al-6B master alloy

The microstructure of the Al-6B master alloy was analyzed by SEM and EDS, and the results are shown in figure 3. The microstructure of the Al-6B master alloy was composed a lot of rod-shaped particles. The largest size of particle is less than 60 μm, as shown in figure 3(a). A rod-shaped area was analyzed through break-even point analysis in EDS, as shown in figure 3(b). It is observed that the main components of the master alloy are Al and B. Based on the quality percentage of each element, it is indicated that the particles in master alloy is AlB_2.

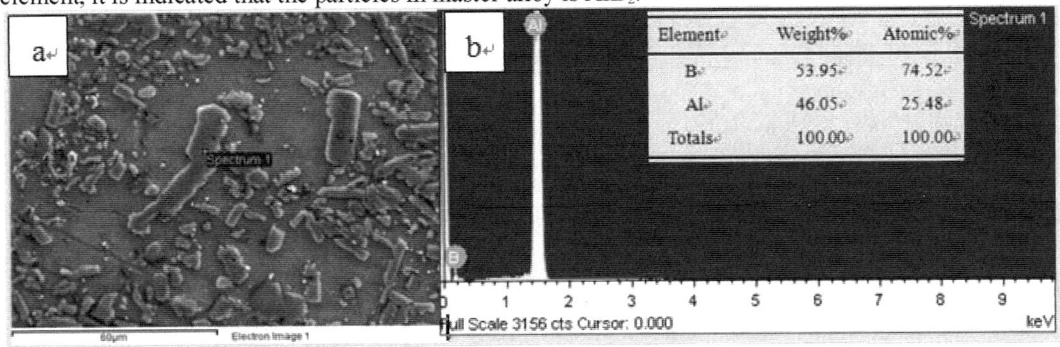

Figure 3. Analysis results of master alloy of Al-6B (a) Microstructure; (b) Energy spectrum analysis.

3.2. Composition and resistivity analysis

The impurity concentrations of each sample were determined by ICP-MS. The concentration of B and Al in different positions of the block are shown in table 1. It was observed that the concentrations of B and Al show an increasing trend from bottom to top of the ingot due to the segregation effect of the elements during the directional solidification process. The segregation effect of Al was more obvious compared with B in the process of directional solidification. The average concentration of B is about 17.20 ppmw in whole ingot. The concentration of Al decreases from 273.00 ppmw to 8.07 ppmw after directional solidification.

Table 1. Concentration of B and Al at different solidified fraction.

Solidified fraction	Concentration of B/ppmw	Concentration of Al/ppmw
0.97	18.84	64.44
0.92	20.36	11.58
0.70	16.95	3.370
0.59	16.98	1.860
0.32	15.56	1.220
0.14	16.42	1.200
0.05	15.14	1.19

Figure 4 shows the resistivity distribution of the silicon ingot. Resistivity was measured by the four point probe. The measured resistivity is P-type along the full ingot height, ranging from 0.002 to 0.030

Ω·cm, which meets the performance requirement for the sputtering silicon target. The measured resistivity shows good uniformity over the ingot's surface and the change trend of resistivity from bottom to top increased firstly and then decreased. The resistivity distribution is associated with the content of element, especially for the content of B.

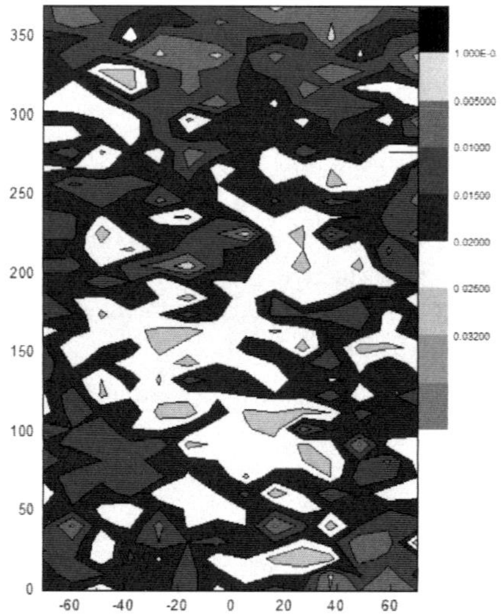

Figure 4. Distribution diagram of the resistivity of the sample.

4. Discussions

4.1. Thermodynamic analysis of AlB_2melting reaction

From the thermodynamic point of view, only the sum of the Gibbs free energies of a series of chemical reactions is less than zero, the total chemical reaction can spontaneously react. The particle of AlB_2 will melt in molten silicon at 1723K. The melting reaction produces the corresponding clusters of B and Al in molten silicon which are represented by [B] and [Al]. The possible reactions for the system are as follows:

$$AlB_2(s) = [Al](in\,Si) + 2[B](in\,Si) \qquad \Delta G \tag{1}$$

$$AlB_2(s) = Al(g) + 2B(g) \qquad \Delta G_1 \tag{2}$$

$$Al(g) = [Al](in\,Si) \qquad \Delta G_2 \tag{3}$$

$$B(g) = [B](in\,Si) \qquad \Delta G_3 \tag{4}$$

The Gibbs free energies values of a series of reaction (2), (3) and (4) in molten silicon at 1723K are as follows:

$$\Delta G_1 = 765.71\,kJ/mol \tag{5}$$

$$\Delta G_2 = -466.809\,kJ/mol \tag{6}$$

$$\Delta G_3 = -290.614\,kJ/mol. \tag{7}$$

Among them, $\Delta G_2 = -335000 - 76.5T(J/mol)$ [10] (8)

Reaction (1) is obtained by (2) + (3) +2(4), and the corresponding Gibbs free energies of reaction (1) is as follows: $\Delta G = \Delta G_1 + \Delta G_2 + 2\Delta G_3 = -282.327\,kJ/mol$ It is observed that the Gibbs free energy of reaction (1) is less than zero, so the melting reaction can spontaneously react. [B] and [Al] are may generated in molten silicon by reaction (1).

4.2. The segregation behaviour of B and Al

During the directional solidification process, impurities segregate at the solid-liquid interface concentrate to the liquid phase due to their lower solubility in solid than in liquid. The concentration of impurity into the solid phase during crystallization, C_S is given by the Scheil equation [11]:

$$C_s = k_{eff} C_0 \ (1 - f_s)^{k_{eff}-1} \tag{9}$$

C_0 is the initial concentration in the liquid at the beginning of solidification. f_s is the solidified fraction. k_{eff} is the effective segregation coefficient.

$$k_{eff} = \frac{k_0}{k_0 + (1 - k_0)exp(-v\,\delta/D)} \tag{10}$$

Among them, k_0 is the equilibrium segregation coefficient. v is the growth rate. D (m^2/s) is the diffusion coefficient of impurity in molten silicon. δ(m) is the solid-liquid interface boundary layer thickness, which can be expressed as [12].

Fitted values of different effective segregation coefficient along the growth direction were calculated by Scheil equation. The results are shown in figure 5 and 6. It is found that the content of B and Al are increasing gradually with the increase of solidified fraction. The effect of segregation of B is small due to the effective segregation coefficient is 0.8 and the control of process. On the contrary, the effect of segregation of Al is obvious because the effective segregation coefficient is about 10^{-3}.

As shown in figure 6, it is observed that the measured values of Al are in good agreement with the curve of the Scheil equation according to different effective segregation coefficient at different solidified fraction. Within the scope of solidified fraction of 0.3 to 0.6, the effective segregation coefficient of 0.00303 is fitting well. Within the scope of solidified fraction of 0.6 to 1.0 and 0.0 to 0.3, the effective segregation coefficient of 0.00378 is fitting well. Overall speaking, fitted values of effective segregation coefficient of 0.00378 are fitting well in the whole height of ingot.

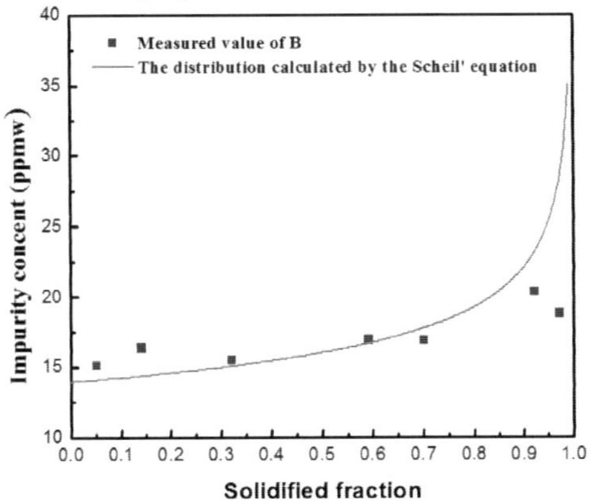

Figure 5. Concentration distribution of B along the growth direction in silicon ingot.

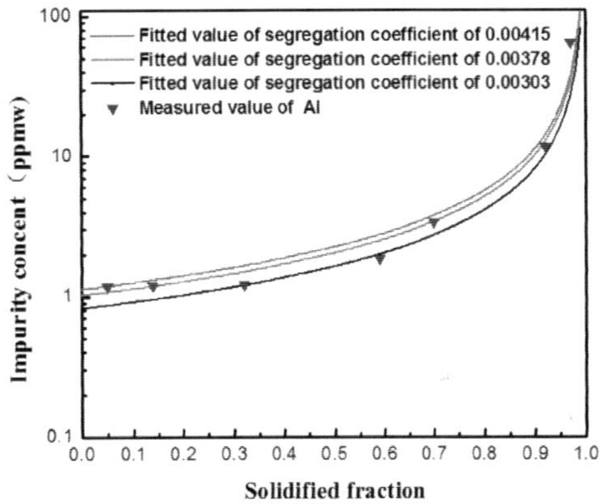

Figure 6. Concentration distribution of Al along the growth direction in silicon ingot.

As a result, the distribution of Al and B can be calculated and predicted by the Scheil equation (9) and (10). Below the 80% of the ingot height, the measured values of B are in good agreement with the curve of the Scheil equation, while the measured values are lower than the value calculated by the Scheil equation in the above of 80% ingot height. The reason is that the environment change of impurity in the last solidified fraction of 20%. The values are affected by complex factors, including concentration gradient, temperature gradient and heat flow during the directional solidification process under a low vacuum condition.

4.3. Effect of Doping Concentration on Resistivity

The measured resistivity along the full ingot height is range from 0.002 to 0.030 Ω·cm, which the mean value is about 0.02 Ω·cm. The actual doping concentration of B is about 17 ppmw. The doping concentration of B can be calculated by formulas according to the test target resistivity, just as shown below:

$$N = \frac{1.33\times10^{16}}{\rho} + \frac{1.082\times10^{17}}{\rho[1+(54.56\rho)^{1.105}]} \tag{11}$$

$$C = \frac{N \times W}{5 \times 10^{22} \times 28 \times 10^{6}} \quad (12)$$

Where N(atom/cm^3)is atomic density, ρ(Ω·cm) is the resistivity, C(ppmw) is the concentration, W is the atomic weight.

According to formulas (11) and (12), when the mean value of resistivity is about 0.020 Ω·cm, the corresponding calculated concentration of B is about 25.01 ppmw. It is observed that the calculated doping concentration of B has a deviation of about 8 ppmw from the actual doping concentration. For the reason that the dopant contains Al and the average value of the Al is about 8.07 ppmw after DS. Al and B are in the same main groups which as the acceptor elements provide vacancy. The outermost layer of Al and B has three electrons. And these elements are presented in the form of substitution in silicon.The effect of the two elements on the resistivity is similar. As a result, the influence of this deviation on resistivity can be compensated by Al.

5. Conclusions

A silicon target material with high purity and low resistivity was prepared by adding Al-6B master alloy. Among them, the purity of silicon target is more than 5N, the resistivity ranges from 0.002 to 0.030 Ω·cm. The content of B and Al are increasing gradually with the increase of solidified fraction. The measured values of Al were fitting well with the curve of values when the effective segregation coefficient is 0.00378. And the mean value of Al is about 8.07 ppmw. Below the 80% of the ingot height, the measured values of B are in good agreement with the curve of the Scheil equation. And the mean value of B is about 17.20 ppmw. The melting reaction of AlB_2 can spontaneously react to generateclusters of [B] and [Al] in molten silicon. The outermost layer of Al and B has three electrons. These elements are presented in the form of substitution in silicon. The influence of the 8 ppmw deviation on resistivity can be compensated by Al.

Acknowledgements

The authors gratefully acknowledge the financial support fromthe China Postdoctoral Science Foundation (GrantNo.2014M551076), National Natural Science Foundation of China (Grant No.51404053).

References

[1] Boumaour M, Sali S, Bahfir A, Kermadi S, Zougar L, Ouarab N and Larebi A 2016 *Journal of Electronic Materials* **45** 3929–3934
[2] Yan X, Li W M, Aberle A G and Venkataraj S 2015 *Journal of Materials Science-Materials in Electronics* **26** 7049–7058
[3] Jung W L and Jae S Y 2011 *Physica Status Solidi A* **208** 2220–2225
[4] Ren S Q, Li P T, Jiang D C, Shi S, Li J Y, Wen S T and Tan Y 2015 *Vacuum* **115** 108–112
[5] Wu J J, Ma W H, Xie K G, Wei K X, Zhou Y, Yang B and Dai Y N 2012 *Journal of Kunming University of Science and Technology* **37** 11–16
[6] Mei X Y, Ma W H, Wei K X and Dai Y N 2009 *Advanced Materials Research* **79-82** 1213-1216
[7] Martorano M A, Neto J B F, Oliveira T S and Tsubaki T O 2011 *Materials Science &Engineering B* **176** 217–226
[8] Jong H W 2005 *Journal of Crystal Growth* **280** 408–412
[9] Huang X M, Arivanandhan M, Gotoh R, Hoshikawa T and Uda S 2008 *Journal of Crystal Growth* **310** 3335–3341
[10] Chase M W 1985 *J.Phys. Chem. Ref. Data.* **14** 1–926
[11] Brown R A and Kim D H 1991 *Journal of Crystal Growth* **109** 50–65
[12] Burton J A, Prim R C and Slichter W P 1987 *Theoretical J. Chem. Phys.* **21**

IOP Conf. Series: Materials Science and Engineering **182**(2017) 012003 doi:10.1088/1757-899X/182/1/01200:

Effect of Co and Pr doping on the properties of solar-reflective $ZnFe_2O_4$ dark pigment

M Suwan, N Sangwong and S Supothina

National Metal and Materials Technology Center, National Science and Technology Development Agency (NSTDA), 114 Thailand Science Park, Phahonyothin Rd., KlongLuang, PathumThani 12120, Thailand

E-mail: sitthis@mtec.or.th

Abstract. High NIR-reflective Co-doped $ZnFe_2O_4$ black pigments were synthesized by a simple solid-state reaction of ZnO and Fe_2O_3 in the presence of 3 to 30 wt.% Co_3O_4 at 1000 and 1100 °C. A series of black pigments with low L^* values in a range of 25.5-26.5 and NIR reflectance of 39.5-48.7% were obtained at 1000 °C while the pigments with comparable L^* values and slightly lower NIR reflectance were obtained at 1100 °C. A change of the pigment property could be attributed to substitution of Co^{2+} for Zn^{2+} as revealed by XRD analysis as well as reflectance spectra. Calcination of ZnO and Fe_2O_3 in the presence of 3 to 30 wt.% Pr_6O_{11} at 1100 °C resulted in dark brown composite pigments consisting of $ZnFe_2O_4$, $PrFe_2O_3$ and unreacted Fe_2O_3 with L^* value around 40 ±1 and NIR reflectance in a range of 48-50%.

Keywords: Solar-reflective pigment; Near-infrared reflectance; Complex inorganic color pigment; Solid-state reaction

1. Introduction

Heat accumulated in surroundings of densely urban areas caused by the heat released by cars, air-conditioning systems as well as buildings and pavements has a significant impact on the 'urban heat island effect'. The increased urban temperature intensifies the use of a large amount of overall cooling load, the demand for peak electricity and the amount of greenhouse gases in the atmosphere. Solar energy is absorbed by construction buildings and cement surfaces, leading to an increase of those surface's temperature as portion of the absorbed energy is radiated back to the atmosphere. It has been reported that energy utilization in the buildings is a substantial portion of total energy consumption [1-3]. In principle, this problem can be reduced by increasing the green areas and selecting a proper design of the buildings for better air circulation. However, these approaches may not be practical for growing ubanized areas. A more practical alternative concerns with the use of suitable building envelopes such as insulating walls and roofs which can retard the heat that is absorbed and built-up inside the buildings. Nevertheless, the heat absorbed by insulated materials will be gradually released into the building.

A new approach of using solar-reflective coating for roof and pavement has been recently introduced to reflect solar radiation, particularly in the near-infrared (NIR) region, thus reducing the heat to be absorbed and transferred into the buildings [4-7]. The coating is formulated by adding an NIR-reflective pigment of desired color in addition to typical colorants. The solar-reflective pigments of various colors, such as green [8], yellow-brown [9-13], and reddish brown and reddish orange [14], have been synthesized. The desired colors were synthesized by doping various dopants into the parent composite oxide. For example, doping the Si^{4+} into the Y_6MoO_{12} resulted in bright yellow color while doping with the Pr^{4+} resulted in red-brown color [13]. Similarly, yellow color was obtained by doping the $Sm_2Ce_2O_7$ with the Mo^{6+} while brown color was obtained by doping with the Pr^{4+} [11].

It is well known that white pigments such as titanium dioxide (TiO_2) has high NIR reflectance of about 90% while black pigments such as carbon black has low NIR reflectance of almost 0. However, the use of TiO_2 white pigment is limited to wall coating. Colored NIR-reflective pigments are preferred for roof and pavement. Moreover, NIR-reflective dark pigments are required for roof coating and glaze in which dark color hue is most popular. Therefore, it becomes a very interesting topic based on academic viewpoint, and there is a challenge of making dark pigment with high NIR reflectance. To our knowledge, there are only few literatures reporting the synthesis of dark-brown/ black NIR-reflective pigments [15-17]. Dark brown Zn, Al-doped $CoCr_2O_4$ pigment has been synthesized by a sol-gel method [15]. The resultant $Co_{1-x}Zn_xCr_2O_4$ pigment had color parameter L^* in a range of approximately 54-60 ($L^* = 0$ for absolute black, $L^* = 100$ for absolute white) and NIR reflectance in a range of 44.5-52.1% depending on the amount of Zn dopant. Similarly, the $CoCr_{2-y}Al_yO_4$ pigment had L^* value in a range of 48.6-51.6 and NIR reflectance in a range of 43.5-43.9% depending on the amount of Al dopant. A sol-gel method was also employed to synthesize dark brown Mg-doped $ZnFe_2O_4$ pigment [16]. Doping of Mg in zinc ferrite results in a change of color hue from brick red to dark brown with color parameter L^* of approximately 48-49 and NIR reflectance in a range of 51-58% depending on the amount of Mg dopant. A series of brownish-black, greenish-black, black blue and black NIR-reflective pigments with L^* of approximately 23-45 and reflectance of 18.0-51.4% in the Vis-NIR regions (300-2500 nm) were obtained from $(CoO)_x(MgO)_{1-x}\cdot(Al_2O_3)_y(Fe_2O_3)_{1-y}$ [17]. The most black pigment ($L^* = 23.5$) has Vis-NIR reflectance of 18.0%.

It can be realized from the previous studies that the pigments with high blackness can be synthesized, however further study should be investigated to enhance their NIR reflectivity. Herein, we report the synthesis of high NIR-reflective Co-doped $ZnFe_2O_4$ black pigment with high blackness ($L^* < 30$) and high reflectance ($R > 40\%$). Effects of adding foreign oxides, including Co_3O_4 and Pr_6O_{11} during solid-state reaction on the pigment's properties are investigated.

2. Experimental procedure

$ZnFe_2O_4$ pigment was synthesized by solid-state reaction of ZnO (Cernic International Co., Ltd.) and Fe_2O_3 (NIC Interchem Co.Ltd.). Stiochiometric amount of ZnO and Fe_2O_3 were homogeneously mixed by wet ball milling. The mixture was calcined in an electrical furnace under ambient atmosphere at 1000 and 1100 °C for 3 h. The resultant products were washed with de-ionized water for 3 times, dried at 105°C, ground into fine powder and finally sieved through a 325-mesh. Effects of adding foreign oxides, i.e. Co_3O_4 and Pr_6O_{11}, were investigated by adding 3 to 30 wt.% Co_3O_4 (Sigma aldrich) or Pr_6O_{11} (C.K. Ceramics) (Sigma aldrich) into the raw material mixture, and synthesized with the same method.

X-ray diffraction analysis was performed on the Rigaku TTRAX III for phase identification. Diffused reflectance spectra were acquired on a Cary 5000 UV-Vis-NIR spectrophotometer (Agilent Technologies) in the wavelength range of 300 to 2500 nm. The NIR reflectance was calculated in the wavelength range of 700 to 2500 nm following the ASTM standard (number E903-96). The CIE 1976 L^* a^* b^* colorimetric method was employed for color measurement using the Spectrophotometer (Konica Minolta, CM-2600d). In this method, the L^* axis represents the lightness (0 = absolute black, 100 = absolute white), the a^* axis represents the green (negative value) to red (positive value), and the b^* axis represents the blue (negative value) to yellow (positive value).

3. Results and discussion

Figure 1 shows XRD patterns of the undoped samples synthesized by calcination of the ZnO and Fe_2O_3 at 1000 and 1100°C for 3 h. The undoped samples show strong reflection peaks which can be well matched with the standard pattern of face-centered-cubic $ZnFe_2O_4$ with a spinel structure (JCPDF file# 22-1012). The reflection peaks of $ZnFe_2O_4$ are sharp and more intense after calcination at higher temperature (1100 □C) due to an increase of crystallinity. No other peaks of minor phase or impurity are observed.

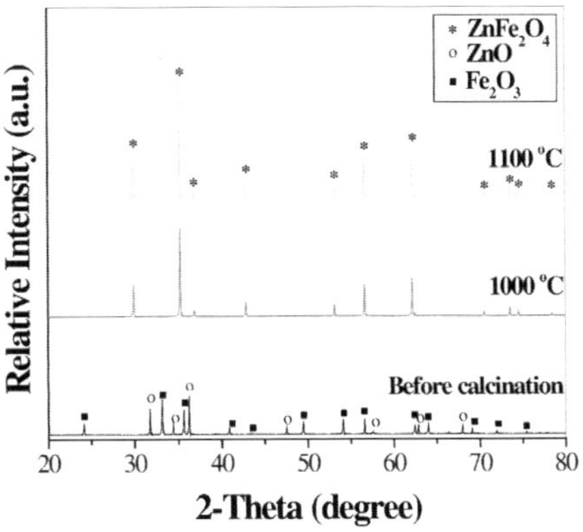

Figure 1. XRD patterns of the undoped samples before and after calcination at 1000 and 1100 °C for 3 h.

Figure 2 shows expanded XRD patterns (for (422), (511) and (440) planes) of Co-doped samples after calcination at 1000 and 1100 □C. It reveals that the addition of Co into the $ZnFe_2O_4$ did not change its crystal structure. However, doping with Co resulted in a gradual shift of XRD peaks towards higher diffraction angles upon the increased Co content compared to that of the parent phase to form the Co-doped compound, $Zn_{(1-x)}Co_xFe_2O_4$. The XRD peak shift can be ascribed to substitution of a smaller Co^{2+} (ionic radius = 0.58 Å [18]) for a larger Zn^{2+} (ionic radius = 0.60 Å [18]) in tetrahedron site of the spinel structure. Such ionic substitution would lead to a reduction of interplanar spacing, and thus a shift towards higher 2θ value. The peak shift increases with the increase of Co doping level. At 30.1 wt.% doping, the maximum doping level employed in this study, the XRD patterns well match with the pattern of $CoFe_2O_4$ phase as a result of complete substitution of Zn^{2+} by Co^{2+} at high doping level. A similar shift of XRD peaks towards higher 2θ value was observed in the substitution of smaller Mg^{2+} for Zn^{2+} in the Mg-doped $ZnFe_2O_4$ [16]. However, only a slight shift was observed in this system, and it appears that the extent of peak shifting does not increase with the Mg doping content. It could be possible that Mg doping became saturated at low doping level. In our study, the Zn^{2+} may be gradually substituted by Co^{2+} to completion, forming $CoFe_2O_4$. Further study on Mo doping into the $ZnFe_2O_4$ is being conducted by Extended X-ray Absorption Fine Structure (EXAFS) analysis.

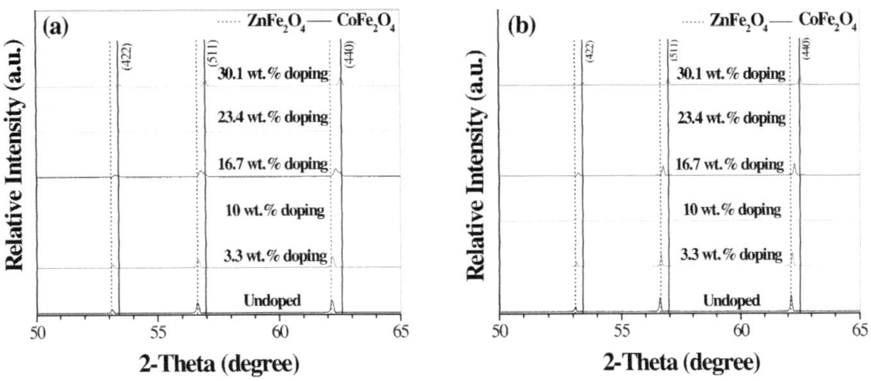

Figure 2. XRD patterns of the Co-doped samples calcined at (a) 1000 and (b) 1100 °C for 3 h at various Co doping levels.

UV-Vis diffused reflectance spectra of the undoped and Co-doped $ZnFe_2O_4$ synthesized at 1000 and 1100 □C are shown in figure 3. The spectra of the undoped samples show two broad absorption bands at around 770 and 1200 nm, a typical characteristics of the zinc ferrite, which are attributed to $^6A_{1g}\rightarrow{}^4T_{2g}$ and $^6A_{1g}\rightarrow{}^4T_{1g}$ for *d-d* electron transition of Fe^{3+} [19]. This undoped samples synthesized at 1000 and 1100 □C have NIR reflectance of 65.9 and 61.2%, respectively. The incorporation of Co into $ZnFe_2O_4$ resulted in a reduction of NIR reflectance and a change of color hue from light brown to dark brown and finally to black, depending on the amount of the Co doping level as summarized in table 1.In addition, the absorption bands at around 770 and 1200 nm are gradually diminished. In general, upon the increase of Co doping level, the L^* value and reflectance gradually decrease due to a darker color hue. For example,the NIR reflectance decreases from 65.9% for the undoped sample to 44.9% for the sample doped with 16.7 wt.% Co_3O_4 after calcination at 1000 °C. Higher doping level results in

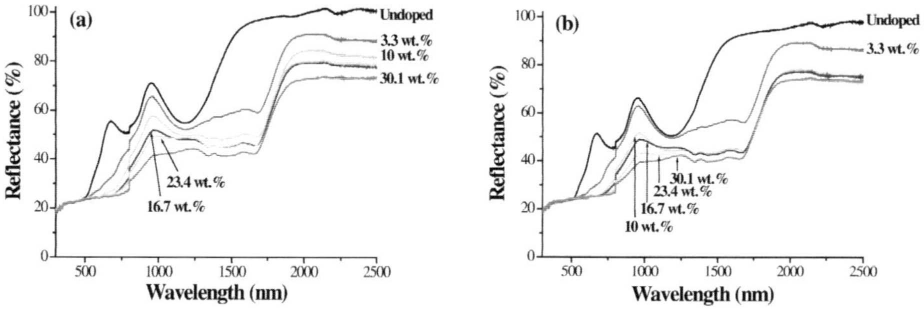

Figure 3. Diffused reflectance spectra of the Co-doped samples calcined at (a) 1000 and (b) 1100 °C for 3 h at various Co doping levels.

Table 1. Properties of NIR-reflective $ZnFe_2O_4$ pigments doped with Co_3O_4 before and after calcination at 1000 and 1100 ℃ for 3 h.

Co_3O_4 (wt.%)	Calcined at 1000 ℃					Calcined at 1100 ℃				
	Color coordinate			Photograph	*R* (%)	Color coordinate			Photograph	*R* (%)
	*L**	*a**	*b**			*L**	*a**	*b**		
0	49.6	25.3	35.5	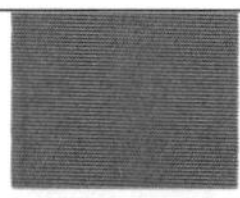	65.9	47.9	27.2	37.5	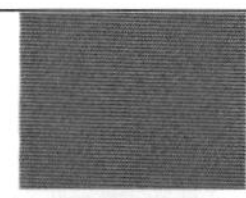	61.2
3.3	34.6	11.8	13.5		55.4	32.6	12.12	14.0		52.8
10.0	25.5	3.9	2.1		48.7	26.2	4.2	2.4		43.8
16.7	25.6	1.6	0.1		44.9	25.7	1.6	0.1		42.8
23.4	26.5	0.5	-0.7	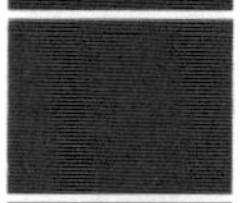	44.0	26.6	0.6	-0.5		39.8
30.1	26.0	0.5	-0.8		39.5	26.8	0.3	-0.8		38.4

only a slight reduction of L^* and reflectance. Thus, no dramatic change of the color hue at Co content higher than 16.7 wt.%. In addition, calcination at 1000 and 1100 □C had no dramatic effect on the NIR reflectance and color hue of the resultant pigments. Note that values of a^* and b^* dramatically decrease from >25 for the undoped samples to approximately 0 for the samples doped with 16.7 wt.% Co or higher, indicating very high black color hue of these pigments. Under the optimum synthesis condition, *i.e.* 1000 □C and 16.7 wt.% doping, the pigment possesses NIR reflectance of 44.9% and L^* value of 25.56. It can be regarded that the pigment synthesized in this work is high in both darkness and reflectance in comparison to the previous works [15-17].

To further investigate the effect of adding foriegn oxide into the zinc ferrite, various amount of Pr_6O_{11} was added to the raw material before subjecting to calcination in the same condition as the Co_3O_4 doping. Figure 4 shows XRD patterns of the samples after calcination at 1100 □C. XRD pattern of the undoped sample well matches with the pattern of $ZnFe_2O_4$ without other reflection peaks of impurity. When the Pr_6O_{11} was added to the raw material, the calcined product consists of $ZnFe_2O_4$, $PrFe_2O_3$, as well as unreacted Fe_2O_3. Upon increasing the Pr_6O_{11} content, reflection peaks of the $ZnFe_2O_4$ decreases while those of the $PrFe_2O_3$ increases. This result indicates that Pr_6O_{11} reacts with Fe_2O_3 during calcination forming $PrFe_2O_3$ compound. It is a competing reaction to the ZnO-Fe_2O_3. The products are composite pigments of $ZnFe_2O_4$, $PrFe_2O_3$ and Fe_2O_3.

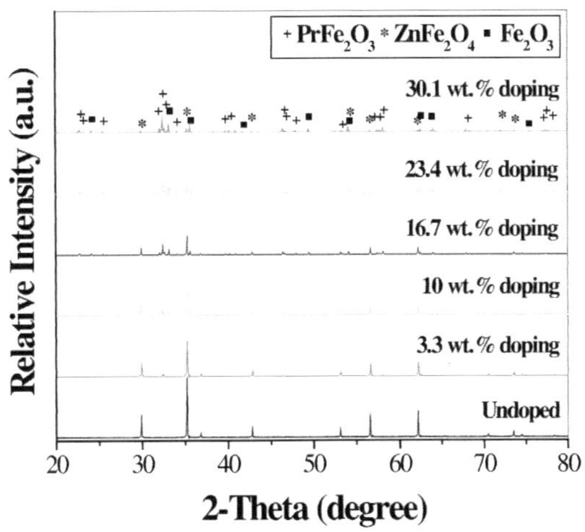

Figure 4. XRD patterns of the NIR-reflective pigments consisting various amount of Pr_6O_{11} in raw material after calcination at 1100 °C for 3 h.

Figure 5 shows UV-Vis diffused reflectance spectra of the composite pigments. In general, their spectra have similar trend with the spectra of the Co-doped $ZnFe_2O_4$ described earlier. Thai is, the reflectance decreases with the increased proportion of Pr_6O_{11} in the raw material. However, the characteristics absorption bands of zinc ferrite at around 770 and 1200 nm slightly decreases with the increased amount of the Pr_6O_{11} but they did not disappear as observed in the case of Co-doped $ZnFe_2O_4$. Such a slight decreased absorption is attributed to the decreased proportion of $ZnFe_2O_4$ in the composite pigment. The NIR reflectance and L^* value gradually decrease from 61.2% and 47.93, respectively, for pure $ZnFe_2O_4$ to 48.1% and 39.59, respectively, for the composite pigment synthesized by adding 30 wt.% Pr_6O_{11} in the raw material as summarized in table 2. Thus, it is obvoius that no doping effect was observed in this case primarily due to its large ionic radius of 0.99 Å (Pr^{3+}) [18]. The composite pigment has dark brown color hue with L^* around 40 ±1 and NIR reflectance in a range of 48-50% which is approximately in the same order with the Al-doped $CoCr_2O_4$ [15] and Mg-doped $ZnFe_2O_4$ [16].

Table 2. Properties of NIR-reflective pigments consisting various amount of Pr_6O_{11} in raw material after calcination at 1100 °C for 3 h.

Pr_6O_{11} (wt.%)	Calcined at 1100 °C				
	Color coordinate			Photograph	*R* (%)
	*L**	*a**	*b**		
0	47.9	27.2	37.5		61.2
5	39.7	26.1	25.9		60.1
10	40.5	25.1	26.6		58.0
15	39.8	23.9	25.2		56.8
20	41.3	22.2	25.2		52.4
30	39.6	18.0	22.2		48.1

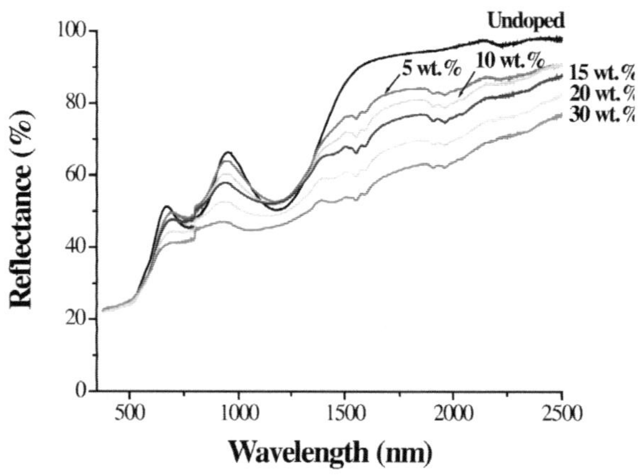

Figure 5. Diffused reflectance spectra of the NIR-reflective pigments consisting various amount of Pr_6O_{11} in raw material after calcination at 1100°C for 3 h.

4. Conclusions

The NIR-reflective Co-doped $ZnFe_2O_4$ black pigment with high darkness (*L** = 25.56) and NIR reflectance (*R*=44.9%) was synthesized by solid-state reaction of ZnO and Fe_2O_3 at 1000°C in the presence of 16.7 wt.% Co_3O_4 dopant. A series of black pigments with a gradual increase of blackness and decrease of reflectance can be synthesized by the addition of 3-30 wt.% Co_3O_4 dopant into the parent $ZnFe_2O_4$. The pigments synthesized at 1100°C had comparable black color hue, but had slightly lower NIR reflectance. The black pigments synthesized in this work are high in both blackness and NIR reflectance in comparison to the previous studies. Addition of the Pr_6O_{11} to the raw material in place of the Co_3O_4 had no doping effect, and thus did not dramatically alter the color hue and NIR reflectance. A composite pigment consisting of $ZnFe_2O_4$, $PrFe_2O_3$ and unreacted Fe_2O_3 with L^* value around 40 ±1 and NIR reflectance in a range of 48-50% was obtained.

Acknowledgements
This work has been finanically supported by the National Metal and Materials Technology Center, the National Science and Technology Development Agency (NSTDA) (Thailand), Grant No. MT-B-59-CER-07-313-I.

References
[1] Hassis S, Santamouris M, Papanikolaou M, Linardi A, Klitsikas N, Georgakis C and Assimakopoulos D 2000 *Energ. Building* **32** 131
[2] Santamouris M, Papanikolaou M and Mihalakakou G 2007 *Climatic Change* **80** 265
[3] Stathopoulou E, Mihalakakou G, Santamouris M and Bagiorgas H 2008 *J. Earth Syst. Sci.* **117** 227
[4] Levinson R, Berdahl P, Akbari H, Miller W, Joedicke I, Reilly J, Suzuki Y and Vondran M 2007 *Sol. Energ. Mat. Sol. Cell* **91** 304
[5] Akbari H and Matthews H D 2012 *Energ. Buildings* **55** 2–6
[6] Santamouris M, Synnefa A and Karlessi T 2011 *Sol. Energy* **85** 3085–3102
[7] Ferrari C, Muscio A, Siligardi C and Manfredini T 2015 *Ceram. Int.* **41** 11106–11116
[8] Thongkanluang T, Kittiauchawal T and Limsuwan P 2011 *Ceram. Int.* **3** 7543
[9] Thongkanluang T, Chirakanohaisarn N and Limsuwan P 2012 *Procedia Eng.* **32** 895
[10] Zou J, Zhang P, Liu C and Peng Y 2014 *Dyes Pigments* **109** 113
[11] Vishnu V S, George G and ReddyM L P 2010 *Dyes Pigments* **85** 117
[12] Vishnu V S and Reddy M L P 2011 *Sol. Energ. Mat. Sol. Cell* **95** 2685
[13] George G, Vishnu V S and ReddyM L P 2011 *Dyes Pigments* **88** 109
[14] Sreeram K J, Aby C P, Nair B U and Ramasami T 2008 *Sol. Energ. Mat. Sol. Cell* **92** 1462
[15] Hedayati H R, Sabbagh A A, Sameie H, Salimi R, Moosakhani S, Tabatabaee F and Zarandi A A 2015 *Dyes Pigments* **113** 588–595
[16] Liu L, Han A, Ye M and Feng W 2015 *Sol. Energy* **113** 48–56
[17] Sanada K, Morisawa Y and Ozaki Y 2014 *J. Ceram. Soc. Japan* **122** 322–328
[18] Shannon R D 1976 *Acta Crystallogr. Sect. A: Found. Crystallogr.* **32** 751–767
[19] Pailhe N, Wattiaux A, Gaudon M and Demourgues A 2008 *J. Solid State Chem.* **181** 1040–1047

Improved electric energy production of solar cell using small silver nanoparticles

X Y Wang[1], H F Li[1], S Shi[1,*] and J L Wang[2,*]

[1] Laboratory of Electromagnetic Protection Technology, Beijing Municipal Institute of Labour Protection, Beijing 100054, China
[2] School of Physics and Nuclear Energy Engineering, Beihang University, Beijing 100191, China

E-mail: [1,*]ss@bmilp.com, [2,*]wangjinliang@buaa.edu.cn

Abstract. The influence of small Ag nanoparticles (NPs) on the all-day electric energy production of solar cells has been investigated. The small Ag NPs were deposited on the cell surface by the magnetron sputtering system. The cell without NPs is used as a reference cell. The external quantum efficiency (EQE) curves indicate that the performance of cell with 8% nanoparticle coverage is better than that of other cells in the long wavelength band. The maximum power-angle curves indicate that the small Ag NPs can effectively improve the all-day electric energy production of solar cell. Comparing with the values of reference cell, the half peak height of maximum power for cell optimized increases by 65%, and the half peak breadth of maximum power for cell optimized broadens by 3%. The enhancements of optimum operating current and optimum operating voltage lead directly to the enhancement of maximum power.

Keywords: small Ag nanoparticle, solar cell, external quantum efficiency, electric energy production

1. Introduction

Photovoltaic is emerging as an important technology for future energy production. Thus, it is necessary to realize a high conversion efficiency of the solar cells [1]. In recent years, surface plasmon has been taken as one of the best solutions to achieve this object [2-4]. Many studies have demonstrated that the NPs coating on solar cell surface can improve the photoelectric performance of solar cells [5-12]. However, most of these studies about the influence of large NPs coating on the optical absorption and conversion efficiency of solar cells are reported. Few studies about the influence of small NPs coating on the all-day electric energy production of solar cells are reported. In our previous work, we have demonstrated that the use of small Ag NPs can greatly enhance the conversion efficiency for solar cells [13]. In the present work, we have applied the small Ag NPs to achieve an enhancement in electric energy production for solar cells. Here, the all-day electric energy production for solar cells is explored by varying the incidence angles.

2. Experiment

In our experiments, the crystalline Si solar cell of 125 mm×125 mm was cut into several square pieces of 20 mm×20 mm, as shown in figure 1. The photovoltaic performance for solar cells was decreased obviously owing to the mechanical damage in cutting process. The performance parameters for solar cells before and after cutting are shown in table 1. To improve nanoparticle adhesion, the solar cells were washed in ethanol, and then blown dry with cold-blast air. Ag NPs were deposited on solar cell surfaces using the magnetron sputtering system (Model: JGP600). The base pressure of vacuum chamber was better than 5×10^{-5} Pa. The working pressure was 0.3 Pa. The deposition rate was 0.05 nm/s. The surface coverage of 5%, 8%, 14% and 23% were obtained respectively by controlling the deposition times.

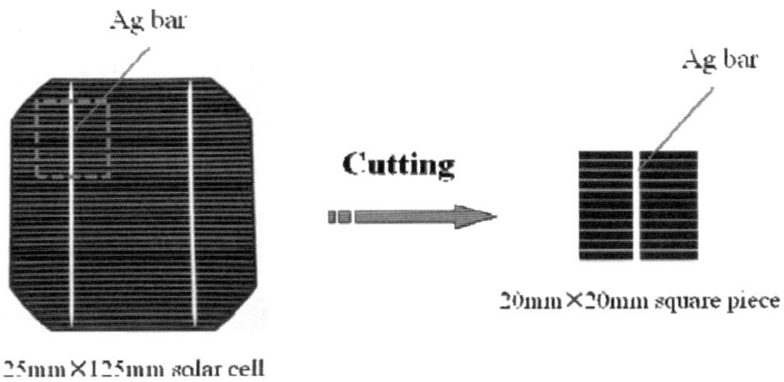

Figure 1. Schematic diagram of solar cell cutting.

Table 1. Performance parameters for solar cells before and after cutting.

Parameter	Short-circuit current (mA)	Open-circuit voltage (V)	Fill factor (%)	Efficiency (%)
125mm×125mm	441.0	9.28	73.0	17.5
20mm×20mm	114.2	0.56	42.2	7.9

Scanning Electron Microscopy (SEM) was used to measure the nanoparticle size. Current-voltage (I-V) characteristics were obtained using the *I-V* characteristic test system (Model: Solar IV-500A). External quantum efficiency (EQE) characteristics were obtained using the quantum efficiency test system (Model: Solar Cell Scan 100).

3. Results and discussion

Figure 2. SEM images of solar cells with different nanoparticle coverage (a) 5 % and (b) 23 %.

Figure 2 shows the SEM images of solar cells with different nanoparticle coverage. Figure 2(a) and figure 2(b) show the SEM images with surface coverage of 5% and 23%, respectively. The coverage can be calculated as

$$\text{coverage} = \frac{S_{\text{NPs}} \times n}{S_{\text{sight}}} = \frac{\pi\left(\frac{d}{2}\right)^2 \times n}{S_{\text{sight}}} \tag{1}$$

Where S_{NPs} is the average vertical projecting area of nanoparticle, S_{sight} is the selected area used for statistics, n is the average amount of nanoparticles in unit area, and d is the average size of nanoparticles. Figure 2 shows that the change of average nanoparticle size is very small when the surface coverage increases, which is due to the similar nucleation rate and growth rate. The nanoparticles are present predominantly as single particles. There are very few nanoparticles in the presence of multiparticle clusters.

Figure 3 shows the EQE curves of solar cells with different nanoparticle coverage. In our experiments, the solar cell without NPs is used as the reference cell. We can find that the EQE of solar cells increases firstly, and then decreases when the nanoparticle coverage increases. This is mainly because the nanoparticle scattering is less than the light absorption of area covered by the nanoparticles. In the short wavelength band, the EQE of solar cells with nanoparticle coverage are decreases obviously in comparison with the reference cell, which indicates that the nanoparticle cover accelerates the front surface recombination. In the long wavelength band, the EQE of solar cells with nanoparticle coverage enhances obviously in comparison with the reference cell, which indicates that the nanoparticle cover can effectively reduce the back surface recombination of solar cell. When the nanoparticle coverage is 8%, the EQE of solar cell reaches the best in the long wavelength band, which is up to 95%. The use of Ag NPs can effectively improve the EQE of solar cells in the long wavelength band.

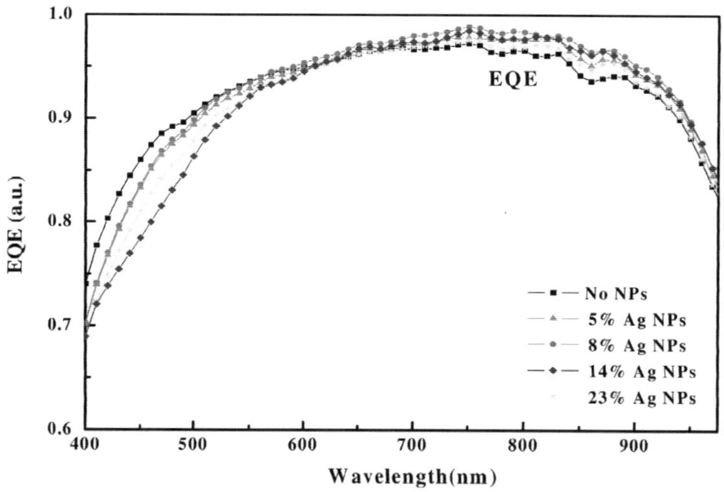

Figure 3. External quantum efficiency curves of solar cells with different nanoparticle coverage.

To simulate the all-day electric energy production of solar cells, the angle of incidence is varied from 0° to 180°. The solar cell without NPs is also used as the reference cell. Figure 4(a) shows the current–voltage and power output curves of the reference cell illuminated at different incidence angles. Figure 4(b) shows the current–voltage and power output curves of the optimized cell illuminated at different incidence angles. Figure 4(c) shows the maximum power (P_{m}) as a function of angle for the reference cell and optimized cell. Figure 4(d) shows the optimum operating current (I_{m}) and the optimum operating voltage (V_{m}) as a function of angle for the reference cell. Figure 4(e) shows the optimum operating current (I_{m}) and the optimum operating voltage (V_{m}) as a function of angle for the optimized cell. In comparison with the reference cell, the half peak height of Pm for the optimized cell increases by 65%, and at the same time broadens by 3%, as shown in figure 4(c). The use of small Ag NPs can effectively improve the all-day electric energy production of the optimized cell. The all-day electric energy production of the optimized cell is much superior to the values of the reference cell. Combining figure 4(d) and figure 4(e), we can find that the enhancements of I_m and V_m lead directly to the enhancement of $P_m = I_m \cdot V_m$. The distribution of peaks for I_m and V_m is reasonable between 30°and 80°, which is confirmed in theory [14]. The enhancement of half peak height for P_m can be calculated as

$$\frac{\Delta P}{P_m(\mathrm{ref})} = \frac{P_m(\mathrm{NPs}) - P_m(\mathrm{ref})}{P_m(\mathrm{ref})} \tag{2}$$

Here, P_m (ref) is the half peak height of P_m for the reference cell. P_m (NPs) is the half peak height of P_m for the optimized cell.

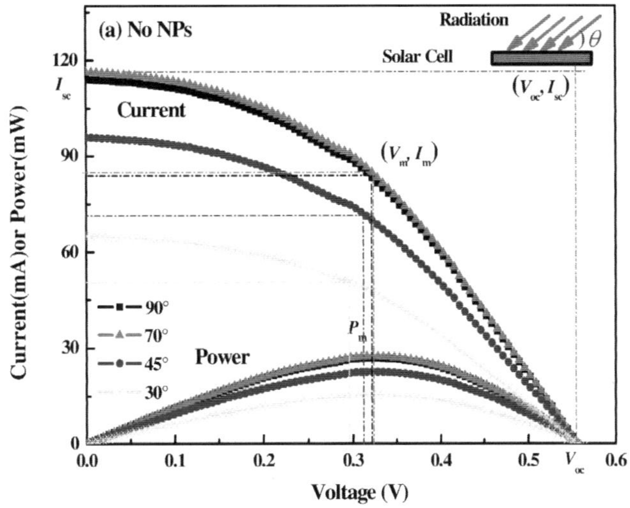

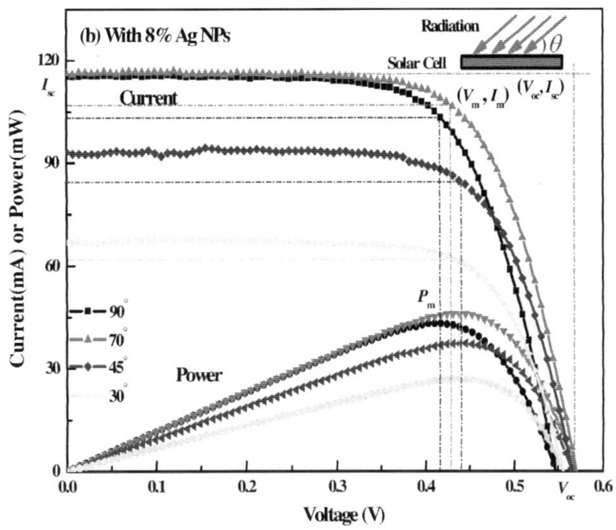

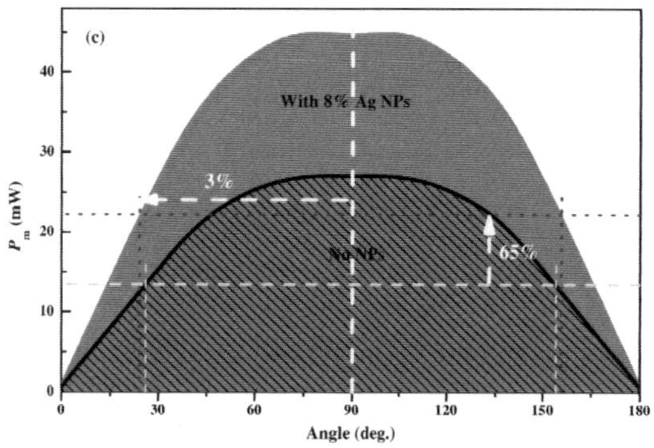

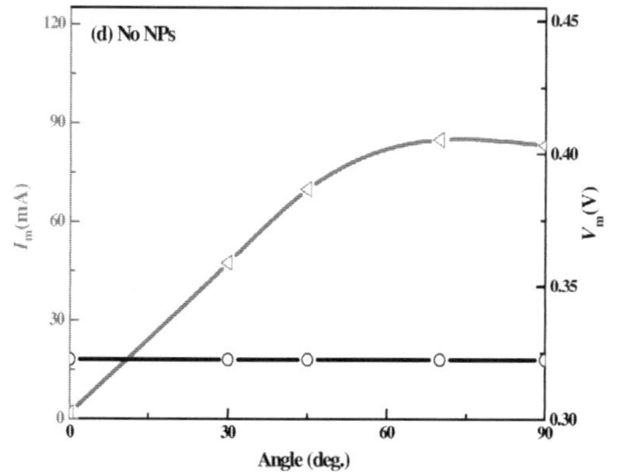

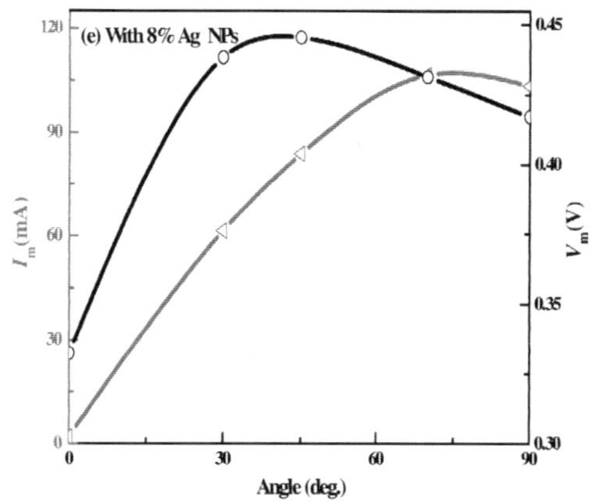

Figure 4. (a) Current-voltage and power curves of the reference cell illuminated at different incidence angles. (b) Current-voltage and power curves of the optimized cell illuminated at different incidence angles. (c) Maximum power as a function of angle for the reference cell (black slash-area) and optimized cell (red area). (d) Optimum operating current (red) and optimum operating voltage (black) as a function of angle for the reference cell. (e) Optimum operating current (red) and optimum operating voltage (black) as a function of angle for the optimized cell.

4. Conclusion

In summary, this study has provided an efficient approach to improve the all-day electric energy production of solar cells via using the small Ag NPs. Comparing with the value of the reference cell, the half peak height of P_m for the optimized cell increases by 65%, and the half peak breadth of P_m for the optimized cell broadens by 3%. The all-day electric energy production of the optimized cell is much superior to the value of the reference cell. The enhancements of I_m and V_m lead directly to the enhancement of P_m.

Acknowledgement

This work is supported by the Beijing Academy of Science and Technology Innovation Team (Grant No. IG201403C2).

References

[1] Xu R, Wang X D, Liu W, Xu X N, Li Y Q, Ji A, Yang F H and Li J M 2012 Chin. Phys. B 21 025202

[2] Shah A V, Schade H, Vanecek M, Meier J, Vallat-Sauvain E, Wyrsch N, Kroll U, Droz C and Bailat J 2004 *Prog. Photovolt. Res. Appl.* **12** 113

[3] Pillai S, Catchpole K R, Trupke T and Green M A 2007 *J. Appl. Phys.* **101** 093105

[4] Catchpole K R and Polman A 2008 *Opt. Express* **16** 21793

[5] Schaadt D M, Feng B and Yu E T 2005 *Appl. Phys. Lett.* **86** 063106
[6] Derkacs D, Lim S H, Matheu P, Mar W and Yu E T 2005 *Appl. Phys. Lett.* **89** 093103
[7] Matheu P, Lim S H, Derkacs D, McPheeters C and Yu E T 2008 *Appl. Phys. Lett.* **93** 113108
[8] Temple T L, Mahanama G D K, Reehal H S and Bagnall D M 2009 *Sol. Energy Mat. Sol. Cells* **93** 1978
[9] Jeong N C, Prasittichai C and Hupp J T 2011 *Langmuir* **27** 14609
[10] Liu W, Wang X D, Li Y Q, Geng Z C, Yang F H and Li J M 2011 *Sol. Energy Mat. Sol. Cells* **95** 693
[11] Deepa K G, Lekha P and Sindhu S 2012 *Sol. Energy* **86** 326
[12] Resta V, Caricato A P, Loiudice A, Rizzo A, Gigli G, Taurino A, Catalano M and Martino M 2013 *J. Nanopart. Res.* **15** 2017
[13] Wang X Y, Wang J L and Wang H 2014 *Sol. Energy* **101** 100
[14] Zhou Y S, Gu B Y, Wang H Y and Zhao L M 2010 *Phys. Rev. A* **81** 035803

The microstructure, mechanical and electrical properties of Niobium pentoxide-doped Titanium oxide ceramic targets

Qi Ling [1, a], TingTing Xu[1], LianZi Wu[1], Wei Peng[1], ZuoXiang Zhu[1], ShengHui Xie[1], HaiPeng Yang[1], JiHua Gao[1], Ling Gao[1] and HaiTao Yang [1*]

[1]Key Laboratory of Functional Materials of Shenzhen, Shenzhen Engineering Laboratory for Advanced Technology of Ceramics, Guangdong Research Center for Interfacial Engineering of Functional Materials, College of Materials Science and Engineering, Shenzhen University, Shenzhen, 518060, China

E-mail: [a]2151120221@email.szu.edu.cn, * yanght63@szu.edu.cn

Abstract. Nb_2O_5 doped TiO_2 (NTO) ceramic targets were prepared by sintering in Ar at 1350℃ for 2 hours. The morphologies structures, densification behavior, mechanical and electrical properties of the sintered ceramic targets with different doping concentration were investigated. The results show that the Nb_2O_5 content of increase from 0 wt% to10 wt% not only enhance the densification but promote the grain growth. The morphologies, compositions, microstructure ,and chemical stateswere characterized by SEMand XRD. The relative density, grain size and resistivity of 7.5 wt% content sintered at 1350℃ in Ar were 94.5%, 13.76 μm, and $7.8 \times 10^{-2}\,\Omega\cdot cm$, respectively.

Keywords: NTO, ceramic targets, sinteing behaviour, transparent conductive films

1. Introduction

Transparent conducting oxide (TCO) thin films have been widely used in liquid crystal displays (LCD), organic light-emitting diodes (OLED) and thin film solar cells. Indium tin oxide(ITO) is currently the most popular commercially available TCO material due to its high transparency (>80%) and low resistivity ($<10^{-3}\Omega\cdot cm$).However, the alternatives to ITO have been extensively developed due to the shortage and high cost of indium. In recent years, worldwide interests have been attracted on TiO_2-based materials which were thought to be most promising in replacing ITO because of its low electrical resistivity with good thermal and chemical stability as well as low cost and non-toxicity [1, 2].As a wide band gap (Eg = 3.2 eV) semiconductor material, intrinsic TiO_2 is highly transparent in the visible region, but relative poor in conductivity. In order to improve its conductivity, it is necessary to induce a proper amount of carriers into it in an adequate manner.

To optimize the performances of TCO films, most researchers have studied the influences of sputtering parameters on the properties of films, including the type of sputter power, sputter power, working pressure, base pressure, substrate temperature, atmosphere, film thickness and post annealing treatment. Unfortunately, only a few studies have attempted to clarify how the TCO target affects the sputtering process and the properties of various films. Many studies were focused on doping of Group III elements into the TiO_2 film, such as B, Al, and Ga [3-13]. In these cases, the plus pentavalent cations of Group V elements Nb as the dopants would substitute the tetravalent Ti cations inside the film, the extra electron is provided as free carrier. Therefore the resistivity was lowered by several orders comparing with pure TiO_2.

For a successful substitution with small lattice deformation, the dopant needs to have a similar radius to that of the host ion Ti (0.56 Å). Based on this consideration, the pentavalent elements of Nb(0.62 Å) have been widely studied as a dopant for fabricating high conductive TiO_2 films. Nb is especially being considered as the most promising candidate because its radius is very similar to that of Ti.

Few researches on pentavalent cation-doped TiO_2 have been published. The pentavalent cation can provide one free electron to contribute the conductivity as it substitutes Ti in TiO_2 films. Ti is a pentavalent cation and has a radius of 0.62Å,which is very close to that of Ti(0.56Å).Thus, it is suitable as a dopant to fabricate high conductive TiO_2 films. Adding Nb_2O_5 dopant to TiO_2film greatly improve the electrical conductivity of the film without significant decrease its visible transmittance. Furubayashi et al. [1] reported the preparation of Nb-doped TiO_2 film with an electrical resistivity of $3.48\times10^{-3}\Omega\cdot cm$ on single crystal substrate.Moreover, Nb-doped $SrTiO_3$ film is also highly conductive.Zhu et al.[14] fabricated a Nb-doped $SrTiO_3$ film with a resistivity of $3.6\times10^{-4}\Omega\cdot cm$.Ohsaki et al.[15] have indicated that a TiO_{2-x} target with a resistivity about 0.3 $\Omega\cdot cm$ can be used to fabricate TiO_2 and NTO film, respectively. Yamada et al.[16] demonstrated that itwas very difficult to manufacture a NTO film with high electrical properties using reactive sputtering of a Ti:Nb metallic target.

It is also widely known that high-quality targets are crucial for the production of high-quality thin films by sputtering. Most research works were focused on thin films and paid little attention to targets. In this paper, the Nb doped TiO_2 (NTO) ceramic targets were prepared by sintering at 1350℃ for 2 hours. The morphologies, structure, densification behavior, mechanical and electrical properties of the sintered ceramic targets with different doping concentration were investigated.

2. Experimental

The selected Nb_2O_5doping concentration was 0wt%, 2.5wt%, 5wt%, 7.5wt% and 10wt% respectively. The selected compositions were mixed ball milled with deionized water. Ball mill for 48 hours with ZrO_2 milling balls in a commercial ball mill, the ball to powder ratio was 2:1. After drying, sieving and granulating, the powder mixtures were dry-pressed into bars in a steel die at 100 MPa.

The compacts were then sintered at 1350℃for 2 hours in Ar. The bulk density was measured using the Archimedean principle. Without being machined, the sintered bar specimens, 5 by 5 by 20 mm in size, were used for three-point bend strength measurements conducted with steel fixtures at a speed of 3 mm/min, with bottom spans of 15 mm. The Vickers hardness was measured hardness tester. The crystal structure, the fracture surface morphology were investigated via X-ray diffraction (XRD) analysis and scanning electron micro-scope (SEM), respectively.

The electrical resistivity of NTO targets was measured by a Physical Properties Measurement System (PPMS).The resistivity, mobility and carrier concentration of the ceramics targets were measured by Hall test while the sheet resistance was obtained through four-probe method.

3. Results and discussion

3.1. Electrical resistivity of NTO targets

Figure 1 shows the electrical resistivity of NTO targets sintered at 1350℃ for 2 hours. The results indicated that the electrical resistivity of TiO_2 without any doping additives is over 10^8 $\Omega\cdot cm$. The electrical resistivity of NTO targets decreased rapidly to 7.8×10^{-2} $\Omega\cdot cm$ with the doping amount increased from 0 to 7.5 wt%.Because the Nb^{5+} can provide one free electron to contribute the conductivity as they substitute the Ti^{4+} in TiO_2 to form NTO solid solution.

We obtain the minimum electrical resistivity at the point of 7.5 wt% doping amount. The electrical resistivity increased to 8.74×10^{-2} $\Omega\cdot cm$ with further increasing total doping amount to 10 wt% due to the formation of excessive Nb^{5+} distributing over the grain boundaries which increases the electrons scattering centers and decreases the mobility of the electrons.

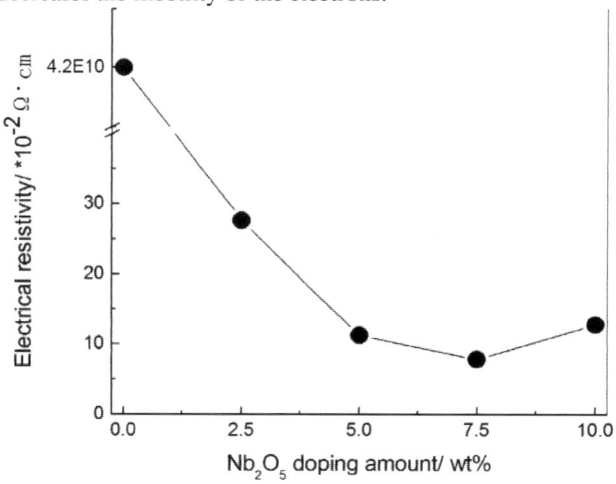

Figure 1. The electrical resistivity of five NTO ceramics as a function of doping content.

3.2. Relative density of NTO targets

Figure 2 shows the relations of relative density versus Nb_2O_5 doping amount for the samples sintered at 1350℃ for 2 hours. The theoretical density of NTO ceramic targets range from 4.23 g/cm^3 to 4.31 g/cm^3,becausethetheoreticaldensitiesof$TiO_2$and Nb_2O_5 are 4.23g/cm^3 (JCPDS 76-0318)and4.98g/cm^3 (JCPDS 30-0873),respectively. The relative density of TiO_2 without any doping additives is only about 89.6 %. The relative density increased rapidly to 93% with the increase of doping amount from 0 to 2.5 wt%, and then the relative density increased slightly to 94 %,the highest pointing, with increasing the doping amount to 5 wt%. The relative density varied slightly decrease from 94 % to 92 % with further increasing the doping amount to 10 wt%.

The results indicated that the porosities of NTO sintered at 1350°C were still high, Nb_2O_5 were not only effective carriers suppliers but also excellent sintering aids for TiO_2.

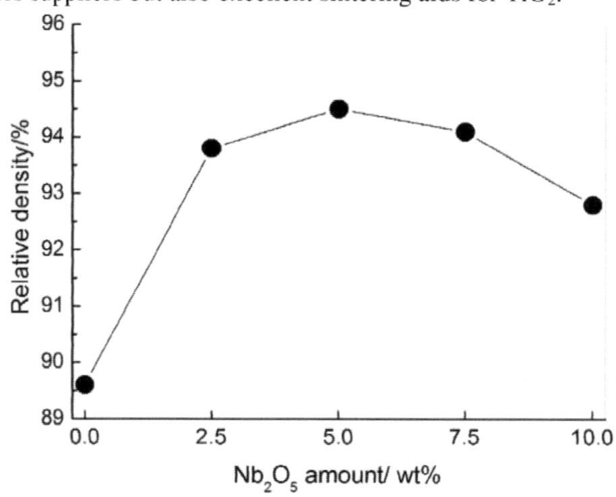

Figure 2. the relative densities of NTO ceramics as a function of doping content.

3.3. grain size of NTO targets

It is well known that the characteristics of the sputtering targets directly affect the quality of the deposited thin films and the stability of the sputtering process [17-20]. In the process of sputtering, targets with high strength and hardness are required to withstand the bombardment of high energy particles. The appropriate size of the grains can get excellent film properties. The average grain sizes of NTO ceramics as a function of the content are shown in figure 3. For the samples sintered at 1350℃ for 2 hours. The results demonstrate that the grain sizes of NTO ceramics increased from 6.4 to 22.79 um, when the Nb_2O_5 content was increased from 0 to 10wt%. The findings on the microstructure and grain size show that increasing the Nb_2O_5 content from 5 to 10 wt% inhibited both the densification , however enhance the grain growth.

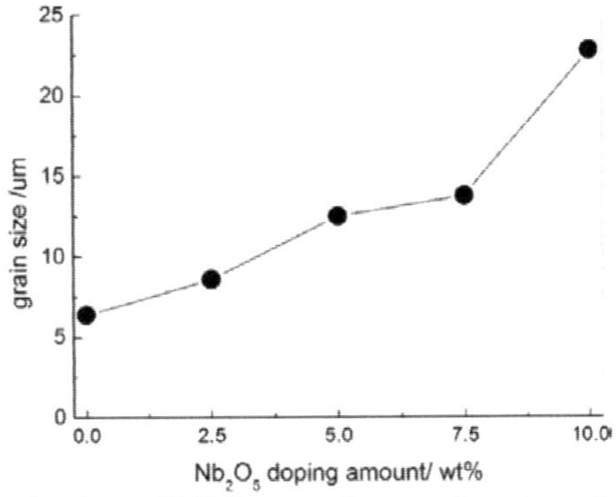

Figure 3. the grain sizes of NTO ceramics as a function of doping content.

3.4. Bending strength of NTO targets

As shown in figure 4, The strength increased from 95.25 MPa to 83.38 MPa with increasing thedoping amount from 0 to 2.5 wt%. The highest strength of 95.25 MPa was obtained at the doping amount of 0wt%, in other words, is pure TiO_2. The results indicated that Nb_2O_5 doping would lower the strength of $TiO_2$and reduced the combination of TiO_2 ceramics targets.The strength decreased to 86.34 MPa within creasing doping amount to 10wt% due to the radius of Nb ion was larger than Ti ion radius, a large number of Ti ions were substituted in the high doping amount, which led to serious crystal lattice distortion and the decrease of binding force. The trend of the change of the bending strength was consistent with the relative density.

Strength is the most important property for targets, which has been neglected by many researchers and producers [21-26]. During sputtering process, the grains and particles with poor bonding strength may be hit off from the target, which contaminates the film being coated, and form new pores on the surface of the targets which deteriorates the quality of the film and shorten the lifetime of the target. High strength targets must be used for the production of large-area high-quality thin films

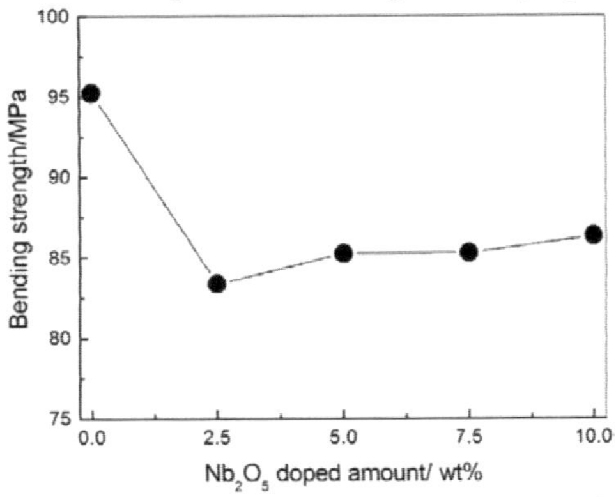

Figure 4. the bending strength of NTO ceramics as a function of doping content.

3.5. The analysis of the XRD on NTO targets

Figure 5 is the XRD pattern of NTO targets with different Nb_2O_5 doping amount from 0 to 10wt% sintered at 1350℃ for 2 hours. It is clear that the peaks of NTO targets are quite identical to the TiO_2 rutile standard patterns. The Nb_2O_5 peaks could not be found, which suggests that the Nb^{5+} has entered into the TiO_2 lattice structure to form a solid solution and provide free electrons to contribute the conductivity after sintering.

With the doping amount increased from 0 to 10wt%., the carrier concentration increased which leads to the electrical resistivity decreased rapidly from 10^8 Ω·cm to 7.8×10^{-2} Ω·cm. Meanwhile, as shown in figure 5. The doping amount reaches 10 wt% without appear the second phase. The second phase would increase the electrons scattering centers and decrease the mobility of the electrons. The results indicated that the solid solubility limit of Nb_2O_5 in TiO_2 was over 10 wt%. That is why the electrical resistivity increased to 12.14×10^{-2} Ω·cm with further increasing doping amount to 10wt% as shown in figure 1. The peak shift indicates the presence of internal stress inside the crystal. With the increase of the doping amount, the peak intensity of (110) lattice plane shows that Nb has a preferential growth effect on the crystal surface.

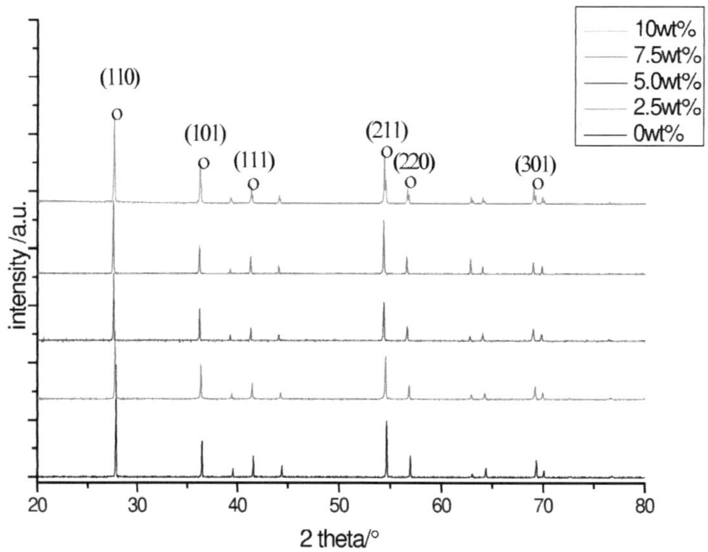

Figure 5. XRD pattern of NTO targets sintered with different doping content.

3.6. The fracture surfaces analysis on targets

Figure 6 shows SEM micrographs of fracture surfaces of the targets with Nb_2O_5 content from 0 to 10wt% sintered at 1350℃ for 2h.

The TiO_2 target without any dopant shows a large amount of big pores duo to its poor density. The fracture surface is simply and smooth. The targets with 5 wt% and 7.5 wt% doping amount show very few small pores, and the fracture surface is rough and complex that often means high strength by absorbing more fracture energy while low strength often leads to simple and smooth fracture surface.

The second phase have not been observed in the samples with the doping amount of 7.5 wt% and 10 wt%, which is corresponding to the results of XRD analysis.

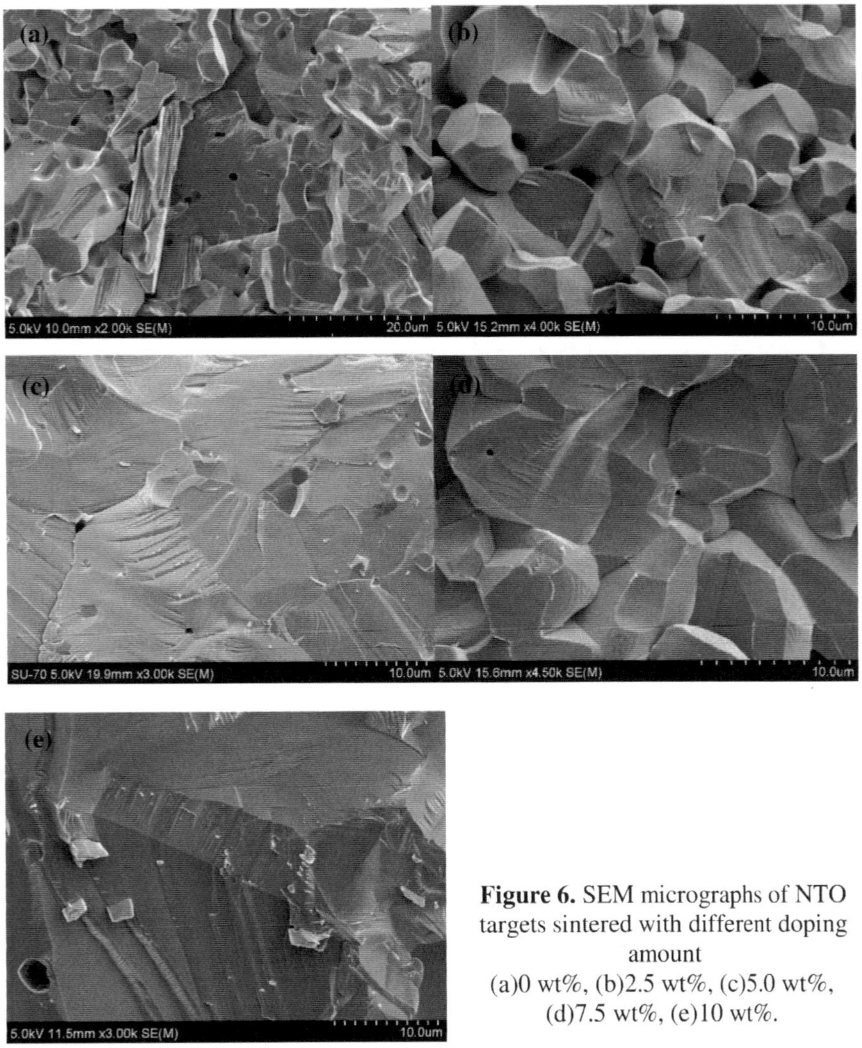

Figure 6. SEM micrographs of NTO targets sintered with different doping amount (a)0 wt%, (b)2.5 wt%, (c)5.0 wt%, (d)7.5 wt%, (e)10 wt%.

4. Conclusions

The doping of Nb_2O_5 was very effective for the densification and the conductivity of TiO_2. The optimum doping amount was 7.5 wt.%. The sintered NTO targets with Nb_2O_5 content achieved an electrical resistivity of 7.8×10^{-2} Ω·cm, a relative density of 94%, and a bending strength of 85.3 MPa.

The solid solubility limit of Nb_2O_5 in TiO_2 was over 10 wt.%. Doping amount of more than 7.5 wt.% in the sintered NTO targets is harmful to the density, and conductivity.

This kind of NTO ceramic target has a potential to be used as a high-quality sputtering target to deposit TiO_2-based transparent conductive films.

Acknowledgements
The research is supported by Shenzhen Innovation and Technology commission under Basic Research Project (Contract NO. JCYJ 2016 0422 1447 51573, JCYJ 2016 0422 1049 21235, JCYJ 2014 0418 1819 58489)

References

[1] Y Furubayashi,T Hitosugi,Y Yamamoto,K Inaba,G Kinoda,Y Hirose,T Shimada and T Hasegawa 2005 *J.Appl.Phys.Lett.* **86** 252101
[2] Ming-Wei Wu and Yan-Chi Chen 2015 *J. Am. Ceram. Soc.* **98(1)** 163–170
[3] S Y Kuo, W C Chen, F I Lai, C P Cheng, H C Kuo, S C Wang and W F Hsieh 2006 *J. Cryst Growth* **287** 78–84
[4] S Benramache, B Benhaoua and O Belahssen 2014 *J. Optik.* **125** 5864–5868
[5] Ming-WeiWu , Pang-HsinLai, Chia-HongHong and Fang-ChengChou 2014 *J. Journal of the European Ceramic Society* **34** 3715–3722
[6] M Nisha, S Anusha, A Antony, R Manoj and M K Jayaraj 2005 *J. Appl Surf Sci.* **252** 1430–1435
[7] Deuk Yong Lee, Ju-Hyun Park, Young-Hun Kim, Myung-Hyun Lee and Nam-Ihn Cho 2014 *J. Current Applied Physics* **14** 421–427
[8] J M Lin, Y Z Zhang, Z Z Ye, X Q Gu, X H Pan, Y F Yang, J G Lu, H P He and B H Zhao 2009 *J. Appl Surf Sci.* **255** 6460–6463
[9] Jinming Liu, Xiaoru Zhao, Libing Duan, Mengmeng Cao, Huinan Sun, Jifeng Shao and Shuai Chen 2011 *J. Applied Surface Science* **257** 10156–10160
[10] J Lee, K N Hui, K S Hui, Y R Cho and H H Chun 2014 *J. Appl Surf Sci.* **293** 55–61
[11] A Shah, Arshad Mahmood, Uzma Aziz, Rashad Rashid, Qaiser Raza and Zahid Ali 2016 *J. Materials Chemistry and Physics* 1–7
[12] Chang-Feng Yua, Shih-Jye Sunb and Jian-Ming Chen 2014 *J. Applied Surface Science* **292** 773–776
[13] T Potlog, P Dumitriua, M Dobromir, A Manole and D Luca 2015 *J. Materials and Design.* **85** 558–563
[14] Y L Zhu, X L Ma, D X Li, H B Lu, Z H Chen and G Z Yang 2005 *J. Acta Mater.* **53** 1277–1284
[15] Linhui Gan, Congcong Wu, Yuan Tan, Bo Chi , Jian Pu and Li Jian 2014 *J. Journal of Alloys and Compounds* **585** 729–733
[16] H Ohsaki, Y Tachibana, A Mitsui, T Kamiyama and Y Hayashi 2001 *J. ThinSolid Films* **392** 169–173
[17] N Yamada, T Hitosugi, J Kasai, N L H Hoang, S Nakao, Y Hirose, T Shimada and T Hasegawa 2010 *J.ThinSolid Films* **518** 3101–3104
[18] J Liu, W J Zhang, D Y Song, Q Ma, L Zhang, H Zhang, X B Ma and H Y Song 2014 *J. Mat Sci Semicon Proc.* **27** 1–11
[19] C S Wu, B T Lin and R Y Yang 2011 *J. Thin Solid Films* **519** 5106–5109
[20] Chaojun Wang, Jian Li and Joonghoe Dho 2014 *J. Materials Science and Engineering.B* **182** 1–5
[21] Liu Lu, MeilanGuo, S Thornley, XiaopingHan, JunhuaHu, M J Thwaites and Guosheng Shao 2016 *J. Solar EnergyMaterials&SolarCells.* **149** 310–319
[22] T Potlog, M Dobromir, D Luca, P Onufrijevs, A Medvids and A Shamardin 2016 *J. Current Applied Physics* **16** 826–829
[23] S K Mukherjee, H W Becker, A P Cadiz Bedini, A Nebatti, C Notthoff, D Rogalla, S Schipporeit, A Soleimani-Esfahani and D Mergel 2014 *J. Thin Solid Films* **568** 94–101
[24] Alireza Sasani, Ardeshir Baktash, Kavoos Mirabbaszadeh and Bahram Khoshnevisan 2016 *J. Applied Surface Science* **384** 298–303
[25] Naoomi Yamada, Taro Hitosugi, Ngoc Lam Huong Hoang, Yutaka Furubayashi, Yasushi Hirose, Seiji Konuma, Toshihiro Shimada and Tetsuya Hasegawa 2008 *J. Thin Solid Films* **516** 5754–5757
[26] Guangmiao Wan, Shenwei Wang, Xinwu Zhang, Miaoling Huang, Yanwei Zhang, Wubiao Duan and Lixin Yi 2015 *J. Applied Surface Science* **357** 622–625

IOP Conf. Series: Materials Science and Engineering **182** (2017) 012006 doi:10.1088/1757-899X/182/1/012006

Dealloying-driven synthesis and characterization of $AgCl/Ag/TiO_2$ nanocomposites with enhanced photocatalytic properties

Tongyang Wang[1,2], Yan Wang[1], Zhengfeng Zhao[1], Lanxiang Zhang[1] and Shifeng Huang[2,*]

[1] School of Materials Science and Engineering, University of Jinan, No. 336, West Road of Nan Xinzhuang, Jinan 250022, China
[2] Shandong Provincial Key Laboratory of Preparation and Measurement of Building Materials, University of Jinan, No. 336, West Road of Nan Xinzhuang, Jinan250022, China

E-mail: *mse_huangsf@ujn.edu.cn

Abstract. The combination of dealloying with acid treatment was used to fabricate mesoporous anatase TiO_2 with high specific surface area of 233 m^2/g. Using anatase TiO_2 as a matrix, a photoreduction strategy was developed to synthesize $AgCl/Ag/TiO_2$ nanocomposites with different Ti/Ag molar ratios.The morphology and properties of $AgCl/Ag/TiO_2$ nanocomposites were investigated by X-ray diffraction, field emission scanning electron microscopy and transmission electron microscopy. The $AgCl/Ag/TiO_2$ nanocomposites showed an enhanced photocatalytic activity for the degradation of methyl orange solution under visible light irradiation. The optimum Ti/Ag molar ratio in the $AgCl/Ag/TiO_2$ nanocomposites was shown to be 6:1, which was attributed to its high specific surface area of 207 m^2/g and the surface plasmon resonance effect.

Keywords: $AgCl/Ag/TiO_2$ nanocomposites, anatase TiO_2, dealloying, specific surface area, photocatalysis.

1. Introduction

For the sustainable development of human society, the solution of both environmental pollution and energy shortage is an urgent task. To solve these problems, semiconductor photocatalysis has emerged as one of the most promising technologies because it represents an easy way to utilize the energy of either natural sunlight or artificial indoor illumination. To date, TiO_2 semiconductors have proved to be one of the best photocatalysts for the purification of water, air and other environment pollutants [1]. And it was widely employed for photocatalytic applications because of its cheapness, non-toxicity, and stability.

Ever since Fujishima and Honda reported that TiO_2 irradiated with ultraviolet radiation could split water [2], discovery of TiO_2 as an effective photocatalyst was recognized as the landmark event that stimulate the investigation of photonic energy conversion by photocatalytic methods. Unfortunately, TiO_2 is not ideal for all purposes due to its wide band-gap of 3.2eV and the recombination of the photogenerated electron-hole pairs. Therefore, the improvement of the photocatalytic activity of TiO_2 is an important issue to be addressed in the photocatalytic field.

In recent years, considerable efforts were made to fabricate semiconductor-noble metal heterojunctions to align the Fermi energy levels for improving the photocatalytic activity [3].Among all noble metals, silver is of particular interest because of its much lower cost as compared with Au, Pd and Pt etc. Recent studies have been shown that nanostructure Ag enhanced the photocatalytic activity of TiO_2 [4-6].In addition, the rapid development of plasmonic Ag@AgCl photocatalysts has offered a new platform to improve the photocatalytic activity of TiO_2 [7-9]. For example, Yu et al. [10] prepared TiO_2/AgCl/Ag with excellent photocatalytic activity for the degradation of methyl orange (MO). Moreover, Li et al. [11] introduced the Ag/AgCl@helical chiral TiO_2nanofibers with enhanced photocatalytic activity for the degradation of 17-β-ethinylestradiol.

Thus, exploration of the nanosized Ag and AgCl to modify nanocrystalline TiO_2 is of great interest. As is known to all, the higher the crystallinity and the higher surface area of components, the better is the photocatalytic performance of the catalysts [12].Therefore, it is essential to investigate the AgCl/Ag/TiO_2nanomaterials with high crystallinity and specific surface area.

There are many reports on photocatalytic properties of AgCl/Ag/TiO_2 nanocomposites under UV-light irradiation. In our work, we have reported the novel AgCl/Ag/TiO_2nanocomposites with high specific surface area and enhanced photocatalytic activity under visible light irradiation. We have found the optimum Ti/Ag molar ratio which could improve the photocatalytic properties of TiO_2 more effectively under visible light. The anatase TiO_2 was firstly prepared by chemical dealloying of Al-Ti precursor and subsequent acid treatment. Dealloying (i.e., selectively removing one or more active elements from alloys) was proven to be effective in producing three-dimensional (3D) nanoporous materials [13-17]. In order to improve the photocatalytic activity of anatase TiO_2, the AgCl/Ag/TiO_2 nanocomposites with different Ti/Ag molar ratios were prepared by a photoreduction process. The photocatalytic activity of the anatase TiO_2 and AgCl/Ag/TiO_2 nanocomposites was evaluated by the degradation of MO solution under visible light irradiation, in order to understand the influence of the noble metal.

2. Experimental

$Al_{95}Ti_5$ (nominal composition, wt.%) alloy was prepared from pure Al (purity, 99.99 wt.%) and Ti (purity, 99.9 wt.%) in a quartz crucible under argon atmosphere using a high-frequency induction furnace. Using a single roller melt spinning instrument, the pre-alloyed ingot was remelted by high-frequency induction heating and then melt-spun onto a copper roller with a diameter of 0.35 m at a speed of 1000 revolutions per minute (rpm) in a controlled argon atmosphere.

The as-spun $Al_{95}Ti_5$ ribbons were dealloyed in a 4 M NaOH aqueous solution under free corrosion conditions. After dealloying for 2 h at ambient temperature, titanate nanomaterials were obtained (Ti+NaOH→Na-titanate). Then the as-dealloyed samples were washed thoroughly with deionized water for several times, in order to remove the residual chemical substances. After drying in air, the as-dealloyed samples were immersed in a 0.1 M HCl aqueous solution in a water bath at 70℃ for 2 h ($\text{Na - titanate} + \text{HCl} \xrightarrow{\text{water bath at } 70^{\circ}\text{C}} TiO_2 + NaCl$) and then washed thoroughly with deionized water. Finally, the anatase TiO_2 was acquired.

AgCl/Ag/TiO_2 nanocomposites with different Ti/Ag molar ratios of 12:1, 6:1 and 3:1 were prepared by the photoreduction method. A typical fabrication process of AgCl/Ag/TiO_2 with the molar ratio of 12:1 was described as follows: 20 mg as-obtained anatase TiO_2 was dispersed in a 10 mL ethanol solution, and 3.5 mg of $AgNO_3$ was added to the suspension. The mixture was stirred magnetically for 30 min. Then, the above mixture suspension was illuminated with UV radiation with the use of a 250 W high-voltage mercury lamp for 10 min. After illumination, the obtained AgCl/Ag/TiO_2 products were thoroughly washed with distilled water and dried at ambient temperature.

All products were characterized by X-ray diffraction (XRD), field emission scanning electron microscopy (FESEM), energy dispersive spectroscopy (EDS), transmission electron microscopy (TEM) with selected area electron diffraction (SAED), and high-resolution transmission electron microscopy (HRTEM). In addition, N_2 adsorption-desorption experiments of the samples were carried out at -196 ℃ by a Gold APP V-Sorb X800 surface area and porosity analyzer. Brunauer-Emmett-Teller (BET) method was used to calculate the specific surface area. The pore size distribution was measured from the desorption branch of isotherm using the corrected form of Kelvin equation by the Barrett-Joyner-Halenda (BJH) method.

The photocatalytic activity of the samples was evaluated by the degradation of MO aqueous solution at ambient temperature [18-21], because MO is a kind of chemically stable and persistent pollutant dye. The photocatalytic experiments were carried out by adding 10 mg TiO_2 or $AgCl/Ag/TiO_2$ photocatalysts into 25 mL MO solution (10 mg/L). A 300 W xenon lamp with a 420 nm cut-filter was used as the visible light source. The suspension was stirred in the dark for 20 min to obtain the saturated adsorption of MO before illumination. The concentration of the MO solution was measured by monitoring the absorbance at 464 nm (the maximum absorbance wavelength of MO) on a UV-Vis spectrophotometer as a function of irradiation time. The photodegradation efficiency, given as a percentage, refers to the difference in the concentration of MO solution before irradiation (C_0) and after light irradiation for M (M= 20, 40 and 60 min) (C_M) divided by C_0 (i.e., $100[C_0\text{-}C_M]/C_0$).

3. Results and Discussion

Figure 1 shows XRD patterns of sodium titanate (Na-titanate) and anatase TiO_2 of the $Al_{95}Ti_5$ ribbons after dealloying process in NaOH solution. It indicats that the titanium atoms in the $Al_{95}Ti_5$ alloy were completely transformed to Na-titanate in figure 1(a) [22-24]. Figure 1(b) presents the XRD pattern of the anatase TiO_2, which was prepared by acid treatment of the Na-titanate samples in the 0.1 M HCl solution. During the course of acid treatment, the phase transformation (from Na-titanate to anatase TiO_2) has occurred on the basis of the typical Bragg reflections of the as-treated samples, according to the standard PDF file (# 21-1272).

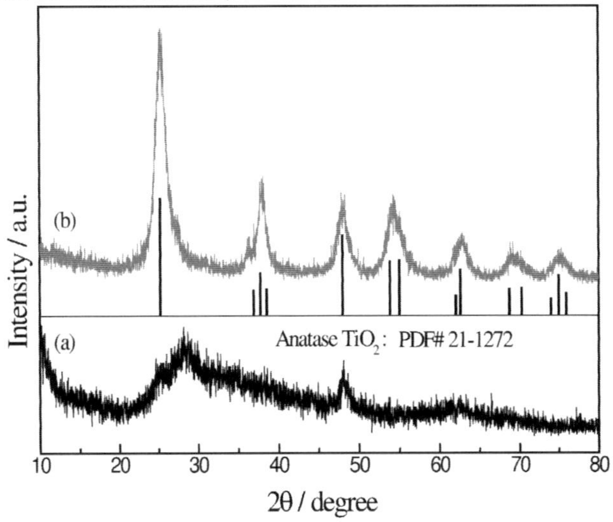

Figure 1. XRD patterns of the (a) Na-titanate samples obtained by dealloying the $Al_{95}Ti_5$ precursor in the 4 M NaOH aqueous solution and (b) the anatase TiO_2 samples acquired via sacrificing the Na-titanate in the 0.1 M HCl aqueous solutionat 70°C.

The Na-titanate samples can be easily prepared by selective leaching of the Al component from Al-Ti alloy in the 4 M NaOH solution in our experiment. Figure 2(a) shows the typical FESEM image (left) and corresponding EDS spectrum (right) of the Na-titanate samples. The Na-titanate samples adopt a uniform 3D network structure and are composed of Na, Ti and O elements. Figure 2(b) presents the morphology and corresponding EDS spectrum of the anatase TiO_2 obtained by acid treatment of Na-titanate in the 0.1 M HCl solution. As can be seen clearly from the EDS spectrum, the anatase TiO_2 samples are mainly comprised of O and Ti elements. In addition, an amount of Cl elements could also be detected due to the strong adsorption of Cl^- ions during the acid treatment in the HCl solution.

The microstructure of the anatase TiO_2 was further characterized by TEM and HRTEM (Figure 3). Figure 3(a) presents the morphology of anatase TiO_2. The corresponding SAED pattern (Figure 3(b)) exhibits a diffuse ring pattern, implying the nature of nanocrystals. The polycrystalline reflection rings can be indexed as (101), (004), (200) and (211) planes of the anatase TiO_2(PDF#21-1272). It is clear that the anatase TiO_2 exhibits a mesoporous structure formed between intra-agglomerated uniform dinky nanoparticlesat higher magnification (Figure 3(c)). The HRTEM (Figure 3(d)) image shows the lattice fringes of one anatase TiO_2 nanoparticle. The lattice fringe with a spacing of 0.35 nm can be assigned to the (101) lattice plane of anatase TiO_2.

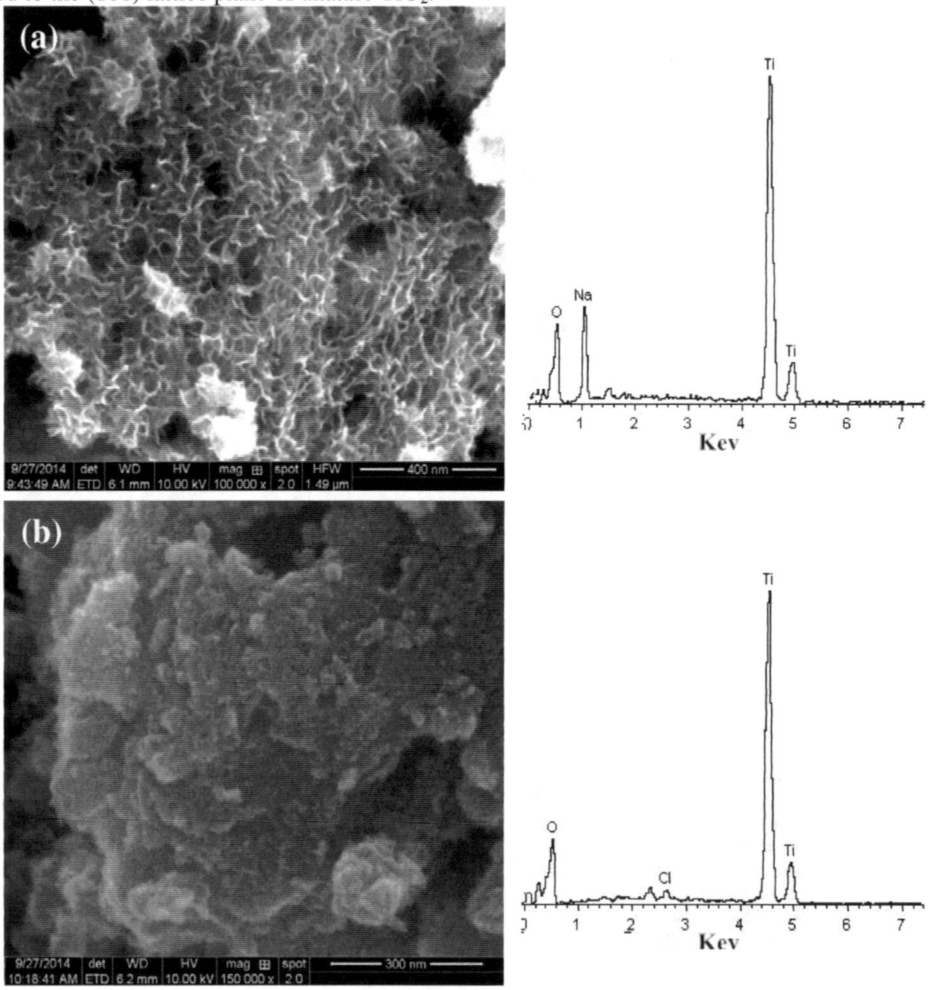

Figure 2. FESEM images and the related EDS spectra of (a) Na-titanate and (b) anatase TiO_2.

Using the mesoporous anatase TiO_2 as the matrix, the $AgCl/Ag/TiO_2$ nanocomposites with different Ti/Ag molar ratios were prepared by the photoreduction process. Figure 4 presents the XRD patterns for the $AgCl/Ag/TiO_2$ nanocomposites with the Ti/Ag molar ratios of 12:1 ($AgCl/Ag/TiO_2$-1), 6:1 ($AgCl/Ag/TiO_2$-2) and 3:1 ($AgCl/Ag/TiO_2$-3). All XRD diffraction peaks can be identified to be anatase TiO_2, metallic Ag and AgCl, according to the standard PDF files (#21-1272, #65-2871 and #31-1238). It implies that the most of the Ag^+ successfully converted to metallic Ag during the photoreduction process. It should be noted that the AgCl phase was obtained in the tested samples, resulting from the substantial Cl^- absorbed on the surface of anatase TiO_2 during the acid treatment of the as-dealloyed Na-titanate in the 0.1 M HCl solution. In addition, AgCl is an important photosensitive semiconductor material due to its self-sensitization process [25-27].

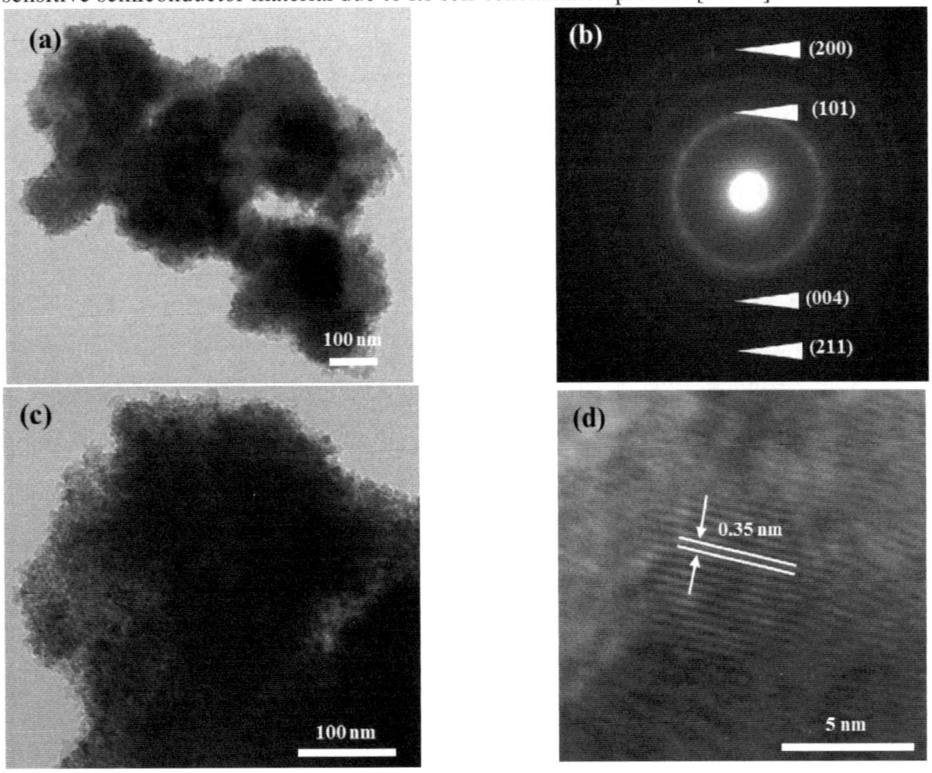

Figure 3. (a, c)TEM and (d) HRTEM imagesshowing the microstructure of the anatase TiO_2. (b) Related SAED pattern corresponding to (c).

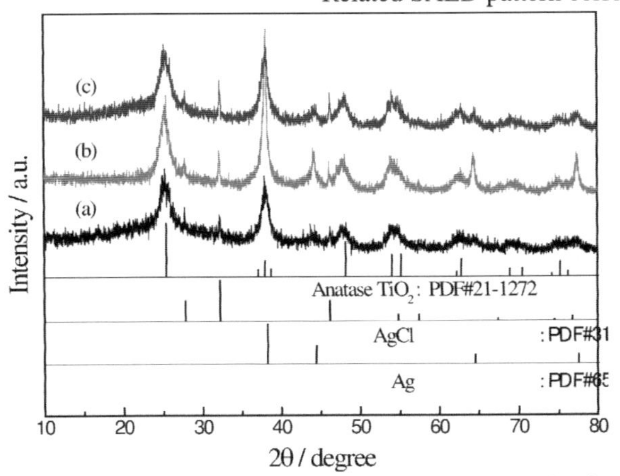

Figure 4. XRD patterns of (a) $AgCl/Ag/TiO_2$-1, (b) $AgCl/Ag/TiO_2$-2 and (c) $AgCl/Ag/TiO_2$-3.

Figure 5 shows the FESEM images (left) and related EDS spectra (right) of the different ratio $AgCl/Ag/TiO_2$ nanocomposites. It can be seen that a certain amount of Ag and AgCl nanoparticles are loaded onto the surface of the anatase TiO_2. Obviously, with the decrease of Ti/Ag molar ratio, the quantities of Ag and AgCl nanoparticles increase in the $AgCl/Ag/TiO_2$ nanomaterials. Moreover, the loading of AgCl/Ag nanoparticles has no obvious influence on the morphology of anatase TiO_2 matrix. The Ag and AgCl nanoparticles can be observed evidently on the surface of anatase TiO_2 for the

AgCl/Ag/TiO_2-1 sample in the circled region (Figure 5(a)). Moreover, the EDS result clearly indicates the existence of Ag and Cl elements in the AgCl/Ag/TiO_2-1 sample (right part of Figure 5(a)). In comparison, more Ag and AgCl nanoparticles can be found in the AgCl/Ag/TiO_2-2 and AgCl/Ag/TiO_2-3 samples, as shown in Figure 5(b and c). In addition, the Cl element is also detected from each related EDS spectrum. Furthermore, the AgCl particles with size of about 100 nm are discovered (marked by arrows), because of their decomposition under electron beam irradiation. Generally, the EDS results are in agreement with the XRD patterns (Figure 4) of the AgCl/Ag/TiO_2 nanomaterials (AgCl/Ag/TiO_2-1, AgCl/Ag/TiO_2-2 and AgCl/Ag/TiO_2-3).

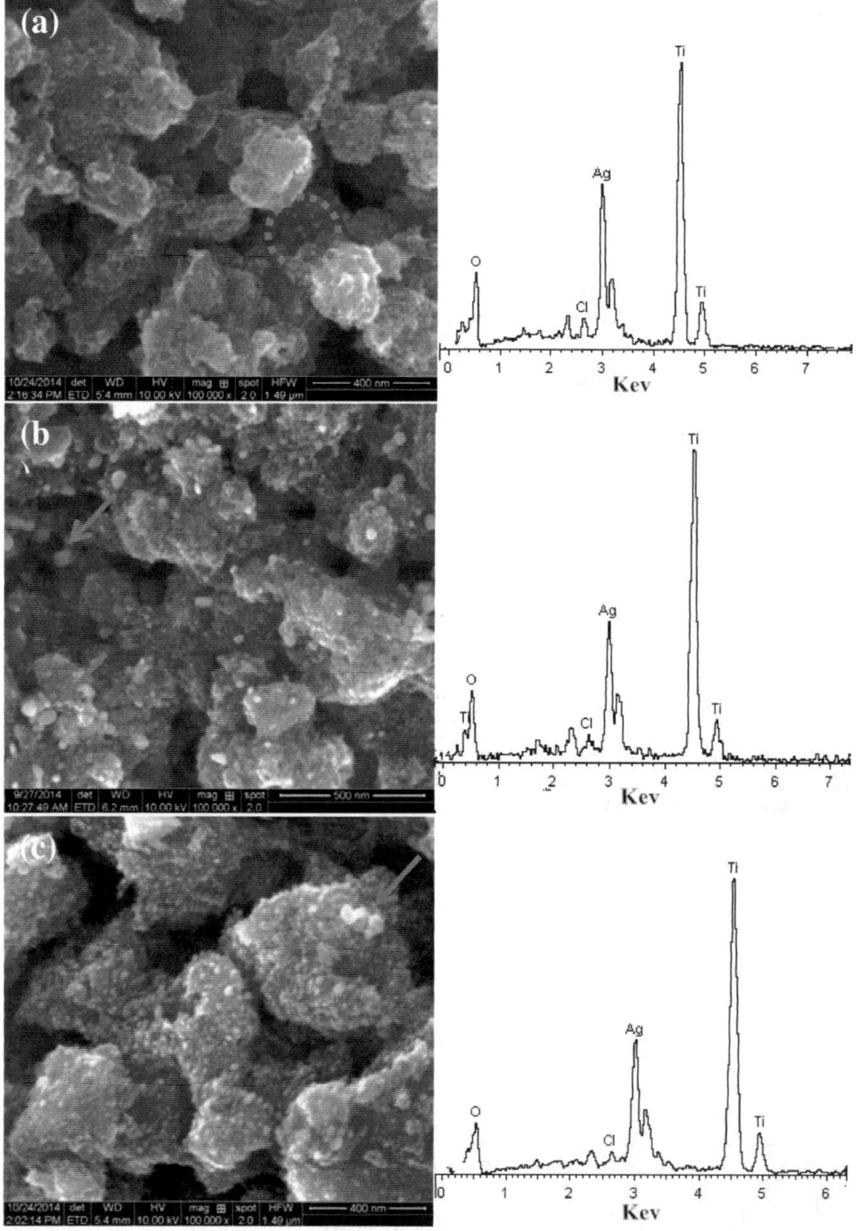

Figure 5. FESEM images and related EDS spectra of (a) AgCl/Ag/TiO_2-1, (b) AgCl/Ag/TiO_2-2 and (c) AgCl/Ag/TiO_2-3.

Figure 6 depicts the microstructure of the AgCl/Ag/TiO_2-2 sample. The TEM image (Figure 6(a)) shows that a large quantity of Ag or AgCl nanoparticles is distributed in the TiO_2 matrix. Furthermore, it can be found by close observation that the Ag and AgCl nanoparticles with size of around 5-50 nm distribute on the surface of TiO_2 matrix (Figure 6(b)). Figure 6(c) shows the corresponding SAED pattern. Besides the diffraction rings of the anatase TiO_2, the strong spots originate from the nanocrystalline Ag and AgCl. As shown in HRTEM image (Figure 6(d)), the clear lattice fringe of 0.35 nm matches the (101) crystallographic plane of the anantase TiO_2. And the fringe spacing of 0.24 nm corresponds to the (111) plane of the Ag phase.

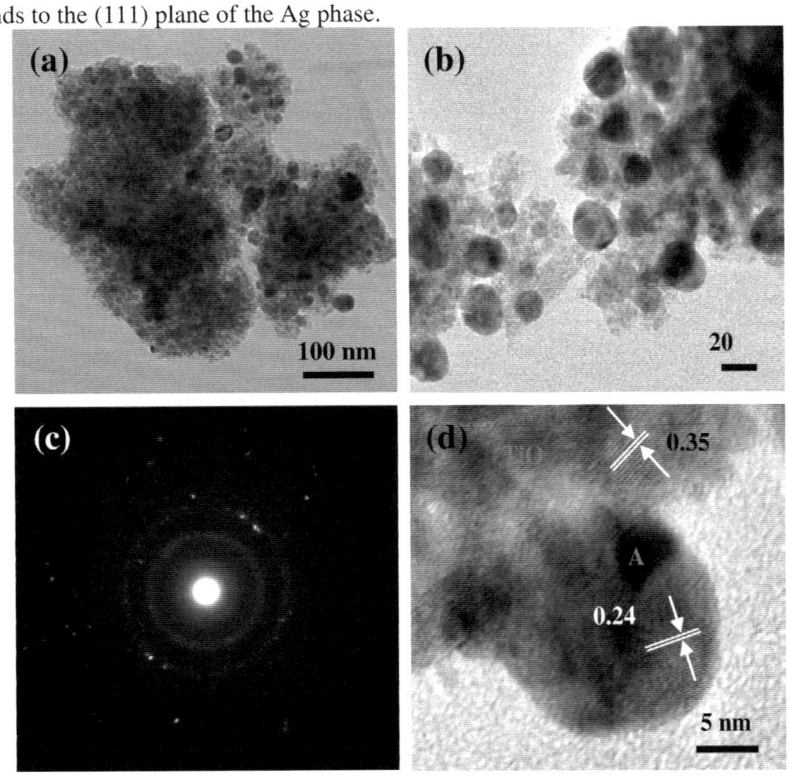

Figure 6. (a, b)TEM and (d) HRTEM imagesshowing the microstructure of AgCl/Ag/TiO_2-2 photocatalyst. (c) SAED pattern corresponding to (a).

It is well known that the specific surface area is an important parameter for enhancing photocatalytic activity. The porous structures of the anatase TiO_2 and AgCl/Ag/TiO_2 samples have also been probed by N_2 adsorption/desorption experiments. Figure 7 presents the N_2 adsorption-desorption isotherms and BJH pore size distribution curves (insets) of the anatase TiO_2 and AgCl/Ag/TiO_2-2samples. Based upon the Brunauer classification [28], Figure 7 displays type IV adsorption-desorption isotherms, which is typical characteristics of mesoporous materials. The loading of AgCl/Ag nanoparticles has a noticeable effect on the adsorption/desorption process of the anatase TiO_2. This result suggests the inter-connected mesoporosity and high pore connectivity in the framework for the anatase TiO_2 and AgCl/Ag/TiO_2 nanocomposites. According to the adsorption-desorption isotherms, the BET surface area of anatase TiO_2 and AgCl/Ag/TiO_2-2 samples was determined to be as high as 233 m^2/g and 207 m^2/g, respectively. The loading of Ag and AgCl nanoparticles slightly decreases the specific surface area of the AgCl/Ag/TiO_2 nanocomposites. The pore size distribution of the tested samples was determined using the BJH model (Figure 7(b) and (d)). The pore size analysis clearly demonstrates that both the anatase TiO_2 and AgCl/Ag/TiO_2 nanomaterials possess mesopores with pore sizes of 2−10 nm. The Ag and AgCl particles grew on the surface of anatase TiO_2, without damaging the pore structure. In addition, the specific surface area of the anatase TiO_2 is much higher than that of TiO_2 nanoparticles reported in the literature, such as Aeroxide P25 (BET 50 m^2/g, 21 nm in diameter, anatase 80% + rutile 20%, Degussa), Aeroxide P90 (BET 90-100 m^2/g, 14 nm in diameter, anatase 90% + rutile 10%, Degussa) [29].

According to above results and analysis, herein we elaborate the formation mechanisms of the anatase TiO_2 and AgCl/Ag/TiO_2 nanocomposites. Based on related literature [30], the Na-titanate has a layered structure as schematically shown in Figure 8. However, during the reaction with the dilute HCl at 70 ℃, the dehydrated titanate rearranges to form anatase lattice with changing the morphology [31]. In the mild condition, the anatase TiO_2 nanocrystals tend to form the nanoparticles, which would agglomerate to shape a mesoporous structure. In addition, the Cl^- ions can be easily absorbed on the surface of anatase TiO_2 because of the large specific surface area of the TiO_2 and the strong adsorption of Cl^- ions [32, 33]. The adsorbed Cl^- ions on the surface of anatase TiO_2 nanomaterials act as active sites for adsorption of Ag^+ ions before photoreduction. The chemical reaction leads to AgCl nanoparticles forming on the surface of the anatase TiO_2. When irradiating with UV, the Ag^+ ions are reduced to metallic Ag by the photo-generated electrons and the Ag nanoparticles nucleate on the surface of the anatase TiO_2. With the increase of Ag content, the surface area of the mesoporous anatase TiO_2 decreases in the AgCl/Ag/TiO_2 nanomaterials.

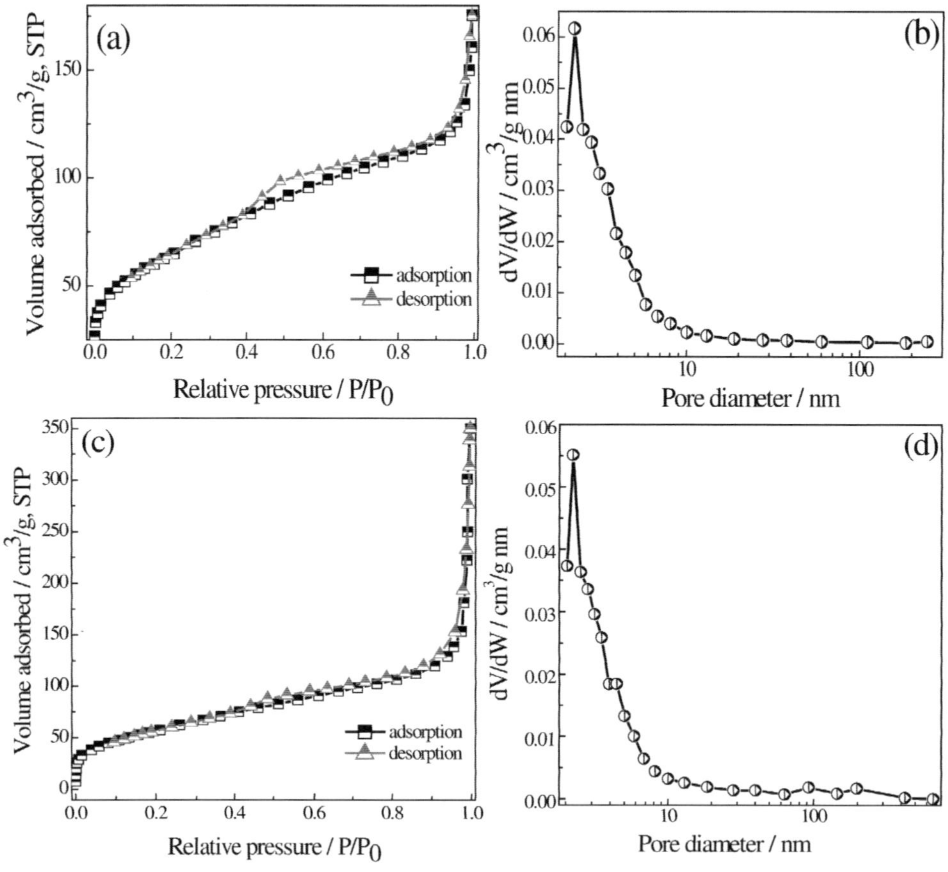

Figure 7. Nitrogenadsorption-desorption isotherms and BJH pore size distribution curves (insets) of the anatase TiO_2 (a),(b) and AgCl/Ag/TiO_2-2(c),(d).

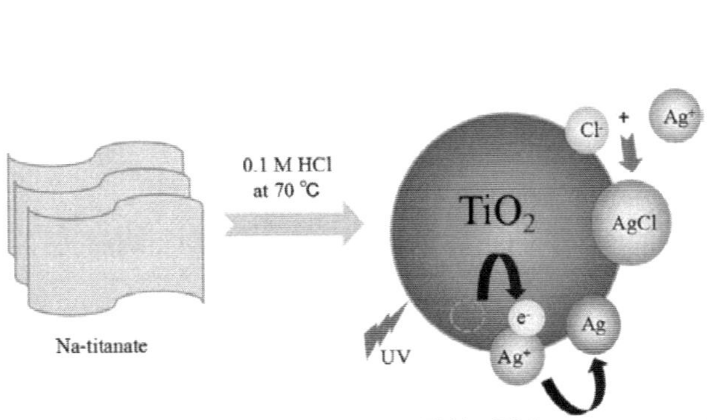

Figure 8. Proposed mechanism to form $AgCl/Ag/TiO_2$ nanocomposites.

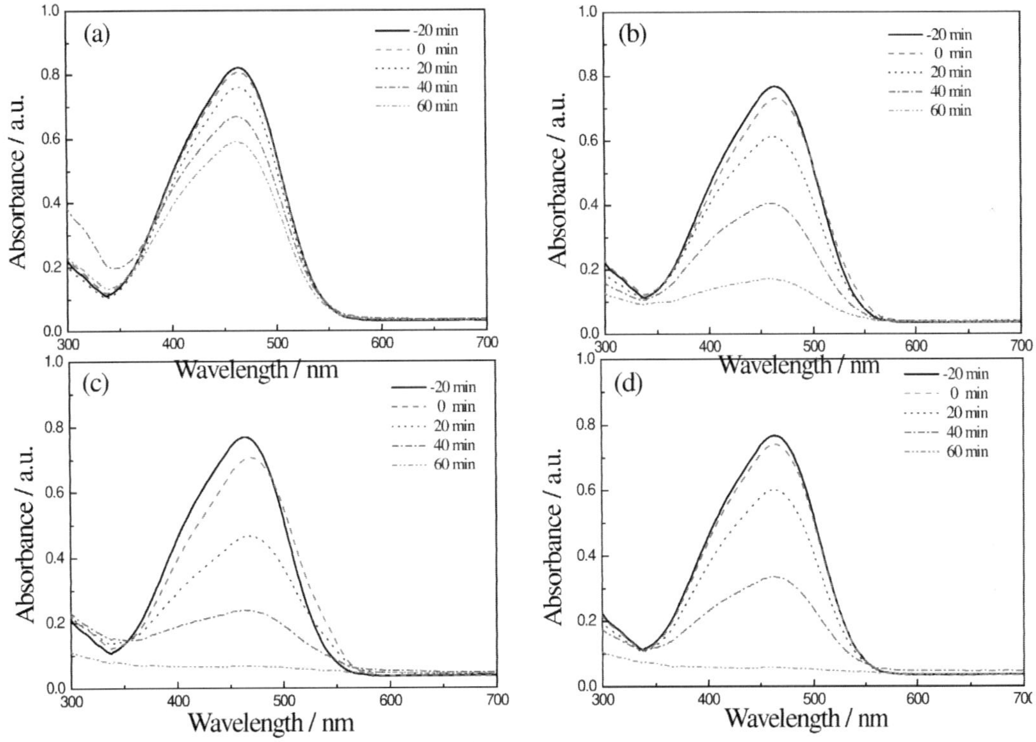

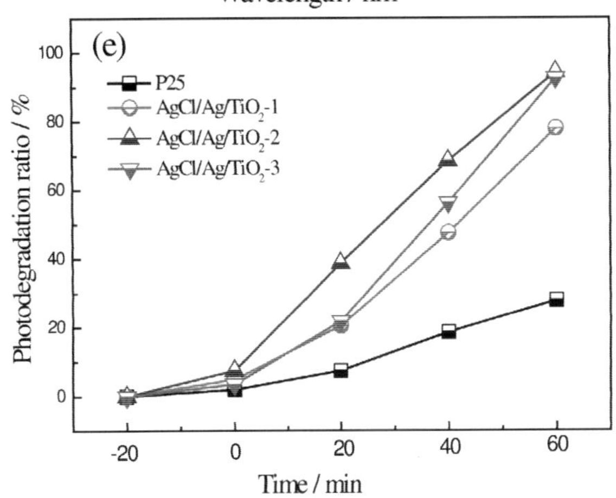

Figure 9. Absorption spectra of MO before and after photodegradation by P25 and $AgCl/Ag/TiO_2$ photocatalysts: (a) P25, (b) $AgCl/Ag/TiO_2$-1, (c) $AgCl/Ag/TiO_2$-2, (d) $AgCl/Ag/TiO_2$-3. (e) Comparison of photocatalytic activities of all the photocatalysts.

The photocatalytic activities of P25 and Ag/AgCl/TiO_2 nanocomposites with different Ti/Ag molar ratios were tested for the photodegradation of MO solution under visible light irradiation. Figure 9(a-d) presents the MO absorbance spectra with different photocatalysts. Obviously, the degradation of MO is slight in the presence of P25, while it is greatly enhanced with the nanocomposites (AgCl/Ag/TiO_2-1,2,3). The photocatalytic efficiencies of P25 and AgCl/Ag/TiO_2 nanomaterials were calculated and the results are shown in Figure 9(e). From Figure 9(e) we found that the saturated adsorption capacity of AgCl/Ag/TiO_2 nanomaterials before illumination is higher than P25, because of its high specific surface area. In addition, the photocatalytic activities of all AgCl/Ag/TiO_2 samples are higher than that of P25. When visible light is irradiated to the Ag/AgCl/TiO_2 photocatalysts, the Ag nanoparticles lead to a surface plasmon resonance effect. The excited electrons in the noble metal nanoparticles transfer to a conduction band of semiconductor, resulting in an interfacial separation of electrons and holes that causes photocatalysis [34-36]. Interestingly, the AgCl/Ag/TiO_2-2 shows the best photocatalytic performance among all the four samples. Obviously, AgCl/Ag/TiO_2-2 degrades the MO solution quickly, and the degradation ratio is almost 100% after visible light irradiation of 60 min. The optimal amount of Ag and AgCl nanoparticles can improve the photocatalytic activity of TiO_2 photocatalysts. Although the degradation ratio of AgCl/Ag/TiO_2-3 sample is also almost 100% after visible light irradiation of 60 min, with the increase of Ag content, the cost will be greatly increased.

4. Conclusions

The mesoporous anatase TiO_2 with large specific surface area of 233 m^2/g can be synthesized through chemical dealloying of the Al-Ti precursor and subsequent acid treatment of the as-dealloyed samples. The AgCl/Ag/TiO_2 nanocomposites with different Ti/Ag molar ratios can be obtained by the photoreduction of $AgNO_3$ using the mesoporous anatase TiO_2 as the matrix. The Ti/Ag molar ratios have a significant influence on the photocatalytic activity of the AgCl/Ag/TiO_2 nanocomposites towards the degradation of MO. And the AgCl/Ag/TiO_2-2 nanocomposite with the Ti/Ag molar ratio of 6:1 displays the best photocatalytic performance. In addition, the enhanced photocatalytic activity of the AgCl/Ag/TiO_2-2 catalyst could be rationalized by the surface plasmon resonance effect and its high specific surface area of 207 m^2/g.

Acknowledgements

The authors acknowledge financial support from the National Natural Science Foundation of China (No. 51171072), and Science and technology development project of Shandong Province (No. 2015GGX102016).

References

[1] H Wang, L Zhang, Z Chen, J Hu, S Li, Z Wang, J Liu and X Wang 2014 *Chem. Soc. Rev.* **43** 5234–5244
[2] A Fujishima and K Honda 1972 *Nature* **238** 37–38
[3] Y Qu and X Duan 2013 *Chem. Soc. Rev.* **42** 2568–2580
[4] M M Khan, S A Ansari, M I Amal, J Lee and M H Cho 2013 *Nanoscale* **5** 4427–4435
[5] Y Liu, C Y Liu, J H Wei, R Xiong, C X Pan and J Shi 2010 *Appl. Surf. Sci.* **256** 6390–6394
[6] H Zhang, X Li and G Chen 2009 *J. Mater. Chem.* **19** 8223–8231
[7] J Tian, R Liu, G Wang, Y Xu, X Wang and H Yu 2014 *Appl. Surf. Sci.* **319** 324–331
[8] P Wang, B Huang, X Qin, X Zhang, Y Dai, J Wei and M H Whangbo 2008 *Angew. Chem., Int. Ed.* **47** 7931–7933
[9] Z F Zhao, Y Z Wang, J Xu, C Y Shang and Y Wang 2015 *Appl. Surf. Sci.* **351** 416–424
[10] J Yu, G Dai, B Huang 2009 *J. Phys. Chem. C* **113** 16394–16401
[11] D Wang, Y Li, G L Puma, C Wang, P Wang, W Zhang and Q Wang 2013 *Chem. Commun.* **49** 10367–10369
[12] H Zhang, G Wang, D Chen, X J Lv and J H Li 2008 *Chem. Mater.* **20** 6543–6549
[13] J Erlebacher, M J Aziz, A Karma, N Dimitrov and K Sieradzki 2001 *Nature* **410** 450–453

[14] Z Qi, C C Zhao, X G Wang, J K Lin, W Shao, Z H Zhang and X F Bian 2009 *J. Phys. Chem. C* **113** 6694–6698
[15] Z Qi, H R Geng, X G Wang, C C Zhao, H Ji, C Zhang, J L Xu and Z H Zhang 2011 *J. Power Sources* **196** 5823–5828
[16] Z Zhang, C Zhang, J Sun, T Kou, Q Bai, Y Wang and Y Ding 2013 *J. Mater. Chem.A* **1** 3620–3628
[17] J Sun, J Shi, J Xu, X Chen, Z Zhang and Z Peng 2015 *J. Power Sources* **279** 334–344
[18] Z Zhao, J Xu, P K Liaw, B Wu and Y Wang 2014 *Corros. Sci.* **84** 66–73
[19] Z F Zhao, J Xu, C Y Shang, R Ye and Y Wang 2015 *Corros. Sci.* **98** 651–660
[20] Z F Zhao, Y Z Wang, J Xu and Y Wang 2015 *RSC. Adv.* **5** 59297–59305
[21] Z F Zhao, X X Zhu, M Zuo, J Xu and Y Wang 2016 *CrystEngComm.* **18** 1636–1644
[22] Y W L Lim, Y Tang, Y H Cheng and Z Chen 2010 *Nanoscale* **2** 2751–2757
[23] R Ma, T Sasaki and Y Bando 2005 *Chem. Commun.* 948–950
[24] F Wu, Z Wang, X Li and H Guo 2011 *J. Mater. Chem.* **21** 12675–12681
[25] T Morimoto, K Suzuki, M Torikoshi, T Kawahara and H Tada 2007 *Chem. Commun.* 4291–4293
[26] M Lanz, D SchuErch and G Calzaferri 1999 *J. Photochem. Photobiol. A* **120** 105–117
[27] Y Nakajima, Q L Jin and H Tada 2008 *Electrochem. Commun.* **10** 1132–1135
[28] M Kruk and M Jaroniec 2001 *Chem. Mater.* **13** 3169–3183
[29] Y Masuda, T Ohji and K Kato 2010 *Cryst. Growth Des.* **10** 913–922
[30] H K Wang, W Shao, F Gu, L Zhang, M K Lu and C Z Li 2009 *Inorg. Chem.* **48** 9732–9736
[31] H Y Zhu, Y Lan, X P Gao, S P Ringer, Z F Zheng, D Y Song and J C Zhao 2005 *J. Am. Chem. Soc.* **127** 6730–6736
[32] R C Newman and K Sieradzki 1994 *Science* **263** 1708–1709
[33] M Hakamada and M Mabuchi 2006 *Nano Lett.* **6** 882–885
[34] W Zhao, Z Zhang, J Zhang, H Wu, L Xi and C Ruan 2016 *Mater. Lett.* **171** 182–186
[35] Y Tian and T Tatsuma 2005 *J. Am. Chem. Soc.* **127** 7632–7637
[36] A Furube, L Du, K Hara, R Katoh and M Tachiya 2007 *J. Am. Chem. Soc.* **129** 14852–14853

IOP Conf. Series: Materials Science and Engineering 182(2017) 012007 doi:10.1088/1757-899X/182/1/01200

Enhanced solar photocurrent of $LaTaON_2$ photoanodes via electrochemical treatment

Huiting Huang[1,2,3] and Zhaosheng Li[1,2,3*]

[1] College of Engineering and Applied Sciences, Nanjing University, Nanjing 210093, China
[2] Collaborative Innovation Centre of Advanced Microstructures, National Laboratory of Solid State Microstructures, Nanjing University, Nanjing 210093, China
[3] Jiangsu Key Laboratory for Nano Technology, Nanjing University, Nanjing 210093, China

E-mail: *zsli@nju.edu.cn

Abstract. Having a theoretical 18.5% solar-to-hydrogen efficiency, $LaTaON_2$ has emerged as a promising photoanode material. However, its performance is crucially limited by low photocurrent in the past reports. To improve its solar photocurrent, a negative electrochemical treatment was applied for the $LaTaON_2$ photoanode. The sample powder was successfully synthesized by a flux assisted-nitridation with precursor derived from solid state sintering. And the $LaTaON_2$ photoanodes were fabricated by electrophoretic deposition with a post-necking procedure. The solar photocurrent of as-fabricated $LaTaON_2$ photoanode has increased to 1.2 mA cm^{-2} at 1.6 V_{RHE} after the negative electrochemical treatment in the dark. The photoanodes with and without the electrochemical treatment were investigated by scanning electron microscopy (SEM), Mott–Schottky test, transient photocurrent and open-circuit photovoltage. The results showed that the enhancement maybe ascribed to the soared carrier density and elimination of surface recombination centre. Therefore it is proposed that the electrochemical treatment eliminates the surface recombination centre of the oxynitrides leading to increased solar photocurrent. It was also found that the $SrTaO_2N$ photoanode had increased photocurrent after the electrochemical treatment. This study provides a facile and general way to improve the solar water-splitting current of photoanodes.

1. Introduction

With the depletion of fossil fuels, a clean way to obtain energy is the only access to our sustainable development. On figuring out this way, photoelectrochemical water splitting seems to be the best choice [1]. It feeds on sunlight and water, produces hydrogen and oxygen. However, the availability of high efficient photoelectrodes is now hampering the realization of this technique [2]. Tremendous amount of effort has been paid to implement the full photoelectrochemical performances of$BiVO_4$, Fe_2O_3, WO_3 and TiO_2 [3]. (Oxy)nitrides, which possess ideal band structures and promising high solar-to-hydrogen (STH) efficiencies, may become the alternative photoanode material to address this dilemma [3]. At present, the solar photocurrents of the majority of (oxy)nitrides are far from their theoretical values.

The slow kinetic process of water oxidation and the existence of surface recombination centre could be the main limiting factors for the improvement of photoanodes. To accelerate the kinetic process of water oxidation, surface modification of oxygen evolution catalyst is adopted to improve thesolar water-splitting current of (oxy)nitride photoanode [4-6]. On the other hand, surface recombination centre on the photoanodes has been revealed by the transient absorption spectroscopy [7]. Therefore, the elimination of surface recombination centre would offer another way to ameliorate the photoelectrochemical performance of the photoelectrodes [8].

Overlayer has been deposited on hematite photoanodes to passivate surface states via atomic layer deposition [8].Due to the high synthetic temperature accompany with aggressive ammonia flow, regular (oxy)nitride films are hard to obtain [3]. (Oxy)nitride photoanodes are usually particle-ensemble films, which may be hard for the deposition of overlayers Recently, the photocurrents of $BiVO_4$, Fe_2O_3, WO_3 and $TiO_2$are enhanced by surface treatment electrochemically [9-12], which may be an alternative viable strategy to applied on the particle-ensemble (oxy)nitride films. Taking $LaTaON_2$ as the study subject [13], the electrochemical treatment is adopted to improve its solar photocurrent for the first time. The sample powders were successfully synthesized by a flux assisted-nitridation with precursor derived from solid state sintering. And the $LaTaON_2$ photoanodes were fabricated by electrophoretic deposition with a post-necking procedure. The solar photocurrent of as-fabricated $LaTaON_2$ photoanode has increased to 1.2 mA cm^{-2} at 1.6 V_{RHE} after the negative electrochemical treatment in the dark.And (photo-)electrochemical measurements are carried out to quest for the cause of the enhancement [14].The results show that the enhancement may be due to the increased carrier density and suppressed surface recombination center. This study has proved that the electrochemical treatment is a facile and general way to improve the solar water-splitting current of (oxy)nitride.

2. Experimental section

2.1. Synthesis of the sample
The $LaTaON_2$ powder sample was synthesized by a two-step method, which combined solid state reaction and the following flux-assisted nitriding. Briefly, 2 mmol La_2O_3 and 2 mmol Ta_2O_5 were thoroughly grounded with an agate mortar and pestle. Then, the mixture was transferred into a corundum boat, and calcined at 1623 K for 10 h. The as-obtained oxide precursor was thoroughly mixed with an equal weight of KCl, after which it was sintered at 1223 K for 10 h under the flowing ammonia (500 mL/min). The sample was washed with deionized water for several times to remove the excess flux, and was dried at 333 K overnight.

2.2. Fabrication of the photoanodes
The $LaTaON_2$ photoanodes were fabricated by an electrophoretic deposition method (EPD) followed by a post-necking procedure. In a typical procedure, 20 mg sample powder and 5 mg iodine were added to 25 mL acetone. The ultrasonic dispersion was applied on the mixture for 30 min, after which a suspension was formed. Two fluorine-doped tin oxide substrates (FTO) with their conductive sides face to face were immersed in the suspension and kept at 10 V for 1 min. The as-obtained electrode film was dried at room temperature and then 30μL 10 mM $TiCl_4$ methanol solution was dripped to the particle-ensemble film with a 5 min heat-treatment at 473 K. After the drip and heat treatment route were repeated for several times, the electrode film was kept at 713 K for 30 min under the flowing ammonia (300 mL/min).

2.3. Characterization of the sample and electrodes
The X-ray diffraction (XRD) pattern of the as-synthesized powder sample was obtained with an Ultima III X-ray diffractor (Rigaku), and the absorption spectrum of the sample by a UV2550 UV-vis spectrophotometer (Shimadzu). The morphology of the powder and photoanodes was observed by a scanning electron microscope (SEM) (Nova NanoSEM).

2.4. Electrochemical treatment and (photo-)electrochemical measurement of the electrodes

IOP Conf. Series: Materials Science and Engineering **182**(2017) 012007 doi:10.1088/1757-899X/182/1/012007

Both electrochemical measurement was carried out with a CHI633c electrochemical workstation. The three-electrode configuration was used with the $LaTaON_2$ photoanode, Pt foil and the saturated calomel electrode (SCE) serving as the work electrode, the counter electrode and the reference electrode. 1M NaOH (pH=13.6) was used as the electrolyte. The applied bias reported here are against the reversible hydrogen electrode (RHE) according to the conversion relationship:

$$E_{RHE} = E_{SCE} + 0.059 \times pH + 0.242 \quad (1)$$

where pH represents the pH value of the electrolyte. For the photoelectrochemical measurement, an OrielSol3A Class AAA Solar Simulator, which was calibrated to 100 $mWcm^{-2}$ (AM1.5G) by a Newport 91150 reference cell, was used as the light source. To ensure the incident areas of different photoanodes, a black mask with a 6 mm hole was pasted on the back of the electrode (glass). Therefore, all the photocurrents were normalized to 1 cm^2.

3. Results and discussion

3.1. Characterization of the sample powder

The XRD pattern (figure 1(a)) of the as-obtained powder matches the standard pattern (PDF#47-1366) well. In other words, the $LaTaON_2$ sample powder, derived from the two-step method, has been successfully obtained [13]. The XRD pattern with sharp peaks indicates the as-obtained sample powder is of high crystallinity without strong orientation. As shown in figure 1(b), the as-obtained orange powder has an optical absorption edge at ca.650nm. The inset of Tauc plot based on the absorption spectrum indicates that the as-obtained powder possesses an indirect band gap of ca. 1.9 eV, which is in consistent with the literature [15, 16].

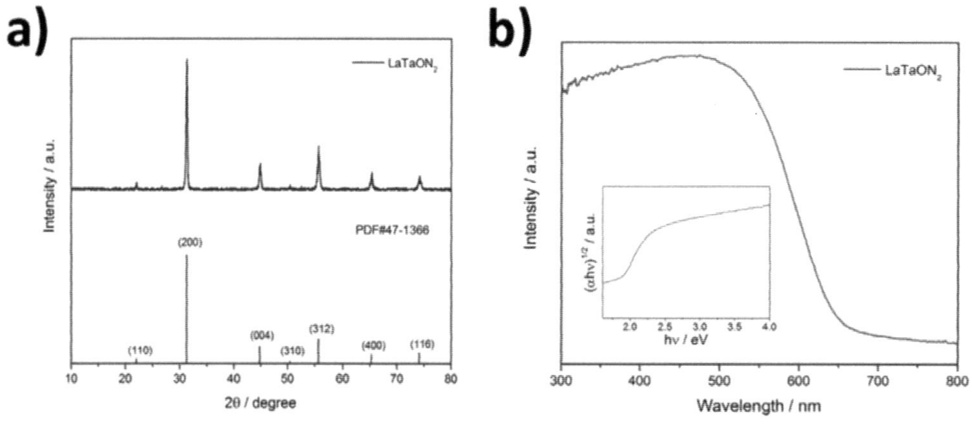

Figure 1. XRD pattern (a) and the absorption spectrum (b) of the as-obtained sample powder, the inset shows the Tauc plot of the sample based on the spectrum.

Figure 2 shows the surface morphology of the as-obtained $LaTaON_2$ powder. The sample powder is a mixture of large particles with several micrometres and small particles with several hundred nanometres. As shown in figure 2(a), it appears that the large particle is formed by the aggregation of small particles. All particles are of smooth surface, which may be originated from the flux-assisted nitriding procedure [17]. It is supposed that voids are formed for most of the (oxy)nitrides during the high-temperature nitriding procedure. As for $LaTaON_2$, three O^{2-} ions are substituted by two N^{3-} ions during the forming process. In a direct nitriding route, the oxide precursor ($LaTaO_4$) would transform to large $LaTaON_2$ particles with voids and stress [13]. In the flux-assisted route, the substance moves more easily in the molten KCl and the stress maybe released. As for the oxide precursor, there is a phase transition of $LaTaO_4$at around 513 K [18], which indicates stress could be formed in the particle during the natural cooling-down process of the solid state reaction and the heating-up process of the nitriding procedure. Thus, the molten KCl may help the transformation to $LaTaON_2$ and releasing the stress, which might force the large oxide precursor particles to change into porous large particles or to break into small ones.

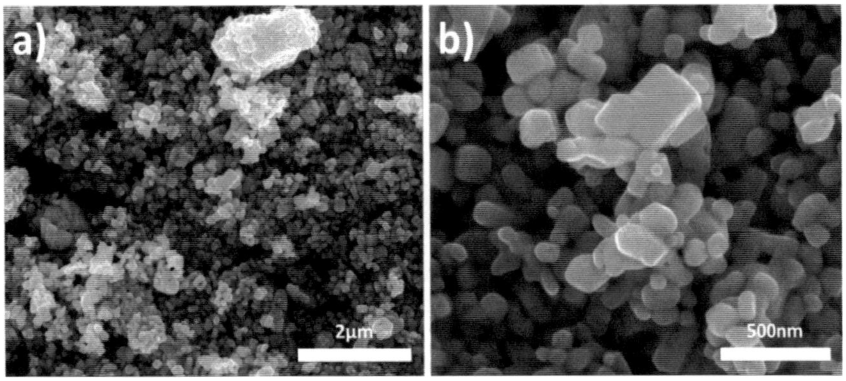

Figure 2. The SEM images of the as-obtained sample powder.

3.2. Electrochemical treatment on the $LaTaON_2$ photoanodes

Typically, the photocurrent of (oxy)nitride without modification of oxygen evolution catalystdecreases severely in minutes during the stability test, even after the first run of the linear voltammetry [6]. But the key factors for the rapid decay have not yet been found. In figure 3(a), the as-fabricated $LaTaON_2$ photoanode shows a saturated ca. 1 $mAcm^{-2}$ at 1.6 V_{RHE} with an onset potential of ca. 0.9 V_{RHE}. In the second run of the linear voltammetry, the photocurrent at 1.6 V_{RHE} sharply decreases to 0.5 $mAcm^{-2}$, while the onset potential becomes ca. 1.1 V_{RHE}. With the electrochemical treatment, the photocurrent at 1.6 V_{RHE} increases to 1.25 $mAcm^{-2}$ with a similar onset potential of ca.0.9V_{RHE}. This enhancement, however, did not maintain at the second run with a photocurrent of ca. 0.65 $mAcm^{-2}$ at 1.6 V_{RHE}. The result shows that it may provide us with a facile and practical way to improve the solar water-splitting current of photoanodes. Thus, it is necessary to figure out the electrochemical treatment and its effect on the $LaTaON_2$ photoanodes.

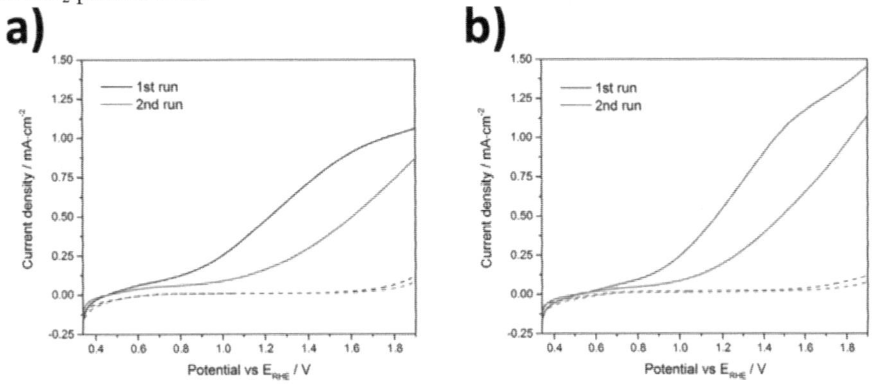

Figure 3. Photocurrent characteristics of the $LaTaON_2$ photoanodes without (a) and with (b) electrochemical treatment. The dash lines show the current of the $LaTaON_2$ photoanodes in the dark. The scan rates were $30mVs^{-1}$.

Before the electrochemical treatment, cyclic voltammograms of the as-fabricated $LaTaON_2$ photoanode in the dark show decreasing current densities during the first few cycles, and the dark current minimized and stabilized after the fifth cycle, as shown in figure 4(a). The initial cyclic voltammogram shows 0.1 $mAcm^{-2}$ at 1.6 V_{RHE}, indicated some reduced species exist on the surface of the photoanode. In the second one, the dark current decreases to 0.03 $mAcm^{-2}$ at 1.6 V_{RHE}, indicated the reduced species may be irreversible or had been removed. In the fifth one, the dark current minimized and stabilized at 0.01 $mAcm^{-2}$. After the cyclic voltammetry in the dark, the photocurrent reaches 0.9 $mAcm^{-2}$ at 1.6 V_{RHE}. Compared with the second scan of the photocurrent test in figure 3(a), the photocurrent of the same photoanode after electrochemical treatment increases to 1.15 $mAcm^{-2}$ at 1.6 V_{RHE}, which is close to the electrochemical treated one in figure 3(b). The cyclic voltammograms of the electrochemical treatment are shown in figure 4(c). During the cyclic scans between -1.48 V_{RHE} and -0.5 V_{RHE}, large amount of hydrogen formed on the surface of the photoanode. It should be noted that bare FTOs could not be immersed in the electrolyte while being conducting the electrochemical treatment, or the tin oxide would dissolve in the electrolyte resulting in insulated substrates. The similar current responds of the five scans shows that the electrochemical treatment mainly acts on the surface of the particle-ensemble film. Someone may argue that some reduced species could be produced during the electrochemical treatment, thus the increased photocurrent would come from these easily oxidized species. After the treatment, a cyclic voltammetry in the dark was performed. As shown in figure 4(d), the reduced species did exist, and exhausted after the first scan, resulting in decreasing dark current. Therefore, the increased photocurrent should not be due to the easily oxidized species formed during the electrochemical treatment.

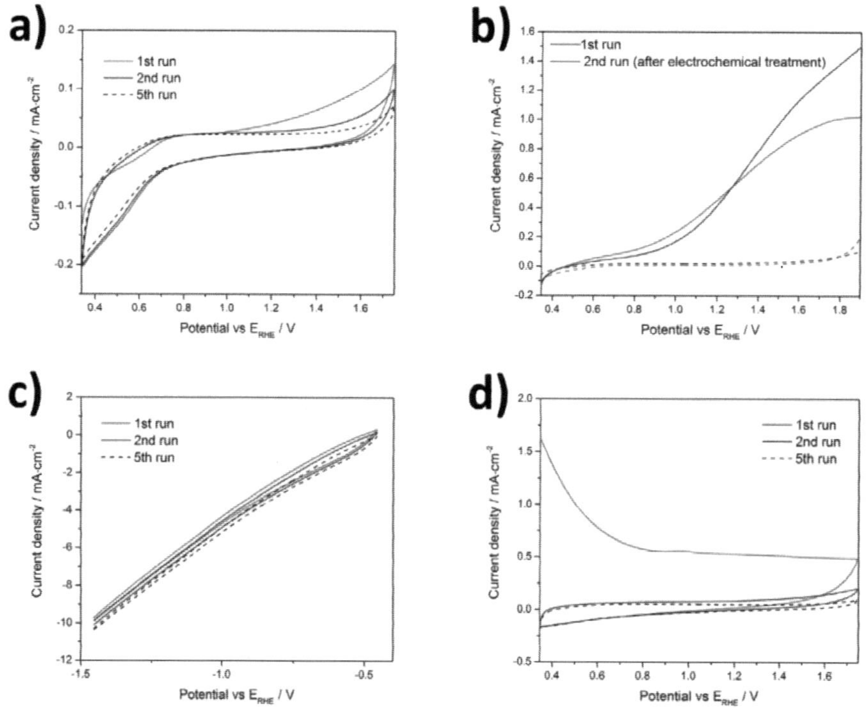

Figure 4. Cyclic voltammograms of the $LaTaON_2$ photoanode before (a) and after (d) the electrochemical treatment in the dark at the scan rate of 100 mVs^{-1}. (b) the photocurrent characteristics of the $LaTaON_2$ photoanodes before and after the electrochemical treatment at the scan rate of 30 mVs^{-1}. (c) Cyclic voltammograms of the electrochemical treatment on the $LaTaON_2$ photoanode at the scan rate of 200 mVs^{-1}.

Figure 5 shows the morphologies and cross sections of the $LaTaON_2$ photoanodes before and after the electrochemical treatment. As shown in figure 5(a) and figure 5(c), the particle-ensemble films are composed of large porous particles and small particles with TiO_x gluing them. The particle-ensemble films are of ca.1μm thick, as shown in figure 5(b) and figure 5(d). The SEM images indicate no obvious difference between the $LaTaON_2$ photoanodes before and after the electrochemical treatment. It is supposed that a few atomic layers of the sample particle would have excess oxygen impurity that deviated from stoichiometric ratio. And it has been confirmed that the removal of the superficial oxygen on the $SrTaO_2N$ particles will arise the electro-conductivity and lead to improved photocurrent. However, it requires high-temperature anneal in 5% hydrogen in argon. The electrochemical treated $LaTaON_2$ photoanodes may go through similar process leading to the decrease of the oxygen content on the surface. However, the electrochemical treated $LaTaON_2$ photoanode was sensitive to the air. After the treatment the $LaTaON_2$ photoanode was placed in air for several minutes, the photocurrent was slightly higher than the as-fabricated one. Therefore, the several electrochemical measurements were performed on the $LaTaON_2$ photoanodes before and after the electrochemical treatment.

Figure 5. SEM images of the $LaTaON_2$ photoanodes before (a,b) and after (c,d) the electrochemical treatment.

In the Mott-Schottky test, the as-fabricated $LaTaON_2$ photoanode shows a flat band position of ca. 0 V_{RHE}, and the flat band position does not change obviously after two scans of photocurrent tests, as shown in figure 6(a). In addition, there is a turning point on the Mott-Schottky plot, which may be due to the intermediate state in the bandgap [19]. After the electrochemical treatment, the photoanode shows a flat band position of ca. 0.4 V_{RHE}, and the flat band position comes closer to 0 V_{RHE} after each scan of photocurrent test, which is the flat band position of the as-fabricated one, as shown in figure 6(b). And the turning point on the Mott-Schottky plot seems to emerge after the second scan of the photocurrent test. The lower slope of the Mott-Schottky plot of the electrochemical treated one indicates higher density of charge carrier [8]. The missing turning point may indicate the elimination of the intermediate state in the bandgap, probably due to the removal of the surface recombination centre. After the second scan of the photocurrent test, the surface recombination centres seem to form, indicating that it may relate to the reaction intermediate of water oxidation. Further analysis should be obtained from the in-situ surface sensitive techniques.

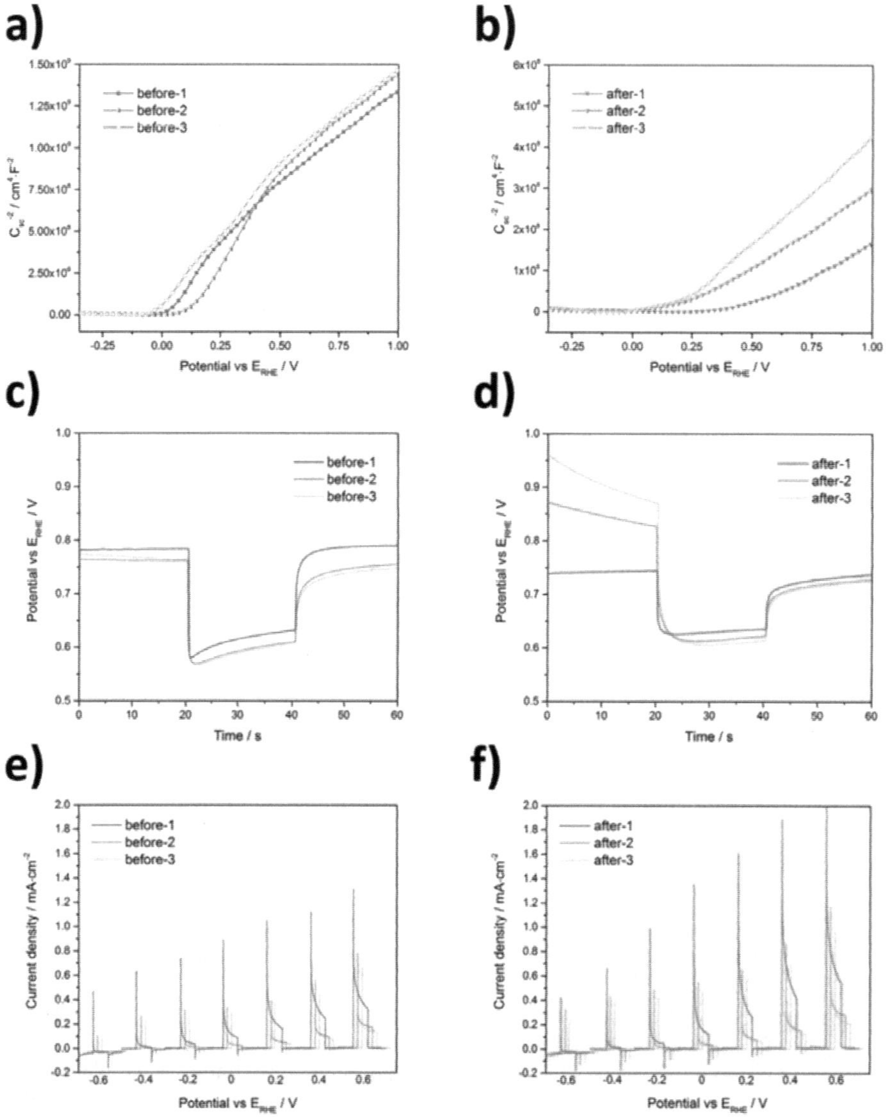

Figure 6. The left column (a, c, e) shows the electrochemical test results of the as-fabricated $LaTaON_2$ photoanodes, while the right column (b, d, f) shows the corresponding electrochemical treated ones. All the electrochemical tests were conducted before the photocurrent tests of the photoanodes, and each photoanode had been tested twice. (a, b) the Mott-Schottky test results of the photoanodes, (c, d) the open circuit voltage of the $LaTaON_2$ photoanodes in the dark and under illumination and (e, f) the transient (photo-)current of the $LaTaON_2$ photoanodes under different bias for 30s.

The as-fabricated $LaTaON_2$ photoanode shows photovoltage of0.15 V with similar responds to the illumination after the photocurrent test, as shown in figure 6(c). The electrochemical treated $LaTaON_2$ photoanode shows smaller photovoltage of ca. 0.1 V in figure 6(d). The responds to the illumination is different from the as-fabricated one, and the open circuit voltage in the dark becomes higher after the following photocurrent tests. As shown in figure 6(e) and figure 6(f), both photoanodes show large transient photocurrent once the light was on, and the transient photocurrent of the electrochemical treated photoanode is larger than the as-fabricated one. Once the light is off, the transient currents of both photoanodes disappear at bias larger than 0.2 V_{RHE}. As discuss previous, the electrochemical treated $LaTaON_2$ photoanode is sensitive to air, keeping it from further experiments, such as XPS. The electrochemical treatment was also conducted on $SrTaO_2N$ photoanodes. It is shown in figure 7 that the photocurrent of $SrTaO_2N$ photoanode increases from 0.3 $mAcm^{-2}$ for the as-fabricated one to 0.6 $mAcm^{-2}$ for the electrochemical treated one at 1.6 V_{RHE}. More (oxy)nitride photoanodes are now being electrochemical treated in our laboratory to test the validityand trying to probe the changes before and after the treatment.

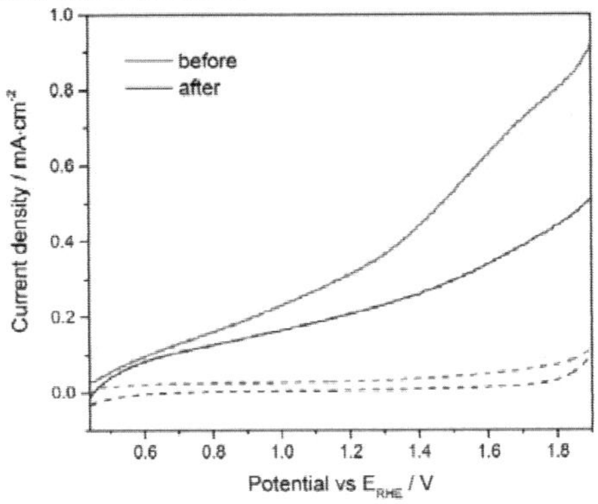

Figure 7. The photocurrent characteristics of the $SrTaO_2N$ photoanodes before (blue bold line) and after (red bold line) the electrochemical treatment. The dash lines show the current of the corresponding $SrTaO_2N$ photoanode in the dark. The scan rate was 30 mVs^{-1}.

4. Conclusion

The negative electrochemical treatment on the $LaTaON_2$ photoanodes boosts the photocurrent from ca. 1 $mAcm^{-2}$ to 1.25 $mAcm^{-2}$ at 1.6 V_{RHE} in 1M NaOH under stimulated sunlight (100 $mWcm^{-2}$). The treatment can eliminate the surface recombination center, resulting in the soaring of charge carrier. No further experiment could be conducted due to the sensitive of the treated surface of the $LaTaON_2$ photoanodes in the electrolyte or air. The negative electrochemical treatment on the $SrTaO_2N$ photoanodes also shows increase of the photocurrent, which indicates that it would be a facile and general strategy to improve the solar water-splitting current of photoanodes. More in-situ characterization techniques should be conducted to monitor the changes during the electrochemical treatment, which might cast light on the construction of high efficient photoelectrodes.

Acknowledgments

We thank Zhan Shi in the National Laboratory of Solid State Microstructures for performing the SEM observation.

This work was financially supported by 973 Program (No. 2013CB632404), a Project Funded by the Priority Academic Program Development of Jiangsu Higher Education Institutions, and National Natural Science Foundation of China (Nos. 51272101, 51272102 and 21473090).

References

[1] Li Z, Feng J, Yan S and Zou Z 2015 Solar fuel production: strategies and new opportunities with nanostructures *Nano Today* **10** 468–486

[2] Bard A J and Fox M A 1995Artificial photosynthesis: solar splitting of water to hydrogen and oxygen *Acc. Chem. Res.* **28** 141–145

[3] Li Z, Luo W, Zhang M, Feng J and Zou Z 2013 Photoelectrochemical cells for solar hydrogen production: current state of promising photoelectrodes, methods to improve their properties and outlook *Energy Environ. Sci.* **6** 347–370

[4] Ueda K *et al.* 2015 Photoelectrochemical oxidation of water using $BaTaO_2N$ photoanodes prepared by particle transfer method *J. Am. Chem. Soc.* **137** 2227–2230

[5] Landsmann S, Maegli A E, Trottmann M, Battaglia C, Weidenkaff A and Pokrant S 2015 Design guidelines for high-performance particle-based photoanodes for water splitting: lanthanum titanium oxynitride as a model *ChemSusChem* **8** 3451–3458

[6] Feng J, Luo W, Fang T, Lv H, Wang Z, Gao J, Liu W, Yu T, Li Z and Zou Z 2014 Highly photo-responsive $LaTiO_2N$ photoanodes by improvement of charge carrier transport among film particles *Adv. Funct. Mater.* **24** 3535–3542

[7] Le Formal F, Pastor E, Tilley S D, Mesa C A, Pendlebury S R, Grätzel M and Durrant J R 2015 Rate Law Analysis of Water Oxidation on a Hematite Surface *J. Am. Chem. Soc.* **137** 6629–6637

[8] Le Formal F, Tétreault N, Cornuz M, Moehl T, Grätzel M and Sivula K 2011 Passivating surface states on water splitting hematite photoanodes with alumina overlayers *Chem. Sci.* **2** 737–743

[9] Luo W, Li Z, Yu T and Zou Z 2012 Effects of surface electrochemical pretreatment on the photoelectrochemical performance of Mo-doped $BiVO_4$ *J. Phys. Chem.*C **116** 5076–5081

[10] Wang G, Yang Y, Ling Y, Wang H, Lu X, Pu Y-C, Zhang J Z, Tong Y and Li Y 2016 An electrochemical method to enhance the performance of metal oxides for photoelectrochemical water oxidation *J. Mater. Chem.*A **4** 2849–2855

[11] Luo W, Yang Z, Li Z, Zhang J, Liu J, Zhao Z, Wang Z, Yan S, Yu T and Zou Z 2011 Solar hydrogen generation from seawater with a modified $BiVO_4$ photoanode *Energy Environ. Sci.* **4** 4046–4051

[12] Zhang M, Luo W, Zhang N, Li Z, Yu T and Zou Z 2012 A facile strategy to passivate surface states on the undoped hematite photoanode for water splitting *Electrochem. Commun.* **23** 41–43

[13] Zhang L, Yan S, Feng J, Fang T, Zhong Y, Li Z and Zou Z 2014 Photoelectrochemical water oxidation of $LaTaON_2$ under visible-light irradiation *Int. J. Hydrogen Energy* **39** 7697–7704

[14] Cao D, Luo W, Feng J, Zhao X, Li Z and Zou Z 2014 Cathodic shift of onset potential for water oxidation on a Ti^{4+} doped Fe_2O_3 photoanode by suppressing the back reaction *Energy Environ. Sci.* **7** 752–759

[15] Jansen M and Letschert H P 2000 Inorganic yellow-red pigments without toxic metals *Nature* **404** 980–982

[16] Liu M, You W, Lei Z, Takata T, Domen K and Li C 2006 Photocatalytic water splitting to hydrogen over a visible light-driven $LaTaON_2$catalyst *Chin. J. Catal.* **27** 556–558

[17] Kim Y I 2014 Effects of KCl flux on the morphology, anion composition, and chromaticity of perovskite oxynitrides, $CaTaO_2N$, $SrTaO_2N$ and $LaTaON_2$ *Ceram. Int.* **40** 5275–5281

[18] Vullum F, Nitsche F, Selbach S and Grande T 2008 Solid solubility and phase transitions in the system $LaNb_{1-x}Ta_xO_4$ *J. Solid State Chem.* **181** 2580–2585

[19] Dean M and Stimming U 1987 Capacity of semiconductor electrodes with multiple bulk electronic states Part I. Model and calculations for discrete states *J. Electroanal. Chem. InterfacialElectrochem.* **228** 135–151

IOP Conf. Series: Materials Science and Engineering **182** (2017) 012008 doi:10.1088/1757-899X/182/1/012008

New Sn-Cu/C composite anode materials with high cyclic stability for Lithium ion battery

Zhaowen Huang[1,*], Wenping Xiao[1], Benqin Tang[1], Shejun Hu[2] and Qiang Ru[2]

[1] College of Electronic and Information Engineering, Shunde Polytechnic, Foshan 528300, China

[2] School of Physics and Telecommunication Engineering, South China Normal University, Guangzhou 510006, China

E-mail: *outstandingmore@163.com

Abstract. Sn-Cu/C anode material with core-shell structure was obtained, in which the crystallization of Sn-Cu alloy was weakened. The results show that the weakening of the crystallization decreases the binding force on the interstitial sites and is benefit for the inserting and escaping of the lithium ions. And "Core-shell" structure design can effectively disperse the Sn-Cu alloy particles and leave room for the volume expansion of Sn-Cu alloy. Furthermore, the volume change is controlled within the "core-shell" structure. Through these methods, Sn-Cu/C composite anode material with high cyclic stability was obtained, whose reversible capacity was 450 mAh g^{-1}, without any capacity loss till the 200th cycle. The volume change and irreversible capacity loss of Sn-based anode materials can be partly solved by the weakening of the crystallization and "core-shell" structure design.

Keywords: Lithium ion battery, anode, Sn-Cu/C composite, modification

1. Introduction

Because of the high-speed development of electronic products and electric vehicle (EV) [1], high energy density and long cycle life are required for new generation lithium ion battery. And current commercial anode material (graphite) is considered as one of the bottlenecks in the development of lithium ion battery for low theoretical capacity [2]. Sn-based alloy materials are one of the most popular fields as the substitution anode material. But for the considerable volume change, Sn-based anode materials exhibit poor cyclic stability [3-5].

To buffer the volume change, synthesizing Sn-M (active-active/inactive) systems is considered as a useful way in improving the cyclic stability [6-12]. But adding another component(active or inactive) will only decrease the volume expansion ratio of Sn, which can't completely solve the volume change. To make the Sn-M alloys become commercial anode materials, more modifications are required [13-14].

In this paper, the Sn-Cu alloy is modified by the weakening of the crystallization. And the cyclic stability is further improved by "core-shell" structure design. The crystallization degree is weakened by ball milling, to decrease the binding force from the crystal on the interstitial sites and benefit for the lithium intercalation/deintercalation. And through "core-shell" structure design, room is left for the possible volume change of Sn-Cu alloy and the volume change will be controlled within the core layer. Through these modifications, we obtain a new Sn-based anode material with high cyclic stability.

2. Experimental

Sn-Cu alloy sample synthesis: Sn-Cu alloy was synthesized by solution chemical reduction method. $SnCl_2$ and chlorides of Cu was reduced by $NaBH_4$, in which Sn:Cu=15:85 (atomic ratio). And NaOH was added (till PH>12) to slow the speed of the reactions. The samples were washed by de-ionized water and vacuum dried at 70 ℃ for 24 h.

Weakening of the crystallization: The Sn-Cu alloy sample was ball-milled in ethanol with ball material ratio of 20:1, and sampled at 5 h, 10 h, 15 h and 20 h.

"Core-shell" structure design: The Sn-Cu alloy (20 h ball-milled) sample was ball-milled again with graphite for 5 h, in which citric acid was added. Then the sample was dried by spray drying and heated under 460℃ for 4 h and under 600℃ for 6 h. The preparation process is shown in figure 1.

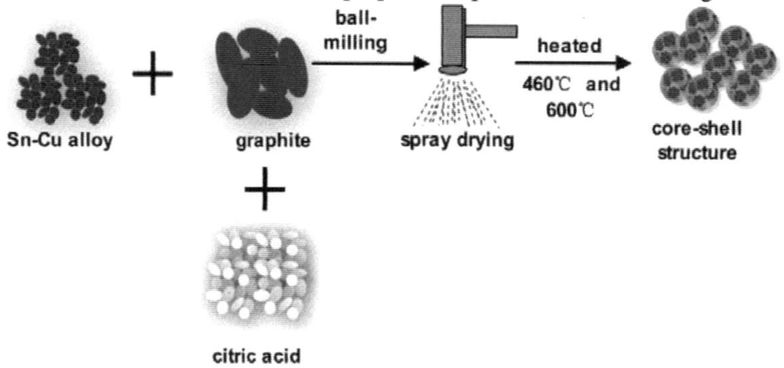

Figure 1. Diagrammatic sketch of the "core-shell" structure design process.

3. Result and Analysis

3.1. Weakening of the crystallization

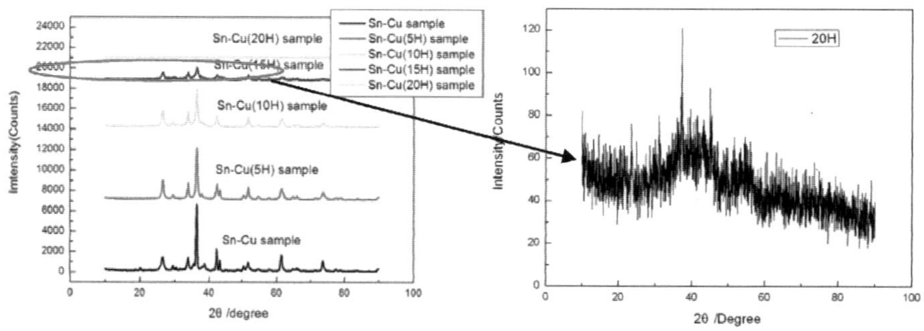

Figure 2. XRD patterns of the Sn-Cu samples after ball-milling.

Figure 2 shows the XRD patterns of the Sn-Cu alloy samples after ball-milling. According to the results, the sites of the characteristic peaks are the same among the samples. Therefore, the components of the Sn-Cu samples haven't changed after ball-milling. For the Sn-Cu (10H) sample, the characteristic peaks (double peaks) at 2θ=42° become single peak, also for the characteristic peaks at 2θ=51°. According to the intensity of the characteristic peaks, as the increasing of ball-milling, the crystallization degree of the sample become weaker and weaker. After ball-milled for 20 h, the Sn-Cu sample become non-crystallization.

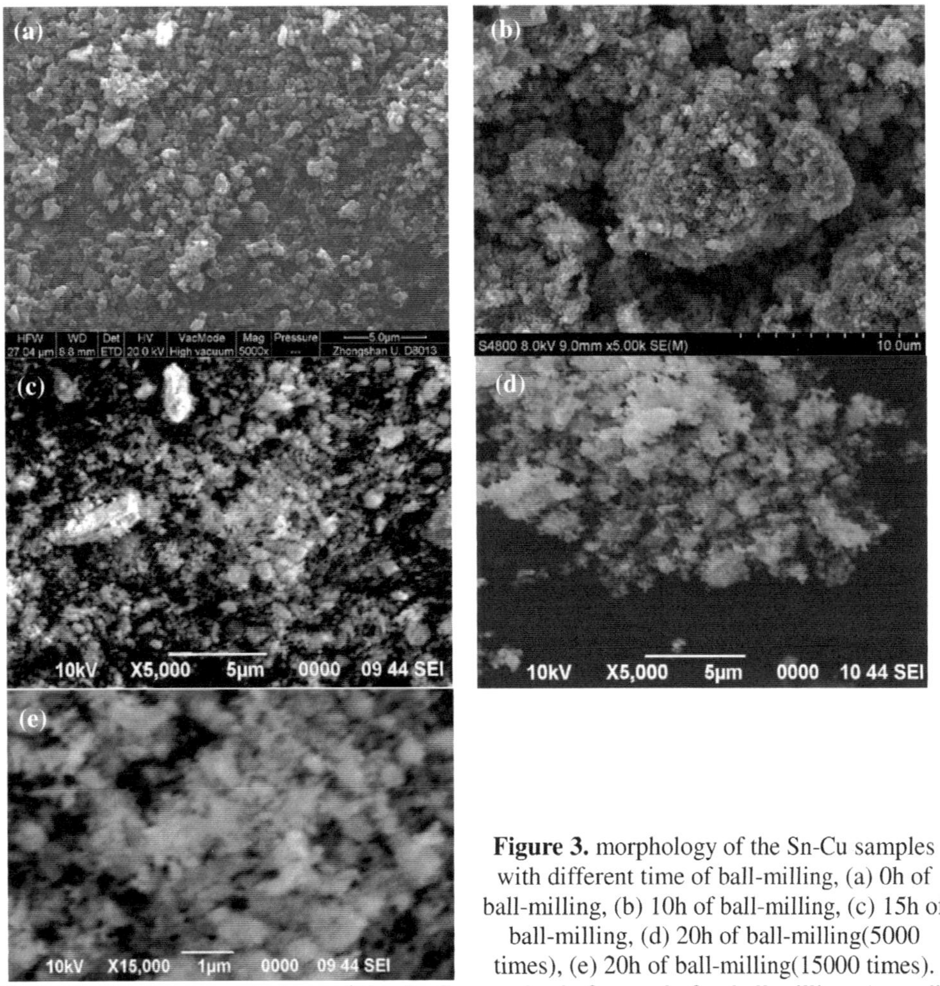

Figure 3. morphology of the Sn-Cu samples with different time of ball-milling, (a) 0h of ball-milling, (b) 10h of ball-milling, (c) 15h of ball-milling, (d) 20h of ball-milling(5000 times), (e) 20h of ball-milling(15000 times).

Figure 3 shows the morphology of the Sn-Cu samples before and after ball-milling. According to figure 3(a), the Sn-Cu sample is nano-scale and reunion into particles of 2 μm. Figure 3(b) shows that after 10 h of ball-milling, the Sn-Cu sample become smaller single-particle size and more serious reunion. However, as the increasing of ball-milling, the particle size of the Sn-Cu sample decrease quickly. As shown in figure 3(c), after 15 h of ball-milling, small part of flocculent morphology even occurr. And after 20 h of ball-milling, the sample all become flocculent.

Figure 4 shows the charge and discharge capacity curve of the Sn-Cu alloy samples after ball milling. According to the results, after 10 h of ball-milling, the charge and discharge specific capacity of the Sn-Cu sample decrease more than 100 $mAh\cdot g^{-1}$, without any improvement of the cyclic stability. However, as the increasing of ball milling, the cyclic stability is gradually improved. And the best cyclic performance is obtained after 20 h of ball-milling.

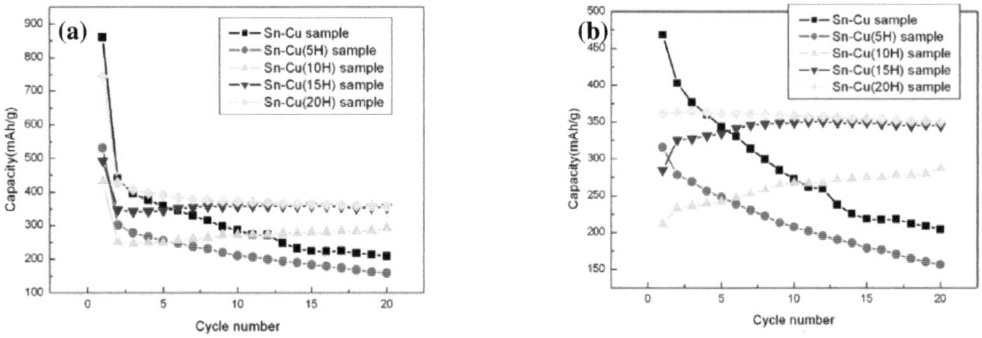

Figure 4. Discharge (a) and charge capacity (b) curves of the Sn-Cu samples after ball milling.

According to the morphology of the Sn-Cu samples, 10 h of ball-milling can only decrease the size of the particles and lead to more serious reunion. As a result, the increasing of reunion is not conducive to the infiltration of the electrolyte and lead to more irreversible capacity loss. This is the reason why 10 h of ball-milling can't improve the cyclic performance and lead to higher capacity loss. According to the XRD patterns and the morphology, the crystallization degree of the sample become weaker and weaker as the increasing of ball-milling. And flocculent morphology occur after 15 h of ball-milling. It should be interpreted that the flocculent morphology provide room for the volume change of Sn-Cu alloy and benefit for inserting and escaping of lithium ions.

To study the relationship between the crystallization degree and lithium intercalation potential, cyclic voltammogram tests are carried out on the Sn-Cu samples. The test results are showed in figure 5. According to figure 5 (a) and (b), the characteristic peaks higher than 1.5V disappear after ball-milling. Meanwhile, several lithium intercalation peaks occurr below 1.0V. It can be interpreted that after the weakening of the crystallization, the binding force on the interstitial sizes decrease, resulting in the decrease of the lithium intercalation formation energy, which present as lower lithium intercalation potential. As shown at figure 5(c), the reversible capacity below 1.0V is activated as the increasing of the cycle number. It should be interpreted that the increasing of reunion is not conducive to the infiltration of the electrolyte and the electrolyte continuously infiltrate with the increasing of the cycle number. According to figure 5(e), after 20h of ball-milling, although the capacity below 0.5V still show trend of increasing, the values difference among the cycles is small and the cyclic voltammogram curves nearly coincide with each other. The Sn-Cu sample after 20 h of ball-milling show the best cyclic stability.

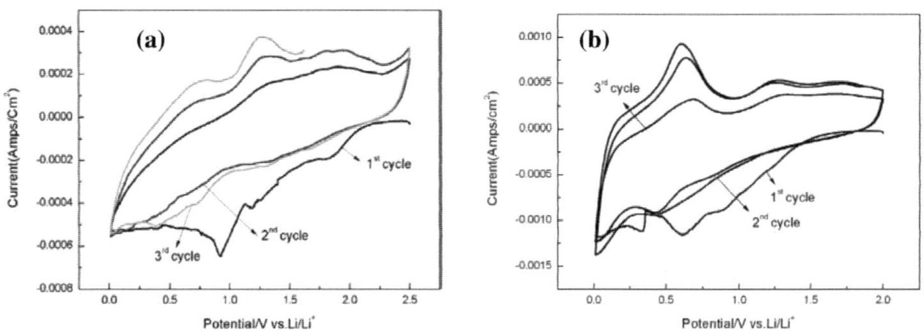

IOP Conf. Series: Materials Science and Engineering **182** (2017) 012008 doi:10.1088/1757-899X/182/1/012008

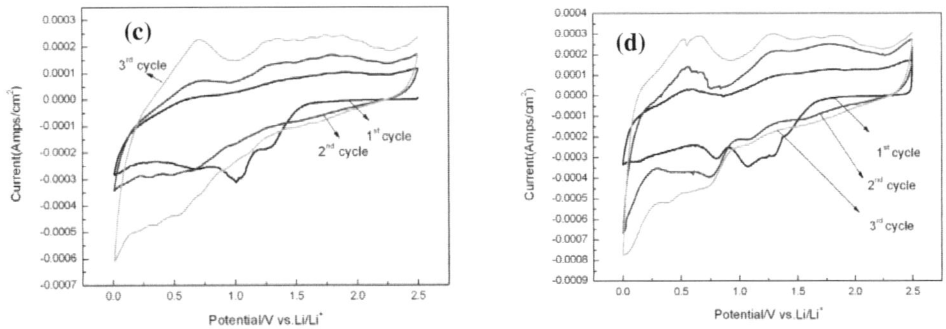

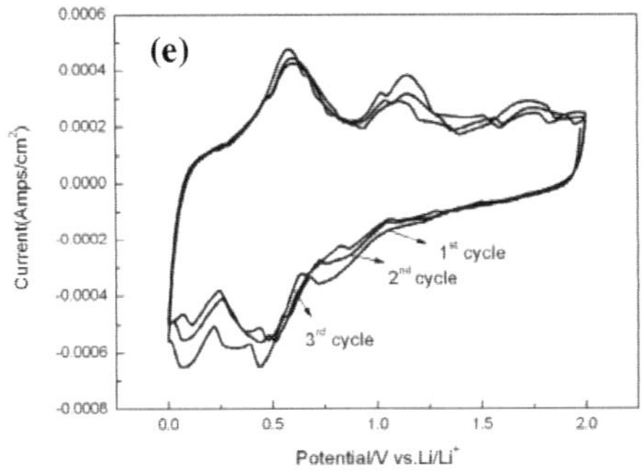

Figure 5. Cyclic voltammogram test results of the Sn-Cu samples with different time of ball-milling, (a) 0 h of ball-milling, (b) 10 h of ball-milling, (c) 15 h of ball-milling, (d) 20 h of ball-milling(5000 times), (e) 20 h of ball-milling(15000 times).

3.2. *"core-shell" structure design*

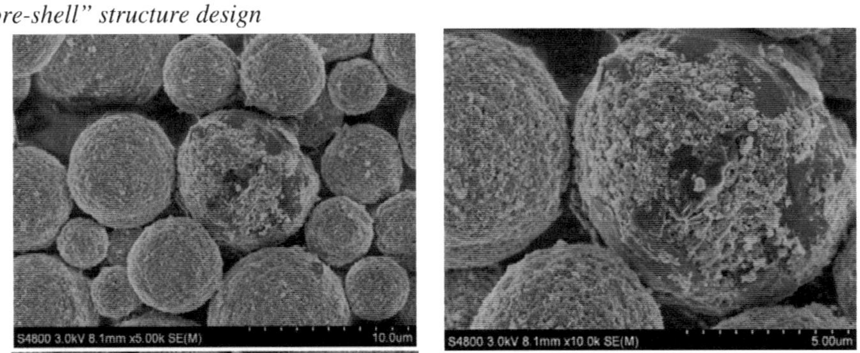

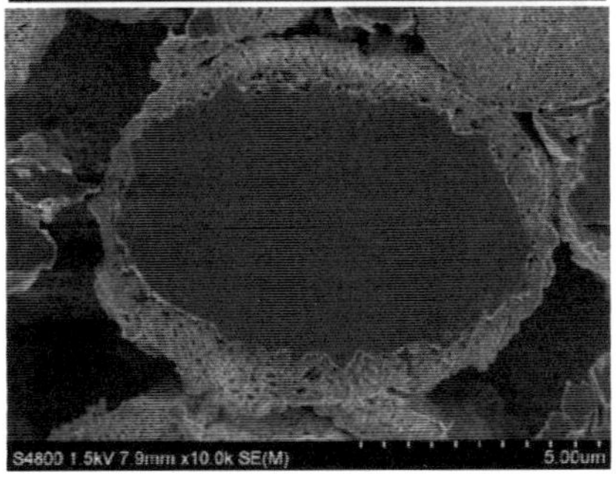

Figure 6. Surface and profile morphology of the Sn-Cu samples with "core-shell" structure.
(a) Surface morphology(5000 times),
(b) Surface morphology(10000 times),
(c) Profile morphology(10000 times).

Figure 6 shows the surface and profile morphology of the sample after "core-shell" structure design. According to figure 6(a) and (b), the shape of the sample is spherical with diameter of less than 10μm. According to figure 6(c), the Sn-Cu alloy particles stay at the surface of the graphite, which is nano-scale. And shells formed by the citric acid cover the whole particle. The structure of the sample accords with that of the target product model. The effective dispersion of Sn-Cu alloy particles will leave room for the volume expansion of the Sn-Cu alloy. And the shell will control the expansion within the "core-shell" structure.

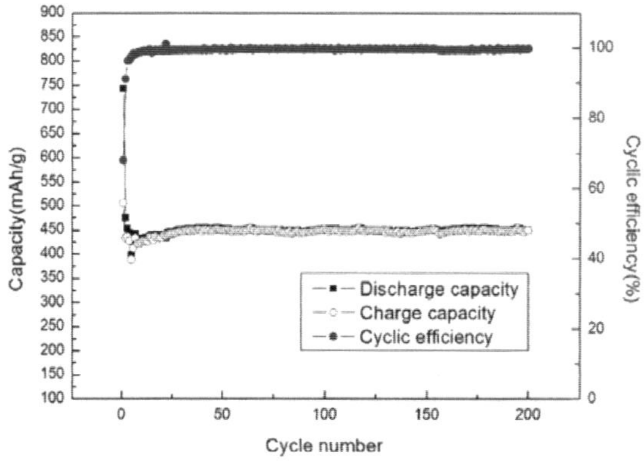

Figure 7. Charge and discharge capacity curve of the Sn-Cu samples with "core-shell" structure.

Figure 7 shows the charge and discharge capacity curve of the Sn-Cu samples with "core-shell" structure. According to the cyclic efficiency curve, although the sample still show large initial irreversible capacity (the initial cyclic efficiency is only 68%), the sample exhibit high cyclic stability. The charge capacity stay at 450 mAh g^{-1} from the 38th cycle and without any capacity loss till the 200th cycle.

After weakening of the crystallization and "core-shell" structure design, the Sn-Cu/C composite anode material exhibit high cyclic stability. The results shown in figure 7 prove that the volume change and irreversible capacity loss of Sn-based anode materials can be partly solved by the weakening of the crystallization and "core-shell" structure design. And the preparation method is simple and low cost.

4. Conclusions

In this paper, by the ways of weakening of the crystallization and "core-shell" structure design, Sn-Cu/C composite with high cyclic stability was obtained. Conclusions are showed as following.

The weakening of the crystallization will decrease the binding force on the interstitial sites and benefit for the inserting and escaping of the lithium ions, which is presented as the decrease of lithium intercalation potential.

"Core-shell" structure design can effectively disperse the Sn-Cu alloy particles and leave room for the volume expansion of Sn-Cu alloy. Furthermore, the volume change is controlled within the "core-shell" structure. Therefore, the way of "core-shell" structure design can dramatically improve the cyclic stability of Sn-Cu alloy anode material.

In this paper, Sn-Cu/C composite anode material with high cyclic stability is obtained, whose reversible capacity was 450 mAh g^{-1}, without any capacity loss till the 200th cycle.

Acknowledgments

This work was financially supported by Guangdong Natural Science Foundation (2015A030310522, 2014A030313436), the Scientific and Technological Plan of Guangdong Province (2016A050503040, 2016B010114002), and the Scientific and Technological Plan of Guangzhou City (201505050909107, 201505050826216)

References

[1] Fotouhi A, Auger D J, Propp K, Longo S and Wild M 2016 *Renewable & Sustainable Energy Reviews* **56** 1008–1021
[2] Bai X, Yu Y, Kung H H, Wang B and Jiang J 2016 *Journal of Power Sources* **306** 42–48
[3] Huang Y G, Pan Q C, Wang H Q, Yan Z X, Yang G H and Chen Y H 2015 *Ceramics International* **42(3)** 4586–4593

[4] Yan Y, Ben L B, Zhan Y and Huang X 2016 *Electrochimica Acta* **187** 186–192
[5] Agubra V A, Zuniga L, Garza D D L, Gallegos L, Pokhrel M and Alcoutlabi M 2016 *Solid State Ionics* **286** 72–82
[6] Wei L, Zhang K, Tao Z and Chen J 2015 *Journal of Alloys & Compounds* **644** 742–749
[7] Uysal M, Cetinkaya T, Alp A and Akbulut H 2015 *Journal of Alloys & Compounds* **645(2)** 235–242
[8] Wu X, Zhang S, Qi T, Fang H, Liu G and Xing Y 2016 *Journal of Power Sources* **307** 753–761
[9] Yui Y, Hayashi M, Hayashi K and Nakamura J 2016 *Solid State Ionics* **288** 219–223
[10] Schmuelling G, Oehl N, Fromm O, Knipper M, Kolny-Olesiak J and Plaggenborg T 2016 *Electrochimica Acta* **196** 597–602
[11] Liu C J, Xue F H, Huang H, Yu X H, Xie C J and Shi M S 2014 *Electrochimica Acta* **129(10)** 93–99
[12] Guo W, Li F, Duan X and Ma J 2016 *Materials Letters* **166** 210–214
[13] Cui C, Liu X, Wu N and Sun Y 2015 *Materials Letters* **143** 35–37
[14] Zhang J, Ma Z, Jiang W, Zou Y, Wang Y and Lu C 2016 *Journal of Electroanalytical Chemistry* **767** 49–55

Thermal performance enhancement of erythritol/carbon foam composites via surface modification of carbon foam

Junfeng Li*, Wu Lu, Zhengping Luo and Yibing Zeng

Aerospace Research Institute of Materials and Processing Technology, Beijing 100076, China

E-mail: *jfli1227@126.com

Abstract. The thermal performance of the erythritol/carbon foam composites, including thermal diffusivity, thermal capacity, thermal conductivity and latent heat, were investigated via surface modification of carbon foam using hydrogen peroxide as oxider. It was found that the surface modification enhanced the wetting ability of carbon foam surface to the liquid erythritol of the carbon foam surface and promoted the increase of erythritol content in the erythritol/carbon foam composites. The dense interfaces were formed between erythritol and carbon foam, which is due to that the formation of oxygen functional groups C=O and C-OH on the carbon surface increased the surface polarity and reduced the interface resistance of carbon foam surface to the liquid erythritol. The latent heat of the erythritol/carbon foam composites increased from 202.0 to 217.2 J/g through surface modification of carbon foam. The thermal conductivity of the erythritol/carbon foam composite before and after surface modification further increased from 40.35 to 51.05 W/(m·K). The supercooling degree of erythritol also had a large decrease from 97 to 54 ℃. Additionally, the simple and effective surface modification method of carbon foam provided an extendable way to enhance the thermal performances of the composites composed of carbon foams and PCMs.

Keywords: erythritol, carbon foam, surface modification, thermal conductivity, supercooling degree

1. Introduction

Phase change materials (PCMs) for thermal energy storage and thermal management have attracted a great deal of attention due to its high thermal energy storage density and small temperature variation. A lot of inorganic and organic PCMs, including paraffin, fatty acids, hydrated inorganic salts, and polyols, have been widely applied in many fields, such as the thermal control of the electronic devices, industrial waste heat recovering, and solar energy utilization [1-6]. Among these various kinds of PCMs, erythritol ($C_4H_{10}O_4$), containing four hydroxyl groups, has high latent heat of 347.7 kJ/kg, is not poisonous and corrosive, and is low volatile. However, erythritol has the same problem as ther other organic PCMs, that is, low thermal conductivity, which decrease the heat transfer rate between a heat sourse and PCMs.

To increase the thermal conductivity of PCMs, Many kinds of high thermal conductivity materials, including metal foams [7-10], fins [11,12], graphite [13-19], carbon fibers [20-24], carbon foam [25-27], are used to form composites with PCMs and increase the heat transfer from a heat sourse to PCMs. Among these high thermal conductivity materials, carbon foam has low bulk density and good chemical inertness as well as high thermal conductivity, which are beneficial to derease the weight of thermal energy storage system or thermal management system, and increase the systems' operational stablity. Many studies using carbon foam as thermal performance enhancer have been done.

Guo et al. [28] used graphite foams to enhance the thermal diffusivity of paraffin. They found that the structure of graphite foam, such as pore sizes, ligament thickness, has important effects on the thermal diffusivity and the thermal capacity, and large pore size and thinner ligament in graphite foam resultes in a larger latent heat but a lower thermal diffusivity. Lafdi et al. [29] prepared the composites composed of paraffin wax and carbon foam matrices with different porosities and thermal properties, and experimentally and numerically investigated the thermal characteristics of the composites. They found that the carbon foam matrix with higher porosity has more stability of thermal performance. Fleischer et al. [27] studied the influences of graphite foam, aluminium foam, and graphite nanofibers on the thermal performance and energy storage capabilities of paraffin wax. The study shows that the graphite foam is a more effective thermal conductivity enhancer for the infiltrated paraffin, and the paraffin melts rapidly, without exhibiting any significant delay to steady-state.

Carbon foam generally has high prosity, more than 70%, and hence has a high interface/volume ratio. Surface driven properties such as surface polarity and bond formation with the stuffing materials are crucial for applicability. However, few work focus on how the chemical affinity of interfaces between carbon foam and PCMs affects the thermal performance of composites. Thus this study is designed to study the thermophysical properties of the erythritol/carbon foam composites before and after the controlled surface modification of carbon foam.

In the present work, the effects of controlled surface modification of carbon foam on the thermal performance of the erythritol/carbon foam composites, including latent heat, thermal conductivity, thermal diffusivity, thermal capacity, and supercooling degree, are investigated. The microstructural changes of carbon foam surface before and after surface modification are also studied by a Fourier-transform infrared (FTIR) spectrometer and an optical microscope.

2. Experimental

The carbon foams were cut into small bulks (30mm×30mm×15mm), and part of them were suface-modified in the hydrogen peroxide at 90 ℃ for 2 h, respectively. The solid erythritol was completely melted at 150℃ in an oven. Then the suface-modified and starting carbon foam bulks were immersed into the liquid erythritol under vaccum for about 20 min until the liquid erythritol solidified. The above immersing process was repeated for three times, which ensured the pores of carbon foam can be adequately filled with the liquid erythritol.

The weight of carbon foam and the erythritol/carbon foam composites were recorded and used to calculate the weight and volume of erythritol filled in per gram of carbon foam. According to the latent heat of pure erythritol, the unit latent heat of the composites was caltulated. Meanwhile, the unit latent heat of the composites was also measured by a differential scanning calometer (DSC, 204F1, Netzsch, Germany) with a heating speed of 5 ℃/min. Optical microscopy images were obtained using an optical microscope (DM2500P, Leica, Germany). The FTIR spectroscopy of carbon foam before and after surface modification were obtained by a Fourier-transform infrared spectrometer (FTS3000, BIO-RAD, USA) with a resolution of 4 cm^{-1}. The thermal parameters were measured at room temperature using the laser flash technique (LFA427, Netzsch, Germany), and the sample sizes were Φ12.6 mm×1.5 mm.

3. Results and discussion

Figure 1 shows the photos and optical microscopy images of carbon foam and erythritol/carbon foam composites. The photos of carbon foam and erythritol/carbon foam composites are given in figure 1(a), (b), and (c), and their optical microscopy images are given in figure 1(d), (e), and (f) respectively. The images show that the pores in surface-modified carbon foam are completely filled with the erythritol (figure 1(f)), while there are lots of pores just partly filled with the erythritol in the no surface-modified carbon foam (figure 1(e)). This result indicates that surface modification of carbon foam has effectively increased the wetting ability of liquid erythritol to the carbon foam surface. Moreover, the liquid erythritol is much easier to get into the carbon foam pores, and hence the dense erythritol/carbon composite is formed.

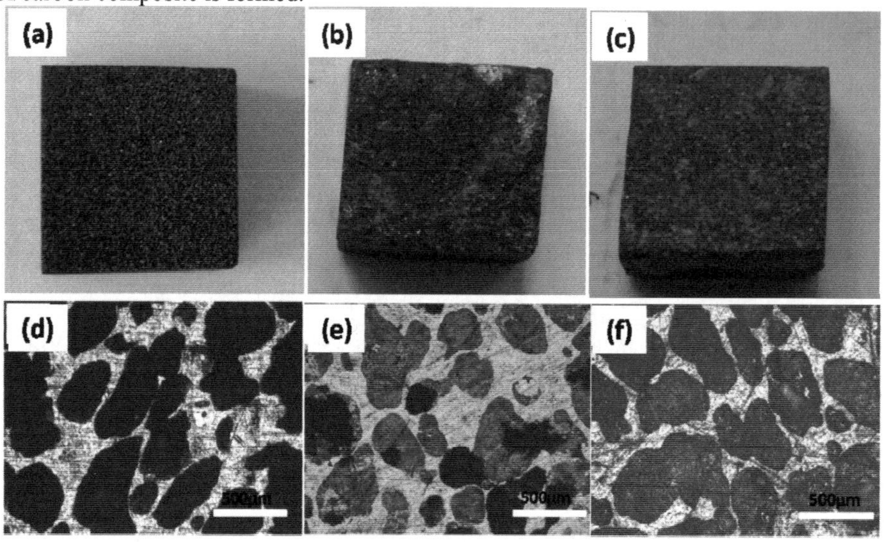

Figure 1. Photos and OM images of (a) photos of starting carbon foam (b) the composite of starting carbon foam and erythritol and (c) the composite of surface-modified carbon foam and erythritol. (d), (e), and (f) are the corresponding microstructures of the samples (a), (b), and (c), respectively.

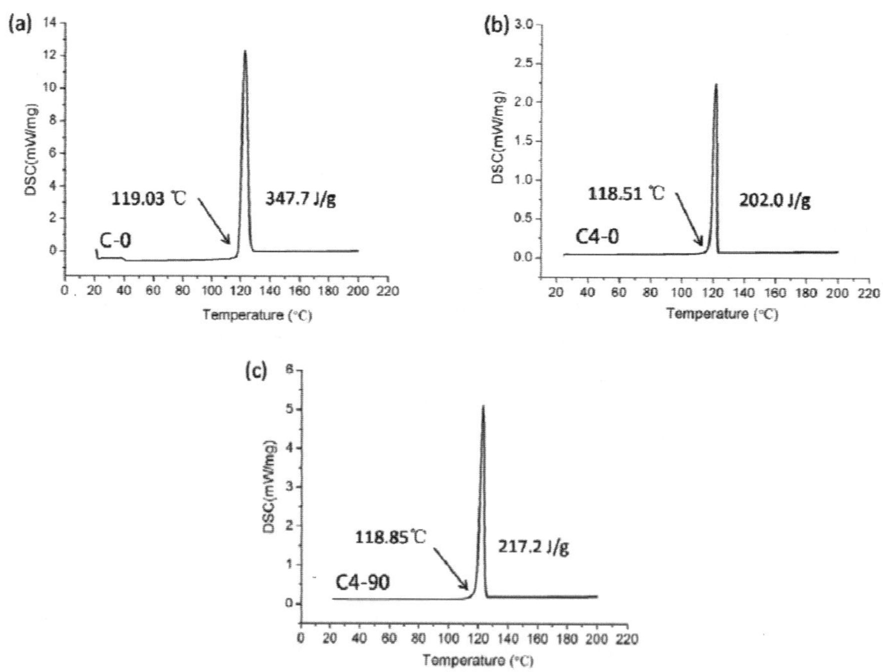

Figure 2. DSC curves of (a) erythritol, (b) starting carbon foam/erythritol composites and (c) surface-modified carbon foam/erythritol composites.

Figure 2 shows the influences of surface modification of carbon foam on the latent heat of the erythritol/carbon foam composites. It is found that the carbon foam has almost no effect on the shape of endothermal peaks and the melting point of erythritol, 119.03 ℃ for pure erythritol, 118.51 ℃ for the sample C4-0, and 118.85 ℃ for the sample C4-90. The latent heat of the samples C-0, C4-0, and C4-90 was 347.7, 202.0, and 217.2 J/g measured by DSC, respectively. For the sample C4-90, there is a increase of 15.2 J/g, increasing by 7.5% compared with that of the sample C4-0. This latent heat increase is accorded with the microstructure changes in figure 1, that is, there is more erythritol in the carbon foam pores after the surface modification of carbon foam

Table 1 shows the thermal diffusivity, thermal capacity, and thermal conductivity of starting carbon foam and erythritol/carbon foam composites. The thermal diffusivity of the erythritol/carbon foam composites increases from 19.749 to 23.892 mm^2/s before and after the surface modification of carbon foam, increasing 20.98%. The thermal capacity of the erythritol/carbon foam composites also increases from 1.287 to 1.346 J/(g·K) with the density increasing. The thermal conductivity correspondingly increases from 40.35 to 53.06 W/(m·K), increasing 31.50% in the condition of no changes in the porosity and pore structure of carbon foam. The pure erythritol has a large supercooling degree of 97℃, while the supercooling degree of the samples C4-0 and C4-90 dramatically decreases to 56 ℃ and 54 ℃, respectively. The result indicates that the carbon foam can promote the crystallization of erythritol in the freezing process and moreover reduce the supercooling degree of erythritol, and the surface modification of carbon foam can further reduce the supercooling degree of Erythritol.

Table 1. Thermal diffusivity, thermal capacity, and thermal conductivity of starting carbon foam and erythritol/carbon foam composites

Sample	Thermal diffusivity (mm^2/s)	Thermal capacity ($J/g{\cdot}K^{-1}$)	Thermal conductivity (W/m·K)	Density (g/cm^3)	Supercooling degree (℃)
C-0	79.565	0.748	29.76	0.50	/
C4-0	19.749	1.287	40.35	1.58	56
C4-90	23.892	1.346	53.06	1.65	54

Note: The carbon foam of the sample C-0 is not surface-modified starting carbon foam. The carbon foam of the sample C4-0 is not surface-modified, and the carbon foam of the sample C4-90 is surface-modified at 90 ℃.

Figure 3 depicts FT-IR spectra of surface-unmodified carbon foam and surface modified carbon foam. The peaks at 1738, 1640, 1461, and 1388 cm^{-1} are attributed to the stretching or bending vibration of C=O, C-C, C-H, C-OH, respectively [30-32]. The peaks (1026~1104 cm^{-1}) corresponding to oxygen functional groups markedly increase after the carbon foam surface is oxided in hydrogen peroxide at 90 ℃. The formation of oxygen functional groups C=O and C-OH on the carbon surface increases the surface polarity. The erythritol contains four C-OH groups, and hence the liquid erythritol is also polar liquid. Therefore, the wetting ability of the liquid erythritol to the oxided carbon foam surface increases, which reduces the interface resistance of carbon foam surface to the liquid erythritol. Finally, the pores of the surface-modified carbon foam are completely filled with the erythritol.

The thermal performance enhancement of erythritol/carbon foam composites are also attributed to the dense interface formation between the erythritol and carbon foam, as shown in figure 4. It is found that the dense interfaces form not only in the flat interface but also in the curve interface. The dense interfaces are beneficial to reduce the interface thermal resistance and enhance the thermophysical properties of the composites.

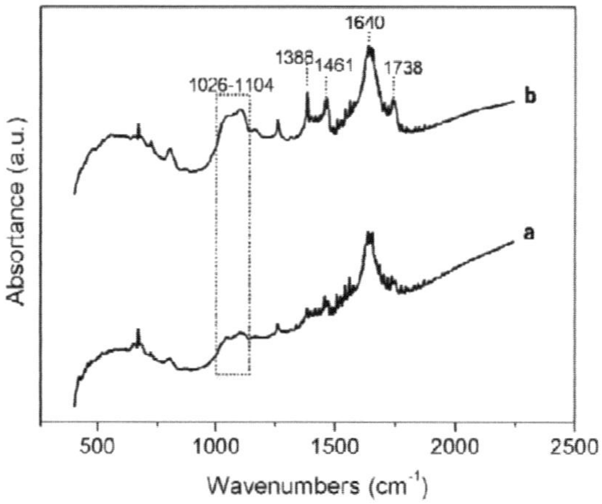

Figure 3. FT-IR spectra of (a) starting carbon foam and (b) surface-modified carbon foam.

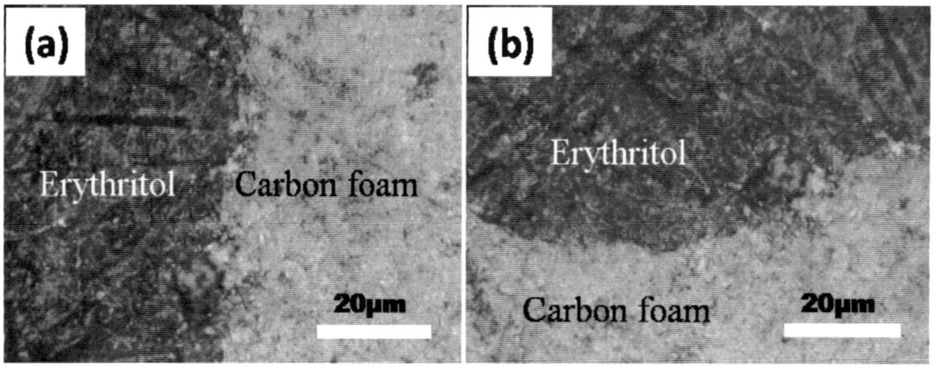

Figure 4. Interfaces between surface-modified carbon foam and erythritol in the sample C4-90, (a) flat interface, (b) curve interface.

4. Conclusions

The thermal performance enhancements of erythritol/carbon foam composites in which carbon foam surface is modified using hydrogen peroxide as oxidizer have been demonstrated. There are carbonyl and hydroxyl groups formed on the carbon foam surfaces after the surface modification, and the polarity of carbon foam surface increases, which enhances the wetting ability of polar liquid erythritol to the carbon foam surface. The pores of surface-modified carbon foam are completely filled with the erythritol, and the dense interfaces form between the carbon foam surface and erythritol. Moreover, the thermal diffusivity, thermal capacity, thermal conductivity, and latent heat of the erythritol/carbon foam composites are markedly improved. There is an increase of 15.2 J/g for the sample C4-90, increasing by 7.5% compared with that of the sample C4-0. The thermal conductivity of the erythritol/carbon foam composite before and after surface modification further increases from 40.35 to 53.06W/(m·K). The supercooling degree of erythritol has a large decrease from 97 to 54 ℃. In addition, the carbon foam, whose surfaces are modified by this simple and effective method, is also able to form dense composites with other polar PCMs. The surface modification of carbon foam provides an extendable route for enhancing the thermal performance of the composites composed of carbon foams and PCMs.

Acknowledgements
These authors would like to express their gratitude for the support provided by the National Natural Science Foundation of China (NSFC, 51502069) and National Key Basic Research Program of China (NKBRP, 2015CB655200).

References

[1] Fan L W and Khodadadi J M 2011 *Renew. Sust. Energ. Rev.* **15** 24–46
[2] Raj V A A and Velraj R 2010 *Renew. Sust. Energ. Rev.* **14** 2819–2829
[3] Baetens R, Jelle B P and Gustavsen A 2010 *Energ. Buildings* **42** 1361–1368
[4] Kuznik F, David D, Johannes K and Roux J J 2011 *Renew. Sust. Energ. Rev.* **15** 379–391
[5] Kenisarin M M 2010 *Renew. Sust. Energ. Rev.* **14** 955–970
[6] Pal D 1998 *J. Thermophys. Heat Tr.* **12(2)** 256
[7] Hong S T and Herling D R 2006 *Scripta Mater.* **55** 887–890
[8] Zhou D and Zhao C Y 2011 *Appl. Therm. Eng.* **31** 970–977
[9] Zhao C Y, Lu W and Tian Y 2010 Sol. Energ. **84** 1402–1412
[10] Chen Z Q, Gu M W and Peng D H 2010 *Appl. Therm. Eng.* **30** 1967–1973
[11] Stritih U 2004 *Int. J. Heat Mass Tran.* **47** 2841–2847
[12] Shatikian V, Ziskind G, Letan R 2005 *Int. J. Heat Mass Tran.* **48** 3689–3706
[13] Zhong Y J, Li S Z, Wei X H, Liu Z J, Guo Q G, Shi J L and Liu L 2010 *Carbon* **48** 300–304
[14] Xia L, Zhang P, Wang R Z 2010 *Carbon* **48** 2538–2548
[15] Kim S, Drzal L T 2009 *Sol. Energ. Mater. Sol. Cells* **93** 136–142
[16] Zhang Z G, Zhang N, Peng J, Fang X M, Gao X N and Fang Y T 2012 *Appl. Energ.* **91** 426–431
[17] Pincemin S, Olives R, Py X and Christ M 2008 *Sol. Energ. Mater. Sol. Cells* **92** 603–613
[18] Xiang J L, Drzal L T 2011 *Sol. Energ. Mater. Sol. Cells* **95** 1811–1818
[19] Zhao J G, Guo Y, Feng F, Tong Q H, Qv W S, Wang H Q 2011 *Renew. Energ.* **36** 1339–1342
[20] Sanusi O, Warzoha R, Fleischer AS 2011 *Int. J. Heat Mass Tran.* **54** 4429–4436
[21] Elgafy A, Lafdi K 2005 *Carbon* **43** 3067–3074
[22] Karaipekli A, Sarıa A, Kaygusuz K 2007 *Renew. Energ.* **32** 2201–2210
[23] Fukai J, Hamada Y, Morozumi Y and Miyatake O 2003 *Int. J. Heat Mass Tran.* **46** 4513–4525
[24] Frusteri F, Leonardi V, Vasta S and Restuccia G 2005 *Appl. Therm. Eng.* **25** 1623–1633
[25] Lafdi K, Mesalhy O and Elgafy A 2008 *Carbon* **46** 159–168
[26] Wierschke K W, Franke M E, Watts R and Ponnappan R 2006 *J. Thermophys. Heat Tr.***20** 865–870
[27] Chintakrinda K, Weinstein R D and Fleischer A S 2011 *Int. J. Therm. Sci.* **50** 1639–1647
[28] Zhong Y, Guo Q, Li S Z, Shi J L and Liu L 2010 *Sol. Energ. Mater. Sol. Cells* **94** 1011–1014
[29] Mesalhy O, Lafdi K and Elgafy A 2006 *Carbon* **44** 2080–2088
[30] Cataldo F 2003 *Fullerenes Nano. Carbon Nanostruc.* **11(1)** 1–13
[31] Zhang J L, Yang H J, Shen G X, Cheng P, Zhang J Yand Guo S W 2010 *Chem. Commun.* **46(7)** 1112–1114
[32] Furer V L 1999 *J. Mol. Struct.* **513** 1–8

Preparation and thermal characterization of oxalic acid dihydrate/bentonite composite as shape-stabilized phase change materials for thermal energy storage

Lipeng Han[1,2], Shaolei Xie[1], Jinhe Sun[1,3,*] and Yongzhong Jia[1,4,*]

[1] Key Laboratory of Comprehensive and Highly Efficient Utilization of Salt Lake Resources, Qinghai Institute of Salt Lakes, Chinese Academy of Sciences, Xining, 810008, China
[2] University of Chinese Academy of Sciences, Beijing, 100049, China

E-mail: [3,*]jinhesun@163.com, [4,*]jiayzh@hotmail.com

Abstract. Oxalic acid dihydrate (OAD) which has very high initial phase transition enthalpy is a promising phase change material (PCM). In this paper, shape-stabilized composite PCMs composed of OAD and bentonite were prepared by a facile blending method to overcome the problem of leakage. FT-IR results indicated the interactions between OAD and bentonite, such as the capillary force and the hydrogen bonding, resulting in the confined crystallization process. As a result, the OAD was confined to be amorphous. The thermogravimetric analysis and scanning electron microscope results showed that sample had the best coating effect when the amount of bentonite was 17.7%. The differential scanning calorimetry analyses demonstrated that a decrease in the OAD content was accompanied by a continuous decrease in the melting point and phase change enthalpy of the composites.

Keywords: phase change materials, thermal properties, oxalic acid dehydrate, bentonite

1. Introduction

Heat storage is used to close the temporal gap between thermal energy production and energy demand. In many cases, it is the prerequisite for the utilization of waste heat in periodic industrial processes and the connection of renewable energy sources (especially solar energy) to continuous thermal loads. The application of phase change materials (PCMs) is one possible heat storage technology. PCMs are functional materials for thermal energy storage by storing and releasing amount of latent heats during their phase change processes [1]. In recent years, due to high energy storage density and latent heat property, small temperature variation from storage to retrieval, and repeatable utilization property, PCMs have become a hotspot in the fields of electronics cooling [2,3], solar power generation [4,5], waste heat utilization [6,7] and aerospace applications [8], etc.

PCMs can be divided into inorganic or organic compounds. The latent heat value of inorganic PCMs is approximately twice as high as the latent heat value of organic PCMs; in addition, inorganic PCMs are commercially available and fairly inexpensive [9-11]. Oxalic acid dihydrate (OAD) is low manufacturing cost and wide usage. In OAD, alternating acid and water molecules that act both as hydrogen-bond donors and acceptors [12]. OAD needs a lot of energy to break the hydrogen-bond in the melting process. Therefore, OAD has a very high heat of fusion of 370 $J \cdot g^{-1}$ [13], which is a very promising PCM in TES for applications such as industrial waste heat recovery or solar energy storage systems [14-17]. This is why authors of this paper have considered the use of OAD as a PCM. However, OAD has large degradations of thermophysical properties over time. The melting point of OAD is 101°C (onset value) higher than the boiling point of water. When OAD is heated to melt, it is easy to produce high pressure caused by the water steam. High pressure not only results in severe leakage but also puts forward advanced requirements of the equipment. Meanwhile, actual applications of OAD inevitably face the enduring problems: relatively strong corrosivity. Some researchers have demonstrated that PCMs can be encapsulated by using porous materials to overcome these problems [18].

Bentonite has strong hygroscopicity and swelling properties, can be dispersed in an aqueous medium into gel shape and slurry. Bentonite is an absorbent aluminum phyllosilicate, and its main uses are for drilling mud, binder, absorbent and as a groundwater barrier. Due to its nano-layer structures, high thermal conductivity and low cost, bentonite also can be used as the supporting material of shape-stabilized PCMs (ss-PCMs) [19,20].

In this paper, we aim to maximize impregnation of OAD between interlayer of bentonite, to reduce leakage problem. Therefore, we prepared OAD and bentonite composites with varying concentration of bentonite and subsequently investigated the effects of a variety of concentration of bentonite on the OAD composites as PCMs.

2. Experiments

2.1. Materials

In this investigation, oxalic acid dihydrate (purity 99.5%) and bentonite were purchased from Aladdin Chemistry Co. Ltd (China). Deionized water was prepared in the laboratory.

2.2. Preparation of OAD composites containing bentonite

The OAD composites containing bentonite were prepared using a simple blending method. 10 g OAD was mixed with different mass bentonite (0.5 g, 1 g, 1.5 g, 2 g or 2.5 g) in a round bottom flask. The components of the OAD/bentonite composites are listed in table 1. The resulting mixture was then heated at 105 °C and stirred for 30 min. 1 mL deionized water was added to make up the water that adhered to the vessel wall in the melting process. Finally, the OAD composites containing bentonite were obtained after cooling at room temperature. It was worth noting that bentonite was too many to form a homogeneous mixture when the addition amount was 2.5 g. Some bentonite particles agglomerated, resulting in non-uniform dispersion. Therefore, the maximum addition amount of bentonite was 2 g.

Table 1. Components of the prepared OAD/bentonite ss-PCMs.

Samples	Components	OAD mass fraction
ss-sample I	10.0 g OAD + 0.5 g bentonite	95.2%
ss-sample II	10.0 g OAD + 1.0 g bentonite	90.9%
ss-sample III	10.0 g OAD + 1.5 g bentonite	87.0%
ss-sample IV	10.0 g OAD + 2.0 g bentonite	83.3%

2.3. Characterization

The morphologies and microstructure of the obtained composite PCMs were investigated using a scanning electron microscope (SEM) instrument (JSM-5610LV). Fourier transform infrared (FT-IR) spectra were recorded using a Thermo Nicolet NEXUS FT-IR spectrometer. A resolution of 4 cm^{-1} and the average of 32 automated scans from 400 to 4000 cm^{-1} were used to obtain the FT-IR spectra. Thermal properties of the composite PCMs, such as melting and latent heats, were measured using a Mettler Toledo differential scanning calorimetry (DSC). Measurements were performed at 10 $K \cdot min^{-1}$ or 2 $K \cdot min^{-1}$ in heating procedure and cooling procedure in nitrogen atmosphere. The temperature accuracy was ±0.01 °C, and the heat flow repeatability was 0.2 μW. The thermal stability of the composite PCMs was determined using a Netzsch STA 449F3 at a scan rate of 5 K·min^{-1} under an atmosphere of nitrogen.

3. Results and discussion

3.1. Micro-morphology analysis of the ss-PCMs

Figure 1 shows the SEM images of pristine bentonite powders and the composites. From figure 1(a), bentonite is a crystal with layer structures. Moreover, one can observe the presence of a large space between discrete layers, which facilitates the intercalation of the solid-liquid PCMs. As illustrated from figure 1(b) to (c), grains are getting smaller. Meanwhile, it is easily to observe a homogeneous and rough morphology with increasing weight percentage of bentonite. OAD molecules can not only crystallize onto the bentonite surface but also in the inner pore of bentonite networks. According to the results of SEM, the ss-sample IV shows the best coating effect. This uniform encapsulation also indicates good cohesion between the inorganic matrix and OAD, which is due to the effect of capillary and surface tension force.

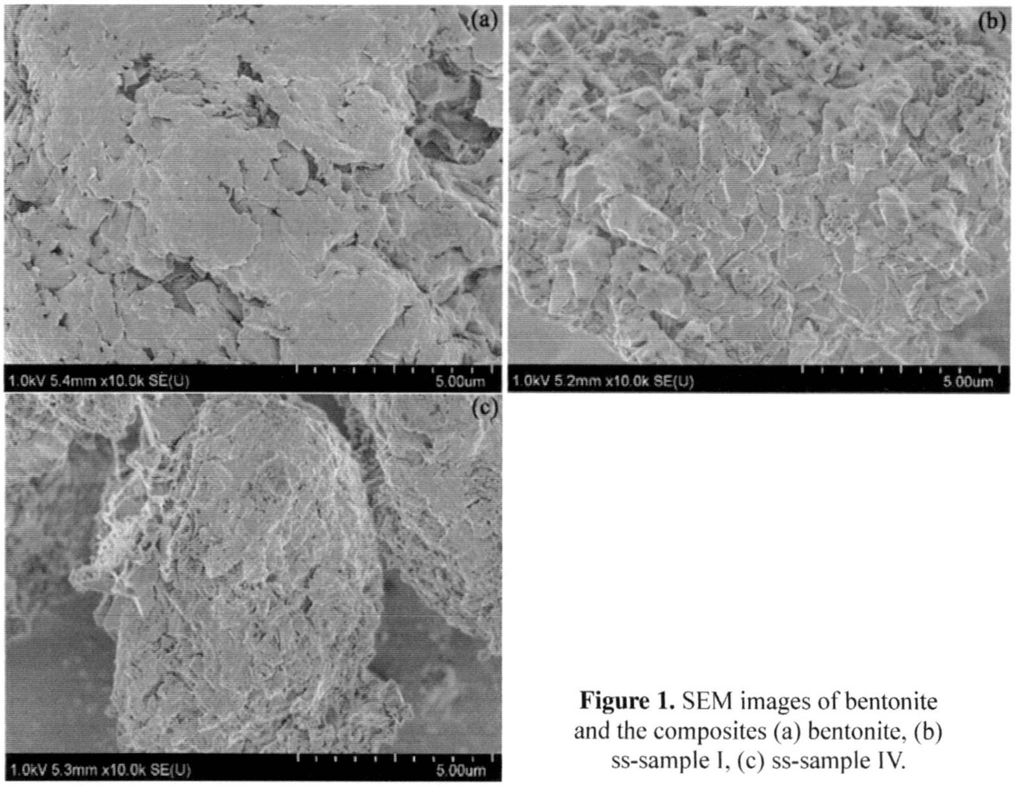

Figure 1. SEM images of bentonite and the composites (a) bentonite, (b) ss-sample I, (c) ss-sample IV.

3.2. FT-IR analysis of the ss-PCMs

FT-IR spectroscopy is usually used to investigate the specific interactions and material compositions in composites. Figure 2 displays the FT-IR spectra of pure OAD, bentonite and ss-PCMs. In the FT-IR spectrum of OAD, the peak at 1689 cm^{-1} represents the stretching vibration of the C=O group. The peaks at 3430 cm^{-1} signify O–H stretching vibration. The peak at 1254 cm^{-1} is assigned to the bending vibration of the −OH group. On the spectra of the bentonite, the peaks at 1043 cm^{-1} and 463 cm^{-1} represent O–H stretching vibration and Si–O–Si bending vibration, respectively.

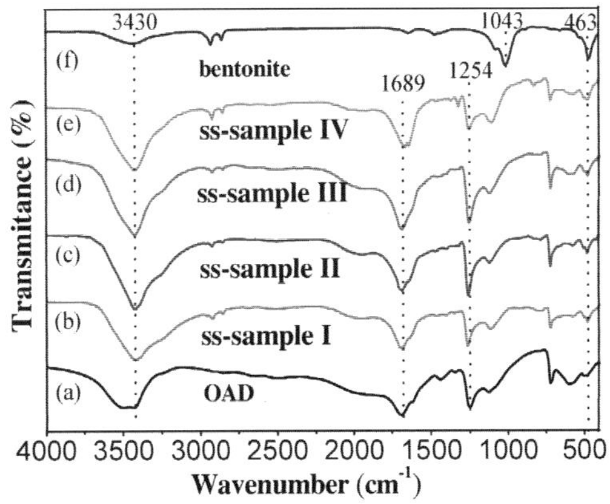

Figure 2. FT-IR spectrum of pure OAD, bentonite and the composite PCMs.

After comparing the spectra of the composite PCMs with those of bentonite and pure OAD, most absorption peaks of the main functional groups of OAD and the corresponding matrix also appear in the spectra of the ss-PCMs with only a slight shift. And no obvious new peaks are observed, which indicates that no new chemical bonds generate between OAD and the matrix. The frequency shift of the functional group indicates that there are some interactions between OAD and the supporting materials, which can prevent the leakage of the melted PCMs from the supporting materials. Due to the nano-layer structure of bentonite, there is a capillary force between melt OAD and bentonite in ss-PCMs [21,22]. Meanwhile, The FT-IR results suggest that there are C=O and O–H groups in both OAD and bentonite. Therefore, the hydrogen bonds infer from the frequency shifts of the functional group also exist between OAD and the porous materials.

3.3. Thermal behavior of OAD/bentonite ss-PCMs

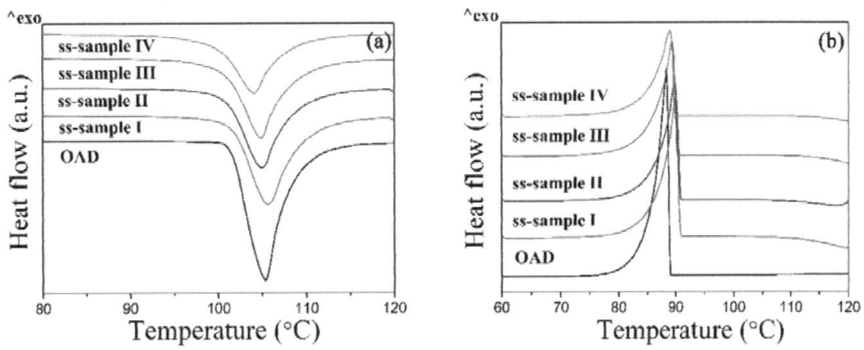

Figure 3. DSC thermograms of OAD and OAD/bentonite composited PCMs with the different amount of OAD components
(a) heating process at a rate of 10 k·min^{-1}, (b) cooling process at a rate of 10 k·min^{-1}.

The phase change temperature and latent heat of composites are measured with the DSC technique. Figure 3(a) and (b) demonstrate the heating and cooling DSC curves of pure OAD and composites with various OAD weight fractions at a heating and cooling rate of 10 K min^{-1}, and the corresponding data deduce from the DSC curves are listed in table 2. Clearly, all the composites exhibit similar curves as pure OAD, indicating that the OAD acts as the PCM during the melting process. However, the phase change characteristics of the OADs are affected by adding bentonite. As listed in table 2, the onset melting (T_{mo}) temperature shifts slightly from 101.2 °C for pure OAD to around 99.5 °C for the composites. This phenomenon can be ascribed to two reasons. On the one hand, the porous bentonite network in the composites provides heat conduction path in the OAD, which improves the thermal conductivity of OAD and consequently accelerates the phase change speed of the composites [23, 24]. On the other hand, bentonite is the impurity for OAD and can destroy the perfect crystallization, so the defects of crystal lattices increase, which cause the crystals can be damaged at low temperature [25].

In addition, the areas of the endothermic and exothermic peaks decrease significantly when OAD is incorporated into bentonite. Compared with pure OAD, the latent heat of OAD/bentonite composites is in the range of 205.8–287.7 $J \cdot g^{-1}$, which is lower than ideal latent heat which is 372.7 $J \cdot g^{-1}$. The decrease of the latent heat of the composite PCMs cannot be attributed to the lower fraction of OAD alone. Another factor is the blending of mesoporous matrices interferes with the crystallization of OAD. The interaction between OAD and its supports may keep OAD from crystallizing and influence the thermal properties of the composites and result in the smaller enthalpy. Perhaps fast heating and cooling rates of 10 $k \cdot min^{-1}$ make the OAD of the composites in the inner pore of bentonite networks fail to fully melt and crystallize. Extra measurements are then completed at a lower heating and cooling rate of 2 $k \cdot min^{-1}$. Figure 4(a) and (b) show that the thermal behaviour of composited PCMs at the heating and cooling rates of 2 $k \cdot min^{-1}$ is close to the former measurements. Onset melting (T_{mo}), degree of supercooling and phase change enthalpy only have a little change. Therefore, the speed of heating and cooling has moderate effect on the thermal behavior of the composites.

Table 2. Thermographic data of OAD and OAD/ bentonite composited PCMs.

	10 $k \cdot min^{-1}$				2 $k \cdot min^{-1}$			
samples	T_{mo} (°C)	ΔH_m (J g^{-1})	T_{co} (°C)	ΔH_c (J g^{-1})	T_{mo} (°C)	ΔH_m (J g^{-1})	T_{co} (°C)	ΔH_c (J g^{-1})
pure OAD	101.2	372.7	89.1	362.0	100.2	371.9	90.4	363.2
ss-sample I	101.2	287.7	91.0	286.7	99.6	291.0	93.4	290.8
ss-sample II	100.6	267.9	91.0	261.3	99.1	266.1	92.8	263.1
ss-sample III	100.4	255.4	90.4	251.0	98.8	250.9	92.6	249.7
ss-sample IV	99.5	205.8	90.0	202.5	98.2	201.0	91.5	195.8

We compare the thermal properties of the prepared composite PCM in this study with other composite PCMs in the literatures in table 3. It can be remarkably noted that the prepared OAD/bentonite composite PCM has high latent heat and appropriate phase change temperature. It indicates that the prepared composite PCM has good thermal energy storage potential.

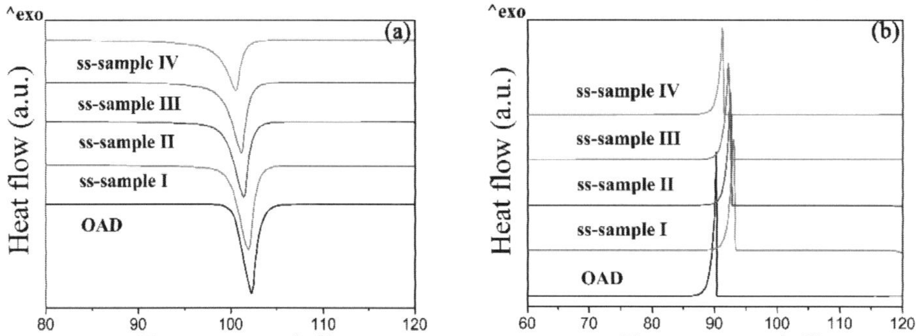

Figure 4. DSC thermograms of OAD and OAD/bentonite composited PCMs with the different amount of OAD components
(a) heating process at a rate of 2 $k \cdot min^{-1}$, (b) cooling process at a rate of 2 $k \cdot min^{-1}$.

Table 3. Thermal characteristics of some composite PCMs in literature.

Material/mass/%	Melting point (°C)	Latent heat ($J \cdot g^{-1}$)	Reference
Myristic acid (50%)/bentonite (50%)	53.2	54.5	[20]
Capric-myristic acid (20 %)/vermiculite (80%)	19.7	27.0	[26]
PEG (45 %)/diatomite (45 %)/EG (10 %)	27.7	88.2	[27]
$CaCl_2 \cdot 6H_2O$ (50%)/EG (50%)	29.9	88.5	[28]
Paraffin (72 %)/SiO_2 (20.8 %)/EG (7.2 %)	27.7	104.4	[29]
Paraffin(60 %)/HDPE(15 %)/(APP/PER/MA)25 %	51.7	81.5	[30]

3.4. Thermal stability of the OAD/bentonite composite PCMs

The OAD and the various OAD/bentonite composites are verified using thermogravimetric (TG) analyses in order to understand the thermal stability of them. Figure 5 shows TG curves of pure OAD and the composite PCMs. As can be seen from the curves, the mass loss of OAD and the composites starts at 70 °C and ends at 92.6–100.0 °C due to the degradation of crystal water in OAD. With the increase of the bentonite content, the mass loss becomes smaller and mass loss rate of composites also becomes lower when temperature is under 80 °C. It implies that the interactions between OAD and porous materials can provide composited PCMs with the good structure stable properties.

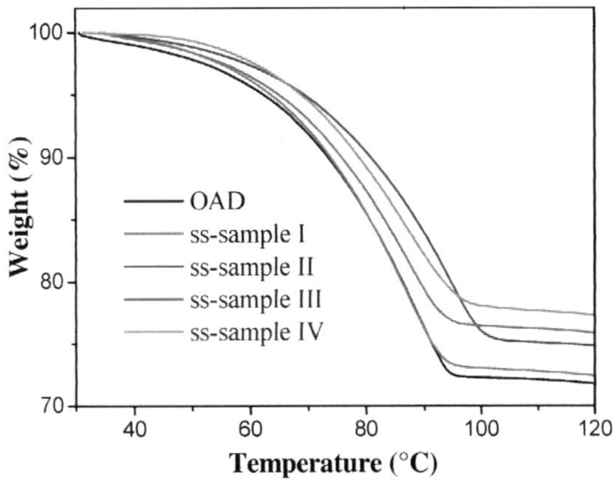

Figure 5. TG curves of OAD and the composites.

4. Conclusions

In this paper, the ss-PCMs composed of bentonite and OAD are prepared by a direct blending method. According to the results of SEM and TG, when the weight percentage of OAD in the composites is 83.3 wt.%, it can be found that OAD is preferably coated with bentonite. DSC data show a continuous lowering of the OAD melting point and phase change enthalpy with increasing bentonite content. According to the FT-IR results, interactions between OAD and bentonite, such as the capillary force and the hydrogen bonding, result in the confine crystallization process. OAD/bentonite composite with mass fraction of 83.3 wt.% OAD is promising candidates for heat storage application due to its moderate melting temperature, significant latent heat storage capacity and form-stable property.

Acknowledgement

Financial support from the National Natural Science Foundation of China (No. U1407205) and Youth Innovation Promotion Association of Chinese Academy of Sciences (No. 2015351) is gratefully acknowledged.

References

[1] Dutil Y, Rousse D, Salah N, Lassue S and Zalewski L 2011 *Renew. Sust. Energ. Rev.* **15** 112–130
[2] Khateeb S, Amiruddin S, Farid M, Selman J and Al-Hallaj S 2005 *J. Power Sources* **142** 345–353
[3] Sabbah R, Kizilel R, Selman J and Al-Hallaj S 2008 *J. Power Sources* **182** 630–638
[4] Steinmann W and Tamme R 2008 *J. Sol. Energy Eng.* **130** 1–5
[5] Pincemin S, Olives R, Py X and Christ M 2008 *Sol. Energy Mater. Sol. Cells* **92** 603–613
[6] Maruoka N and Akiyama T 2006 *Energy* **31** 1632–1642
[7] Kaizawa A, Kamano H, Kawai A, Jozuka T, Senda T, Maruoka, N and Akiyama T 2008 *ISIJ Int.* **48** 540–548
[8] Lafdi K, Mesalhy O and Elgafy A 2008 *Carbon* **46** 159–168
[9] Kenisarin M and Mahkamov K 2007 *Renew. Sust. Energ. Rev.* **11** 1913–1965
[10] Farid M, Khudhair A, Razack S and Al-Hallaj S 2004 *Energy Conv. Manag.* **45** 1597–1615
[11] Shin H, Park M, Kim H and Park S 2015 *Appl. Therm. Eng.* **75** 978–983
[12] Stare J and Hadži D 2014 *J. Chem. Theory Comput.* **10** 1817–1823
[13] Haillot D, Bauer T, Kröner U and Tamme R 2011 *Thermochim. Acta* **513** 49–59
[14] Naumann R and Emons H 1989 *J. Therm. Anal. Calorim.* **35** 1009–1031
[15] Gil A, Medrano M, Martorell I, Lázaro A, Dolado P, Zalba B and Cabeza L 2010 *Renew. Sust. Energ. Rev.* **14** 31–55
[16] Gutierrez A, Ushak S, Galleguillos H, Fernandez A, Cabeza L and Grágeda M 2015 *Appl. Energy* **154** 616–621
[17] El-Sebaii A, Al-Heniti S, Al-Agel F, Al-Ghamdi A and Al-Marzouki F 2011 Energy Conv. Manag. **52** 1771–1777
[18] Kadoono T and Ogura M 2014 *Phys. Chem. Chem. Phys.* **16** 5495–5498
[19] Li M, Wu Z, Kao H and Tan J 2011 *Energy Conv. Manag.* **52** 3275–3281
[20] Chen C, Liu X, Liu W and Ma M 2014 *Sol. Energy Mater. Sol. Cells* **127** 14–20
[21] Wang C, Feng L, Li W, Zheng J, Tian W and Li X 2012 *Sol. Energy Mater. Sol. Cells* **105** 21–26
[22] Sari A and Biçer A 2012 *Sol. Energy Mater. Sol. Cells* **101** 114–122
[23] Xia L and Zhang P 2011 *Sol. Energy Mater. Sol. Cells* **95** 2246–2254
[24] Wang X, Guo Q, Zhong Y, Wei X and Liu L 2013 *Renew. Energy* **51** 241–246
[25] Yuan Y, Li T, Zhang N, Cao X and Yang X 2016 *J. Therm. Anal. Calorim.* **124** 881–888
[26] Karaipekli A and Sari A 2009 *Sol. Energy* **83** 323–332
[27] Karaman S, Karaipekli A, Sarı A and Bicer A 2011 *Sol. Energy Mater. Sol. Cells* **95** 1647–1653
[28] Duan Z, Zhang H, Sun L, Cao Z, Xu F, Zou Y and Zhou H 2014 *J. Therm. Anal. Calorim.* **115** 111–117
[29] Li M, Wu Z and Tan J 2012 *Appl. Energy* **92** 456–461
[30] Zhang P, Hu Y, Song L, Ni J, Xing W and Wang J 2010 *Sol. Energy Mater. Sol. Cells* **94** 360–365

Aluminum alloy anode materials for Li-ion batteries

Z H Sun[1,2], Z F Chen[1], Q W Fu[1] and X Y Jiang[1,3,*]

[1] School of materials and engineering, Tianjin University of Technology, 391 Binshui Xidao, Xiqing District, Tianjin 300384, China

E-mail: [2] sunzhh_977@163.com, [3,*] jiangxunyong@tjut.edu.cn

Abstract. Aluminum has larger theoretical capacity of 2235 mAh/g than that of graphite (372 mAh/g), but it has big disadvantages including shorter cycle life and higher irreversible capacity loss. Improving cycle performance can be obtained via alloying of aluminum. In this paper, two ternary aluminum alloy, Al_7Cu_2Fe and $Al_{73}Cu_5Fe_{22}$ were prepared. The main phase of Al_7Cu_2Fe alloy was Al_7Cu_2Fe. The heat treatment increased the proportion of Al_7Cu_2Fe. The main phase of $Al_{73}Cu_5Fe_{22}$ alloy was $Al_{60}Cu_{30}Fe_{10}$. The heat treatment reduced the proportion of $Al_{60}Cu_{30}Fe_{10}$. For two alloys, the heat treatment could increase discharge capacity compared with cast alloy. The discharge capacity was improved by 50%. The content of aluminum in alloys has little effect on improving cycle performance, and it has obvious influence on the phase structure of alloy with heat treatment.

Keywords: aluminum alloy, AlCuFe; anode, lithium storage

1. Introduction

In recent years, lithium-ion batteries (LIBs) have been most important energy devices because of high energy density, high output voltage, long life and low pollution, etc [1]. In order to meet people's needs for high capacity batteries, looking for a new anode material with great performance becomes research focus [2]. The energy density of LIBs is proportional to the capacity and the operating voltage, thus developing electrode materials with high reversible capacity is one important approach to improve the energy density of LIBs. Aluminum has theoretical capacity of 2235 mAh/g if the Al_4Li_9 alloy is formed [6], which is much larger than that of graphite (372 mAh/g). Pure aluminum anode has big disadvantages including short cycle life and high irreversible capacity loss because of the large volume expansion during lithiation and delithiation, followed by the loss of electrical contact between individual particles and particles and current collector, resulting in a rapid capacity fade [7]. Improved cycle ability can be obtained by accommodating the volume change via alloying of aluminum. Al-based intermetallic materials (Al_2Cu, Fe_2Al_5) have been studied [10]. The results show that the cycling ability of aluminum alloy anode is improved comparing with pure aluminum electrode.

Here, we report a new aluminum alloy electrode to improve the cycling ability of aluminum anode. The composition of alloy is Al_7Cu_2Fe and $Al_{73}Cu_5Fe_{22}$. The purpose of addition of third element is to reduce the volume change during the charging and discharging process. To investigate the influence of different microstructure on the capacity of alloy electrode, subsequent heat treatment is applied on the casting alloy powder. The results show that heat treatment can increase the capacity of alloy electrode.

2. Experimental

Al_7Cu_2Fe and $Al_{73}Cu_5Fe_{22}$ ingot were prepared by art melting with Fe (chip 99.9%), Cu (shot 99.9%) and Al (shot 99.9%). The ingot were crushed into powder by two routes, 1) One ingot was heat treated at 800 °C in the protection of argon for 12 hours, then cooled down to room temperature. After heat treatment, the alloy was put into ball mill in the ball material ratio 10:1 (stainless steel ball) and proper filling coefficient until the fines were prepared to 300 meshes with molecular sieve. 2) Other ingot was directly milled by mechanical ball milling under the same conditions, without heat treatment. The electrodes for next test were prepared as follow: the alloy powder, the SBR, CMC and acetylene black were mixed with the mass ratios of 85: 2: 3: 10. And then the mixture after a suitable amount of deionized water were stirred with a magnetic whisk for 1 h, and pasted onto the copper foils. The materials were punched into circular electrode slices (15 mm in diameter) after they were dry for 4h at 25 °C. Then the electrode slices were placed on a wild mouthed petri dish to be heated for 10 h at 115 °C in the vacuum drying oven. The average weight of the electrodes was approximately 18 mg. The mental Lithium was used as anode in the cells. And a porous polymeric membrane (Celgard 2400) was used as diaphragm. 1 M LiBF4 dissolved in EC, DEC and EMC mixture of equal volumes was used as electrolyte. The cells were assembled in an argon-filled glove box with the concentrations of moisture and oxygen below 0.1 ppm. Charge-discharge cycles of the electrodes were carried out on the NEWARE tester at about 25 °C. The current was carried 100 times at 0.1 C rate. The structure of materials were characterized by X-ray diffraction (XRD, Rigaku D/Max 2500) with Cu Kα radiation in the range of 10°~80°.

3. Results and discussion

3.1. Microstructure and electrochemical characteristics of Al_7Cu_2Fe

The results of XRD in figure 1(a) show that the main phase of casting Al_7Cu_2Fe alloy is Al_7Cu_2Fe. There are small content of $Al_{23}CuFe_4$ and $Al_{60.3}Cu_{30}Fe_{9.7}$, Al_2Cu in the ingot. The multiphase structure comes from the rapid cooling rate during the arc melting process, which makes incomplete peritectic reaction [13]. Heat treatment and mechanical milling only change the proportion of different phase. After mechanical milling and heat treatment, the intensity of diffraction peak on Al_2Cu decrease, while the peak on Al_7Cu_2Fe increase. These may also suggest heat treatment could increase the content of main phase [14].

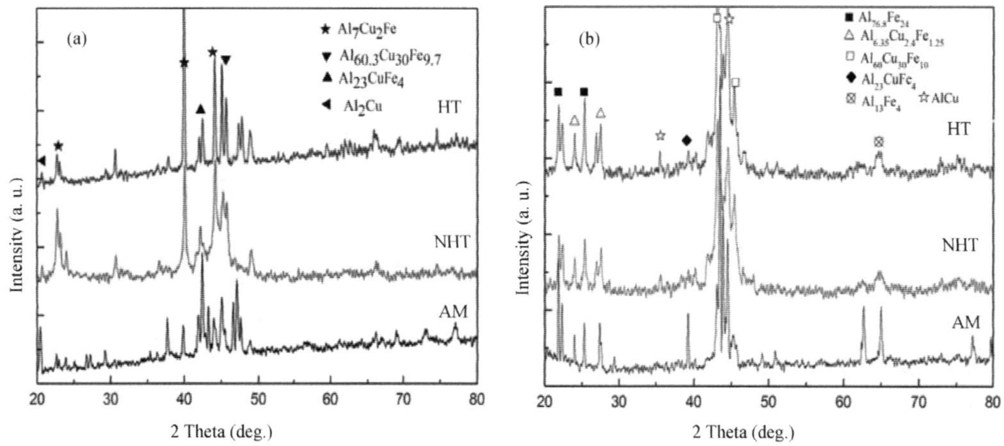

Figure 1. XRD spectra of alloys (a) Al_7Cu_2Fe and (b) $Al_{73}Cu_5Fe_{22}$ under different treatment processes, (HT: heat treatment + balling, NHT: ball milling, AM: as cast).

The electrochemical performance of Al_7Cu_2Fe is studied using the prepared Al_7Cu_2Fe electrode as working electrode and metallic lithium as the counter electrode. The results are shown in figure 2. Cast

Al_7Cu_2Fe without heat treatment performed a small discharging platform is about 0.3 V and the charging plateau is about 0.45 V. The charge-discharge cycling of the cells at 0.1 C rate is carried 50 times. It can be seen that the capacity reduce from about 60 mAg/h to about 35 mAh/g after first cycle. The capacity of electrode keeps constant. The materials with heat treatment have a higher initial discharge capacity which is double higher than cast electrode. After first cycle, the electrode shows the similar large irreversible capacity as casting alloy, but it shows a better cycling property comparing with casting alloy.

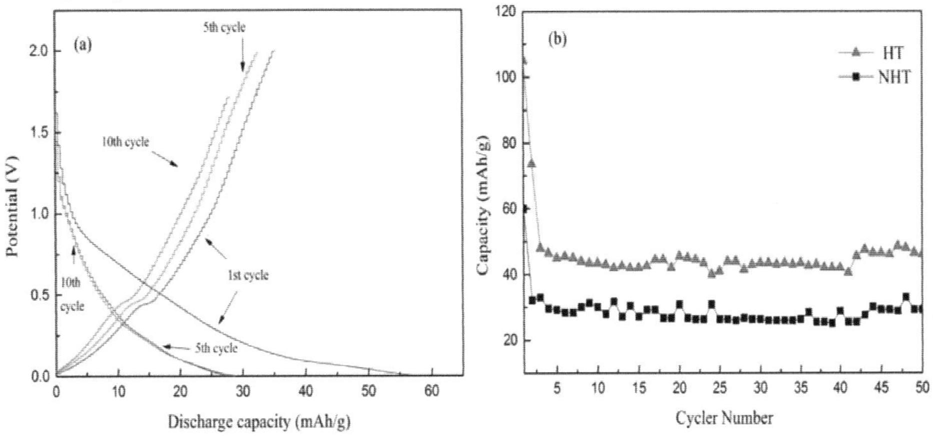

Figure 2. (a) Voltage-capacity of Al_7Cu_2Fe without heat treatment at first, fifth, tenth cycle, (b) Discharge capacity of Al_7Cu_2Fe electrodes and change with heat treatment, (HT: heat treatment + balling, NHT: ball milling).

3.2. Microstructure and electrochemical characteristics of $Al_{73}Cu_5Fe_{22}$

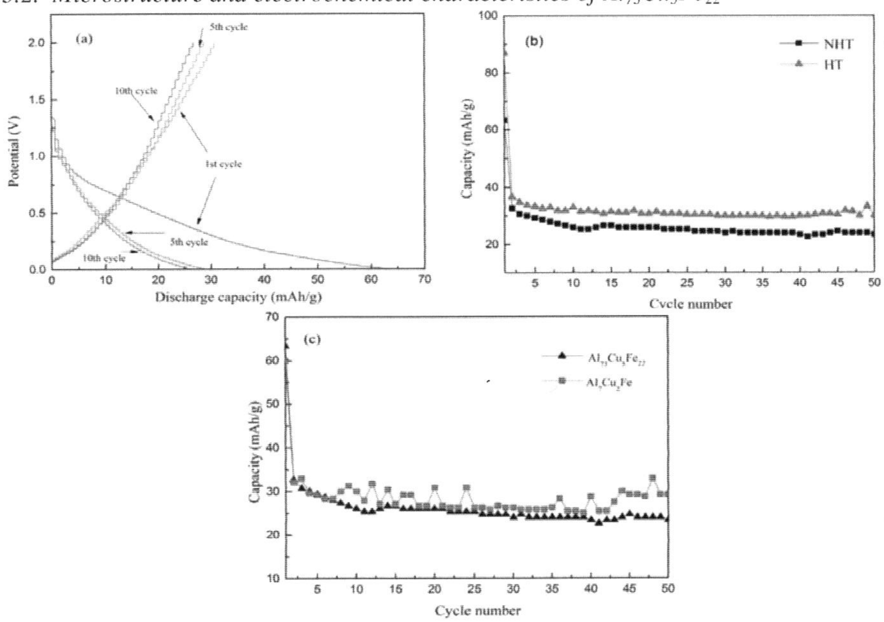

Figure 3. (a) Voltage-capacity of $Al_{73}Cu_5Fe_{22}$ without heat treatment at first, fifth, tenth cycle, (b) Discharge capacity of $Al_{73}Cu_5Fe_{22}$ electrodes and change with heat treatment, (c) Discharge capacity of different alloys without heat treatment, (HT: heat treatment + balling, NHT: ball milling).

Although the content of aluminum in $Al_{73}Cu_5Fe_{22}$ is a little higher than Al_7Cu_2Fe, the main phase of cast $Al_{73}Cu_5Fe_{22}$ is $Al_{60}Cu_{30}Fe_{10}$, which is different from Al_7Cu_2Fe in Al_7Cu_2Fe alloy in figure 1 (b). There are some AlCu and a small amount of $Al_{6.35}Cu_{2.4}Fe_{1.25}$, $Al_{76.8}Fe_{24}$, $Al_{23}CuFe_4$ and $Al_{13}Fe_4$ in the $Al_{73}Cu_5Fe_{22}$ alloy. Heat treatment and mechanical milling only change the proportion of main phase. After mechanical milling and heat treatment, the intensity of diffraction peak of $Al_{6.35}Cu_{2.4}Fe_{1.25}$ and $Al_{76.8}Fe_{24}$ increase, instead. This means that the content of main phase of $Al_{73}Cu_5Fe_{22}$ decrease with heat treatment.

In figure 3, the electrochemical performance of $Al_{73}Cu_5Fe_{22}$ is studied using the prepared $Al_{73}Cu_5Fe_{22}$ electrodes as working electrode and metallic lithium as the counter electrode. In figure 3 (a), cast $Al_{73}Cu_5Fe_{22}$ doesn't perform a small discharging platform like Al_7Cu_2Fe in figure 2(a). The charge-discharge cycling of the cells at 0.1 C rate is carried 50 times. The materials with heat treatment have a higher discharge capacity and a better cycling life than materials without heat treatment in figure 3(b).

The figure 3(c) shows the discharge capacity of two alloys with cycle number. It can be seen that the discharge capacity of two alloys is similar. Although content of aluminum of Al_7Cu_2Fe is lower than $Al_{73}Cu_5Fe_{22}$, its capacity is a little higher than $Al_{73}Cu_5Fe_{22}$. The two alloys are composed by different crystalline phase. The main phase of Al_7Cu_2Fe is Al_7Cu_2Fe, but the main phase of $Al_{73}Cu_5Fe_{22}$ is $Al_{60}Cu_{30}Fe_{10}$. The excess aluminum in $Al_{73}Cu_5Fe_{22}$ precipitates as other phase. This is the reason why Al_7Cu_2Fe has better capacity than $Al_{73}Cu_5Fe_2$.

4. Conclusion

Two ternary aluminum alloy, Al_7Cu_2Fe and $Al_{73}Cu_5Fe_{22}$ have been prepared, and the lithium storage property of two alloys studied. The experimental results show that: the main phase of cast Al_7Cu_2Fe is Al_7Cu_2Fe. The heat treatment increase the proportion of Al_7Cu_2Fe. The main phase of cast $Al_{73}Cu_5Fe_{22}$ is $Al_{60}Cu_{30}Fe_{10}$. The heat treatment reduces the proportion of $Al_{60}Cu_{30}Fe_{10}$. For two alloys, the heat treatment can increase discharge capacity comparing with cast alloys. The discharge capacity is improved by 50%. The content of aluminum in alloys has little effect on improving cycle performance.

References

[1] Zhang W J 2011 A review of the electrochemical performance of alloy anodes for lithium-ion batteries *Journal of Power Sources* **196** 13–24

[2] Winter M, Besenhard J O and Spahr M E 1998 Insertion electrode materials for rechargeable lithium batteries *Advanced Materials* **10** 725–763

[3] Larcher D, Edstroem K, Beattie S and Larcher D 2007 Recent findings and prospects in the field of pure metals as negative electrodes for Li-ion batteries *J. Mater. Chem* **17** 3759–3772

[4] Thackeray M M, Vaughey J T and Fransson L M L 2002 Recent developments in anode materials for lithium batteries *JOM*,**54** 20–23

[5] Wachtler M, Winter M and Besenhard J O 2002 Anodic materials for rechargeable Li-batteries *Journal of Power Sources* **105** 151–160

[6] Gayle F W, Sande J B V and Mcalister A J 1984 The Al-Li (aluminum-lithium) system *Journal of Phase Equilibria* **5** 21

[7] Liu Y *et al.* 2011 In situ transmission electron microscopy observation of pulverization of aluminum nanowires and evolution of the thin surface Al_2O_3 layers during lithiation-delithiation cycles *Nano Letters* **11** 4188–4194

[8] Kasavajjula U, Wang C and Appleby A J 2007 Nano- and bulk-silicon-based insertion anodes for lithium-ion secondary cells *Journal of Power Sources* **163** 1003–1039

[9] Winter M, Brodd R J 2004 What are Batteries, fuel Cells and supercapacitors? *ChemInform*, **104** 4245–69

[10] Lindsay M J, Wang G X and Liu H K 2003 Al-based anode materials for Li-ion batteries *Journal of Power Sources* **119** 84–87

[11] Su Y C, Yan J and Lu P T 2006 Thermodynamic analysis and experimental research on Li

IOP Conf. Series: Materials Science and Engineering **182** (2017) 012011 doi:10.1088/1757-899X/182/1/012011

intercalation reactions of the intermetallic compound Al_2Cu *Solid State Ionics* **177** 507–513
[12] Hamon Y, Brousse T and Jousse F 2001 Aluminum negative electrode in lithium ion batteries *Journal of Power Sources* **97–98** 185–187
[13] Jian X *et al.* 2003 Study on $NiSb_2$ as a new lithium-ions batteries anode material *Rare Metal Materials & Engineering* **32** 283–286
[14] Wang Y F and Zhuang X W 1992 Amorphous materials and their application (Beijing, China: Higher Education Press) pp 21–30

Superstructure and physical properties of skutterudite-related phase $CoGe_{1.5}Se_{1.5}$

Y Liang, B Fang, X M Zhu and M M Liang

Institute of Nuclear Technology and Application, School of Science, East China University of Science and Technology, Meilong Road 130, Shanghai 200237, China

E-mail: yliang@ecust.edu.cn

Abstract. $CoGe_{1.5}Se_{1.5}$ skutterudite-related phase with a homogeneity range has been synthesized by solid-state reaction. The phase purity, homogeneity range, crystal structure, thermal stability and electrical resistivity were studied. XRD data indicates that $CoGe_{1.5}Se_{1.5}$ crystallized in a modification of the skutterudite $CoAs_3$ type structure with space group $R\bar{3}$ ($a = b$ = 11.751(1) Å, c = 14.36(1) Å). HRTEM-SAED shows more information about the superstructure to confirm the rhombohedral symmetry with space group $R\bar{3}$. The lattice parameter of this skutterudite-related phase was found to be dependent on the concentration of Ge and Se. $CoGe_{1.5}Se_{1.5}$ decomposed between 1073 K and 1173 K under argon atmosphere investigated by in-situ XRD, suggesting a good thermal stability. $CoGe_{1.49}Se_{1.42}$, $CoGe_{1.43}Se_{1.34}$ and $CoGe_{1.50}Se_{1.15}$ dense bulk samples were obtained by hot-press technique. The chemical composition detected by FESEM/EDS suggests the homogeneity range and the existence of voids at framework positions. The electrical resistivity of the compounds decreases with increasing temperature, acting as a semiconductor. The chemical composition has a big influence on the value of electrical resistivity and energy gap.

Keywords: related skutterudite, $CoGe_{1.5}Se_{1.5}$, superstructure, electrical resistivity

1. Introduction

Since the natural mineral $CoAs_3$ was first found in 1845 by Wilhelm von Haidinger at Skutterud, Norway, this structure class of MPn_3 (M = Co, Rh, Ir; Pn = P, As, Sb) was named as skutterudite [1]. The skutterudite compounds crystallize in a body-centered cubic with space group $Im\bar{3}$. The pnictogen atoms framework located at the 24g sites offers several types of voids which can be filled by guest atoms. As shown in figure 1, the octahedral voids are filled by transitional metal atoms at 8c sites, forming tilted corner-sharing octahedrons [$MPn6$]. In some cases, the icosahedral voids can be filled by electropositive atoms at 2a sites, giving rise to filled skutterudites of general formula, $R_xM_4Pn_{12}$ (R = alkali metal, alkaline earth metal and rare earth metal). In addition to the unfilled binary and filled ternary skutterudites, another group of materials with skutterudite-related structure exists. These compounds, also known as ternary but unfilled skutterudites, are obtained by simultaneous substitution of group 14 and 16 atoms on the pnictogen sites with general formula of $MT_{1.5}Ch_{1.5}$ (T = Si, Ge and Sn; Ch = S, Se and Te). The skutterudite-related phase behaves complex crystal structure because of a distortion of the anion sublattice, as well as various physical properties due to its large variety of compositions.

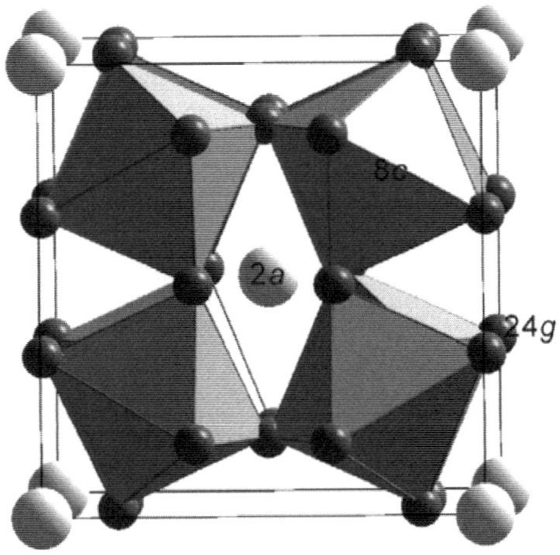

Figure 1. Crystal structure of skutterudite.

For the crystallographic studies, an ordered variant of the skutterudite structure was discussed for $CoGe_{1.5}S_{1.5}$ by Wold et al. from additional superstructure reflections in the X-ray powder diffraction pattern [2]. Then they suggested a short-range ordering of group 14 and 16 anions in primitive cubic model for $IrGe_{1.5}S_{1.5}$, $IrGe_{1.5}Se_{1.5}$, $IrSn_{1.5}S_{1.5}$ and $RhGe_{1.5}S_{1.5}$ [3]. Lutz et al. proposed the space group *R*3 for $CoGe_{1.5}Ch_{1.5}$ (*Ch* = S, Se) from single crystal data of a pseudo-merohedral twin [4] and their model explained the superstructure. Later, Vaqueiro et al. found a center of inversion for the proposed crystal structure from the X-ray powder study on $CoGe_{1.5}Te_{1.5}$ [5]. Since then, all anion substituted ternary skutterudites, $MT_{1.5}Ch_{1.5}$ (*M* = Co, Rh and Ir; *T* = Ge and Sn; *Ch* = S, Se and Te), have been refined in the space group $R\overline{3}$ from X-ray or neutron powder diffraction data [6-11]. Recently, Weihrich et al. refined the crystal structure of $IrSn_{1.5}Se_{1.5}$ in space group $R\overline{3}$ from single crystal data [12]. For electronic properties, the related skutterudites should be semiconductors, as they are isoelectronic with $CoSb_3$. In fact, the electrical resistivity data of $CoGe_{1.5}S_{1.5}$, $RhGe_{1.5}S_{1.5}$ [7], $CoGe_{1.5}Se_{1.5}$ [13], $RhGe_{1.5}Se_{1.5}$ [10] and $CoGe_{1.5}Te_{1.5}$ [11] showed semiconducting temperature dependences. $CoSn_{1.5}Se_{1.5}$ [8], $RhSn_{1.5}Te_{1.5}$ and $IrSn_{1.5}Te_{1.5}$ [6, 14] behaved as semiconductors at low temperature, while showed metal-like behavior above 300 K. In addition, Nolas et al. reported that $CoGe_{1.452}Se_{1.379}$ and $CoGe_{1.431}Se_{1.385}$ is n-type and p-type semiconductor, respectively, because of the different concentration of Ge and Se [15]. Liang et al. found that the Ge/Se ratios significantly affected the electrical resistivity of $RhGe_xSe_{3-x}$ [10]. However, no detailed information concerning the homogeneity and superstructure of $CoGe_{1.5}Se_{1.5}$ is available.

Herein, $CoGe_{1.5}Se_{1.5}$ skutterudite-related phase with a homogeneity range has been synthesized by solid-state reaction and hot-press treatment. The phase purity, homogeneity range, superstructure, thermal stability and electrical resistivity were studied.

2. Synthesis and Characterization

The samples of $CoGe_xSe_{3-x}$ ($1.0 \leq x \leq 2.0$) were prepared from powdered Co (Chempur, 99.9 %), Ge (Chempur, 99.999 %) and Se (Chempur, 99.99 %). Stoichiometric amounts of starting materials about 2 g were mixed and pressed to pellets. The pellets were sealed in vacuum silica tubes (10^{-3} Pa), annealed at 873 K for 5 d and quenched in water to room temperature. The products were stable in air and water. The so obtained samples after solid-state reaction were ground to fine powder and loaded into graphite dies of inner diameter 10 mm for hot-press treatment. Densification was achieved by heating to 823 K at 30 K/min under a uniaxial pressure of 100 MPa in argon atmosphere. After a reaction time of 30 minutes, the samples were cooled down to room temperature. The pellets obtained were cut with a diamond wire-saw for the electrical resistivity measurements.

For phase analysis and determination of unit cell parameters, data were collected with a D/MAX-2550 diffractometer (Rigaku, Cu *K*α, $10° \leq 2\theta \leq 80°$, $\Delta 2\theta = 0.02°$). Unit cell parameters were calculated from reflection positions determined by profile deconvolution with the program package WinCSD [16] and corrected with internal LaB_6 standard (NIST, $a = 4.15692(1)$ Å). The in-situ X-ray powder diffraction was performed on a DX-2700A diffractometer (Cu *K*α, $10° \leq 2\theta \leq 80°$, $\Delta 2\theta = 0.02°$). The sample was heated stepwise from 293 K to 1273 K (heating rate ≈ 50 K/min, holding time at each temperature before collecting data ≈ 5 min) under argon. Specimens for TEM investigations were prepared by dispersing and crushing small amounts of powder in ethanol, and then deposited on holey carbon support films on copper grids. A JEOL JEM-2100F electron microscope was used. The selected area electron diffraction (SAED) method was performed on thin particles for unit cell characterization. Energy-Dispersive X-ray spectroscopy (EDS) was performed with a Hitachi S-4800 field emission scanning electron microscope. Electrical resistivity was measured on hot-pressed samples in a PPMS in the temperature range 2.5 K – 400 K using a four-point method.

3. Results and Discussion

The X-ray powder diffraction pattern of $CoGe_{1.5}Se_{1.5}$ revealed the pure skutterudite-related phase (figure 2). Comparing with the calculated X-ray diffraction pattern in body-centered cubic model with space group $Im\bar{3}$, in the XRD pattern of $CoGe_{1.5}Se_{1.5}$ in fact superstructure reflections with low intensities were visible. The superstructure reflections at $2\theta = 32.30°$, 35.84°, 44.96°, 47.67°, 50.28°, 59.94°, 64.36° and 72.91° correspond to the (131), (223), (413), (241), (333), ($\bar{1}65$), (351) and (437) planes of the rhombohedral crystal structure with space group $R\bar{3}$, respectively. The lattice parameter of $CoGe_{1.5}Se_{1.5}$ is $a = b = 11.751(1)$ Å, $c = 14.36(1)$ Å.

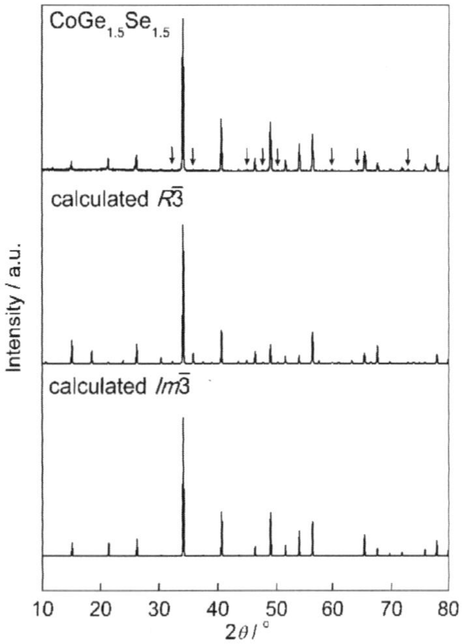

Figure 2. XRD pattern of synthesized $CoGe_{1.5}Se_{1.5}$ powder, calculated patterns for $CoGe_{1.5}Se_{1.5}$ with space group $Im\bar{3}$ and $R\bar{3}$, respectively. Arrows mark weak superstructure reflections.

Figure 3. HRTEM and FFT (inset) images of $CoGe_{1.5}Se_{1.5}$ view along [001] direction.

HRTEM investigations of $CoGe_{1.5}Se_{1.5}$ reveal particles with large crystalline domains of the skutterudite-related phase (figure 3). To further confirm the superstructure and obtain more information about the symmetry, SAED patterns along different directions were studied (figure 4). At first, the analysis was performed in the original body-centered cubic model with space group $Im\bar{3}$. In this case, Figure 4(a) matches the SAED patterns along [111] direction. The electron diffraction spots with the same spacing distance of d = 5.876 Å correspond to the ($\bar{1}01$), ($10\bar{1}$), ($\bar{1}10$), ($1\bar{1}0$), ($01\bar{1}$) and ($0\bar{1}1$) planes. While, there are not planes match the typical electron diffraction spots with spacing distances of d = 8.303 Å, 5.867 Å, 4.796 Å and 4.787 Å, as shown in figure 4(b). As a consequence, the rhombohedral model with space group $R\bar{3}$ was applied. According to this model, figure 4(a) is the SAED patterns along [001] direction. The electron diffraction spots with the spacing distance of d = 5.876 Å correspond to the (110), ($\bar{1}\bar{1}0$), ($2\bar{1}0$), ($\bar{2}10$), ($1\bar{2}0$) and ($\bar{1}20$) planes. Along [100] direction, the electron diffraction spots with spacing distances of d = 8.303 Å, 5.867 Å, 4.796 Å and 4.787 Å correspond to the planes of ($01\bar{1}$), (012), (021) and (003), respectively (figure 4(b)). Figure 4(c) presents the SAED patterns along [$1\bar{1}\bar{1}$] direction. The typical electron diffraction spots with spacing distances of d = 8.303Å, 5.876 Å and 5.867 Å represent the (101), (110) and ($\bar{1}1\bar{2}$) planes, respectively. In the SAED patterns along [$0\bar{2}1$] direction (figure 4(d)), the marked spots confirm the existence of (012), (300) and ($\bar{3}12$) planes with spacing distances of d = 5.867Å, 3.392 Å and 3.391 Å, respectively. The results indicate that the SAED patterns along (001), (100), ($1\bar{1}\bar{1}$) and ($0\bar{2}1$) directions consist well with the rhombohedral crystal structure.

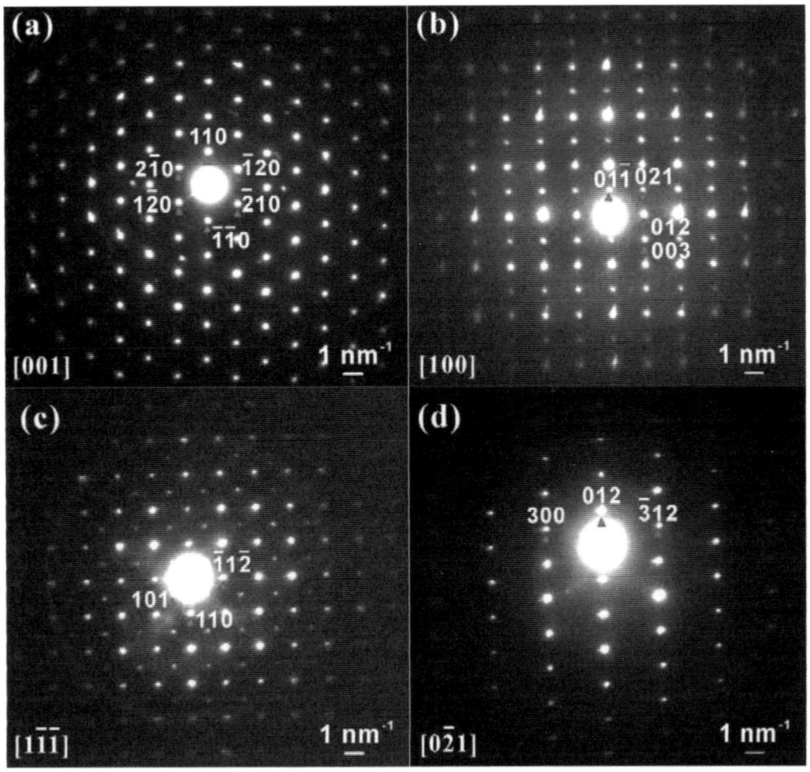

Figure 4. HRTEM-SAED patterns of synthesized $CoGe_{1.5}Se_{1.5}$ powder along (a) [001], (b) [100], (c) [$1\bar{1}\bar{1}$] and (d) [$0\bar{2}1$] directions.

In order to determine the homogeneity range, samples with different nominal compositions of $CoGe_xSe_{3-x}$ ($1.0 \leq x \leq 2.0$) were prepared by solid-state reaction. XRD patterns showed only reflections of the skutterudite-related phase for $1.3 \le x \le 1.8$ samples. Besides skutterudite-related main phase, some elemental Ge in $x \ge 1.9$ samples, or $CoSe_2$ in $x \le 1.2$ samples were also investigated in XRD patterns (not shown). Despite of the large range of nominal compositions, unit cell refinement showed only small differences in the lattice parameters (table 1).

Table 1. Nominal composition and lattice parameters of $CoGe_xSe_{3-x}$ （$1.3 \le x \le 1.8$） skutterudite-related phase.

Nominal composition	Lattice parameters / Å
$CoGe_{1.3}Se_{1.7}$	$a = b = 11.750(2)$, $c = 14.40(2)$
$CoGe_{1.4}Se_{1.6}$	$a = b = 11.744(2)$, $c = 14.35(3)$
$CoGe_{1.5}Se_{1.5}$	$a = b = 11.751(1)$, $c = 14.36(1)$
$CoGe_{1.6}Se_{1.4}$	$a = b = 11.741(3)$, $c = 14.43(3)$
$CoGe_{1.7}Se_{1.3}$	$a = b = 11.749(2)$, $c = 14.32(3)$
$CoGe_{1.8}Se_{1.2}$	$a = b = 11.750(3)$, $c = 14.28(4)$

The in-situ XRD performed on the sample with nominal composition $CoGe_{1.5}Se_{1.5}$ exhibited the cell expansion with increasing temperature of the skutterudite-related phase. On heating from 293 K to 1173 K a shift of the reflection positions to lower diffraction angles was observed (figure 5). The lattice parameter of the skutterudite-related phase in this temperature range increased linearly (figure 6). At 1173 K, additional reflections corresponding to some impurities were observed and became clearer at 1273 K, indicating that the decomposed temperature was between1073 K and 1173 K.

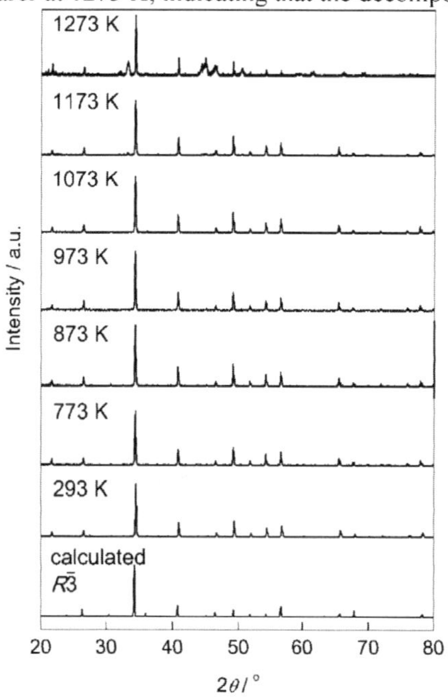

Figure 5. XRD patterns collected in-situ on heating $CoGe_{1.5}Se_{1.5}$ powder.

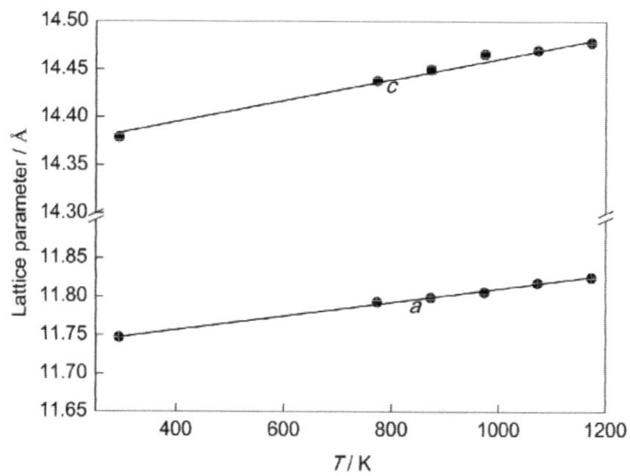

Figure 6. Temperature dependence of the unit cell parameters of $CoGe_{1.5}Se_{1.5}$. Error bars lie within the points.

The obtained three samples with nominal compositions of $CoGe_{1.4}Se_{1.6}$, $CoGe_{1.5}Se_{1.5}$ and $CoGe_{1.6}Se_{1.4}$ were further treated by hot-press technique. The preparation resulted in dark-gray bulk with at least 95 % of the theoretical density. XRD patterns reveal the skutterudite-related main phase with some binary GeSe or elemental Ge byproducts (figure 7). The composition of the main phase is $CoGe_{1.49}Se_{1.42}$, $CoGe_{1.43}Se_{1.34}$ and $CoGe_{1.50}Se_{1.15}$, respectively, established by EDS analysis, whereby it was assumed that the 8*c* positions were completely occupied by cobalt atoms (table 2). The results suggest the homogeneity range of chemical composition, as well as the existence of vacancies at the framework. For all the three samples $CoGe_{1.49}Se_{1.42}$, $CoGe_{1.43}Se_{1.34}$ and $CoGe_{1.50}Se_{1.15}$, the resistivity decreases with increasing temperature from 2.5 K to 400 K confirming the semiconducting behavior of materials (figure 8). The value of electrical resistivity at 300 K is 1.8×10^{-3} Ωm, 6.6×10^{-3} Ωm and 3.0×10^{-2} Ωm, respectively. The difference increases with decreasing temperature. The much higher electrical resistivity observed for $CoGe_{1.50}Se_{1.15}$ perhaps is due to its lower carrier mobility or concentration than that of $CoGe_{1.49}Se_{1.42}$ and $CoGe_{1.43}Se_{1.34}$. The data at higher temperature (300 K – 400 K) can be fitted by the Arrhenius activation law $\ln(\rho)$ = constant + $E_g/2k_BT$, where E_g is the energy gap, resulting in E_g = 0.194 eV, 0.182 eV and 0.538 eV for $CoGe_{1.49}Se_{1.42}$, $CoGe_{1.43}Se_{1.34}$ and $CoGe_{1.50}Se_{1.15}$, respectively (table 2), comparable with that for $CoGe_{1.452}Se_{1.379}$ (0.336 eV) in *ref.* 15. The results indicate that the estimated intrinsic band gap strongly depends on the concentrations.

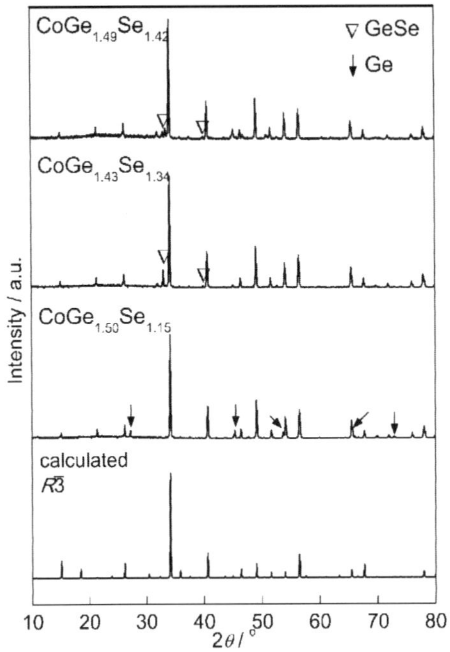

Figure 7. XRD patterns of hot-press treated $CoGe_{1.49}Se_{1.42}$, $CoGe_{1.43}Se_{1.34}$ and $CoGe_{1.50}Se_{1.15}$ samples, calculated pattern with space group $R\overline{3}$.

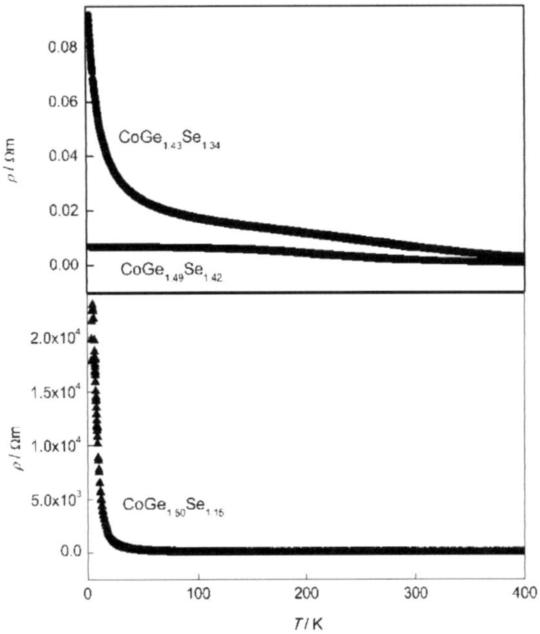

Figure 8. Temperature dependence of the electrical resistivity of $CoGe_{1.49}Se_{1.42}$, $CoGe_{1.43}Se_{1.34}$ and $CoGe_{1.50}Se_{1.15}$.

Table 2. FESEM/EDS results and energy gap (E_g) of hot-press treated samples.

Nominal composition	Composition of main phase from FESEM/EDS	E_g (eV)
$CoGe_{1.4}Se_{1.6}$	$Co_{1.00(5)}Ge_{1.49(6)}Se_{1.42(3)}$	0.194
$CoGe_{1.5}Se_{1.5}$	$Co_{1.00(1)}Ge_{1.43(4)}Se_{1.34(3)}$	0.182
$CoGe_{1.6}Se_{1.4}$	$Co_{1.00(4)}Ge_{1.50(6)}Se_{1.15(9)}$	0.538

4. Conclusions

The skutterudite-related phase $CoGe_xSe_{3-x}$ with a homogeneity range has been synthesized by solid-state reaction. XRD shows weak superstructure reflections for all concentrations. More information about superstructure obtained from SAED confirms the rhombohedral crystal structure with space group $R\bar{3}$. The concentration of Ge and Se detected by EDS suggests the existence of vacancies at the framework positions. The electrical resistivity of $CoGe_xSe_{3-x}$ for different concentrations increases with decreasing temperature, indicating a semiconducting behavior of the material. The band gap significantly depends on the concentration.

Acknowledgments
This work was supported by the National Natural Science Foundation of China (Grant number: 51301071), the Innovation Program of Shanghai Municipal Education Commission (Grant number: 14ZZ057).

References

[1] Haidinger W 1845 *Handbuch der Bestimmenden Mineralogie* (Wien: Braumüller und Seidel)
[2] Korenstein R, Soled S, Wold A and Collin G 1977 *Inorg. Chem.* **16** 2344
[3] Lyons A, Gruska R P, Case C, Subbarao S N and Wold A 1978 *Mater. Res. Bull.* **13** 125
[4] Partik M, Kringe C and Lutz H D 1996 *Z. Kristallogr.* **211** 304
[5] Vaqueiro P, Sobany G G, Powell A V and Knight K S 2006 *J. Solid State Chem.* **179** 2047
[6] Bos J W G and Cava R J 2007 *Solid State Commun* **141** 38
[7] Vaqueiro P, Sobany G G and Stindl M 2008 *J. Solid State Chem.* **181** 768
[8] Laufek F, Navrátil J, Plášil J, Plecháček T and Drašar Č 2009 *J. Alloys Compd.* **479** 102
[9] Vaqueiro P, Sobany G G and Powell A V 2010 *Dalton Trans.* **39** 1020
[10] Liang Y, Schnelle W, Oeschler N, Budnyk S and Grin Yu 2011 *Z. Kristallogr.* **226** 62
[11] Kaltzoglou A, Powell AV, Knight K S and Vaqueiro P 2013 *J. Solid State Chem.* **198** 525
[12] Yan W, Pielnhofer F, Tragl S A and Weihrich R 2015 *Z. Anorg. Allg. Chem.* **641** 543
[13] Wei K, Dong Y, Puneet P, Tritt T M and Nolas G S 2014 *J. Alloys Compd.* **614** 330
[14] Zevalkink A, Star K, Aydemir U, Snyder G J, Fleurial J P, Bux S and Vo T von Allmen P 2015 *J. Appl. Phys.* **118** 035107
[15] Nolas G S, Yang J and Ertenberg R W 2003 *Phys. Rev. B* **68** 193206
[16] Akselrud L and Grin Yu 2014 *J. Appl. Cryst.* **47** 803

Activated porous carbon wrapped sulfur sub-microparticles as cathode materials for lithium sulfur batteries

Y Wang[1], Y L Yan[1,*], B Ren[1], R Yang[2], W Zhang[1] and Y H Xu[1]

[1] School of Materials Science and Engineering, Xi'an 710048, China
[2] School of Science, Xi'an University of Technology, Xi'an 710048, China

E-mail: *yyl3550@xaut.edu.cn

Abstract. The lithium-sulfur batteries holds a high theoretical capacity and specific energy, which is 4-5 times larger than that of today's lithium-ion batteries, yet the low sulfur loading and large particles in the cathode greatly offset its advantage in high energy density. In the present paper, a liquid phase deposition method was introduced to synthesize sub-micro sulfur particles, which utilized as cathode materials after composed with activated porous carbon. Compared with common sublimed sulfur cathodes, as-obtained composite cathode shows an enhanced initial discharge capacity from 840.7 mAh/g to 1093 mAh/g at C/10. The reversible specific capacity after 50 cycles increased from 383 mAh/g to 504 mAh/g. The developed method has the advantages of simple process, convenient operation and low cost, and is suitable for the industrial preparation of lithium/sulfur batteries.

Keywords: lithium sulfur batteries, liquid phase deposition, sub-micro sulfur particles, porous carbon

1. Introduction

With the rapid development of electric vehicles and mobile electronic devices, there is an urgent need to develop higher energy density batteries. The largest theoretical specific capacity of conventional lithium ion battery cathode material was only 300mAh/g [1], however, the actual specific capacity is lower than 300 mAh/g. It has become increasingly difficult to meet the development of the market demand for the high capacity secondary batteries. The theoretical specific capacity and specific energy of lithium sulfur secondary batteries are up to 1675 mAh/g and 2600 Wh/kg [2,3], respectively. Besides, the cathode active material of element sulfur is very abundant in nature, low cost and environmental friendly [4]. It shows good performance and bright future in application, which is the research topic in the field of chemical power source.

However, in spite of these advantages, the practical application of Li-S batteries is still facing with several problems, including the stability of lithium metal and the insulating nature of pure sulfur with extraordinary low conductivity of 5×10-30 $S \cdot cm^{-1}$ [5] at room temperature. It is difficult to have sufficient electrochemical contact between the cathode active materials of sulfur, conductive carbon and the current collector, thus resulting in the insufficient utilization of sulfur [6,7]. Various attempts to overcome the above-mentioned problems appeared, most of which focus on enhancing the electrical conductivity of the cathode and trapping the polysulfides in a conducting matrix during cycling. Conducting matrix includes porous carbon [8,9], graphene [10,11], carbon nanotubes [12,13], metal oxides [14,15], and conductive polymers [16,17]. In addition, it can start from the size of the elemental sulfur. However, at present, the particle size of sulfur usually distributes in micron level, such as sublimed sulfur. If sublimed sulfur is use directly to be the cathode in lithium batteries, it is not benefit to develop the potential performance of battery. It is found that the construction of cathode material with micro/nano structure is an effective way to improve the electrochemical performance of lithium sulfur battery, sub-micro sulfur [18] has the advantages of small size, large specific surface area, rich pore volume, which can improve the electrochemical performance of the battery, so it is more suitable for application in lithium/sulfur batteries.

At present, the preparation of micro/nano sulfur materials mainly are used by high temperature gas phase solidification method, which needs to be carried out under high temperature and inert gas protection. It is not suitable for industrial production due to its harsh preparation conditions, high energy consumption and high cost. Guo [19] prepared nano sulfur material with regular spherical structure by water-oil two phase micro emulsion method. However, this method required a large amount of organic solvent and surfactant, and the subsequent cleaning process was more complicated. Recently, liquid phase deposition method [20,21] was used to prepare micro/nano materials efficiently, which was the moderate chemical reaction in the solution. Compared with the preparation method above, it is easier to control the chemical reaction process of liquid phase deposition method with the simple process and low cost. For example, Shin [22] used liquid phase deposition method to synthesize colloidal micron sulfur.

Herein, in this paper, we prepared sub-micro sulfur/activated porous carbon (AC) composites by a simple method as the cathode material in lithium sulfur batteries. Using sub-micro S prepared by the liquid phase deposition as an active material. AC was used as a conductive matrix with excellent chemical adsorption capacity of lithium sulfides and polysulfide, which enhanced electrochemical performances remarkably. Because of the smaller particle size of sub-micro S, it is more uniformly distributed in activated carbon matrix. Compared with sublimed S, the use of smaller scale sub-micro S indirectly enhance the conductivity of the active material, and greatly improves the charge and discharge specific capacity of lithium sulfur batteries. In addition, adsorption effect of activated carbon for smaller sub-micro S particles is stronger, and the distribution of sulfur particle is expanding in the activated carbon. It will significantly reduce the shuttle effect [23-25] and control polysulfide dissolved into the electrolyte during cycle process, thus improving sulfur utilization and cycle stability.

2. Experimental

2.1. Synthesis of sub-micro sulfur

The synthesis of sub-micro S by liquid phase deposition. In a typical reaction, first, sodium thiosulfate pentahydrate ($Na_2S_2O_3 \cdot 5H_2O$, analytical grade, 7.75 g) was dissolved in deionized water (a laboratory made, 78 mL) with magnetic stirring (0.4M of $Na_2S_2O_3 \cdot 5H_2O$) at the ice bath (0-5 ℃), named solution A. Secondly, PEG-400(5 ml) was added into 10 mL hydrochloric acid (HCl, 1 M), named solution B. Third, solution B was added dropwise (60 drops/min) into solution A. In order to react completely, the reaction process remained for 2 h at room temperature with magnetic stirring and ultrasonic treatment, after which the product was filtered and washed several times with deionized water until the value of pH was reached 7. Finally, the yellow suspension was collected and dried in an air-oven at 55 ℃ for 24 h. Sub-micro S powers were obtained. The reaction equation in the liquid phase occurred as equation (1).

$$Na_2S_2O_3 \cdot 5H_2O + 2HCl \rightarrow S \downarrow + SO_2 \uparrow + 2NaCl + 6H_2O \quad (1)$$

2.2. Synthesis of sulfur/activated carbon composites

Sub-micro S or sublimed S (analytical grade)/activated carbon (AC, Shaanxi XinHua Carbon Industry) composites were prepared via melt infusion. Typically, sulfur powders and AC were taken in a 6:4 weight ratio and mixed well. The mixture was then transferred into a sealed stainless steel vessel (50 ml) and placed in an oven heated at 155 ℃ for 8 h to obtain the S/C composites.

2.3. Materials characterization

To evaluate the characteristic of the S, S/AC composite, X-ray diffraction (Shimadzu XRD-7000, with Cu Ka radiation, λ =1.5406 Å) was conducted to investigate the compositions of the as-prepared product. The morphology and energy dispersive spectroscopy (EDS) mapping of samples were obtained with JEOL JSM-6610 scanning electron microscopy (SEM) and JEOL 3100 transmission electron microscope (TEM), respectively. The particle size of sulfur was test by laser particle analyzer (LPA, BT-2003).

2.4. Electrochemical testing

The cathode was prepared by making a slurry of the S/C composite, KS-6 conductive graphite, and polyvinyledene fluoride (PVDF) binder in a weight ratio of 80:10:10, using N-methyl-2- pyrrolidone (NMP) as the solvent. The slurry was then cast onto an aluminum foil using a glass rod to make ~20 micron thick films followed by drying at 50 ℃ in an oven for 12 h. Lithium metal was used as the anode. A microporous polypropylene Celgard 2300 membrane was used as the separator. 1 M Lithium Bis (Trifluoromethanesulfonyl) Imide (LiTFSi) and 1wt% lithium nitrate ($LiNO_3$) dissolved in a mixture of 1, 3-Dioxolane (DOL) and Dimethoxyethane (DME) (1:1 by volume) was used as the electrolyte. CR2025 type coin cells were assembled in a glove box under inert argon atmosphere. Galvanostatic charge and discharge tests were carried out using a Neware battery tester in the voltage range of 1.5-3.0 V (vs Li^+/Li).

3. Results and discussion

XRD patterns (figure 1) were recorded to investigate the state of the synthesized sub-micro S and AC in the S/C materials. Firstly, pure sulfur displays sharp diffraction peaks that agree well with the characteristic pattern of S8 which matches with the standard value (JCPDF NO. 08-0247), while carbon materials only show two broad diffraction peaks at ~ 24° and ~43° , indicating amorphous characteristics. Furthermore, compared with the pure sub-micro S sample, the sub-micro S/AC composites shows the weaker diffraction peaks than that of bulk crystalline sulfur, indicating that there is some sulfur distributed on the surface of the activated carbon, and the other sulfur exists as highly dispersed fine particles in the pores of carbon matrix. It may alleviate the insulation issue of sulfur, and is good for electron transport in the electrode to be fabricated.

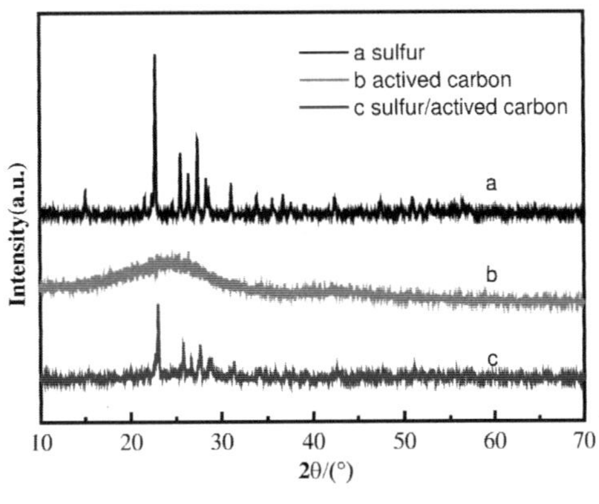

Figure 1. XRD patterns of sub-micro S, AC, sub-micro S/AC.

Figure 2(a) is the SEM image of sub-micro sulfur precipitated in aqueous solution. The particle size of as-prepared sulfur particles is below 1 μm uniformly, and agglomeration phenomenon appear slightly. According to the TEM image of sulfur particles (figure 2(b)), the size of larger particles is about 250 nm. Moreover, some smaller sulfur particles below 100 nm are also presented at the upper right corner in figure 2(b).

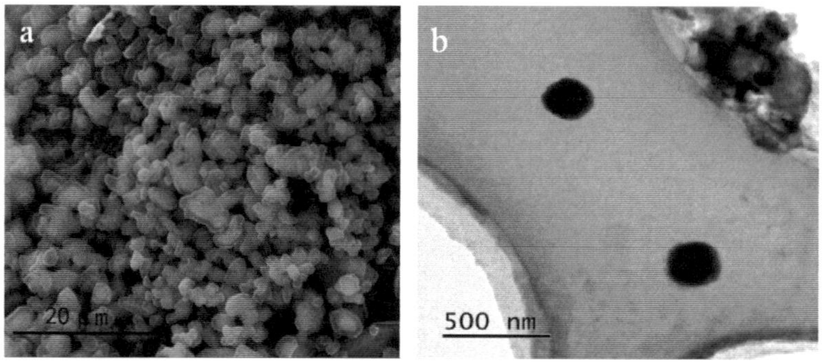

Figure 2. (a) SEM images ,and (b) TEM images of the sub-micro S.

In order to precisely determine the distribution of particle size of sulfur, the sample was tested by laser particle size analyzer and the test result in exhibited in figure 3. As can be seen, the size of the sulfur particles mainly distributes in 150nm ± 10 nm, and the minority of sulfur particles' size is distributed in 100nm ± 10 nm, which confirms the particle size of sulfur is at the sub-micro level. The size control of sub-micro sulfur particles attributes into several reasons, such as (1) the effect of surfactant (PEG) in B solution, (2) slow adding rate of B into A, and (3) the inhibited growing rate caused by ice bathing. As a result, the particle size of sulfur can be reduced to sub-micro, even close to the nano scale.

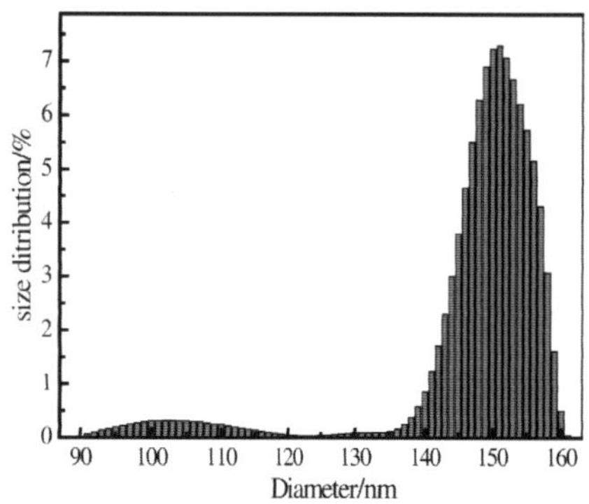

Figure 3. The laser particle-size distribution of sub-micro S.

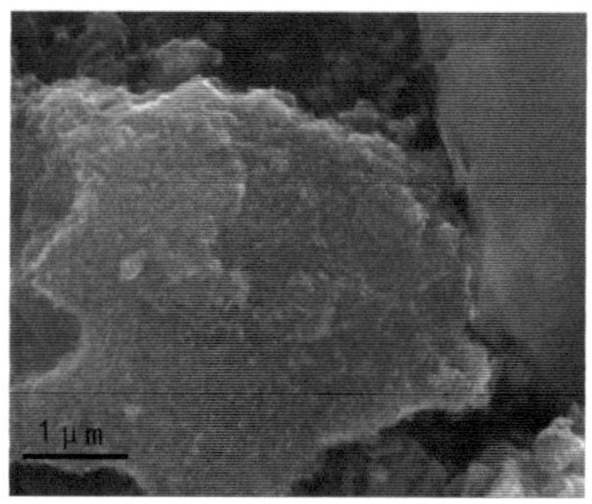

Figure 4. SEM image of AC.

Figure 4 is the SEM image of AC particles. It shows that the AC particles possess a irregular shape and uneven surface. It also has a large number of pores, generally for mesoporous scale (2-50 nm) [26].

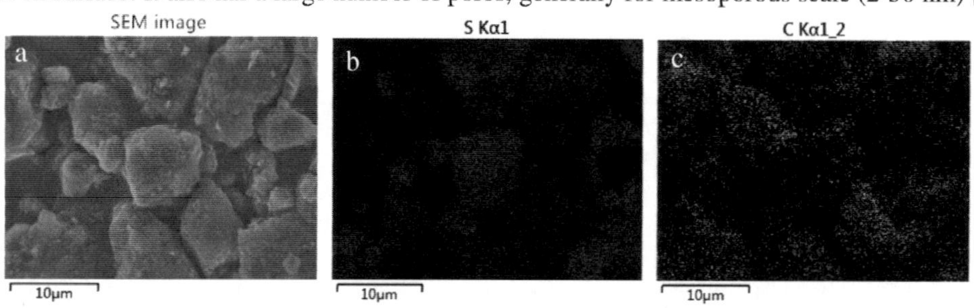

Figure 5. SEM images of (a) the sub-micro S/AC composite, (b) EDS sulfur and (c) EDS carbon mapping of the region shown in (a).

The SEM image and EDS analysis of sub-micro S/AC composite is shown in figure 5. After sub-micro S and AC was compounded, the number of surface pores decreased significantly(figure 5(a)). It suggests that part of sulfur particles have been filled into the pores of AC. It is consistent with the decreasing of diffraction intensity of sulfur element in the XRD result of the previous figure 1. The distributions of S and C elements are shown in figure 5(b) and 5(c), respectively. As can be seen that the majority of sulfur particles distribute uniformly in carbon matrix.

The sub-micro S/AC composites and the sublimed S/AC composites were both assembled into Li/S battery as cathode materials and the charge and discharge performances were tested at 0.1 C. The testing results are shown in figure 6. The discharge platforms are at the vicinity of 2.3 and 2.1 V, and the former corresponds to elemental sulfur is oxidized to sulfur compounds (Li_2S_n, n⩾4) process, while the later corresponds to the high valence sulfides (Li_2S_n, n = 4) reduced to low valence sulfides (Li_2S_n, n < 4) and the Li_2S process. The first discharge capacity of sub micron S/AC battery is up to 1097 mAh/g, higher than that of the sublimed S/AC battery which is only 870 mAh/g. It shows that the prepared sub-micron S can promote the sulfur capacity in battery. Due to the smaller size of sub micron S, shortening the ion diffusion paths in the transmission and promoting the transfer of charge at the same time.

The corresponding cycle properties of sub micron S/AC battery and sublimed S/AC battery at 0.1C for 50 times are also shown in figure 6(b). It can be seen that the sub-micro S/AC composite assembled battery in the first 15 cycles, the capacity decay rapidly, which is mainly caused by internal complex irreversible electrochemical reaction in the battery [27]. Since the sixteenth cycle, the discharge specific capacity tends to be stable. After 50 cycles, it can still maintain the capacity of 503.7 mAh/g, about 46.1% of the initial capacity. At the same time, the charge and discharge efficiency is almost up to 100%. However, the specific capacity of sublimed S/AC battery decreases quickly in the previous 4 cycles. After 50 cycles, it can still maintain the initial capacity of about 44% (382.9 mAh/g), and the charge and discharge efficiency is about 98.8%. In summary, the discharge specific capacity and capacity retention rate of sub-micro S/AC composites battery is better than that of sublimed S/AC composites.

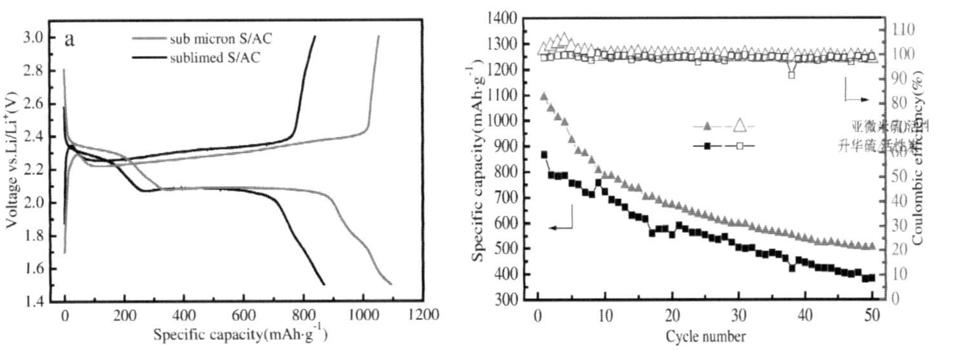

Figure 6. Electrochemical performance of sub-micro S/AC and sublimed S/AC composites. (a) Charge/discharge voltage profiles of electrodes at 0.1C between 1.5 and 3 V, (b) Comparison of the cycling performance over 50 cycles at 0.1C.

The EIS measurements were performed on coin cells with the same loading of the sulfur and the results are shown in figure 7. The resistance (R_Ω), corresponding to the intercept at the real axis, is determined by the ionic resistance of the electrolyte, the intrinsic resistance of the active materials, and the contact resistance at the active material/current collector interface. The charge transfer resistance (R_{ct}), corresponding to the semicircle in the high-frequency region, represents the kinetic resistance of the electrochemical reaction at the electrode-electrolyte boundary. The Warburg impedance, corresponding to the inclined line at low frequency, is associated with Li-ion diffusion in the electrode [28, 29].

Before cycling, both the EIS results from sub micron S/AC battery and sublimed S/AC battery exhibit typical semicircles at high frequencies and short inclined lines in the low frequency region as shown in figure 7(a). The R_{ct} is 221 Ω for sub-micro S, which is lower than that for sublimed sulfur (330 Ω). As all known, the R_{ct} is significantly affected by the area of interface for the electrochemical reaction, as well as the transfer rate of Li^+, and the conductivity of electrons. In the sub micron S/AC composites, as the reduction of sulfur particle size, the attachment sites between the sulfur particles and the pore walls of AC matrix are increased. Thereby the reaction active sites between cations, anions and electrolyte during the charging and discharging process are improved. Thus, larger reaction zones and better electron transfer rate result a smaller R_{ct}. Moreover, the Li-ion diffusion rate of sub micron S/AC composites cathode materials is as shown in the Nyquist plots in the low- frequency region.

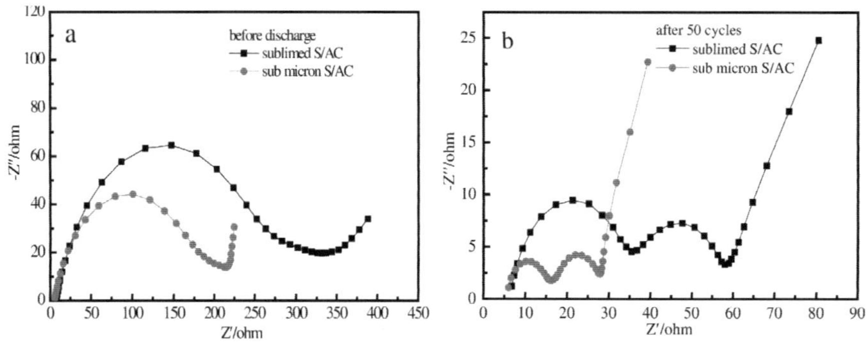

Figure 7. Electrochemical impedance response of Li/S batteries.
(a) before discharge, and (b) after 50 cycles.

After 50 cycles (figure 7(b)), there are two different semicircles followed by a sloping line in the EIS. Generally, the high-frequency semicircle attributes to from the inter-facial charge transfer process, and the semicircle in the medium frequency range is related to the solid-electrolyte-interface (SEI) film resistance caused by the formation of Li_2S or Li_2S_2 on the surface of cathode, and the straight line is ascribed to the diffusion of Li ions in the matrix. As can be seen, both the resistance (R_{ct}) and Li-ion diffusion rate of sub micron S/AC battery are still better than that of sublimed S/AC battery after 50 cycles. The R_{ct} after cycling are decreased remarkably in comparison with that before cycling (from 221 Ω to 28 Ω for sub micron S/AC battery and from 330 Ω to 58 Ω for sublimed S/AC battery), which may be due to the infiltration of the electrolyte and chemical activation process for the dissolution and redistribution of the active materials [30].

4. Conclusions

In summary, a sub-micro S particles, particle size distributed in 150nm±10nm, was successfully fabricated through the liquid phase deposition method. Then it was used to prepare a sub-micro S/AC composite by ball-milling and melt diffusion process. The sulfur element dispersed uniformly in AC matrix from the EDS map. The the sub-micro S/AC cell presented higher initial discharge capacity 1097 mAh/g at 0.1 C and better cycle performance after 50 cycles than the sublimed S/AC cell. This is due to the smaller resistance and faster Li^+ transfer rate in the sub-micro S/AC cell, which is verified by EIS testing. Consequently, the smaller particle size of sulfur played a important role in enhancing the electrochemical performance of sub-micro S/AC composite cathode. The reaction active sites between the positive ions, negative ions and the electrolyte were improved significantly. It is believed that reducing the sulfur particle size is good for high performance Li-S batteries applications.

Acknowledgment

The authors thank the the International Science and Technology Cooperation Program of China (Nos. 2015DFR50350) and Scientific Research Program Funded by Shaanxi Provincial Education Department (Nos. 15JK1536, 14JS047) for the financial support through a laboratory-directed research and development. The synthesis and structural characterization of the bimodal porous materials and the S/C composites were conducted at Xi'an University of technology (Innovation project, Nos. 2015CX002, 2014CX024).

References

[1] Nagata H and Chikusa Y 2014 *J. Power Sources* **264** 206–210
[2] Bruce P G, Freunberger S A, Hardwick L J and Tarascon J M 2012 *Nat. Mater.* **11** 19–29
[3] Wang W, Yu Z B, Yuan K G, Wang A and Yang 2011 *Prog. Chem.* **23**(2/3) 540–546
[4] Goodenough J B and Kim Y 2010 *Chem. Mater.* **22** 587–603
[5] Tarascon J M and Armand M 2001 *Nat.* **414** 359–367
[6] Ji X and Nazar L F 2010 *J. Mater. Chem.* **20** 9821–9826
[7] Cheon S E, Ko K S, Cho J H, Kim S W, Chin E Y and Kim H T 2003 *J. Electrochem. Soc.* **150** A796–A799
[8] Qu Y, Zhang Z, Zhang X, Ren G D, Wang X W, Lai Y Q, Liu Y X and Li J 2014 *Electrochim. Acta* **137** 439–446
[9] Liang C, Dudney N J and Howe J Y 2009 *Chem. Mater.* **21** 4724–4730
[10] Wang H, Yang Y, Liang Y, Robinson J T, Li Y, Jackson A, Cui Y and Dai H 2011 *Nano Lett.* **11** 2644–2647
[11] Jiang Y, Lu M N, Ling X T, Jiao Z, Chen L L, Chen L, Hu P F and Zhao B 2015 *J. Alloys Compd.* **645** 509–516
[12] Guo J, Xu Y and Wang C 2011 *Nano Lett.* **11** 4288–4294
[13] Dörfler S, Hagen M, Althues H, Tübke J, Kaskel S and Hoffmann M J 2012 *Chem. Commun.* **48** 4097–4099
[14] Wei S Z, Li W, Cha J J, Zheng G, Yang Y, McDowell M T, Hsu P C and Cui Y 2013 *Nat. Commun.* **4** 1331
[15] Zhang Y, Bakenov Z, Zhao Y, Konarov A, The Nam Long Doan, Sun K E K, Yermukhambetova A and Chen P 2013 *Powder Technol.* **235** 248–255
[16] Xiao L, Cao Y, Xiao J, Schwenzer B, Engelhard M H, Saraf L V, Nie Z, Exarhos G J and Liu J 2012 *Adv. Mater.* **24** 1176–1181
[17] Li G C, Li G R, Ye S H and Gao X P 2012 *Adv. Energy Mater.* **2** 1238–1245
[18] Chen J J, Jia X, She Q J, Wang C and Zhang Q 2010 *Electrochimi. Acta* **55** 8062－8066
[19] Guo Y, Zhao J, Yang S, Yu K and Wang Z 2006 *Powder Technol.* **162** 83–86
[20] Ni H and Fan L Z 2012 *J. Power Sources* **214** 195–199
[21] Guo W L and Wang X K 2004 *Technol. Dev. Chem. Ind.* **33** 1–3
[22] Shin J H and Cairns E J 2008 *J. Electrochem. Soc.* **155** A368–A373
[23] Diao Y, Xie K, Xiong S and Hong X 2013 *J. Power Sources* **235** 181–186
[24] Elezari R, Salitra G, Garsuch A, Panchenko A and Aurbach D 2011 *Adv. Mater.* **23** 5641–5644
[25] Wang Q, Jin J, Wu X, Ma G and Yang J 2014 *Phys. Chem. Chem. Phys.* **16** 21225–1229
[26] Yang C S, Yun S J and Jeong H K 2014 *Curr. Appl. Phys.* **14** 1616–1620
[27] Diao Y, Xie K, Hong X B and Xiong S Z 2013 *Acta Chim. Sinica* **71** 508–518
[28] Deng Z F, Zhang Z A, Lai Y Q, Liu J, Li J and Liu Y X 2013 *J Electrochem. Soc.* **160** A553－A558
[29] Kolosnitsyn V S, Kuzmina E V, Karaseva E V and Mochalov S E 2010 *J. Power Sources* **196** 1478–1482
[30] Zhang Z Y, Lai Y Q, Zhang Z A, Zhang K and Li J 2014 *Electrochim. Acta* **129** 55–61

Plasma assisted fabrication of multi-layer graphene/nickel hybrid film as enhanced micro-supercapacitor electrodes

Q Ding[1], W L Li[3,4], W L Zhao[1], J Y Wang[1], Y P Xing[1], X Li[2], T Xue[2], W Qi[5], K L Zhang[1], Z C Yang[1,6,*] and J S Zhao[1,7,*]

[1] School of Electronics Information Engineering, Tianjin Key Laboratory of Film Electronic & Communication Devices, Tianjin University of Technology, 391 West Binshui Road, Tianjin, 300384, China
[2] Center for Analysis, Tianjin University, Tianjin 300072, China
[3] State Key Laboratory of Transient Optics and Photonics, Xi'an Institute of Optics and Precision Mechanics, Chinese Academy of Sciences, Xi'an, Shaanxi,710119, China
[4] University of Chinese Academy of Sciences, Beijing 100049, China
[5] Beijing Key Laboratory of Energy Nanomaterials, Advance Technology & Materials Co., Ltd, China Iron & steel Research Institute Group, Beijing, 100081, China

E-mail: [6,*] yangzhengchuntjut@163.com, [7,*] jinshi58@163.com

Abstract. A facile synthesis strategy has been developed for fabricating multi-layer graphene/nickel hybrid film as micro-supercapacitor electrodes by using plasma enhanced chemical vapor deposition. The as-presented method is advantageous for rapid graphene growth at relatively low temperature of 650 °C. In addition, after pre-treating for the as-deposited nickel film by using argon plasma bombardment, the surface-to-volume ratio of graphene film on the treated nickel substrate is effectively increased by the increasing of surface roughness. This is demonstrated by the characterization results from transmission electron microscopy, scanning electron microscope and atomic force microscopy. Moreover, the electrochemical performance of the resultant graphene/nickel hybrid film as micro-supercapacitor working electrode was investigated by cyclic voltammetry and galvanostatic charge/discharge measurements. It was found that the increase of the surface-to-volume ratio of graphene/nickel hybrid film improved the specific capacitance of 10 times as the working electrode of micro-supercapacitor. Finally, by using comb columnar shadow mask pattern, the micro-supercapacitor full cell device was fabricated. The electrochemical performance measurements of the micro-supercapacitor devices indicate that the method presented in this study provides an effective way to fabricate micro-supercapacitor device with enhanced energy storage property.

Keywords: plasma enhanced chemical vapor deposition, graphene; thin film, micro-supercapacitor

1. Introduction

As a kind of energy storage devices with high power density, high efficiency, long life expectancy and environmental friendliness, supercapacitors have received considerable research attention for the past two decades [1, 2]. Moreover, due to the increasing needs of small-scale energy storage units within the micro/nanosystems of electronic devices, the studies about micro-supercapacitors have become focus topics in recent years [3, 4]. In general, based on the electrical double layer energy storage mechanism, carbon-based materials have been approved as the promising candidates for electrode materials in supercapacitor applications [5, 6]. Recently, different types of carbon-based materials, such as activated carbons, carbon nanotubes (CNTs) and graphene, have been employed to fabricate high performance micro-supercapacitor electrodes [7-12]. As a two-dimensional material with atomic thickness, graphene has attracted significant attention for use as supercapacitor electrodes, owing to its unique structure and physical behaviors, such as large surface area and high in-plane electrical conductivity [11-14]. In addition, due to the requirement of microelectrodes with two-dimensional permeable channels, graphene sheets are ideally suitable to fabricate micro-supercapacitor electrodes, which can be integrated on a substrate [15, 16].

For the synthesis of graphene, mechanical exfoliation, chemical exfoliation, epitaxial growth over SiC, and chemical vapor deposition (CVD) are mostly used methods [17-26]. Comparing with other methods, the CVD technique is mostly promising for large-scale graphene growth and meets the projected demand of graphene production for its single-step process [27, 28]. Metallic substrate, which is commonly copper (Cu) or nickel (Ni), is essential in the CVD method because it would decrease the decomposition energy of raw reactants [29, 30]. However, the synthesis temperature of graphene by using CVD method is relatively high, while the consuming time of graphene growth by using CVD method is relatively long [31, 32]. Therefore, for large-scale applications of graphene materials on micro-supercapacitors, there is still a challenging to develop a simple and effective method for synthesizing graphene sheets at relatively low temperature. Since plasma energy could effectively help to dissociate hydrocarbons, plasma enhanced CVD (PECVD) technique has been widely utilized for low temperature preparation of graphene sheets (graphene layer >3 layers) with reasonable growth rates [33, 34].

However, micro-supercapacitors made of conventional graphene electrodes exhibit limited specific capacitance due to the ultra flat surface of graphene films. Since the surface-to-volume ratio for the electrode of micro-supercapacitors is the essential parameter related to the specific capacitance, considerable efforts should be made to increase the surface roughness of the as-prepared graphene films in order to improve the specific capacitance of micro-supercapacitors.

Thus, in this study, the graphene film is grown on the Ni film deposited on the silicon dioxide/silicon (SiO_2/Si) substrate by using PECVD method. In order to increase the surface roughness, the as-deposited Ni film is pre-treated by using argon (Ar) plasma bombardment to form the rough surface. As a result, the as-synthesized graphene film on the plasma treated Ni film could correspondingly have rough surface. The electrochemical performances of the graphene synthesized by PECVD method are investigated, demonstrating that the graphene/Ni hybrid film with high surface roughness exhibits the expected higher specific capacitance.

2. Experimental

2.1. Chemicals

Iron trichloride ($FeCl_3$), sodium sulfate (Na_2SO_4) and poly (ethylene oxide) (PEO) are purchased from Sigma-Aldrich Company.

2.2. Preparation of Ni film

In this study, there were two types of Ni film which was deposited on the SiO_2/Si substrate. One type of Ni film was deposited on the SiO_2/Si substrate by using electron beam evaporation (EB500-I, Shenyang Scientific Instrument Limited, China). On the other hand, to create the rough surface, the as-deposited Ni film was further conducted by Ar plasma treated at 250 W for 35 min. The thicknesses of all of the as-prepared Ni films were 150 nm.

2.3. Synthesis of graphene/Ni hybrid film

In a typical synthesis procedure according to the previous reports [35], the graphene was synthesized by using PECVD (400 model, Shenyang Scientific Instrument Limited, China) on the as-prepared Ni film. The as-prepared Ni films were placed inside the PECVD system chamber. After placing the Ni films inside the PECVD chamber, the base pressure of the PECVD system vacuum chamber was pumped down to 8.5×10^{-4} Pa. Then, hydrogen gas (H_2) was piping into the chamber at the flow rate of 80 mLmin^{-1}, and then the Ni films were preheated at 650 °C for 40 min at a pressure of 220 Pa. After this process, the mixed gases with H_2, Ar and methane (CH_4) were piped into the chamber at the flow rate of 4, 80 and 160 mLmin^{-1}, respectively. The pressure of vacuum was fixed at 250 Pa. Finally, graphene was synthesized by applying radio frequency power of 200 W on the Ni films at the temperature of 650 °C for 200 s. After turning off the radio frequency power, the heater and CH_4 gas were correspondingly shut down. In order to cool down the reaction system, Ar and H_2 were kept flowing into the chamber at the rate of 160 and 80 mLmin^{-1}, respectively. In this study, the two types of graphene/Ni hybrid film, which were made from graphene film grown on the Ni film and the plasma treated Ni film, were named as sample 1 and sample 2, respectively.

2.4. *Characterization*

To evaluate the quality of the as-prepared graphene, the transfer of graphene was conducted by dissolving Ni film in 1 M $FeCl_3$ solution, followed by rinsing graphene with de-ionized (DI) water for 3 times. And then graphene was transferred to SiO_2/Si substrate for further characterization. The morphology and structure of the as-prepared graphene was studied by using transmission electron microscopy (TEM; JEM-2100 JEDL). In addition, the surface morphology of the hybrid film of sample 1 and 2 was characterized by using field-emission scanning electron microscope (FESEM; X-650 HITACHI) and atomic force microscopy (AFM; 5600LS Agilent). Raman spectra of the as-prepared graphene were acquired by using inVia Renishaw Raman Microscope with 532 nm laser radiation source.

2.5. *Electrochemical measurement the graphene/Ni hybrid films*

Electrochemical behaviors of the graphene/Ni hybrid films as electrodes in micro-supercapacitor were investigated. Both of the two kinds of as-fabricated graphene/Ni hybrid film were used as working electrodes of micro-supercapacitor. In addition, a Pt foil and saturated calomel electrode (SCE) were used as the electrolyte, counter electrode, respectively. A series of electrochemical measurements of the as-fabricated graphene/Ni hybrid films were tested, including the galvanostatic charge-discharge and cyclic voltammetry (CV), by using a three-electrode compartment on a CHI660D Electrochemical System with 1 M Na_2SO_4 aqueous solution as the electrolyte. Nitrogen gas (N_2) was pushed into the electrolyte solution to remove oxygen (O_2) in DI water.

2.6. *Electrochemical measurement the micro-supercapacitor device*

By using the same procedure of synthesizing sample 2, the micro-supercapacitor device was fabricated. In particular, the Ni film was deposited on the SiO_2/Si substrate by using electron beam evaporation with comb columnar shadow mask pattern. Then as-fabricated comb columnar Ni film was further conducted by Ar plasma treated at 250 W for 35 min. Finally, the graphene film was grown on the surface of Ni film by using the same procedure mention above. The as-fabricated comb columnar electrode was shown in figure 1. The gel electrolyte used for micro-supercapacitor device encapsulation was made from PEO and Na_2SO_4. Typically, 4 g PEO was dissolved into 50 mL Na_2SO_4 solution of 1 M. After stirring for a while, the syrupy gel electrolyte was formed. The as-fabricated comb columnar electrode was encapsulated by using polymethyl methacrylate (PMMA) cap with gel electrolyte inside the cap. Then, a series of electrochemical measurements of the encapsulated micro-supercapacitor device were tested, including the galvanostatic charge-discharge and CV, by using CHI660D electrochemical system.

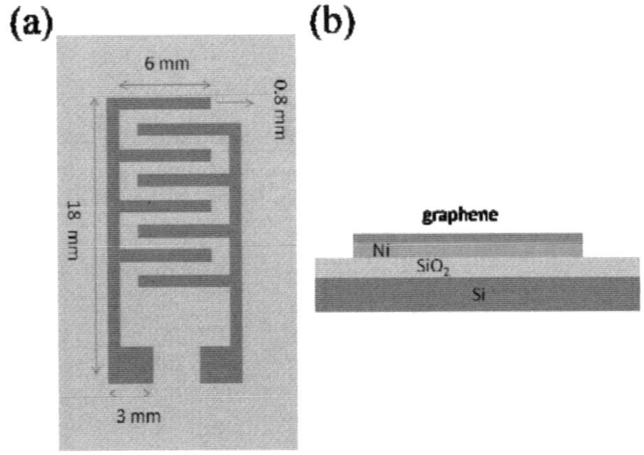

Figure 1. Schematic diagram of the top view (a) and the side view (b) of the as-fabricated micro-supercapacitor device with comb columnar electrode.

3. Results and discussions

3.1. Characterizations of graphene

Figure 2 shows the Raman spectra of graphene transferred from the as-prepared two types of Ni film onto SiO_2/Si substrate, where a and b represent graphene sheets from sample 1 and sample 2, respectively. As shown in figure 2, there are three characteristic bands of graphene materials in both Raman spectra of a and b, which are the D band (1350 cm^{-1}), G band (1580 cm^{-1}) and 2D band (2700 cm^{-1}). The intensity ratio between D and G band (I_D /I_G ratio) in Raman spectra of a and b are 1.6 and 1.9, respectively, which indicates that the defect degree of graphene grown on plasma treated Ni film is higher than that of graphene grown on the non-deposited one [36, 37]. Since the pits on the as-deposited Ni film are formed after plasma treating, the as-formed pits are turned into the nucleation center of graphene and prevent the movement of carbon atoms on the surface of Ni film. As a result, the number of graphene flakes tends to be increased, while the edges of these flakes attributed to the high disorder structure. Therefore, it can be concluded that a rougher surface of Ni film could increase the defect degree of graphene films grown on it. To further estimate the surface morphology of graphene sheets from sample 1 and 2, TEM observation is conducted. As shown from the typical TEM images in Figure 3, the as-prepared graphene sheets from sample 1 have flat surface while the as-prepared graphene sheets from sample 2 have rough surface. In addition, from the high resolution TEM in Figure 3, there are crystallized carbon structure with typical interlayer spacing of 0.34 nm in both two graphene samples, which is closely matched with the (002) lattice spacing of graphene [38, 39].

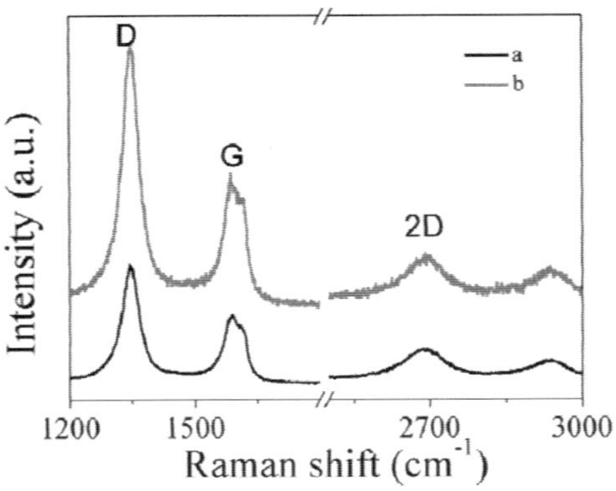

Figure 2. Raman spectra of graphene sheets obtained from sample 1 (a) and sample 2 (b) [35].

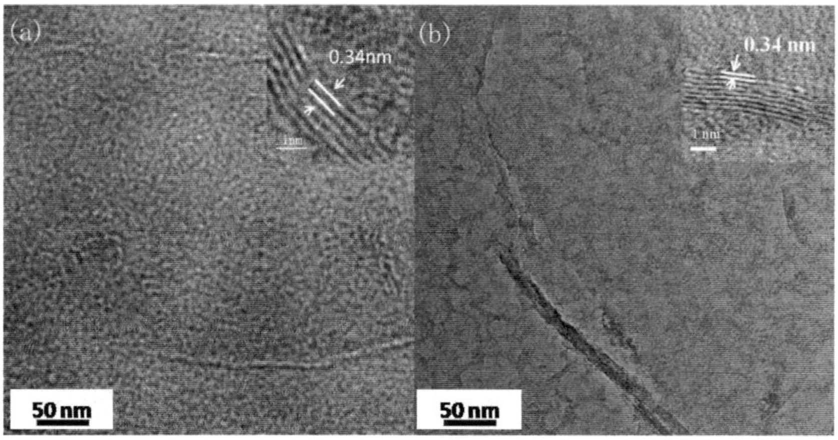

Figure 3. Typical TEM images of graphene sheets obtained from sample 1(a) and sample 2(b) [35]. The inset shows corresponding high resolution TEM image.

3.2. Characterizations of the as-fabricated electrodes

Since electrochemical measurements are conducted on the electrodes fabricated by using the graphene films on the Ni films, the surface morphology of the final electrode was investigated by using SEM and AFM. By comparing the SEM images of sample 1 and 2 in Figure 4, the black dots in Figure 4 (b) indicate that there are obvious pits on the surface of sample 2, while the surface of sample 1 is flat. Furthermore, the cross section view of the hybrid film in the inset images of Figure 4 (b) indicates the surface roughness of sample 2, which shows the up and down pits on the surface. In addition, as shown in Figure 5, a homogeneous film of sample 1 with a tiny roughness on the surface is observed, while the surface of sample 2 is rougher due to the plasma treated of Ni film. In addition, from the three dimension (3D) view of sample 1 and 2, which is shown in Figure 5 (c) and (d), respectively, sample 2 displays sharper surface in 3D view of AFM images. The morphology of the ups and downs of sample 2 agrees well with the TEM result in Fig 3 (b), which indicates that the graphene films grown on the plasma treated Ni film have rough surface with more pits.

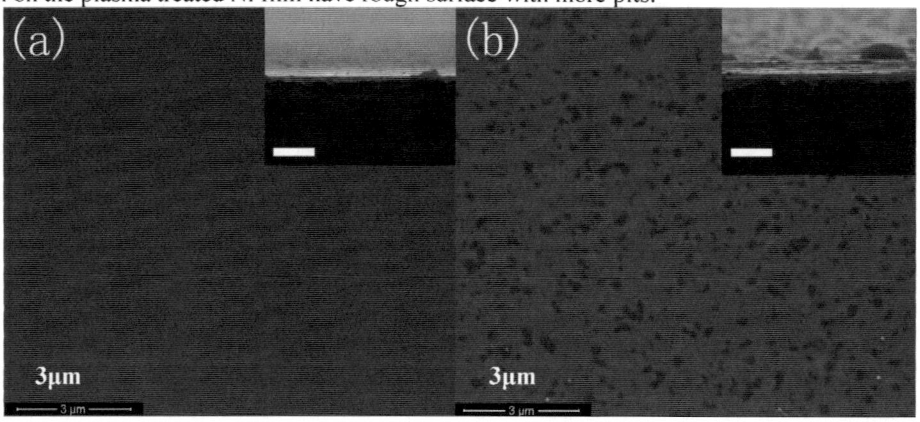

Figure 4. SEM images of sample 1(a) and sample 2(b) [35]. The inset shows the corresponding cross section images, where the scale bar is 1μm.

IOP Conf. Series: Materials Science and Engineering **182** (2017) 012014 doi:10.1088/1757-899X/182/1/012014

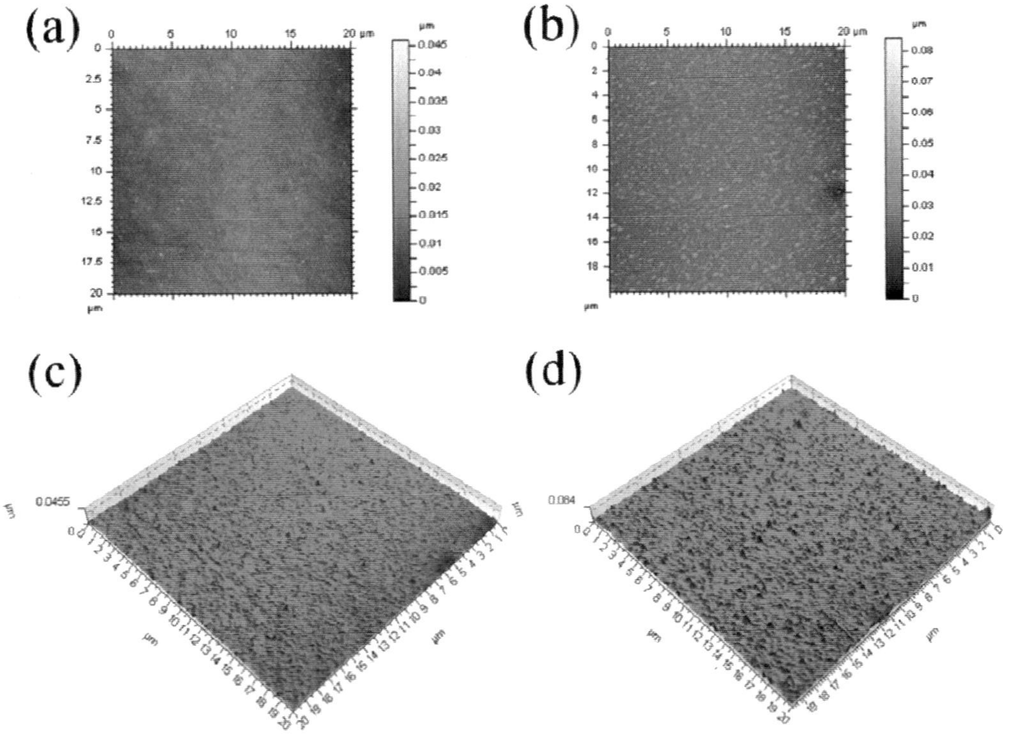

Figure 5. AFM images of sample 1(a) and sample 2(b) with the corresponding 3D view of sample 1 (c) and sample 2 (d).

CV and galvanostatic charge-discharge measurements are used to investigate the capacitive performances of sample 1 and sample 2 as the working electrode of micro-supercapacitors in 1 M Na_2SO_4 solution. The electrochemical performances are characterized by using a three-electrode configuration. Figure6 (a) and (b) show the CV curves for sample 1 and sample 2 as the working electrodes under different scan rates from 1 to 100 mVs^{-1}, respectively. The exhibition of quasi-rectangular CV curves in the potential window of -0.3-0.7 V demonstrates that the excellent reversibility of the as-synthesized micro-supercapacitor electrodes [40, 41]. Moreover, a higher current density is observed in Figure 6 (b) than that in Figure 6 (a), suggesting that specific capacitance of sample 2 is higher than that of sample 1. The higher specific capacitance of sample 2 is further confirmed by the galvanostatic charge discharge profiles in Figure 6 (c), which are obtained at the current density of 0.5 Am^{-2}. As shown in Figure 6 (c), the nearly linear profile and rather symmetric shapes imply that the as-fabricated hybrid films demonstrate the typical double layer capacitive behavior and excellent electrochemical reversibility [42, 43]. Besides the galvanostatic charge discharge behavior of sample 1 and 2 at current density of 0.5 Am^{-2}, the charge storage capacity of the as-fabricated two types of electrode is further evaluated by the specific capacitance value obtained from galvanostatic charge discharge test at different current density. As shown in Figure6 (d), the overall specific capacitance of sample 2 is higher than that of sample 1 at each scan rate, which is due to the larger surface-to-volume ratio of graphene provided by the plasma treated Ni film with rough surface. At low current density of 0.1 Am^{-2}, the specific capacitances of sample 1 and sample 2 as the working electrode are 0.5 and 9 Fm^{-2}, respectively. Even at the current density of 2 Am^{-2}, the specific capacitances of sample 1 and sample 2 are 0.45 and 2.6 Fm^{-2}, respectively. Based on the analysis above, the graphene/Ni hybrid film with a rough surface gives rise to about 10 times enhancement of specific capacitance due to the increasing of surface-to-volume ratio. This enhancement indicates that the synthesis method mentioned in the present work provides a facile and effective way to fabricate micro-supercapacitor electrodes.

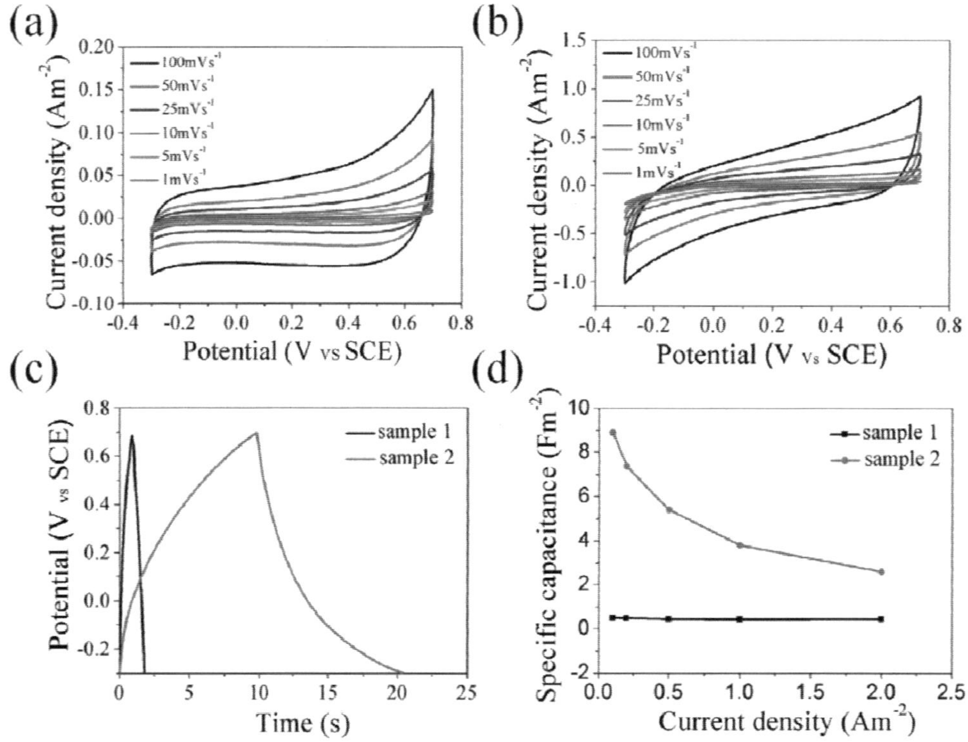

Figure 6. Electrochemical behavior of the micro-supercapacitor electrode made of the as-fabricated graphene/Ni hybrid film. (a) CV curves of the micro-supercapacitor electrode made of sample 1; (b) CV curves of the micro-supercapacitor electrode made of sample 2; (c) galvanostatic charge-discharge of the micro-supercapacitor electrode at a current density of 0.5 Am^{-2}; (d) variation in specific capacitance for the micro-supercapacitor electrode against different current densities.

3.3. Electrochemical behavior of the micro-supercapacitor device

To further investigate the electrochemical performance of the as-obtained graphene/Ni hybrid film in a full symmetrical cell, micro-supercapacitor device based on sample 2 with comb columnar electrodes were fabricated and shown in Figure 7 (a). Due to the advantages of easy fabrication and wide working temperature, gel electrolyte of PEO/Na_2SO_4 system was used in the micro-supercapacitor device package. As shown in Figure 7 (b), the as-obtained micro-supercapacitor device exhibits quasi-rectangular CV curves at various scan rates from 1 to 100 mVs^{-1} a potential window range from -0.3 to 0.7 V, which indicates the fast charge transfer between the electrodes and typical double layer electrochemical behavior. According to Figure 7 (c), the galvanostatic charge discharge curve with a linear and symmetrical shape at the current density of 0.1 Am^{-2} indicates the capacitor-like characterizations of the as-obtained micro-supercapacitor device. In addition, the micro-supercapacitor device shows a specific capacitance of 0.114 Fm^{-2} at 0.1 Am^{-2}, and maintains 90% of this value when operated at current density of 1 Am^{-2}, which is shown in Figure 7 (d). Due to the low ionic conductivity of the gel electrolyte utilized in this work, the specific capacitance of the as-fabricated micro-supercapacitor device is unexpected considerable. Nevertheless, the synthesis method in this study provides one effective way to fabricate micro-supercapacitor devices with improved capacitance.

In addition, some ionic liquid or organic electrolytes are attempted to be employed in the future to optimize the performance of the micro-supercapacitor devices.

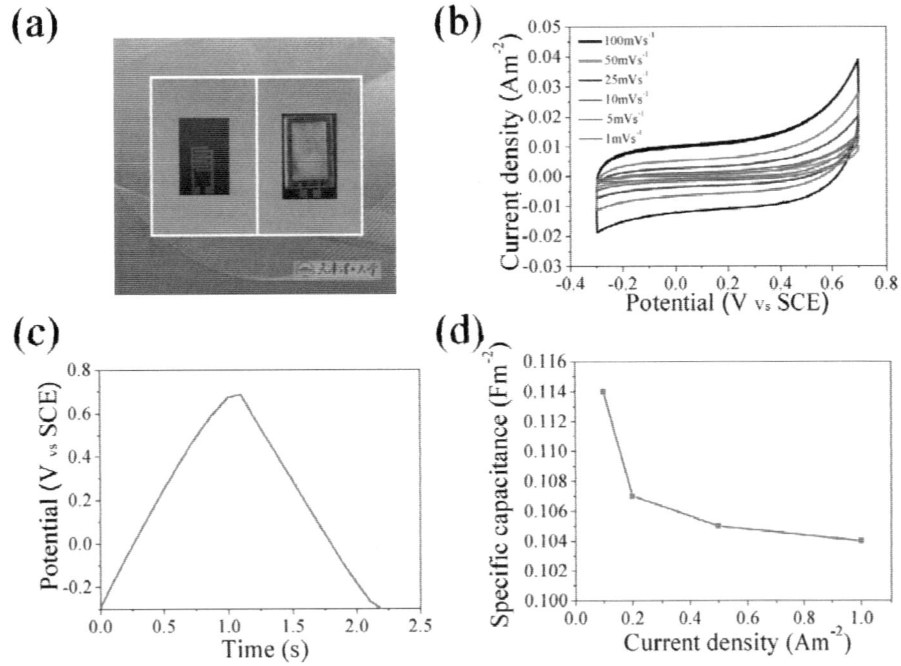

Figure 7. (a) Digital photograph of the as-fabricated micro-supercapacitor device. (b) CV profiles of micro-supercapacitor device at various scan rates from 1 to 100 mVs^{-1}. (c) galvanostatic charge-discharge curve of micro-supercapacitor device at the current density of 0.1 Am^{-2}. (d) specific capacitance of micro-supercapacitor device obtained from different current densities.

4. Conclusions

In summary, graphene with multi layers is successfully synthesized on Ni film by using PECVD. In addition, the surface of the as-synthesized graphene/Ni hybrid film is tuned by using Ar plasma, giving rise to generate rough surface with improved surface-to-volume ratio. It is found that the electrochemical properties of graphene/Ni hybrid film with rough surface are dramatically improved. When the graphene/Ni hybrid film with improved surface-to-volume ratio is used as the electrode in micro-supercapacitor, the specific capacitance is 10 times higher than the untreated graphene/Ni film. Furthermore, the electrochemical performance of the micro-supercapacitor device with comb columnar electrodes shows that the synthesis method presented in this work provides an effective way to fabricate micro-supercapacitor devices and can be served as guidance for the future design of high performance micro-supercapacitor devices.

Acknowledgments

This work is supported by the National High Technology Research and Development Program (863 Program) (Grant No. 2015AA034601) and Student's Platform for Innovation and Entrepreneurship Training Program (Grant No. 201510060030).

References

[1] Wang G, Zhang L and Zhang J 2012 *Chem. Soc. Rev* **41** 797–828
[2] Zhang L and Zhao X S 2009 *Chem. Soc. Rev* **38** 2520–2531
[3] Beidaghi M and Gogotsi Y 2014 *Energy Environ. Sci* **7** 867–884

[4] Chmiola J, Largeot C, Taberna P L, Simon P and Gogotsi Y 2010 *Science* **328** 480–483
[5] Pandolfo A G and Hollenkamp A F, 2006 *J. Power Sources* **157** 11–27
[6] Zhang Z, Wang W, Li C, Wei L, Chen X, Tong Y, Mai K and Lu X 2006 *J. Power Sources* **248** 1248–1255
[7] Jiang S, Shi T, Zhan X, Long H, Xi S, Hu H and Tang Z 2014 *J. Power Sources* **272** 16–23
[8] Hu S, Zhang S, Pan N and Hsieh Y L 2014 *J. Power Sources* **270** 106–112
[9] Brown B, Swain B, Hiltwine J, Brooks D B and Zhou Z G 2014 *J. Power Sources* **272** 979–986
[10] Cao Z and Wei B Q 2013 *Energy Environ. Sc.* **6** 3183–3201
[11] Liu C, Yu Z, Neff D, Zhamu A and Jang B Z 2010 *Nano Lett.* **10** 4863–4868
[12] Zhao B, Liu P, Jiang Y, Pan D, Tao H, Song J, Fang T and Xu W 2012 *J. Power Sources* **198** 423–427
[13] Sun D, Yan X, Lang J and Xue Q, 2013 *J. Power Sources* **222** 52–58
[14] Zhang S, Li Y and Pan N, 2012 *J. Power Sources* **206** 476–482
[15] Byon H R, Lee S W, Chen S, Hammond P T and Shao-Horn Y 2010 *Carbon* **49** 457–467
[16] Sarker A K and Hong J D 2012 *Langmuir* **28** 12637–12646
[17] Geim A K and Novoselov K S 2007 *Nat. Mater.* **6** 183–191
[18] Pang S, Englert J M, Tsao H N, Hemandez Y, Hirsch A, Feng X and Mullen K 2010 *Adv. Mater.* **22** 5374–5377
[19] Hummers Jr W S and Offeman R E 1958 *J. Am. Chem. Soc.* **80** 1339
[20] Zhang L, Liang J J, Huang Y, Ma Y F, Wang Y and Chen Y S, 2009 *Carbon* **47** 3365–3368
[21] Kim K S, Zhao Y, Jang H, Lee S Y, Kim J M, Kim K S, Ahn J H, Kim P, Choi J Y and Hong B H 2009 *Nature* **457** 706–710
[22] Reina A, Jia X, Ho J, Nezich D, Son H, Bulovic V, Dresselhaus M S and Kong J 2009 *Nano Lett.* **9** 30–35
[23] Munoz R and Gomez-Aleixandre C 2013 *Chem. Vap. Deposition* **19** 297–322
[24] Coraux J, N' Diaye AT, Busse C and Michely T 2008 *Nano Lett.* **8** 565–570
[25] Sutter P W, Flege J I and Sutter E A 2008 *Nat. Mater.* **7** 406–411
[26] Pan Y, Zhang H G, Shi D X, Sun J T, Du S X, Liu F and Gao H J 2009 *Adv. Mater.* **21** 2777–2780
[27] Sun Z, Yan Z, Yao J, Beitler E, Zhu Y and Tour J M 2010 *Nature* **468** 549–552
[28] Lee S, Lee K and Zhong Z 2010 *Nano Lett.* **10** 4702–4707
[29] Bae S, Kim H, Lee Y and Xu X 2010 *Nat. Nanotechnol.* **5** 574–578
[30] Obraztsov A N, Obraztsova E A, Tyurnina A V and Zolotukhin A A 2007 *Carbon* **45** 2017–2021
[31] Yu Q, Lian J, Siriponglert S, Li H, Chen Y P and Pei S S 2008 *Appl. Phys. Lett.* **93** 113103
[32] Vazquez de Parga A L, Calleja F, Borca Jr B, Passeggi M C G, Hinarejos J J, Guinea F and Miranda R 2008 *Phys. Rev. Lett.* **100** 056807
[33] Shashurin A and Keidar M 2008 *Carbon* **46** 1826–1828
[34] Levchenko I, Ostrikov K, Long J D and Xu S 2007 *Appl. Phys. Lett.* **91** 113115
[35] Yang Z, Zhao W, Wang J, Chen R, Li X, Xue T, Zhang K and Zhao J 2016 *Sci. Adv. Mater.* **8** *Preprint*
[36] Qi J L, Zheng W T, Zheng X H, Wang X and Tian H W 2011 *Appl. Surf. Sci.* **257** 6531–6534
[37] Malesevic A, Vitchev R, Schouteden K, Volodin A, Zhang L, Tendeloo G V, Vanhulsel A and Haesendonck C V 2008 *Nanotechnology* **19** 305604
[38] Yang Z, Li X and Wang J 2011 *Carbon* **49** 5207–5212
[39] Yang Z, Tang C, Zhang Y, Gong H, Li X and Wang J 2013 *Sci. Rep.* **3** 2925
[40] Wang J, Yang Y, Huang Z and Kang F 2012 *J. Power Sources* **204** 236–243
[41] Yang Z, Tang C, Gong H, Li X and Wang J 2013 *J. Power Sources* **240** 713–720
[42] Li Q, Li Z, Lin L, Wang X, Wang Y, Zhang C and Wang H 2010 *Chem. Eng. J.* **156** 500–504
[43] Xu Y, Hennig I, Freyberg D, Strudwick A J, Schwab M G, Weitz T and Cha K C P 2014 *J. Power Sources* **248** 483–488

Microencapsulation of phase change materials with carbon nanotubes reinforced shell for enhancement of thermal conductivity

Weiwei Cui[1,2], Yongpeng Xia[1,2], Huanzhi Zhang[1,2*], Fen Xu[1,2], Yongjin Zou[1,2,3,*], Cuili Xiang[1,2], Hailiang Chu[1,2], Shujun Qiu[1,2] and Lixian Sun[1,2,4,*]

[1] School of Materials Science and Engineering, Guilin University of Electronic Technology, Guilin 541004, China,
[2] Guangxi Key Laboratory of Information Materials, Guilin University of Electronic Technology, Guilin 541004, China.

E-mail:[3,*] zhanghuanzhi@guet.edu.cn, [4,*] sunlx@guet.edu.cn

Abstract. Novel microencapsulated phase change materials (micro-PCMs) were synthesized via *in-situ* polymerization with modified carbon nanotubes(CNTs) reinforced melamine-formaldehyde resin as shell material and CNTs reinforced *n*-octadecane as PCMs core. DSC results confirm that the micro-PCMs possess good phase change behavior and excellent thermal cycling stability. Melting enthalpy of the micro-PCMs can achieve 133.1 J/g and has slight changes after 20 times of thermal cyclings. And the incorporation of CNTs supplies the micro-PCMs with fast thermal response rate which increases the crystallization temperature of the micro-PCMs. Moreover, the thermal conductivity of the micro-PCMs has been significantly enhanced by introducing CNTs into their shell and core materials. And the thermal conductivity of micro-PCMs with 1.67 wt.% CNTs can increase by 25%. These results exhibit that the obtained micro-PCMs have a good prospect in thermal energy storage applications.

Keywords: phase change materials, microencapsulation, thermal conductivity, carbon nanotubes reinforced shell

1. Introduction

Energy is well recognized to be a significant role in the rapid development of worldwide economics [1]. However, conventional fossil energy sources are limited and non-renewable. Thus, their increasing consumption inevitably results in the shortages of energy resources [2]. Nowadays, the well-designed thermal energy storage unit is an important system for reducing dependency on the fossil fuels, which is mainly attributed to the rational use of thermal energy by "peak shaving" [3, 4]. Phase change materials (PCMs) are recognized as potential candidates for thermal energy storage units due to their very high density of latent heat [5, 6]. PCMs are materials which can absorb or release large amount of heat during phase transformation process within a unit change of temperature [7, 8]. Therefore, they have opened a prospective application in various fields and have been widely used as thermal energy storage materials. However, practical application of PCMs in thermal systems is not easy to manipulate because of their poor chemical stability, erosion and super-cooling problems [9]. Microencapsulation is a commonly used technique to overcome the inherent shortcomings of PCMs by which a core material can be protected by a shell. Consequently, it makes PCMs easy to be manipulated in commercial refrigeration, solar heating systems, textile and buildings [10]. Microencapsulated PCMs (Micro-PCMs) not only have high heat transfer rates and good durability, but also reduce interactions of PCMs with the external environment [11]. Generally, Micro-PCMs exhibit a well-defined core-shell structure via self-assembly to protect interior specific functional materials from leakage [12]. As a result, their chemical and thermal stability, thermal response rate and compatibility are depended on the shell structure, so the selection of shell materials is very crucial to the application of Micro-PCMs [13]. Traditionally, the shell materials of micro-PCMs consist of polymer such as melamine–formaldehyde (MF) resin, urea–formaldehyde (UF) resin, polymethyl methacrylate (PMMA), styrene–butadiene-styrene (SBS) copolymer and polystyrene (PS) [14,15]. Although these micro-PCMs with polymer shell materials have good performance, there are still some defects such as poor thermal stability, low thermal conductivity [16]. Moreover, the inherent thermal conductivity of PCMs is quite low. Accordingly, in order to improve comprehensively thermal properties and mechanical properties for micro-PCMs, many researchers rely on high thermal conductive fillers mixing with PCMs such as metal nanowires and carbon nano-particles [17]. Nevertheless, poor compatibility between organic PCMs and these inorganic nano-fillers always induces low thermal stability and durability for thermal energy storage systems. In this case, the extremely high interface area between the fillers (nanoparticles or nanofibres) and the polymer shell or core materials may solve these problems and confer the micro-PCMs excellent comprehensive properties.

Carbon nanotubes (CNTs) are a kind of beneficial nano-reinforcement materials with low density, high aspect ratios, extraordinary mechanical properties and distinguished intrinsic thermal conductivity (~10^3 W/(m·K)) due to their unique structure [18,19]. Furthermore, the high interface of CNTs makes them susceptible to bonding with different materials and contributes to the enhancement of properties for composites [20]. Hence, CNTs have stimulated significant interest in the nano-reinforcement material world, especially in the field of polymer-based composites. And the composites impregnated with CNTs exhibit significantly higher thermal conductivity than that of pure polymers [21]. Moreover, some researchers have filled CNTs into pure PCMs to achieve high thermal conductivity [22-25]. Nevertheless, there are little reports on the micro-PCMs with CNTs reinforced shell materials and PCMs core at the same time.

Therefore, in this paper, novel micro-PCMs with CNTs reinforced shell and core materials are synthesized via *in-situ* polymerization. In order to improve the thermal conductivity of traditional micro-PCMs, amount of CNTs are mixed with PCMs core and as well modified CNTs are pre-polymerized with melamine–formaldehyde pre-polymers forming shell materials, respectively. Furthermore, the influence of CNTs on the properties of micro-PCMs is also investigated. It can be found that the obtained novel micro-PCMs exhibit excellent comprehensive properties which have a potential prospect as thermal energy storage materials.

2. Experimental

2.1. Materials

Carbon nanotubes (CNTs) (purity of 95%, Shenzhen Nanotech Port Co., Ltd) were used to reinforce micro-PCMs. *n*-Octadecane (Analytical reagent, purity of 90%, Alfa Aesar) was used as the phase change material (PCM). Formaldehyde aqueous solution (Analytical reagent, purity of 37%, Xilong Chemical Co., Ltd) and Melamine (Analytical reagent, Sinopharm Chemical Reagent Co., Ltd) were used to prepare the shell materials of micro-PCMs. Citric acid monohydrate (Analytical reagent, purity of 99.5%, Xilong Chemical Co., Ltd) and triethanolamine (Analytical reagent, purity of 98%, Xilong Chemical Co., Ltd) were used to control pH value. The sodium salt of styrene–maleic anhydride copolymer (SMA) (19% of aqueous solution, Shanghai Leather Chemical Plant) was used as an emulsifier. Distilled water was used as a solvent. All materials are used as obtained without further purification.

2.2. Sample preparation

2.2.1. Modification of Carbon nanotubes. In order to enhance the compatibility and bonding energy between CNTs and the polymer shell of micro-PCMs. The purchased CNTs were modified by using concentrated acid to receive hydrophilic property which makes sure the reaction and bonding between CNTs and shell materials. And the synthetic process is presented as follows: 1.2 g CNTs and 50 ml of concentrated sulfuric acid (98 wt%) were mixed in a 250 ml three-neck flask with magnetic stirring for 15 min at room temperature. Then, the mixed solution was ultrasonicated for 1 hour. After that, 50 ml of concentrated nitric acid (65 wt%) was added into the solution and stirred for 15 min. Next, the mixed solution reacted at 140 ^{o}C in an oil thermostat bath with refluxing for 2 hours. Finally, the resulting suspension was cooled and washed with distilled water for several times until the solution became neutral. And then the product was dried at room temperature.

2.2.2. Preparation of novel CNTs reinforced micro-PCMs. Modified CNTs are subjected to ultrasonication in a 100 ml stand-up flask for 10 min to obtain a homogeneous solution. Then, 3 g melamine was added and ultrasonicated for another 10 min. Subsequently, 6 ml formaldehyde aqueous solution and 10 ml distilled water were added. And the mixture was reacted at 70 ^{o}C with stirring until it changed into transparent solution getting the CNTs reinforced shell pre-polymer.

Novel CNTs reinforced Micro-PCMs were synthesized via *in-situ* polymerization. First, 9 g *n*-octadecane and 50 mg CNTs were mixed homogeneously by ultrasonication for 15 min in a 250 ml three-neck flask. Then, 5 g SMA and 150 ml distilled water were adding into the mixture. And the mixed solution was emulsified at 50 ^{o}C with a stirring rate of 1200 rpm for 2 hours. Next, the CNTs reinforced shell pre-polymer was slowly added into the emulsion. Then, pH value of the mixed solution was adjusted to 4-5 by citric acid monohydrate. After that, the polymerization was conducted for 6 hours at 70 ^{o}C with a stirring rate of 600 rpm. At last, pH value of the mixture was adjusted to 8-9 by triethanolamine to terminate the reaction. The product was washed with 60 ^{o}C distilled water for several times till the solution became neutral and then dried at room temperature. Series of novel micro-PCMs were prepared with modified CNTs content ranging from 0.22 wt% to 1.67 wt%. These samples are summarized in table 1.

Table 1. The constituent of the CNTs reinforced micro-PCMs.

Microcapsule name	Core materials	Shell materials	The amount of CNTs for core materials	The amount of modified CNTs for shell materials
Sample 1	*n*-octadecane	MF resin	0	0
Sample 2	*n*-octadecane	MF resin	50 mg	0
Sample 3	*n*-octadecane	MF resin	50 mg	20 mg
Sample 4	*n*-octadecane	MF resin	50 mg	50 mg
Sample 5	*n*-octadecane	MF resin	50 mg	100 mg
Sample 6	*n*-octadecane	MF resin	50 mg	150 mg

2.3. Characterization techniques

Morphologies of the obtained samples were observed by using scanning electron microscopy (SEM) (FEI, Quanta 450 FEG). Fourier transform infrared (FTIR) (Thermo Fisher Scientific, Nicolet 6700) spectra was used to survey the chemical structures of the samples with wavenumber range between 500 cm^{-1} and 4000 cm^{-1} by KBr sampling method. X-ray diffraction (XRD) (Bruker, D8 Advance) is used to detect the crystallization properties of the obtained micro-PCMs with copper Ka radiation in the 2θ range of 10-90°. Differential scanning calorimeter (DSC) (Setaram, Sensys) is used to measure the thermal storage properties of the samples in the range of -30–50 ^{o}C under a nitrogen atmosphere. DSC analysis was carried out with the sample mass of 5–10 mg. Thermal conductivity analysis (TCA)

(Setaram, TCi) was used to measure the thermal conductivity of the obtained micro-PCMs at room temperature. Reported results demonstrated the average value of three measurements for each sample to ensure the repeatability.

3. Results and discussion

3.1. Synthetic mechanism of the CNTs reinforced micro-PCMs

In order to enhance the thermal conductivity and mechanical properties of micro-PCMs, CNTs and the modified CNTs are filled into the PCMs core and shell materials, respectively. The high interface area of CNTs makes sure the strong bonding energy among CNTs, core and shell materials of micro-PCMs. And high thermal conductivity and good mechanical properties of CNTs may provide excellent comprehensive thermal properties for the obtained CNTs reinforced micro-PCMs. The corresponding synthetic mechanism of the CNTs reinforced micro-PCMs (as seen from figure 1) is illustrated as follows: first, the mixture of *n*-octadecane and CNTs is emulsified into homogeneous micellar solution. The added CNTs can improve the thermal conductivity and thermal stability of core materials due to their good lipophilicity. Then, the modified CNTs which possess hydrophilic groups through covalent modification are reacted with melamine and formaldehyde by hydrogen bond forming into CNTs reinforced shell pre-polymers. Subsequently, the pre-polymers are dropped into the micellar solution and adsorbed around the micellar liquids by the hydrogen bond of different hydrophilic groups between pre-polymers and micelles. And the resulted polymers after reaction are deposited gradually onto the surface of the micellar liquids forming the shell materials of micro-PCMs. Consequently, CNTs and modified CNTs are successfully filled into the core and shell materials of micro-PCMs. As a result, CNTs reinforced micro-PCMs are successfully synthesized.

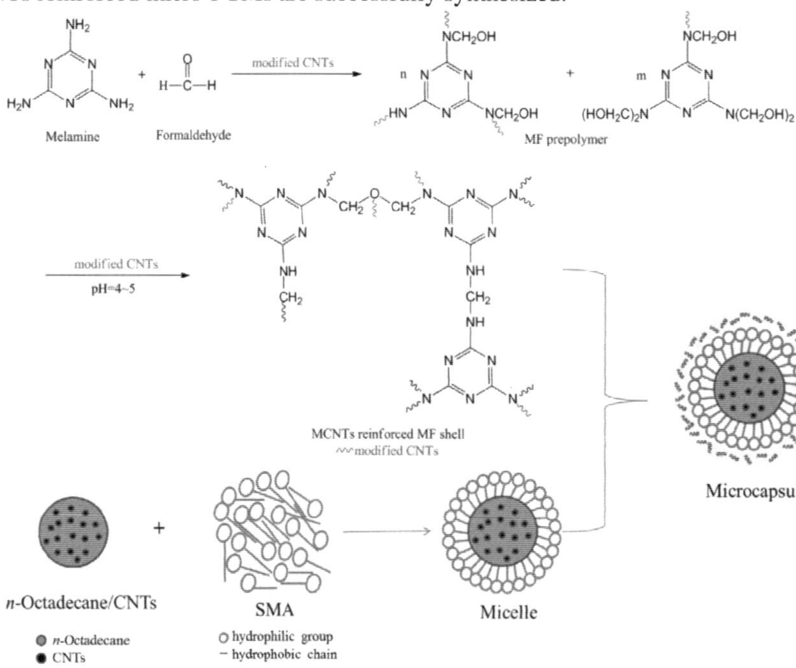

Figure 1. Schematic formation of CNTs reinforced micro-PCMs.

3.2. Morphology of the CNTs reinforced micro-PCMs

The micrographs of the obtained micro-PCMs detected by SEM are shown in figure 2. It is obvious that the synthesized micro-PCMs have a regular spherical structure with fairly uniform morphology in each case. And the diameter of the microcapsules is about 5 μm. This phenomenon indicates that the addition of modified CNTs has no effect on the formation of micro-PCMs and the hydrophilic groups of modified CNTs are successfully combined with the shell materials during the polymerization process. It also demonstrates that modified CNTs are homogeneously dispersed into the polymer chains, instead of being placed on the surface of the microcapsule. As a result, the surfaces of the obtained micro-PCMs are smooth and compact. Nevertheless, there are some particles accumulating on the surface of spherical micro-PCMs. It is attributed to the fact that some polymers are not forming into the micro-PCMs shell and sticking to the surface of micro-PCMs. These results mean that series of micro-PCMs doped with different content of modified CNTs are successfully synthesized. Furthermore, the maximum addition amount of the modified CNTs is 150 mg to synthesize the reinforced micro-PCMs as known from the preparation process.

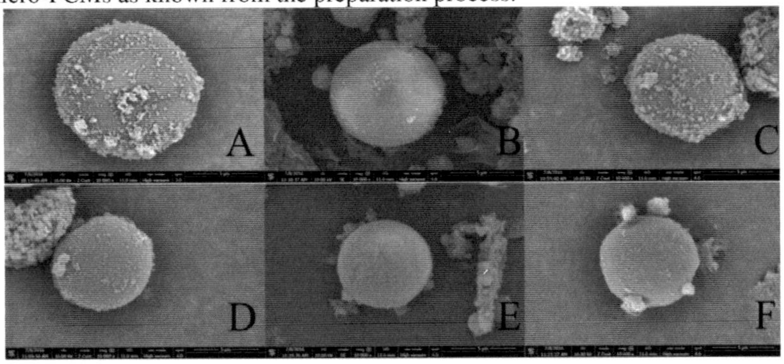

Figure 2. SEM images of the CNTs reinforced micro-PCMs: A-Sample 1, B-Sample 2, C-Sample 3, D-Sample 4, E-Sample 5, F-Sample 6.

3.3. Chemical structure of the CNTs reinforced micro-PCMs

FTIR spectra are performed to investigate the chemical compositions of modified CNTs, micro-PCMs without CNTs and the CNTs reinforced micro-PCMs. The corresponding spectra are presented in figure 3. It is distinct that there is a wide peak at approximately 3400 cm^{-1} in the spectrum of modified CNTs, which is corresponding to the absorption of O–H and –COOH. As seen from the spectrum of micro-PCMs without CNTs, the strong peaks at 2922 cm^{-1} and 2852 cm^{-1} are associated with the asymmetric and symmetric stretching vibration of aliphatic C-H. The wide absorption peaks at approximately 3400 cm^{-1} are attributed to the stretching vibrations of O–H and N–H in MF resin. Furthermore, the peak at approximately 1015 cm^{-1} is resulted from the stretching vibration of C–OH. It is clear that all the characterized peaks of modified CNTs and micro-PCMs without CNTs are appearing on the spectra of the obtained micro-PCMs with different content of modified CNTs. This result further indicates that the CNTs reinforced micro-PCMs are successfully synthesized.

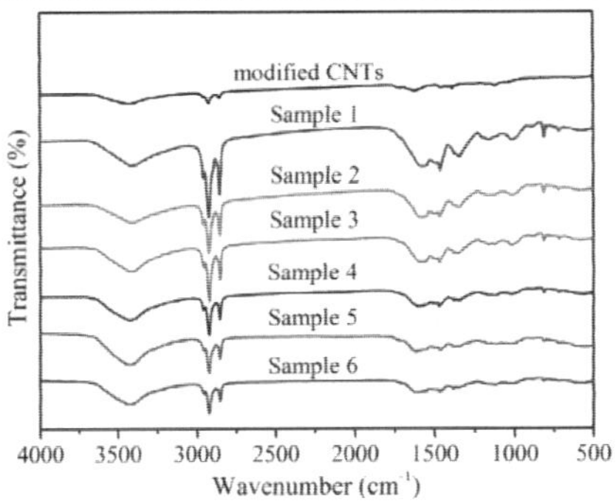

Figure 3. FT-IR spectra of the modified CNTs and the CNTs reinforced micro-PCMs.

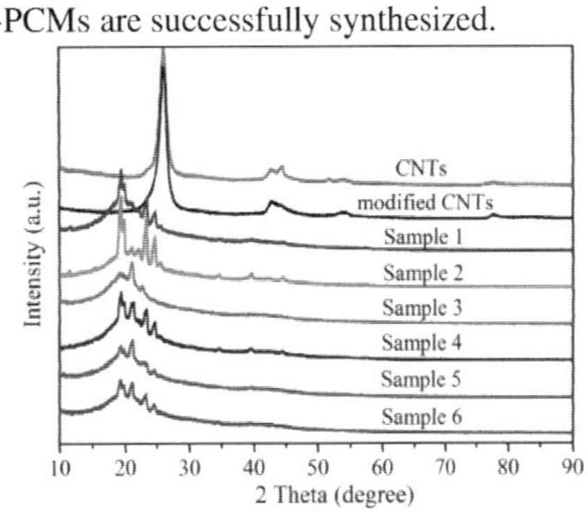

Figure 4. XRD patterns of CNTs, modified CNTs and the CNTs reinforced micro-PCMs.

3.4. Crystallization properties of the CNTs reinforced micro-PCMs

The crystallinity of CNTs, modified CNTs, micro-PCMs without CNTs and the CNTs reinforced micro-PCMs are determined by XRD. And the influence of CNTs and modified CNTs on crystallization features of micro-PCMs is also investigated. The received XRD patterns are shown in figure 4. A diffraction peak for CNTs and modified CNTs can be remarkably observed at the 2θ value of 26°, which is well corresponding to (002) plane of CNTs. It is clearly observed that micro-PCMs without CNTs reveal series of diffraction peaks at approximately 2θ of 19.4°, 19.9°, 23.5° and 24.8°, which can be indexed as (011), (012), (101) and (102) reflections of *n*-octadecane, respectively. These diffraction peaks also appear on the XRD patterns of CNTs reinforced micro-PCMs with different modified CNTs content, indicating that crystallization structure of micro-PCMs almost have no change after adding CNTs. It demonstrates that the incorporation of CNTs and modified CNTs does not affect the crystallization structure of micro-PCMs and they are just uniformly dispersed in the core and shell materials by interaction between molecules. As a result, the novel CNTs reinforced micro-PCMs are successfully synthesized and they keep a good crystallization structure which may ensure them excellent phase change behaviors.

3.5. Thermal storage properties of the CNTs reinforced micro-PCMs

Figure 5 shows the DSC curves of micro-PCMs without CNTs and the CNTs reinforced micro-PCMs, and the date calculated from DSC curves are summarized in table 2. For the CNTs reinforced micro-PCMs, the phase change characteristics during phase process are quite similar to those of micro-PCMs without CNTs because of the fact that there is no significant influence of CNTs additional content on the formation of microcapsules. It is clearly seen that there is a single peak in the melting process for the micro-PCMs, while there are two phase change peaks in the cooling thermograms. Some researchers pointed out that *n*-alkane probably exist several stable phases and partially ordered phases [26]. The minor peak and the mainly wide peak in figure 5 are corresponding to the solid–solid phase transition (the order–disorder transitions in PCMs) and solid–liquid phase change (the melting-cooling process of PCMs), respectively. In this case, the temperature of the solid–solid transition is lower than that of the solid–liquid one [27]. These results imply that the CNTs reinforced micro-PCMs maintain a good phase change behavior. Furthermore, the melting peaks of the CNTs reinforced micro-PCMs shift to a lower temperature with the increasing of CNTs content, and their crystallization peaks move to a higher temperature correspondingly. It suggests that the addition of CNTs improves the thermal conductivity of the obtained micro-PCMs which supplies a quickly thermal response rate to the ambient temperature.

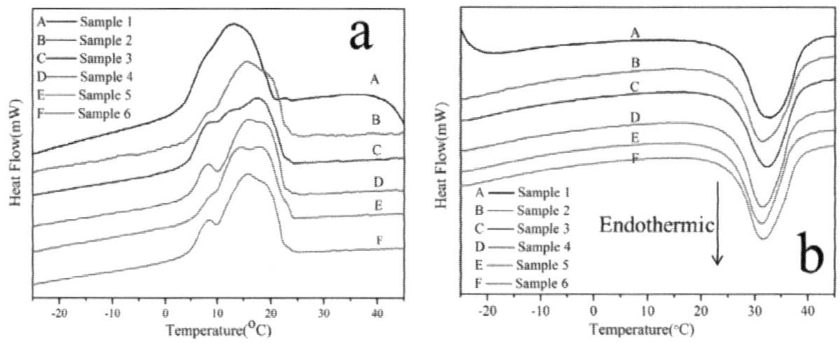

Figure 5. DSC curves of the obtained micro-PCMs.

Moreover, as seen from table 2, the crystal latent heats of the samples reinforced with CNTs are approximately the same as that of micro-PCMs without CNTs, while the melting latent heats are obviously enhanced after reinforced with CNTs. This result also shows that the CNTs are dispersed uniformly in the micro-PCMs and form a continuous thermal conductive network so that the interior PCMs can adequately change their phase from solid to liquid. In addition, the melting latent heat of sample 3 is the highest among all the samples. And the melting latent heat decrease slightly with the increase of the CNTs meaning that the addition content of the CNTs has little influence on the phase change behavior of the obtained micro-PCMs. That is to say, the CNTs reinforced micro-PCMs possess a quite good thermal energy- storage property.

Table 2. DSC data of the CNTs reinforced micro-PCMs.

Sample name	Solidifying		Melting	
	Temperature (°C)	Latent heat (J/g)	Temperature (°C)	Latent heat (J/g)
Sample 1	13.13	-131.93	32.85	116.89
Sample 2	15.73	-132.08	31.34	118.54
Sample 3	17.28	-133.11	32.13	121.22
Sample 4	16.01	-132.40	31.49	120.10
Sample 5	14.60	-130.04	31.25	120.44
Sample 6	15.89	-131.92	31.73	122.71

3.6. Thermal durability of the CNTs reinforced micro-PCMs

The change in thermal properties of the CNTs reinforced micro-PCMs is also evaluated by using sample 3 tested for 20 times of DSC thermal cycles and is shown in figure 6. Melting and crystallization temperatures of the sample 3 almost keep the same till the 20th thermal cycling. Meanwhile, the peak areas of melting and crystallization process change slightly after 20 times of thermal cycling, suggesting that there is no significant change in latent heats for the sample after 20 thermal cyclings. Based on these results, it is notable that the CNTs reinforced micro-PCMs exhibit high thermal reliability in its phase change process. The phase change temperatures and latent heats can keep almost constant in each heating-cooling cycle. It implies that the synthesized samples can maintain repetitious phase change behavior, meaning that the samples perform a good thermal durability in heat storage capacity.

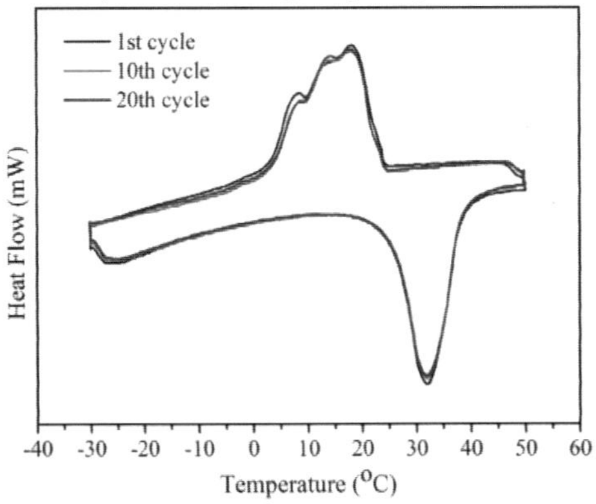

Figure 6. DSC curves of the CNTs reinforced micro-PCMs for 20 times of thermal cycles.

3.7. Different heating rates test of the CNTs reinforced micro-PCMs

The DSC curves are performed as well to investigate the phase change behavior and thermal energy-storage property of the micro-PCMs without CNTs and the CNTs reinforced micro-PCMs at different heating rates of 5 K/min, 10 K/min, 15 K/min and 20 K/min. The phase transformation behaviors and the phase change characteristic parameters are plotted in figure 7 and table 3, respectively. The sample exhibits comparatively high enthalpy which almost is unchanged during both heating and cooling phase change process at different heating rates. It is also notable that the crystallization peaks of the samples shift to a lower temperature with the increasing heating rate, and the melting peaks move to a higher temperature. This phenomenon is ascribed to the truth that a faster scanning rate can cause more serious super-cooling for a phase change system [28]. However, the crystallization temperatures

IOP Conf. Series: Materials Science and Engineering **182** (2017) 012015 doi:10.1088/1757-899X/182/1/012015

of the CNTs reinforced micro-PCMs are higher than those of the micro-PCMs without CNTs. This result indicates that the thermal response rate of the CNTs reinforced micro-PCMs is faster than that of the pure micro-PCMs. Such a feature is attributed to the highly thermal conductive CNTs which provide a superior thermal response rate for the CNTs reinforced micro-PCMs.

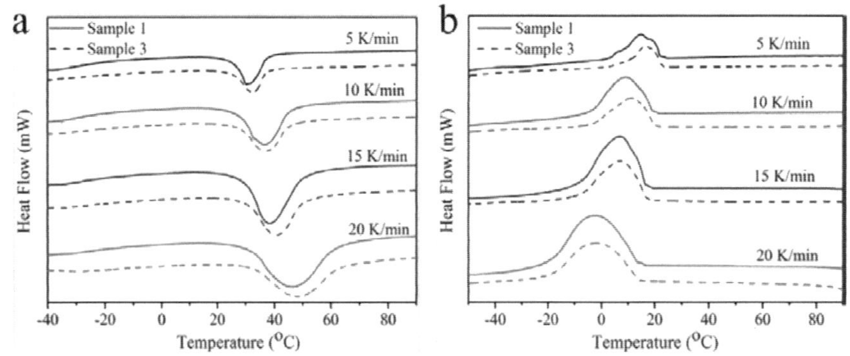

Figure 7. DSC curves of the CNTs reinforced micro-PCMs tested at different heating rates.

Table 3. DSC data of the CNTs reinforced micro-PCMs at different heating rates.

Heating rate	Solidifying Temperature (°C)		Melting Temperature (°C)	
	Sample 1	Sample 3	Sample 1	Sample 3
5 K/min	14.65	16.37	30.21	32.51
10 K/min	8.99	11.07	36.73	36.83
15 K/min	6.82	7.07	38.47	40.76
20 K/min	-2.59	-1.89	46.41	48.22

3.8. Thermal conductivity of the CNTs reinforced micro-PCMs

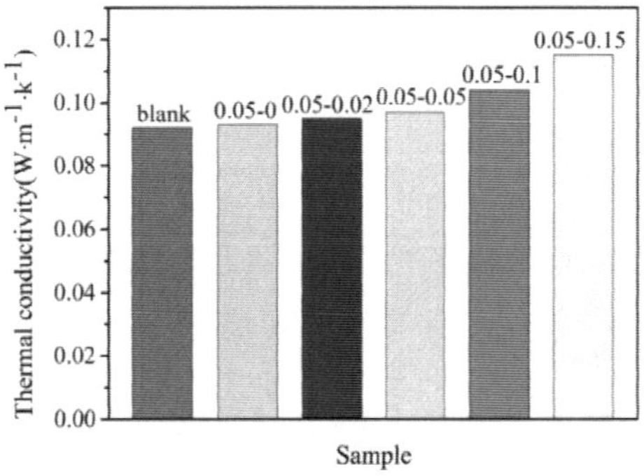

Figure 8. Thermal conductivity of the CNTs reinforced micro-PCMs.

To further evaluate the thermal energy storage and thermo-regulating capacity of PCMs, thermal conductivity is considered as an important contributing factor, as it may delay or promote the thermal response to the storage and release of latent heat. The thermal conductivity of each sample is shown in figure 8. The effect of different mass fractions of modified CNTs on the thermal conductivity of micro-PCMs is also provided. The thermal conductivities of sample1-6 are 0.092, 0.093, 0.095, 0.097, 0.104 and 0.115 W/(m·K) at room temperature, respectively. It is evident that thermal conductivity of the samples show a linear increase with the increment of modified CNTs in shell materials, owing to the internal high thermal conductivity of CNTs. Compared to the sample 1, the thermal conductivity of sample 6 performs a remarkable improvement (25%). These phenomena suggest that CNTs and the modified CNTs are excellently combined with *n*-octadecane and shell materials, respectively. And they are dispersed uniformly in the core and shell materials. Therefore, the CNTs can perfectly improve the thermal conductivity of the obtained micro-PCMs.

4. Conclusions

Novel micro-PCMs with good phase change behavior and high thermal conductivity are successfully synthesized through *in-situ* polymerization with modified CNTs reinforced melamine-formaldehyde resin as shell material and CNTs reinforced *n*-octadecane as PCMs core. The comprehensive performance of the CNTs reinforced micro-PCMs is determined by SEM, FT-IR, XRD, DSC, TGA and TCA. SEM results confirm that the samples are successfully prepared with smooth and compact spherical surface. FTIR spectra and XRD patterns reveal that the incorporation of CNTs and modified CNTs are just uniformly dispersed in the core and shell materials by interaction between molecules. On the basis of the DSC data, all of the samples present good phase change behavior and thermal cycling stability. The TCA implies that thermal conductivity of the samples shows a linear increase with the increment of CNTs mass fractions, owing to the internal high thermal conductivity of the CNTs. As a result, the CNTs reinforced micro-PCMs have excellent phase-change latent heat and high thermal conductivity, which provide a good prospect in thermal energy storage applications.

Acknowledgements

The authors greatly appreciate the financial support by the National Natural Science Foundation of China (Grant No. 51102230, 51462006), the Guangxi Natural Science Foundation (No. 2014GXNSFAA118401, 2013GXNSFBA019244), Innovation Project of GUET Graduate Education (2016YJCX21) and Program for Postgraduate Joint Training Base of GUET-CJYRE (No. 20160513-14-Z).

References

[1] S A Memon 2014 Phase change materials integrated in building walls: A state of the art review *Renewable Sustainable Energy Rev* **31** 870–906

[2] F He, X D Wang and D Z Wu 2014 New approach for sol-gel synthesis of microencapsulated n-octadecane phase change material with silica wall using sodium silicate precursor *Energy* **67** 223–233

[3] Y Allouche, S Varga, C Bouden and A C Oliveira 2016 Validation of a CFD model for the simulation of heat transfer in a tubes-in-tank PCM storage unit *Renewable Energy* **89** 371–379

[4] L F Cabeza, C Castellon, M Nogues, M Medrano, R Leppers, O Zubillaga, R Leppers and O Zubillaga 2007 Use of microencapsulated PCM in concrete walls for energy savings *Energ Buildings* **39** 113–119

[5] S P Zhan, S H Chen, L Chen and W M Hou 2016 Preparation and characterization of polyurea microencapsulated phase change material by interfacial polycondensation method *Powder Technol* **292** 217–222

[6] L Zhao, J Luo, H Wang, G L Song and G Y Tang 2016 Self-assembly fabrication of microencapsulated n-octadecane with natural silk fibroin shell for thermal-regulating textiles *Appl. Therm. Eng.* **99** 495–501

[7] Z Chen and M H Qin 2016 Preparation and hygrothermal properties of composite phase change humidity control materials *Appl. Therm. Eng.* **98** 1150–1157

[8] S Demirbag and S A Aksoy 2016 Encapsulation of phase change materials by complex coacervation to improve thermal performances and flame retardant properties of the cotton fabrics *Fibers Polym.* **17** 408–417

[9] H Zhang, F Xing, H Z Cui, D Z Chen, X Ouyang, S Z Xu, J X Wang, Y T Huang, J D Zuo and J N Tang 2016 A novel phase-change cement composite for thermal energy storage: Fabrication, thermal and mechanical properties *Appl. Energy* **170** 130–139

[10] F M Bezerra, O G Carmona, C G Carmona, M J Lis and F F de Moraes 2016 Controlled release of microencapsulated citronella essential oil on cotton and polyester matrices *Cellulose* **23** 1459–1470

[11] X L Qiu, W Li, G L Song, X D Chu and G Y Tang 2012 Microencapsulated n-octadecane with

different methylmethacrylate-based copolymer shells as phase change materials for thermal energy storage *Energy* **46** 188–199

[12] M Kaseem, K Hamad and Y G Ko 2016 Fabrication and materials properties of polystyrene/carbon nanotube (PS/CNT) composites: A review *Eur Polym J.* **79** 36–62

[13] P Zhang, Z W Ma and R Z Wang 2010 An overview of phase change material slurries: MPCS and CHS *Renewable Sustainable Energy Rev* **14** 598–614

[14] C Alkan, A Sari and A Karaipekli 2011 Preparation, thermal properties and thermal reliability of microencapsulated n-eicosane as novel phase change material for thermal energy storage *Energy Convers. Manage.* **52** 687–692

[15] L X Chai, X D Wang and D Z Wu 2015 Development of bifunctional microencapsulated phase change materials with crystalline titanium dioxide shell for latent-heat storage and photocatalytic effectiveness *Appl. Energy* **138** 661–674

[16] T Y Wang, S F Wang, R L Luo, C Y Zhu, T Akiyama, Z G Zhang, T Akiyama and Z G Zhang 2016 Microencapsulation of phase change materials with binary cores and calcium carbonate shell for thermal energy storage *Appl. Energy* **171** 113–119

[17] J F Wang, H Q Xie and Z Xin 2009 Thermal properties of paraffin based composites containing multi-walled carbon nanotubes *Thermochim. Acta.* **488** 39–42

[18] X C Gui, Z Q Lin, Z P Zeng, K L Wang, D H Wu and Z K Tang 2013 Controllable synthesis of spongy carbon nanotube blocks with tunable macro- and microstructures *Nanotechnology* **24** 085705

[19] P Ciambelli, L Arurault, M Sarno, S Fontorbes, C Leone, L Datas, D Sannino, P Lenormand and S L B D Plouy 2011 Controlled growth of CNT in mesoporous AAO through optimized conditions for membrane preparation and CVD operation *Nanotechnology* **22** 265613

[20] J S Ye and F S Sheu 2006 Functionalization of CNTs: New routes towards the development of novel electrochemical sensors *Curr. Nanosci.* **2** 319–327

[21] M Naffakh, A M Diez-Pascual, C Marco and G Ellis 2012 Morphology and thermal properties of novel poly(phenylene sulfide) hybrid nanocomposites based on single-walled carbon nanotubes and inorganic fullerene-like WS2 nanoparticles *J. Mater. Chem.* **22** 1418–1425

[22] I Burmistrov, N Gorshkov, I Ilinykh, D Muratov, E Kolesnikov, S Anshin, I Mazov, J P Issi and D Kusnezov 2016 Improvement of carbon black based polymer composite electrical conductivity with additions of MWCNT *Compos. Sci. Technol.* **129** 79–85

[23] R Nadiv, M Shtein, M Refaeli, A Peled and O Regev 2016 The critical role of nanotube shape in cement composites *Cem. Concr. Compos.* **71** 166–174

[24] H Wang, A S Tazebay, G Yang, H T Lin, W Choi and C H Yu 2016 Highly deformable thermal interface materials enabled by covalently-bonded carbon nanotubes *Carbon* **106** 152–157

[25] H D Yun, J Kwak, S Y Kim, H Seo, I C Bang, S Y Kim, S Kang and S Y Kwon 2016 High performance all-carbon composite transparent electrodes containing uniform carbon nanotube networks *J. Alloys Compd.* **675** 37–45

[26] S Sari-Bey, M L Fois, I Krupa, L Ibos, B Benyoucef and Y Candau 2014 Thermal characterization of polymer matrix composites containing microencapsulated paraffin in solid or liquid state *Energy Convers. Manage.* **78** 796–804

[27] X L Qiu, L X Lu, P J Han, G Y Tang and G L Song 2016 Fabrication, thermal property and thermal reliability of microencapsulated paraffin with ethyl methacrylate-based copolymer shell *J. Therm. Anal. Calorim.* **124** 1291–1299

[28] F Y Jiang, X D Wang and D Z Wu 2016 Magnetic microencapsulated phase change materials with an organo-silica shell: Design, synthesis and application for electromagnetic shielding and thermal regulating polyimide films *Energy* **98** 225–239

A damage mechanics based method for fatigue life prediction of the metal graded materials

Yang Tong [*]**, Weiping Hu and Qingchun Meng**

School of Aeronautic Science and Engineering, Beihang University, Beijing 100191, China

E-mail: [*]tongyang0909@buaa.edu.cn

Abstract. Based on the continuum damage mechanics theory, the fatigue life prediction for TC4-TC11 graded material was conducted. At first, the damage evolution equation was derived, then the method to calibrate material parameters for TC4-TC11 graded material was proposed, and all the material parameters were obtained. A beam model with TC4-TC11 graded material was established by using the stratified method and finite element method. Finally, the fatigue life of TC4-TC11 graded beam was predicted.

Keywords: damage mechanics, fatigue life prediction, metal graded material, finite element

1. Introduction

In the year of 1984, Japan's MasyuhiNINO, ToshioHIRA and other materials scientists proposed the concept of "functionally graded materials" (FGM) [1]. Scientists from China and many other countries have also started to study the preparation technology, material properties and practical value of FGM. FGM has become a new star in the material world, which is with a strong development trend. The metal graded material is a kind of FGM, which is composed of two or more metal materials with the rational distribution and uniform transition [2]. Due to the special mechanical properties of metal graded material, it can meet the requirement that different location has different mechanical property. So the metal graded material has a wide application in various fields, especially in the aerospace field [3-8]. However, because of its special inhomogeneity, the method to predict the life of uniform materials is not adapted to this new material. Therefore, a new method to predict the fatigue life of metal graded material is presented, which is based on the theory of continuum damage mechanics [9-11]. Firstly, according to the principle of thermodynamics, the damage evolution equation of metal graded materials is derived. Secondly, the damage evolution parameters for TC4-TC11 metal graded materials are calibrated according to the experimental data. Finally, a model of beam is established and the stress analysis and life prediction are carried out. The simulation results show that this method is feasible.

2. Theoretical model

In this paper, the TC4-TC11 metal graded materials are studied. The corresponding damage evolution equation is derived, based on the continuum damage mechanics theory. For isotropic materials, the damage variable D is introduced to describe the deterioration of materials, which can be given as the reduction of stiffness [12-13].

$$D = \frac{E - E_D}{E} \tag{1}$$

where E is Young's Modulus of material and E_D is the effective Young's Modulus with damage. As E_D ranges from E to 0, D varies between 0 and 1.

Using Kronecker delta

$$\{x\} = \begin{cases} x & x > 0 \\ 0 & x \le 0 \end{cases}$$

the damage evolution equation is expressed by[14]

$$\frac{dD}{dN} = \alpha \frac{\left\{ Y_{max}^{\frac{1}{2}} - Y_{th}^{\frac{1}{2}} \right\}^m}{(1-D)^m} \tag{2}$$

where α, m, Y_{th} are parameters related to the properties of the material and Y is strain energy release rate. Y_{max} is the maximum value and Y_{th} is the threshold value, which means the damage occurs when $Y_{max} > Y_{th}$. According to thermodynamics of damaged materials, Y is defined as

$$Y = -\rho \frac{\partial \psi}{\partial D} \tag{3}$$

where ρ is density and Ψ is Helmholtz free energy function. For the case of isotropic damage, Equation (4) can be derived when stress is constant.

$$Y = \frac{1}{2} C_{ijkl}^0 \varepsilon_{ij}^e \varepsilon_{kl}^e = \frac{W^E}{1-D} \tag{4}$$

where W^E is strain energy density.

In the condition of uniaxial loading,

$$W^E = \frac{1}{2} \sigma \varepsilon^e, \quad \sigma = E(1-D)\varepsilon^e \tag{5}$$

then,

$$Y = \frac{\sigma^2}{2E(1-D)^2} \tag{6}$$

In the case of multiaxial loading,

$$W^E = \frac{1}{2} \sigma_{ij} \varepsilon_{ij}^e, \quad \sigma_{ij} = C_{ijkl}^0 \varepsilon_{ij}^e \varepsilon_{kl}^e (1-D) \tag{7}$$

then the strain energy release rate is

$$Y = \frac{W^E}{1-D} = \frac{(\sigma_{eq} (R_v)^{1/2})^2}{2E(1-D)^2} \tag{8}$$

where

$$R_v \equiv \frac{2}{3}(1+v) + 3(1-2v)\left(\frac{\sigma_H}{\sigma_{eq}}\right)^2 \tag{9}$$

In order to obtain the uniform equation, the concept of damage equivalent stress is introduced. The equivalent stress can be denoted as:

$$\sigma_e = \sigma_{eq} (R_v)^{\frac{1}{2}} = [\frac{3}{2} s_{ij} s_{ij}]^{1/2} (R_v)^{\frac{1}{2}} \tag{10}$$

So the uniform equation of the strain energy release rate can be written as:

$$Y = \frac{1}{2} (\frac{1}{1-D})^2 \frac{1}{E} \sigma_e^2 \tag{11}$$

From Equation (11), Y_{max} can be expressed by the following formula:

$$Y_{max} = \frac{1}{2} (\frac{1}{1-D})^2 \frac{1}{E} \sigma_{e,max}^2 \tag{12}$$

IOP Conf. Series: Materials Science and Engineering **182** (2017) 012016 doi:10.1088/1757-899X/182/1/012016

Here the relationship between the threshold value Y_{th} and the D is assumed to satisfy Equation (11). Then,

$$Y_{th}=\frac{1}{2}\left(\frac{1}{1-D}\right)^2\frac{1}{E}\sigma_{e,th}^2 \tag{13}$$

where $\sigma_{e,th}$ is the threshold value of the equivalent stress.

Substituting Equation (12) and Equation (13) into Equation (2), the damage evolution equation can be obtained as:

$$\frac{dD}{dN}=\alpha\left(\frac{1}{2E}\right)^{\frac{m}{2}}\frac{\left(\sigma_{e,max}-\sigma_{e,th}\right)^m}{(1-D)^{2m}} \tag{14}$$

Then Equation (14) could be rewritten as:

$$\frac{dD}{dN}=\beta\frac{\left(\sigma_{e,max}-\sigma_{e,th}\right)^m}{(1-D)^{2m}} \tag{15}$$

where

$$\beta=\alpha\left(\frac{1}{2E}\right)^{\frac{m}{2}} \tag{16}$$

Separating the Equation (15), Equation (17) and Equation (18) can be obtained as follows

$$\int_0^{N_f}\left(\sigma_{e,max}-\sigma_{e,th}\right)^m dN=\left(\sigma_{e,max}-\sigma_{e,th}\right)^m N_f \tag{17}$$

$$\frac{1}{\beta}\int_{D_0}^{1}(1-D)^{2m}dD=\frac{1}{\beta(2m+1)}(1-D_0)^{2m+1} \tag{18}$$

where D_0 is the initial damage.

Equation (17) is equal to Equation (18) with the same value C_1. It is clear that different fatigue curves can be obtained when D_0 has different values. By choosing $D_0 = D_{0,m}$, the mean fatigue S-N curve can be obtained as:

$$\left(\sigma_{e,max}-\sigma_{e,th}\right)^m N_{f,m}=C_{1,m} \tag{19}$$

where $N_{f,m}$ is the mean fatigue life.

$$C_{1,m}=\frac{1}{\beta(2m+1)}(1-D_{0,m})^{2m+1} \tag{20}$$

Finally, the fatigue failure criterion of metal graded materials under load spectrum is studied. Here the simplified load spectrum (as shown in figure 1) is taken as the research object. The simplified load spectrum is composed of several load blocks with the constant amplitude, which have different maximum stress and corresponding cycles.

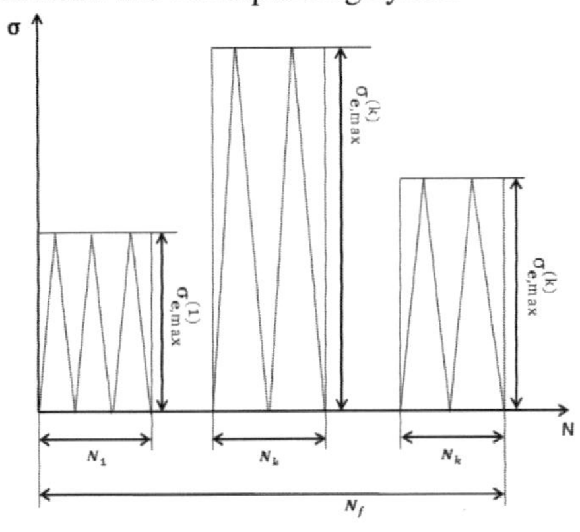

Figure 1. load spectrum composed of different load blocks.

Processing Equation (15), it can be obtained as follows:

$$\int_0^{N_f}\left(\sigma_{e,max}-\sigma_{e,th}\right)^m dN=\frac{1}{\beta}\int_{D_0}^{1}\left(1-D\right)^{2m} dD \tag{21}$$

For the load spectrum, it can be calculated as:

$$\int_0^{N_f}\left(\sigma_{e,max}-\sigma_{e,th}\right)^m dN=\sum_{k=1}^{k}\left(\sigma_{e,max}^{(k)}-\sigma_{e,th}\right)^m N_k \tag{22}$$

where,

$$N_f=\sum_{k=1}^{k}N_k \tag{23}$$

And,

$$\frac{1}{\beta}\int_{D_0}^{1}\left(1-D\right)^{2m} dD=\frac{1}{\beta\left(2m+1\right)}(1-D_0)^{2m+1}=C \tag{24}$$

The following fatigue failure criterion can be derived by combining Equation (22) and Equation (24):

$$\sum_{k=1}^{k}\left(\sigma_{e,max}^{(k)}-\sigma_{e,th}\right)^m N_k=C \tag{25}$$

For each load block, the corresponding fatigue life can be calculated as:

$$N_{f,k}=\frac{C}{\left(\sigma_{e,max}^{(k)}-\sigma_{e,th}\right)^m} \tag{26}$$

where $N_{f,k}$ is the corresponding fatigue life of metal graded material under the stress $\sigma_{e,max}^{(k)}$.

In summary, the following form of fatigue failure criteria can be obtained [9,12]:

$$\sum_{k=1}^{k}\frac{N_k}{N_{f,k}}=1 \tag{27}$$

3. Parameter calibration for TC4-TC11 medal graded material

3.1. Calibration method

According to the characteristics of metal graded materials, it can be divided into two parts: the substrate region and the transition region. In this paper, it is assumed that the substrate zone and transition zone have the same form of damage evolution equation, but the specific parameters of the different regions are different. For parameter calibration in the substrate region, the conventional damage evolution model for homogeneous material is used. Then for parameter calibration in the transition region, the interpolation method is used for the rest part, after the calibration of the part with specific proportion has been done. In this paper, the parameters of the damage evolution equation have been identified according to the fatigue test results. Taking TC4-TC11 medal graded material as an example, the basic steps are as follows:

(1) Fitting the S-N curve of two substrate regions materials and fitting the specific region which is made up of two base materials each 50%.

(2) Deriving the damage evolution equation after determined the force situation as a uniaxial, biaxial or multiaxial force.

(3) Parts of the parameters can be calibrate by comparing the S-N curve equation and the damage evolution equation and the rest material parameters can be obtained by a method of selecting the farthest point.

3.2. Results of calibration

Firstly, the S-N curve is fitted by the experimental data. The results of three kinds of materials are shown in figure 2:

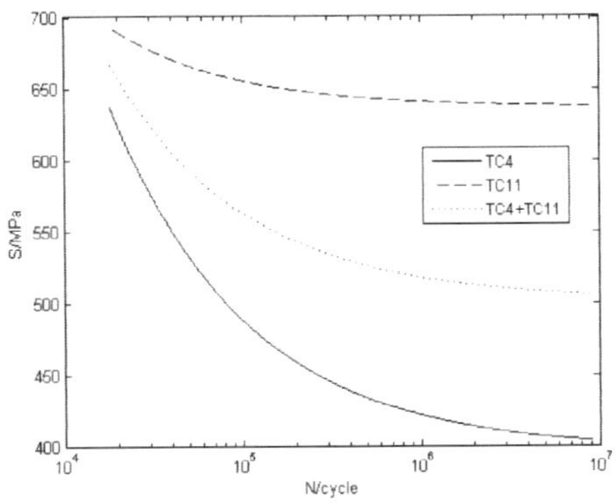

Figure 2. S-N curves of fatigue test for standard parts of three kinds of materials (K_T=1, R=-1).

The fitting equations are:

- TC4: $S = 396.28 + 58173.465 \times N^{-0.56}$
- TC11: $S = 636.31 + 30053.545 \times N^{-0.64}$
- TC4+TC11: $S = 501.106 + 43958.758 \times N^{-0.57}$

This paper takes the simplest situation of K_T=1 as an example.

According to Equation (17) and Equation (18), the stress-life relationship of standard specimens under K_T=1 can be written as follows:

$$\left(\sigma_{k,max} - \sigma_{k,th}\right)^{m_k} N_f = \frac{1}{\beta_k\left(2m_k+1\right)}\left(1-D_{0,k}\right)^{2m_k+1} \tag{28}$$

where $\sigma_{k,th}$, m_k, β_k, $D_{0,k}$ are material parameters.

For simplifying Equation (28), another parameter C_1 is defined as:

$$C_1 = \frac{1}{\beta_k\left(2m_k+1\right)}(1-D_{0,k})^{2m_k+1} \tag{29}$$

Then the Equation (28) can be written by:

$$N_f = \frac{C_1}{(\sigma_{k,\max} - \sigma_{k,th})^{m_k}}$$

Logarithm of the upper equation:

$$lgN_f = lgC_1 - m_k lg(\sigma_{k,max} - \sigma_{k,th}) \tag{30}$$

The equation form of the S-N curve is:

$$lgN = C - mlg(\sigma_{max} - \sigma_{th}) \tag{31}$$

Comparing the Equation (30) with Equation (31), the following equations can be obtained:

$$\begin{aligned} lgC_1 &= C \\ m_k &= m \\ \sigma_{k,th} &= \sigma_{th} \end{aligned} \tag{32}$$

Then two material parameters of the damage evolution equation $\sigma_{k,th}$ and m_k can be identified. Meanwhile, an equation about β_k and $D_{0,k}$ can be obtained.

In order to identify the value of β_k and $D_{0,k}$, the point which has the longest fatigue life has been picked out, assuming the initial damage $D_{0,k} = 0$.

Then,

$$\left(\sigma_{k,max} - \sigma_{k,th}\right)^{m_k} N_f = \frac{1}{\beta_k\left(2m_k + 1\right)} \tag{33}$$

β_k can be obtained. Substituting β_k into Equation (29), $D_{0,k}$ can be obtained.

At this point, all the material parameters of the damage evolution equation can be obtained.

The material parameters of other regions in transition zone can be obtained by interpolation with the identified data.

The results of the calibrated parameters in transition zone are shown in table 1.

Table 1. The results of the calibration of damage evolution equation parameters.

Percentage of TC4	σ_{th}	m_k	$D_{0,k}$	β_k
100	396.2765	1.7857	0.0163	0.6228
90	414.8123	1.7923	0.0466	0.6056
80	434.5631	1.7924	0.0739	0.5993
70	455.5289	1.7862	0.0983	0.6039
60	477.7099	1.7735	0.1196	0.6193
50	501.1059	1.7544	0.1380	0.6457
40	525.7171	1.7289	0.1535	0.6829
30	551.5432	1.6969	0.1659	0.7310
20	578.5845	1.6585	0.1754	0.7901
10	606.8409	1.6137	0.1819	0.8600
0	636.3123	1.5625	0.1854	0.9408

4. Fatigue life prediction of TC4-TC11 metal graded material

4.1. Damage mechanics-finite element method

When predicting the fatigue life of components by using damage mechanics method [15-17], the stress field and damage field must be calculated. The process of calculation is summarized as follows

(1) Calculating the increment of the damage ΔD_i for element after ΔN cycles

$$\Delta D_i = \alpha_k \frac{\left\{Y_{max}^{\frac{1}{2}} - Y_{k,th}^{\frac{1}{2}}\right\}^{m_k}}{\left(1 - D_{i-1}\right)^{m_k}} \Delta N \tag{34}$$

(2) Updating the damage and element stiffness as shown in Equations (35) and (36).

$$D_i = D_{i-1} + \Delta D_i \tag{35}$$

$$E = E_0 \times \left(1 - D_i\right) \tag{36}$$

(3) Repeating Steps1-2 until the damage extent at any element reaches 1 and the fatigue life is obtained.

4.2. Stratified method and finite element analysis

4.2.1. Stratified method

The properties of the metal graded materials vary uniformly in the transition region, thus in order to obtain a more precise stress field, the stratified method is used in the finite element analysis. The stratified method is referred to that the model is divided into many layers and assuming that the same layer has the same parameters, but the parameters are changed in different layers which abide by some certain rules. According to this assumption, the traditional finite element method is applied to the stress analysis in each layer. Based on the finite element analysis, it can be known that the more numbers of layers, the higher the accuracy. However, at the same time, too many layers will result in a waste of computation time. So the proper number of layers is important. And in order to ensure continuity and alleviate the mutation between different layers, the material properties are assumed to change by a certain function.

4.2.2. Finite element analysis

A cantilever beam is designed with a size of 100mm*20mm*6mm, which is under the applied force of 2650N in the free end, as shown in figure 3.

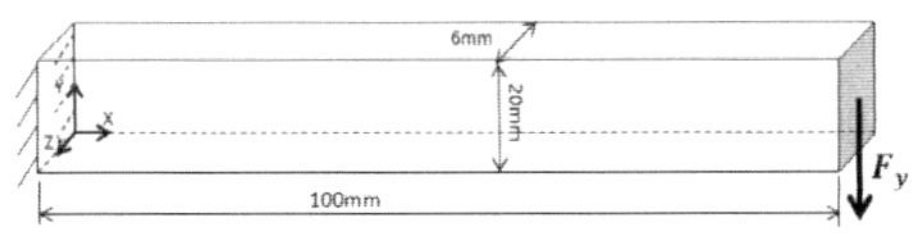

Figure 3. The design and applied force of cantilever beam.

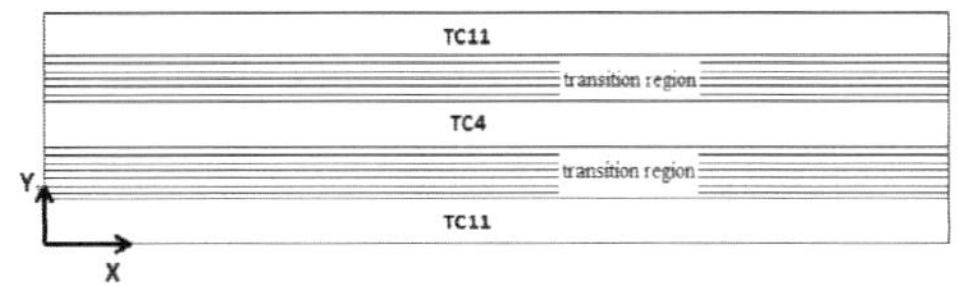

Figure 4. Component distribution of metal graded beam.

From the Equation (28), it can be found that the material fatigue life is not only related to the stress, but also to the Young's modulus E. If E is small, the increment of each cycle is larger. Thus, the crack initiates may not occur at the position of maximum stress.

Based on the above consideration, a design that the upper and lower ends of the beam adopting TC11 with a higher Young's modulus and the middle part adopting TC4 with a lower Young's modulus is put forward as shown in figure 4.

To calculate the stress distribution, the stratified method is used [18-21]. The transition region is divided into ten layers with different layers having different material properties. And the assumption that the properties vary following the power exponential function is taken as the calculated results show it has a higher accuracy than following a linear function.

Solid185 element is chosen to carry out the mesh generation and stress analysis, and the calculated results are shown in figure 5 to figure 7.

As a contrast, the finite element model of TC4 homogeneous beam is established. The structure size and loading condition are the same as that of the graded beam. The results are shown in figure 8.

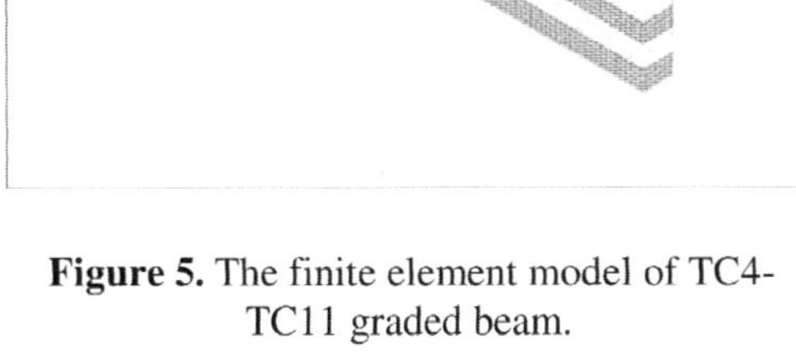

Figure 5. The finite element model of TC4-TC11 graded beam.

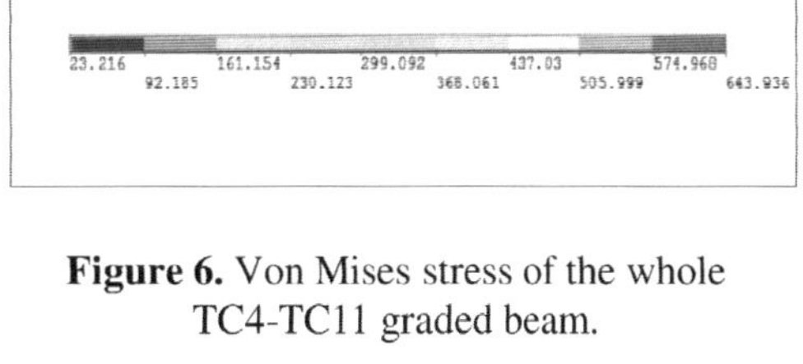

Figure 6. Von Mises stress of the whole TC4-TC11 graded beam.

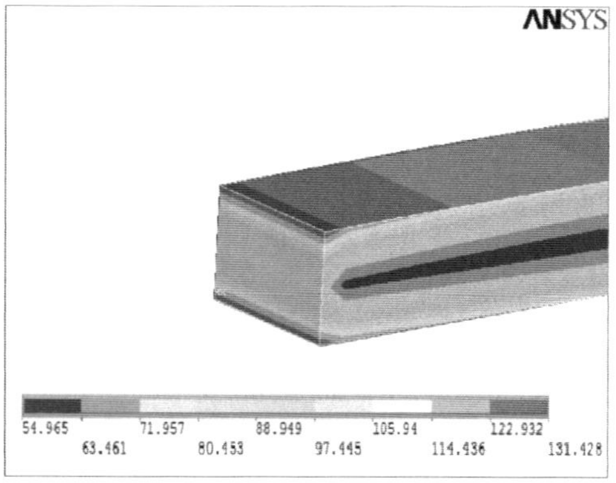

Figure 7. Von Mises stress of the middle layer of TC4-TC11 graded beam.

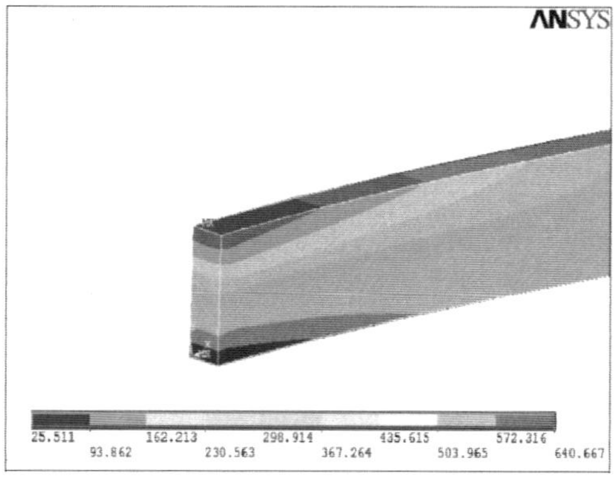

Figure 8. Von Mises stress of the whole TC4 homogeneous beam.

4.3. Life prediction results

The life prediction for homogeneous beam and graded beam is conducted, respectively. The results show that the maximum stress of homogeneous beam appears at the top and bottom, whose value is $\sigma = 640.667MPa$. And the predicted life is $N_f = 28800$

The results of the maximum stress and fatigue life of graded beam are shown in table.2. The table only lists the results of the lower part of the beam considering its symmetry.

Table 2. Maximum stress and fatigue life of each regions of graded beam.

Region	The lowest region	The bottom of the lower transition region	The upper of the lower transition region	The bottom of the middle layer region
Threshold value (MPa)	636.3	609.5	413.1	396.3
Maximum stress(MPa)	643.9	376.8	145.6	131.4
Fatigue life	668000	$N_f > 10^7$(because the maximum is under the threshold value)		

From table 2, the fatigue life of the graded beam is $N_f = 668000$.

Comparing the calculated results of the two beams, it shows that although the maximum stress of metal graded beam is larger than homogeneous beam, the fatigue life of metal graded beam is longer than the homogeneous one. The calculated results are consistent with the practical situation, which indicates that the method is feasible.

5. Conclusions

Based on the continuum damage mechanics theory, the fatigue life prediction of metal graded materials is carried out, and the main conclusions are summarized as follows:

According to the principle of thermodynamics, the damage evolution equation for the fatigue life prediction of metal graded materials is derived.

The calibration method of the material parameters in the damage evolution equation is proposed. Taking the TC4-TC11 metal graded material as an example, the specific process of material parameter calibration is given.

The fatigue life prediction of TC4-TC11 graded beam is carried out by using the damage mechanics-finite element method and the results show that the method is feasible.

In this paper, the influence of environment, temperature, etc. hasn't been considered when calculating the stress and damage. Besides, in the practical application, most of the metal graded materials are in the complex environment. Therefore, in the following research, analysis considering the environmental temperature and mechanical environment synthetically will be conducted, which can make the results more close to the actual situation.

References

[1] Niino M, Hrai T and Watanabe R 1987 *Joural of Japan Society of Composite Materials* **13** 257–264

[2] Jing Shikai, Zhang He, Zhou Jingtao and Song Guohua 2015 *Chinese Journal of Mechanical Engineering.* **6** 1186–1193

[3] C AtkinsonR and D List 1987 *International Journal of Engineering Science* **10**

[4] Shanmugavel P *et al.* 2012 *European Journal of Scientific Research* **68(3)** 412–439

[5] Atai A A, Nikeranjbar A and Kasiri R 2012 *roceedings of the Institution of Mechanical Engineers, Part C: Journal of Mechanical Engineering Science* **226(3)** 607–614

[6] Mahamood R M *et al.* 2012 Functionally graded material: an overview *Proceedings of the World Congress on Engineering (London, UK)* 1593–1597

[7] Zhang Zhengyan *et al.* 2013 *Chinese Journal of Mechanical Engineering* **49(3)** 163–173

[8] Nadeau J C and Ferrari M. 1999 *Mechanics of Materials* **10**

[9] Lemaitre J and Chabocha J L 1974 *Proc. IUTAM, Symp. Of Mechanics of Viscoelastic Media and Bodies, Springer-Verlay, Gethenbury*

[10] Zhan Z *et al.* 2015 Revised damage evolution equation for high cycle fatigue life prediction of aluminum alloy LC4 under uniaxial loading *Applied Mathematics and Mechanics* **36(9)** 1185–1196

[11] Zhan Z *et al.* 2016 Continuum damage mechanics-based approach to the fatigue life prediction for 7050-T7451 aluminum alloy with impact pit *International journal of Damage Mechanics* **25(7)** 943–966

[12] Zhan Z *et al.* 2015 A study on the effect of surface defect on the fatigue performance of metal component based on damage mechanics *Mechanics* **21(1)** 5–10

[13] Zhan Z *et al.* 2013 Experimental method for and theoretical research on defect tolerance of fixed plate based on damage mechanics *Chinese Journal of Aeronautic* **26(5)** 1195–1201

[14] Jean, Lemaitre, Rodrigue and Desmorat 1934 *Continuum damage mechanics* (Berlin:Jean. Lemaitre)

[15] Zhan Z *et al.* 2015 The fatigue life prediction for structure with surface scratch considering cutting residual stress, initial plasticity damage and fatigue damage *International Journal of Fatigue* **74** 173–182

[16] Zhan Z *et al.* 2015 A damage mechanics-based fatigue life prediction approach for 30crmnsia alloy steel with impact defect *Reliability Systems Engineering, 2015 First International Conference on. IEEE* 1–5

[17] Zhan Z X *et al.* 2012 The Fatigue Life Prediction Problem for Fixed Plate of Helicopter *Advanced Materials Research. Trans Tech Publications* **452** 1164–1168

[18] Reddy J N 2000 Analysis of functionally graded plates *International Journal for Numerical Methods in Engineering* **47(1.3)** 663–684

[19] Sankar B V 2001 An elasticity solution for functionally graded beams *Composites Science and Technology* **61(5)** 689–696

[20] Zhong Z and Shang E T 2003 *International Journal of Solids and Structures* **40(2)** 5335–5352

[21] Ding H J, Huang D Jand Chen W Q 2007 Elasticity solutions for plane anisotropic functionally graded beams *International Journal of Solids and Structures* **44(1)** 176–196

Effect of citric acid dosage and sintered temperature on the composition, morphology and electrochemical properties of lithium vanadium oxide prepared by a sol–gel method

C R Zhong[1,2], X J Su[1], G L Hou[1], Z H Liu[1], F S Yu[3], S Bi[1] and H Li[1]

[1] Xi'an Research Institute of High-tech, Xi'an, Shaanxi 710025, China
[2] Baoji Research Institute of High-tech, Baoji, Shaanxi 721013, China
[3] Xi'an Qinghua Cooperation, North Special Energy Group Limited Company, Xi'an Shaanxi 710025, China

E-mail: zhong_0807@163.com

Abstract. A lithium vanadium oxide cathode material was synthesized via sol-gel processing using citric acid as the chelating agent. Different dosage of citric acid and sintered temperature were introduced to investigate their effects on the products composition, morphology and electrochemical properties. The results showed that the V_2O_3 yield was inhibited and the crystallization of grain was accelerated with the increasing dosage of citric acid. Furthermore, V_2O_3 was oxidized to LiV_3O_8 and $Li_{0.3}V_2O_5$ with the increase of sintered temperature.

Keywords: sol-gel method, lithium vanadium oxide, thermal battery, citric acid, electrochemical performance

1. Introduction

The vanadium-based oxides have become one of the most attractive candidates in the field of lithium batteries in the past several decades. Among the vanadium-based oxides, LiV_3O_8 has been extensively investigated as potential and promising substituted cathode materials due to its attractive electrochemical properties such as high specific discharge capacity, good rate capacity, long cycle life, good safety, low cost and facile preparation [1,2]. LiV_3O_8 has a layered structure composed of VO_6 octahedra and VO_5 distorted trigonal bipyramid units to form stable V_3O_8 sub-lattices [3], accommodating additional three lithium-ions per formula unit without any structural change in the tetrahedral sites [4].

It is generally known that the synthesized routes and processing factors have significant influences on the crystal structure and electrochemical performances of LiV_3O_8 [5]. Various methods have been developed to prepare the LiV_3O_8 materials such as the sol–gel method, solid-state reaction, combustion synthesis, hydrothermal reaction and flame spray pyrolysis method [6]. Among these methods, sol–gel process is supposed to be an excellent method for the synthesis of cathode active materials due to its unique advantages in the synthesis of oxide materials. The raw materials can be mixed at the molecular level via sol–gel methods so that the products possess homogeneous composition and higher capacity [7]. Many works have been done to fabricate LiV_3O_8 crystallites by using sol–gel methods in which citric acid was mostly used as chelating agent [6-9].

The vanadium-based oxides also have been used for thermal battery cathode in the past years [10-11]. In the present work, we utilized a sol–gel process to synthesize the LiV_3O_8 cathode material used in thermal battery which using citric acid as chelating agents. The influences of citric acid dosage and sintered temperature on the phase composition, morphology and electrochemical properties of the products were particularly investigated.

2. Experimental

Analytical pure $LiOH \cdot H_2O$, V_2O_5 and citric acid ($C_6H_8O_7 \cdot H_2O$) were used as raw materials without any purification. Stoichiometric amounts of $LiOH \cdot H_2O$ and V_2O_5 (Li: V=1.05: 1.5, molar ratio) were mixed in deionized water. Then the saturation solution of citric acid was slowly dripped into the reaction system with continuous magnetic stirring. The molar ratio of total metal ions to citric acid was 1:1, 1:1.5, 1:2, respectively. After stirred for 1 h and dried the sol at 80°C in air, the gel was obtained. The three piece of obtained gel was dried in a vacuum oven at 120 °C for 6 h to form the bulky honeycombed precursor. Subsequently each precursor was grinded through 100 mesh sieve and was divided into three portions. One portion sieved powder of each sample was handled as one group and heated at 400 °C, 450 °C and 500 °C for 12 h in a muffle furnace respectively. The samples obtained were designated LVO-1-400, LVO-1-450, LVO-1-500, LVO-1.5-400, LVO-1.5-450, LVO-1.5-500, LVO-2-400, LVO-2-450 and LVO-2-500, respectively.

The crystal structures of the samples were determined using Oxford Instruments Company type 7718 X-ray spectroscopy in the range of 10°~80°. The morphology of the samples was observed with a Tescan VEGA 2 XMU SEM.

The cathode powder was fabricated by a mixture of 80 wt% active materials and 20wt% eutectic salt which is made up of LiCl and KCl (45:55, wt %). Every cell was assembled from a pelletized cathode material having a diameter of 20mm, Li-Si anode, eutectic salt and MgO (65:35, wt %) separator, nickel collector. The assemblage and measurement were conducted under a dry air atmosphere (1% humidity). The cells were introduced into a home-built instrument set to the test temperature (500 °C). While testing, the peak value of open circuit potential (OCP) and elevated temperatures discharge measurements were recorded on CorrTest CS 350 electrochemical workstation.

3. Results and discussion

The composition and crystallinity of the materials synthesized using a sol-gel method followed by different heat treatments (400, 450 or 500 °C) were determined using XRD analysis. As shown in figure 1, the powder shows diffractions peaks at 2θ of 13.9° and 28.5° indicating the gradual crystallization of LiV_3O_8 (JCPDS No. 72-1193) phase. The presence of the V_2O_3 (JCPDS No. 34-0187) and $Li_{0.3}V_2O_5$ (JCPDS No. 18-0755) was also detected in the XRD pattern. The impurity of V_2O_3 has rarely been reported in previous literature. From figure 1, it can be found that the diffraction peaks of V_2O_3 decreased with increasing dosage of citric acid. The dosage of citric acid has a great influence on the decomposition of the precursor. The more citric acid used, the less V_2O_3 deoxidized. In addition, the increased citric acid obviously accelerates crystallization of LiV_3O_8 and $Li_{0.3}V_2O_5$ grain. The effect of sintered temperature on the composition of the sample was analyzed at the same dosage of citric acid. The V_2O_3 produced was decreased gradually even disappeared with the increasing sintered temperature may because the V_2O_3 was oxidized to LiV_3O_8 and $Li_{0.3}V_2O_5$. Moreover, the crystallinity of LiV_3O_8 and $Li_{0.3}V_2O_5$ enhances with increasing sintered temperature.

The SEM images of as-synthesized LiV_3O_8 powders are shown in figure 2. It can be seen that the citric acid dosage and sintered temperature has a visible influence on the morphology of the samples. The SEM images reveal the process of the grain emerged and grew with the increasing citric acid at the same sintered temperature. Besides, the citric acid acts as an organic fuel during the sintering process [8] that is the reason why the increasing citric acid accelerates crystallization of grain. Review the images from vertical orientation; it reveals that the sintering temperature has obvious effect on the morphology of the as-synthesized material. The crystallites grain just appeared at 400°C, and then grew with the increasing temperature. In these samples, LVO-1-500, LVO-1.5-500 and LVO-2-450 reveal the rod-shaped crystallites with regular geometric shape and even size distribution. In addition, the volatilized citric acid also causes much interspace in sintering process so that the samples are propitious to infiltration of melting eutectic salt, which may lead to good electrochemical performance at the same dosage of citric acid.

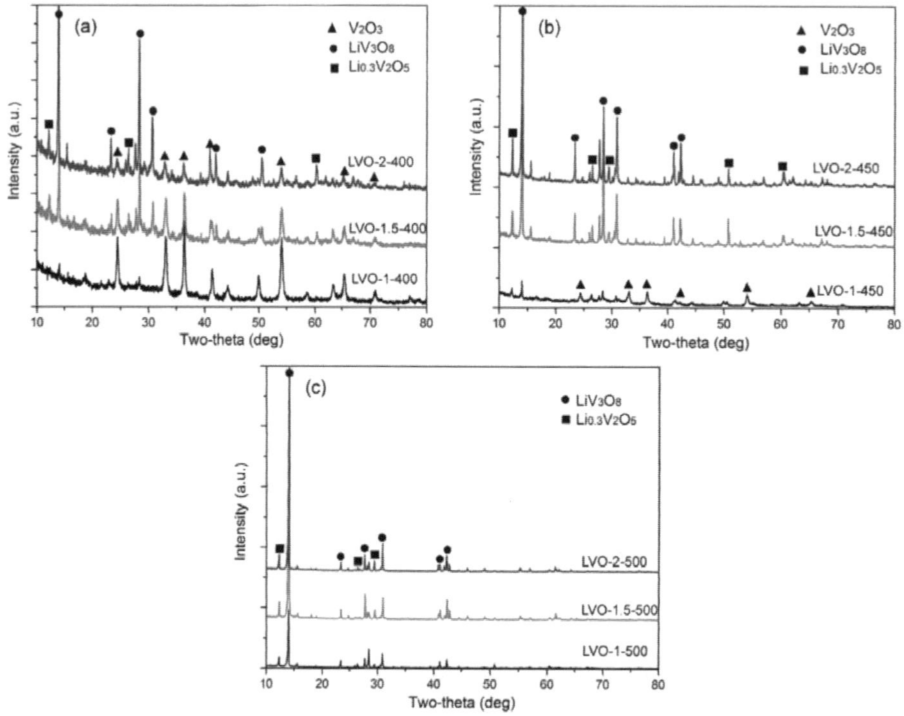

Figure 1. XRD pattern of specimen sintered at (a) 400°C; (b) 450°C; (c) 500°C.

IOP Conf. Series: Materials Science and Engineering **182** (2017) 012017 doi:10.1088/1757-899X/182/1/012017

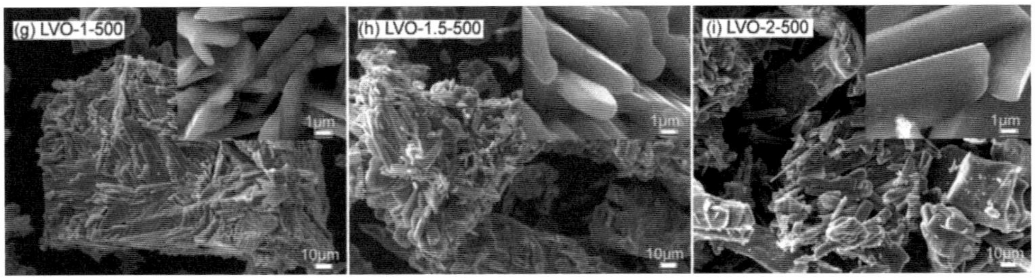

Figure 2. SEM images of prepared specimen (a) LVO-1-400, (b) LVO-1.5-400, (c) LVO-2-400, (d) LVO-1-450, (e) LVO-1.5-450, (f) LVO-2-450, (g) LVO-1-500, (h) LVO-1.5-500, (i) LVO-2-500.

The peak value of open circuit potential (OCP) is shown in figure 3(a). Combining with XRD analysis, we find that the peak value is seriously affected by the composition of material. The presence of $Li_{0.3}V_2O_5$ and V_2O_3 decrease the OCP of the cathode materials. When the dosage of citric acid is equal, the V_2O_3 is oxidized to LiV_3O_8 and $Li_{0.3}V_2O_5$ with the increasing temperature, so the peak value of OCP increase. At the same temperature, the V_2O_3 yield was inhibited and the content of LiV_3O_8 and $Li_{0.3}V_2O_5$ increase with the increasing dosage of citric acid. It has proved that the percentage of LiV_3O_8 is augmented which is according with the XRD analysis.

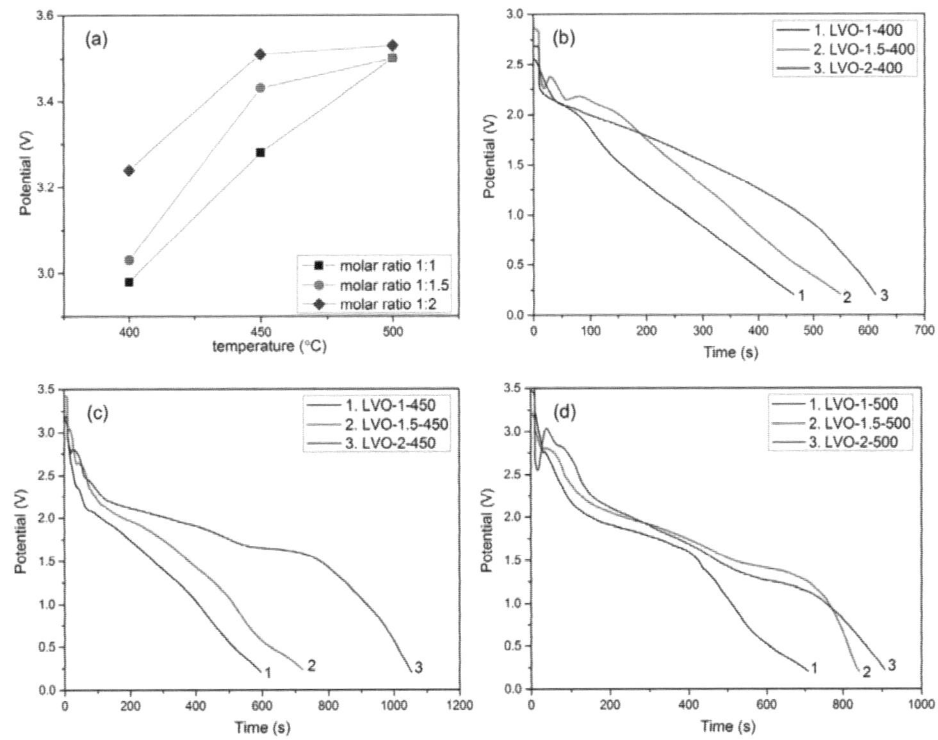

Figure 3. (a) The peak value of OCP, and the galvanostatic discharge curve of samples sintered at (b) 400°C; (c) 450°C; (d) 500°C.

The voltage-time curves of the samples at a current density of 100 mA/cm^2 are illustrated in figure 3(b), (c), (d) (the terminated voltage is 0.2 V). The specific capacity of the electrode material can be calculated according to the current density and the discharge time. Analyzing the discharge curves and specific capacity, we find that the two specimens, LVO-2-450 and LVO-1.5-500, have stable discharge platform which can be used as cathode material of thermal battery. They have a main discharge plateaus in the potential range of 1.65–2.2 V (maintaining for 500~700 s), and separately delivers a high initial discharge capacity of 312.2 mAh/g and 241.5 mAh/g at a current density of 100 mA/cm^2. Compared with the traditional cathode of FeS_2 material, the LiV_3O_8 material exhibits higher discharge potential and shorter working time, which can be used as promising candidates for both medium-high voltage and relatively low currents in thermal batteries applications.

4. Conclusion

The lithium vanadium oxide powder were synthesized via sol–gel process with citric acid as chelating agent. The results show that the samples exhibits better electrochemical properties with increasing dosage of citric acid. The increasing citric acid inhabits the produce of V_2O_3 and accelerates the emergence and growth of LiV_3O_8 and $Li_{0.3}V_2O_5$ grains. While the molar ratio of total metal ions to citric acid is 1:2, the as-synthesized samples exhibits better electrochemical performance.

Acknowledgements
This work was supported by the National Natural Science Foundation of China (51302312, 51502341).

References

[1] Z J Wu and Y Zhou 2012 Effect of Ce-doping on the structure and electrochemical performance of lithium trivanadate prepared by a citrate sol–gel method *J. Power. Source.* **199** 300–307
[2] K Kim, S H Parka, T H Kwon, J E Park, H Ahn and M J Lee 2013 Characterization of Li–V–O nanorod phases and their effect on electrochemical properties of $Li_{1+x}V_3O_8$ cathode materials synthesized by hydrothermal reaction and subsequent heat treatment *Electrochim. Acta.* **89** 708–716
[3] L A D Picciotto, K T Adendorff, D C Liles and M M Thackeray 1993 Structural characterization of $Li_{1+x}V_3O_8$ insertion electrodes by single-crystal X-ray diffraction *Solid State Ionics* **62** 297–307
[4] A Q Pan *et al.* 2011 Nanosheet-structured LiV_3O_8 with high capacity and excellent stability for high energy lithium batteries *J. Mater. Chem.* **21** 10077–10084
[5] D Q Wang, L Y Cao, J F Huang and J P Wu 2012 Synthesis and electrochemical properties of submicron sized sheet-like LiV_3O_8 crystallites for lithium secondary batteries *Mater. Lett.* **71** 48–50
[6] L Liu 2009 An Investigation on the Synthesis and Properties of LiV_3O_8 Cathode Material for Lithium-ion Battery (Tianjin, China: Nankai Universuty Press)
[7] S Sarkar, H Banda and S Mitra 2013 High capacity lithium-ion battery cathode using LiV_3O_8 nanorods. *Electrochim. Acta.* **99** 242–252
[8] D Q Wang, L Y Cao, J F Huang and J P Wu 2013 Effects of different chelating agents on the composition, morphology and electrochemical properties of LiV_3O_8 crystallites synthesized via sol–gel method *Ceram. Int.* **39** 3759–3764
[9] M Shui, W D Zheng, J Shu, Q C Wang, S Gao, D Xu, L L Chen, L Feng and Y L Ren 2013 Synthesis and electrochemical performance of $Li_{1+x}V_3O_8$ as cathode material prepared by citric acid and tartaric acid assisted sol-gel processes *Curr. Appl. Phys.* **13** 517–521
[10] D M Cao, Z Y Li and K C Zhou 2005 Study on cathodical discharge behaviors of $Li_{1+x}V_{3-y}M_yO_8$ (M=Mo, P) at elevated temperature *Rare Metal Mater Eng.* **34** 1550–1553
[11] Z Y Li, B Y Huang, C F Tang, Z J Liu and X H Qu 2001 Synthesis of LiV_3O_8 by sol-gel method and its cathodical discharge behavior at 500°C *Funct. Mater.* **32** 181–183

Influence of Mn contents in 0Cr18Ni10Ti thin wall stainless steel tube on TIG girth weld quality

Bo Liu

CNNC Jianzhong Nuclear Fuel Co., Ltd., Sichuan 644000, China

E-mail: liu2china3@163.com

Abstract. Three kinds of cold worked 0Cr18Ni10Ti thin wall stainless steel tubes with the manganese contents of 1.27%, 1.35% and 1.44% and the cold worked 0Cr18Ni10Ti stainless steel end plug with manganese content of 1.35% were used for TIG girth welding in the present investigation. The effect of different manganese contents in stainless steel tube on weld quality was studied. The results showed that under the same welding conditions, the metallographic performance of the girth weld for the thin wall stainless steel tube with the manganese element content 1.44% welded with end plug was the best. Under the appropriate welding conditions, the quality of the girth weld increased with the increase of the manganese content till 1.44%. It was found that in the case of the Mn content of 1.44%, and under the proper welding condition the welding defects, such as welding cracks were effectively avoided, and the qualified weld penetration can be obtained.. It is concluded that the appropriate increase of the manganese content can significantly improve the TIG girth weld quality of the cold worked 0Cr18Ni10Ti stainless steel tube.

Keywords: thin-wall stainless steel tube, Tungsten-arc Inert-Gas (TIG) welding, metallographic properties

1. Introduction

Austenitic stainless steel has excellent corrosion resistance and high temperature mechanical properties, and the radiation performance is stable. Therefore, it is widely applied for the parts of the nuclear industry, such as the reactor vessel, reactor components, cladding material of PWR associated core components and rods. Especially, the domestic developed austenitic stainless steel has been successfully applied in Qinshan nuclear power plant 300MW PWR as fuel cladding materials.

For a long time, in the manufacture of nuclear fuel elements the cracks and other welding defects easily generate during the tungsten electrode argon arc welding of girth seam for cold worked 0Cr18Ni10Ti thin-wall stainless steel, to produce which seriously affects the quality of product manufacturing. In order to solve the problems in the welding of thin-wall stainless steel tube, the effect of manganese content on the weld quality is discussed in detail in this study. The effect of manganese on the welding quality was analyzed by means of the tungsten electrode argon arc welding tests of three kinds of 0Cr18Ni10Ti thin wall stainless steel tube and the end plug.

2. Experimental

2.1. Welding requirements

The tungsten electrode argon arc welding was used to weld the three kinds of cold worked 0Cr18Ni10Ti thin wall stainless steel tube with the manganese contents of 1.27%, 1.35% and 1.44% and the cold worked 0Cr18Ni10Ti stainless steel end plug with manganese content of 1.35%. The weld and the heat affected zone shall not have cracks and other defects, only to allow the existence of a diameter of less than or equal to 0.25mm of the dispersed porosity [1]. The effective penetration should be no less than 90% of the minimum theoretical wall thickness of the cladding tube, and the melting depth should not be more than 1mm, the melting zone, the heat affected zone and the microstructure of the base metal. The tensile strength of welded joint shall not be lower than the lower limit specified by the tube. X ray inspection was used for metallographic examination.

2.2. Welding techniques

Stainless steel tube and end plug are both made of 0Cr18Ni10Ti. The stainless steel tube and the end plug are connected together by tungsten electrode argon arc welding, and the structure is schematically shown in figure 1. The outer diameter of stainless steel tube is 10±0.03mm, the wall thickness is 0.48mm to 0.56mm, and the surface diameter of end plug is 9mm. The interference fit between the end plug and the stainless steel tube and the amount of interference shall be less than or equal to 0.015mm. Arc stability of non-melting tungsten electrode argon arc welding can be stable under the condition of less than 10A and the welding process is nothing more than splashing, weld forming beautiful. The model is suitable for the welding of cold worked stainless steel tubular material. Due to the thin wall thickness of stainless steel tube, the technical requirements of the weld penetration is not less than the minimum theoretical wall thickness of the stainless steel tube, therefore the TIG welding method is used to manufacture products.

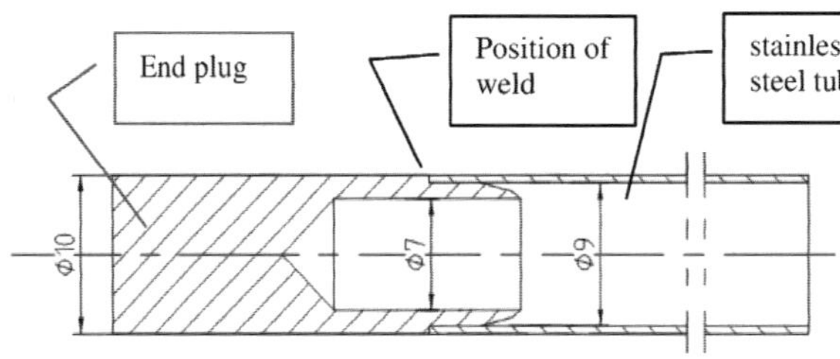

Figure 1. Schematic diagram of welding structure.

The selection of the welding parameters is important for the welding of thin-walled stainless steel material. It has been found that although TIG welding has the advantages for the welding of thin-wall stainless steel material, the tensile strength of weld joints, weld appearance, weld intergranular corrosion resistance can usually meet the requirements, however, in the inspection of the metallographic properties of weld, the welding cracks, lack of penetration and other defects have been often observed if the welding process is not appropriate. If the welding parameters are smaller, although the probability of crack can be reduced, it is easy to produce the problem of low penetration and low tensile strength. On the other hand, it is possible to increase the weld penetration depth, but the probability of crack is very high. Therefore, the actual production of the choice of welding parameters is very small. In the process of welding, the welding defects can be caused by any fluctuation, which seriously affects the quality of the products and the yield of the products.

2.3. Composition of the welding parts material

The stainless steel tube and the end plug used in the product are made of 0Cr18Ni10Ti stainless steel. The main chemical composition is listed in table 1. The addition of Manganese can improve the mechanical strength and medium temperature performance of the material, in addition to increase the solubility of N in metallographic phase. However, the effect of manganese content on the weld properties has been rarely studied. In order to solve this problem, considering the defects such as the welding cracks, no penetration and so on, the influence of the manganese content on the metallographic properties of the welded joint the stainless steel is studied.

Table 1. Chemical composition of stainless steel tube (mass fraction,%).

C	Si	P	S	Mn	Ni	Cr	Ti	N	Co
≤0.08	≤1.00	≤0.035	≤0.020	≤2.00	9.00~ 12.00	17.00~ 19.00	≥5× C	≤0.037	≤0.05

2.4. Treatment for the welding parts

The end plug shall be cleaned by ultrasonic in NaOH solution,. After cleaning, the end plug is dried in the electric heating oven with drying temperature control in the temperature range of 80°C to 100 °C for 30 - 60min, followed by wiping with the silk cloth.

The welding equipment with DC pulsed TIG welding power, and the output current is between 5A ~ 80A. The welding parameters for test are shown in table 2. The welding quality was examined by X-ray test and metallographic examination.

Table 2. Parameters of welding test.

Welding current (A)	Rotary speed of workpiece (r/min)	Welding time (s)	Cooling time (s)
45	12	5	10
50	12	5	10
55	12	5	10
60	12	5	10
65	12	5	10

3. Results Discussion

3.1. Effect of welding current on weld quality

Table 3. Effect of welding current on weld quality.

Mn content in stainless steel tube	Welding current	Quantity of sample	Weld appearance quality	X-Ray inspection	Metallographic quality of weld
1.27%	45A	2	ok	ok	lack of penetration
	50A	2	ok	ok	1 sample is ok,1 sample is lack of penetration
	55A	2	ok	ok	ok
	60A	2	ok	ok	1 sample is ok,1 sample is crack
	65A	2	ok	ok	ok
1.35%	45A	2	ok	ok	lack of penetration
	50A	2	ok	ok	ok
	55A	2	ok	ok	1 sample is ok,1 sample is lack of penetration
	60A	2	ok	ok	1 sample is ok,1 sample is crack
	65A	2	ok	ok	1 sample is ok,1 sample is crack
1.44%	45A	2	ok	ok	lack of penetration
	50A	2	ok	ok	ok
	55A	2	ok	ok	ok
	60A	2	ok	ok	ok
	65A	2	ok	ok	ok

The effect of welding current on the welding quality for three kinds of stainless steel pipe with the end plug is shown in table 3. The results in table 3 indicate that the girth weld of stainless steel pipe with 1.44% manganese content shows the best welding quality, while that of the stainless steel tubes with the manganese contents of 1.27% and 1.35% are not satisfied.

3.2. Discussion on weld metallographic properties

In technical requirements, it is specified that the effective penetration of weld shall not be less than 90% of theoretic wall thickness of cladding tube, while the penetration not more than 1mm; and the cladding the minimum theoretic wall thickness is 0.48mm and the effective penetration is acceptable when it is between 0.423mm and 1.0mm after welding cladding tube and end plug. Figure 2 shows the weld metallographic examination result, the weld penetration is increased with the welding current increasing, especially in the small current, three kinds of manganese content of different thin wall stainless steel pipe and the end plug weld penetration increase significantly, when the current is increased to 55A and more high numerical, weld penetration increases gently, tends to steady state, maintaining 0.513 to 0.545mm.

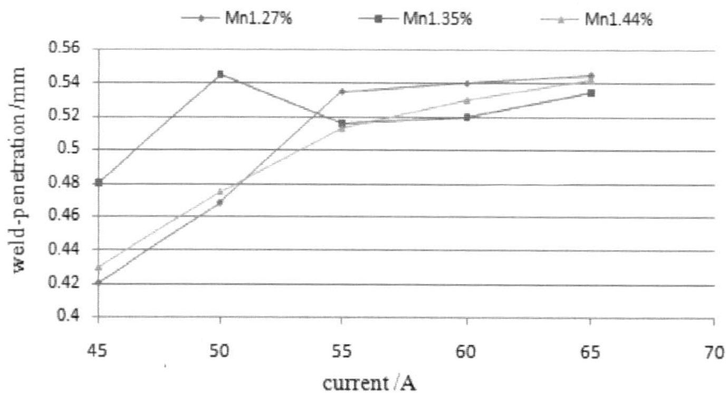

Figure 2. Welding penetration of three kinds of thin-wall stainless steel with different manganese contents with the welding current.

The typical weld penetration is shown in figure 3. Through the in-depth analysis of the microstructure of weld seam it is found that the partial values of the weld penetration meet 0.432 ~ 1.0mm, but no molten metal element in the stainless steel tube and the end plug joint root, forming a cavity, the weld penetration reaches 0.432 ~ 1.0mm in stainless steel tube and the end plug joint by stainless steel tube side, which leads to weld penetration index does not meet the requirements, leading to weld lack of penetration. Such phenomenon in the stainless steel tubes with manganese contents 1.27% and 1.35% is easy to cause lack of penetration when welding, but for the tube with manganese content 1.44%, the lack of penetration takes place only in the welding current 45A, due to the weld penetration smaller than the smallest penetration index 0.432mm.

In the process of girth welding cold worked 0Cr18Ni10Ti thin-wall stainless steel tube and end plug, it is very easy to have cracking in the weld. If the welding material is given, the welding energy plays a decisive role for the welding crack, while the welding energy is given, and the trend generating crack is different for the thin-wall stainless steel tube with different manganese contents. According to the test results, when girth welding thin-wall stainless steel tube with 1.27% manganese content and end plug within lower energy range, only incomplete penetration defect takes place and transverse and longitudinal cracks occur in the weld root. For welding thin-walled stainless steel tube with 1.35% manganese content and end plug, the trend of weld cracking is the same as that of the tube with 1.27% manganese content. When welding the thin wall stainless steel tube with 1.44% manganese content, the probability generating weld crack is much smaller. Only incomplete penetration takes place with smaller welding energy and no crack is produced with a larger range of welding energy.

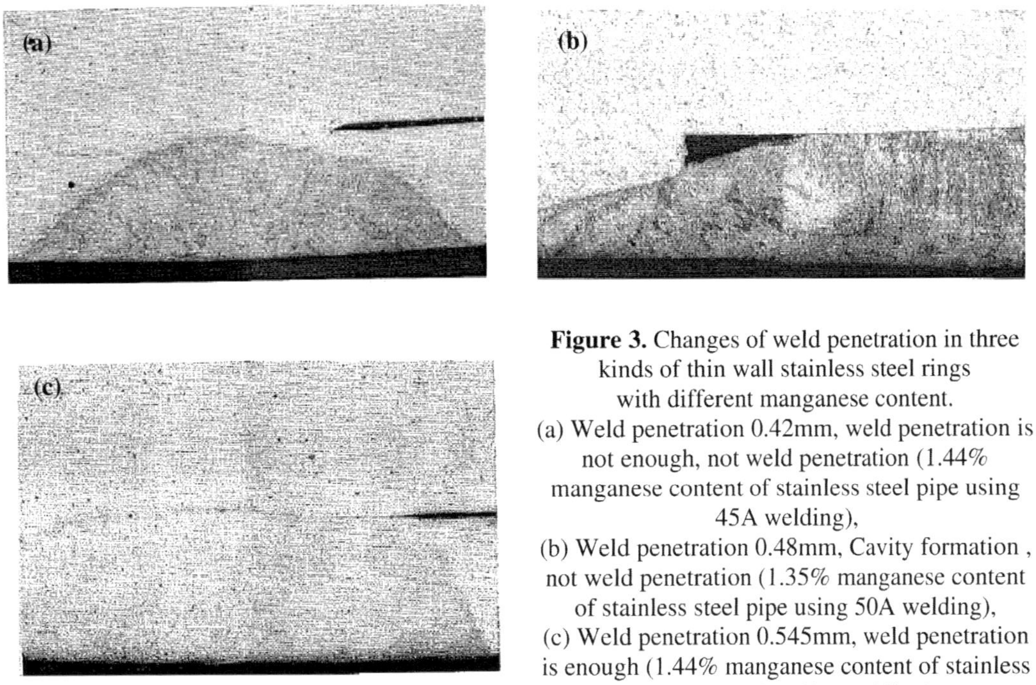

Figure 3. Changes of weld penetration in three kinds of thin wall stainless steel rings with different manganese content.
(a) Weld penetration 0.42mm, weld penetration is not enough, not weld penetration (1.44% manganese content of stainless steel pipe using 45A welding),
(b) Weld penetration 0.48mm, Cavity formation , not weld penetration (1.35% manganese content of stainless steel pipe using 50A welding),
(c) Weld penetration 0.545mm, weld penetration is enough (1.44% manganese content of stainless steel pipe using 60A welding).

When welding the thin wall stainless steel tubes with 1.27% and 1.35% manganese contents and end plugs by 60-65A welding current, the weld crack is very easy to generate. When welding the thin wall stainless steel tube with 1.44% manganese content and end plug by 60-65A welding current, no weld cracking generates. So, it shows that the different Mn content has an effect on the occurrence of welding crack. Under the condition that the content of the chemical composition of 0Cr18Ni10Ti thin wall stainless steel tube is less than or equal to 2%, the Mn content increased to 1.44%, which will significantly reduce the probability of cracking defect in the weld. Fig. 4 is a typical photo of welding crack.

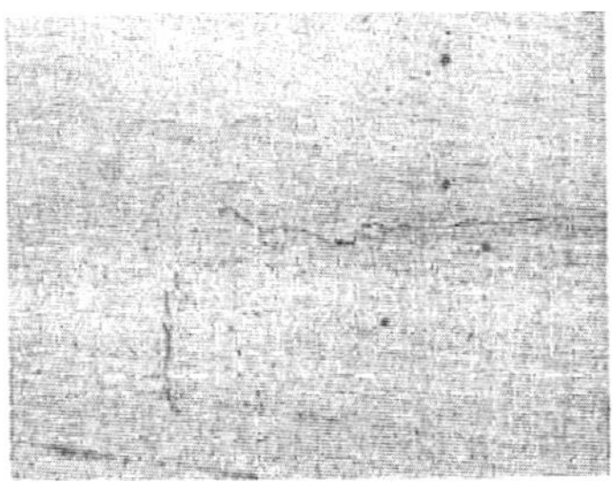

Figure 4. Weld crack defect.

Figure 5. Weld appearance.

3.3. Weld appearance

The typical weld appearance is shown in figure 5. The appearance of thin-wall stainless steel tubs with 1.27%, 1.35% and 1.44% manganese contents and end plugs welded by the same welding energy is equivalent, with the smooth, silver-white and about 3mm uniform width weld and free of any superficial crack defect.

3.4. Weld X - ray examination

In order to check the existence of incomplete penetration, incomplete fusion and the porosity of diameter greater than 0.25mm, the whole girth weld is examined three times by X ray at 120 degree. The X-ray examinations show that no presence of porosity with diameter greater than 0.25mm is observed in the girth welds of the thin wall stainless steel tubes with the three manganese contents and end plugs. Test shows no porosity of diameter greater than 0.25mm is generated when welding the thin-wall stainless steel tubes with the different manganese contents and end plugs.

4. Conclusion

The content of manganese in the 0Cr18Ni10Ti thin-wall stainless steel tubes has a significant effect on the quality of the welds between welding of the stainless steel tube and the end plug. The 1.44% manganese content in thin-wall stainless steel tube is ideal for performing girth welding to end plug. Therefore in order to meet the welding quality requirement, the TIG welding current shall be controlled in the range of 50 - 65A and the Mn content in thin-wall stainless steel tube shall be 1.44%.

References

[1] Test method for weld seam of PWR fuel rods Metallographic examination and X ray inspection GB/T 11809

Microstructure and corrosion behavior of a Zr-Sn-Nb-Fe-Cu-O alloy fabricated by α+β quenching processing

Liangyu Chen [1,2], Lina Zhang [1,*], Linjiang Chai [3], Wenlong Yang [4], Liqiang Wang [2] and Weijie lu [2]

[1] School of Mathematics and Science, Jiangsu University of Science and Technology, Zhenjiang, Jiangsu 212003, China
[2] State Key Laboratory of Metal Matrix Composites, Shanghai Jiao Tong University, Shanghai 200240, China
[3] College of Materials Science and Engineering, Chongqing University of Technology, Chongqing 400054, China
[4] Department of Applied Science, Harbin University of Science and Technology, Harbin 150080, China

E-mail: * lina-zhang@outlook.com

Abstract. In this work, the microstructure of a Zr-Sn-Nb-Fe-Cu-O alloy fabricated by α+β quenching processing (ABQ sample) was investigated by the combined utilization of scanning electron microscopy, electron backscatter diffraction and transmission electron microscopy. The results showed that the polygonal grains evenly distributed in ABQ sample and triangular grains distributed at triple junctions of polygonal grains with densely second phase particles (SPPs). The textures of $<0002>$ directions tilted approximately 25° from normal direction and the rolling direction spreading along the $<11\bar{2}0>$-$<10\bar{1}0>$ were found in the sample, which was also detected in the regularly cold rolled and annealed sample. The occurrence of corrosion kinetics transition of ABQ sample was sight earlier than that of RAX sample. Due to a higher solid solubility of β-Zr, SPPs re-dissolved into β-Zr during α+β annealing and precipitated out afterward in those areas. Finally, discrepant corrosion rate of metal matrix and SPPs led to the formation of protrusions of matrix, which could concentrate stress and generate cracks in the oxide.

Keywords: zirconium alloys, microstructure, corrosion, α+β quenching, second phase particles

1. Introduction

Because of low thermal neutron absorption cross-section, excellent waterside corrosion resistance, and appropriate mechanical properties, zirconium alloys have been successfully utilized as structural materials in nuclear industry [1-7]. The corrosion resistance of zirconium alloys has been considered as one of the most important properties to control the performance and the safety of nuclear reactors [8-10]. During corrosion, a thin oxide film forms on the metal surface by reacting with coolant [10]. This oxide can protect metal from the ingression of oxidant. Thus, the corrosion rate decreases with the ongoing of corrosion [9]. Once the oxide reaches to a certain thickness, it splits into two layers and the corrosion kinetic transition occurs [11]. Then the corrosion resistance of alloys degrades.

The characteristics of oxide film are significantly influenced by the microstructures of alloys which are based on the processing procedure [12-15]. Ly et al. [15] suggested that a "soft" metal substrate could release the in-plane stress of oxide via creep, thus recrystallized Zricaloy-4 possessed better corrosion resistance. Park et al. [16, 17] found that HANA-4, HANA-6 and Zr–1.1Sn–0.2Nb–0.35Fe–0.15Cr–0.1Cu (in wt.%) alloys with lower final annealing temperature had better corrosion resistance and lower oxide growth. Kim et al. [14] studied the microstructure of Zr–0.4Sn–1.5Nb–0.2Fe–0.1Cr fabricated by α/β annealing after β-quenching and discovered that the increased corrosion resistance was caused by a decreased the supersaturated Nb content in the matrix by formation of β-Zr. Chen et al. [18, 19] also presented the quality of oxide is highly depend on the microstructure which is on the basis of processing procedures. Therefore, to optimize performance of zirconium alloys, it is necessary to explore new processing procedure.

In our previous work, we were successful in improving the corrosion resistance of two Zr-Sn-Nb alloys by the second β-quenching processing [18-20]. In this work, α+β quenching processing was added into the regularly cold rolled and annealed processing procedure, and then the microstructure and corrosion behavior of as-received sample were investigated. Autoclave corrosion tests were conducted up to 200 days and weight gain changes were observed as the measure of corrosion resistance of alloy. To study the corrosion mechanism, oxide cross-sectional microstructure of sample was also characterized. This work aimed to under the microstructure and corrosion behavior of a Zr-Sn-Nb-Fe-Cu-O alloy fabricated by α+β quenching processing.

2. Materials and methods

2.1. Samples preparation

Zr-Sn-Nb-Fe-Cu-O alloy was prepared by vacuum arc re-melting for three times to obtain chemical homogeneity. The alloy contains 0.55% Sn, 0.32% Fe, 0.35% Nb, 0.05% Cu, and 0.079% O and with Zr as the balance (in wt. %), where the contents of Sn, Fe, Nb and Cu were analyzed by inductively coupled plasma atomic emission spectrometry and the content of O was determined by Nitrogen/Oxygen Exterminator (TC600, LECO Corporation). The disk-shape ingot was 110 mm in diameter and 50 mm in height. After forged at 1000 °C, the height of ingots was reduced to 10 mm and then the ingots were β-quenched at 1050 °C. Afterward the sample was processed by the regularly rolled and annealed procedure [19]. Before the sample was rolled into 0.6 mm, the intermediate annealing was replaced by α+β quenching. This sample is denoted as ABQ in current work and was compared with the normal regularly cold rolled and annealed processing procedure.

2.2. Microstructure analysis

Microstructures were characterized using a FEI Sirion 200 field emission gun scanning electron microscope (SEM) and a JEOL-2100F transmission electron microscope (TEM) equipped with an energy dispersive X-ray spectroscopy (EDS) detector with an accelerated voltage of 200 kV. Samples for SEM observations were polished according to standard metallographic techniques and then etched in the solution containing H_2O : HNO_3 : HF of 9 : 9 : 2 (in vol.%). Specimens for TEM observations were prepared using a twin-jet polishing with a solution (C_2H_5OH : $HClO_4$ = 9 : 1 in vol.%) at a voltage of 20 V and a temperature of −30 °C. In order to obtain the oxide cross-section fracture surface, the corroded samples were etched in the solution for 15 min to dissolve the metal substrates. Then, the oxide films extruding the metal substrates were gently fractured using a tweezer. The cross-sectional oxides and their fracture surface were coated with a gold film to improve the electrical conductivity before SEM observation.

Electron back scattering diffraction (EBSD) measurements were carried out with an AZtec HKL Max system, installed on a NOVA NanoSEM 230 scanning electron microscope. Samples prepared for EBSD measurements were polished by colloidal silica after etching. EBSD patterns were analyzed by the channel 5 software. The grain boundary angle condition used to define a grain was fixed to 10°.

The channel 5 software can also identify the degree of recrystallization *via* comparing the collected crystallographic data with the existing data in database.

2.3. Corrosion tests

Samples used for autoclave corrosion testing were 15 mm × 20 mm in size and chemically polished using a pickling solution (H_2O : HNO_3 : H_2SO_4 : HF = 3 : 3 : 3 : 1 in vol.%) at 30~50 °C and finally cleaned by deionized water. The corrosion resistance of the sample was evaluated by measuring their weight gain per unit surface area in relation to the exposure time (i.e. 3, 7, 14, 42, 70, 100, 130, 160, and 200 days). The corroded samples were named after their procedure number and exposure time (e.g. sample 1#-100 stands for the sample 1# after 100 days exposure, and so on). According to the ASTM-G2/88 (Standard Test Method for Corrosion Testing of Products of Zirconium, Hafnium, and Their Alloys in Water at 680°F or in Steam at 750 °F) [21], corrosion tests were conducted with a static isothermal autoclave in deionized water and 0.01 mol/L LiOH aqueous solution at 360 °C under a saturation pressure of 18.6 MPa, respectively.

3. Results and discussion

3.1. General microstructural characteristics

Figure 1 shows the general microstructural characteristics of the sample. Two shapes of grains could be found. Polygonal grains are evenly distributed and have high proportion in the sample. Triangular grains distribute at triple junctions of polygonal grains with densely SPPs as indicated by white solid arrows (figure 1(a)). Figure 1(b) is the magnified image of the dash line rectangle in figure 1(a). It is rather clean in the interior of the polygonal grains (α-grains), none of SPPs can be discovered.

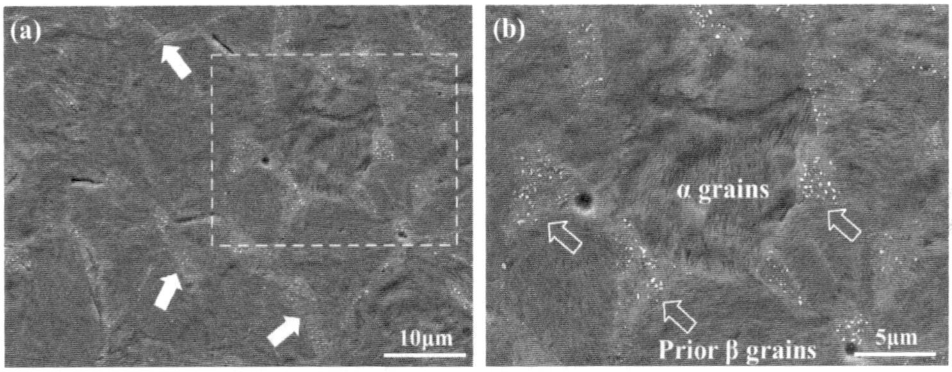

Figure 1. SEM micrographs of (a) the experimental sample, (b) is the magnified micrograph for the square regions marked in (a). The SPPs located areas are indicated as white solid arrows. The prior β grains are indicated by white hollow arrows.

Figure 2 presents the TEM image for SPP in the sample studied. The particle possess face-center cubic (FCC) structure via the selected area diffraction pattern (SADP) analysis (the inset in figure 2). The chemical compositions of SPPs obtained by EDS are in a range of 3.0–8.7% Nb, 21.6–25.7% Fe, 3.5–5.9% Cu (in wt.%) and the balance is Zr. Therefore, the particles are identified as C15 type $(Zr, Nb, Cu)_2Fe$ precipitates, which are coincide with our previous work [19, 22]. The type of SPPs has not been changed in the event of more alloying elements concentrated in the triangular grains.

β-Zr is produced at triple junctions of polygonal grains when sample is annealed at α+β region [23]. Due to larger solid solubility of β-Zr, SPPs which precipitate in the former processing steps dissolve into β-Zr grains again [24, 25]. Thus the contents of solid solution alloying elements are unbalance in different grains when α-Zr and β-Zr grains coexist. After quenching, solid solution alloying elements are kept in the prior β-Zr grains (transformed α-Zr grain). However, solid solubility of prior β-Zr is much lower than that of β-Zr, SPPs precipitate once again at afterward processing steps. On the other hand, it is speculated that the content of alloying elements in α-Zr grains is no more than solid solution limit of α-Zr since no SPP redissolves into them while the sample is annealed at α+β region. Therefore, none of SPPs precipitate out after final annealing. This is the main reason accounting for areal distribution of SPPs.

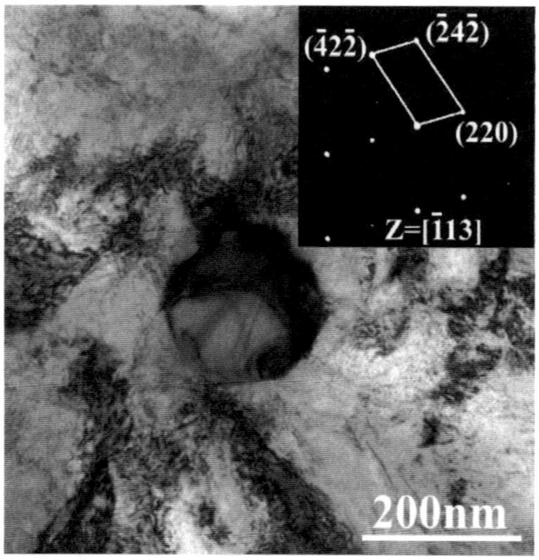

Figure 2. TEM images of the experimental sample. The inset is the SADP of SPP.

3.2. EBSD result and texture
Figure 3 reveals EBSD inverse pole figure (IPF) map of sample. This result is good agreed with the SEM observation of microstructure. The polygonal grains with grain size of about 20~40 μm are surrounded by imperfectly identified regions. These regions are triangular grains with intensively SPPs in figure 1. Due to the small size of SPPs, EBSD analysis could not work well in distinguishing them. Nevertheless, the imperfectly analysis confirms the features of microstructure observed by SEM, and also demonstrates that the SPPs are areal distribution.

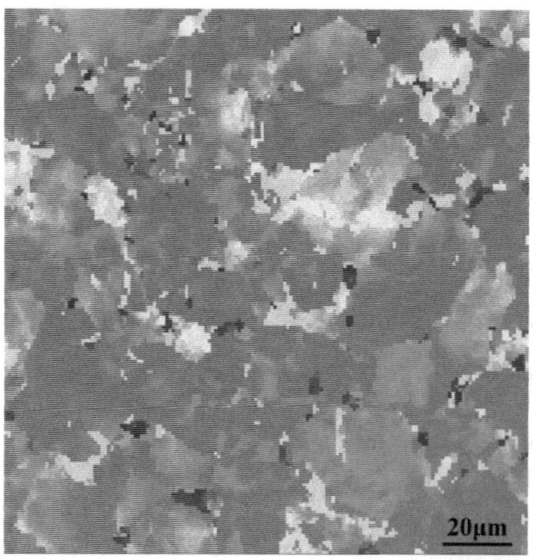

Figure 3. EBSD inverse pole figure maps for the experimental sample.

Figure 4 presents the EBSD inverse pole figure of sample. It can be easily found that strongly $<0002>$ directions are tilted approximately 25° from the normal direction of sample (ND) and the rolling direction (RD) tends to spread along the $<11\bar{2}0>-<10\bar{1}0>$ symmetry boundary of the stereographic triangle. Such kind of texture is always found in cold rolling or at a low temperature annealing (below 600 °C) after cold rolling zirconium alloys [19, 26-28]. In present work, cold rolling and a low temperature final annealing were carried out after α+β quenching. It implies that compared to the regularly cold rolled and annealed processing procedure, α+β quenching processing added into processing procedure before the last cold rolling and final annealing has not significant effect on the texture of sample. Therefore, the influence of textural variation on corrosion behavior of sample can be neglected. Furthermore, the quantitative relationship between texture and corrosion resistance has not been impeccably built up to now [28-30].

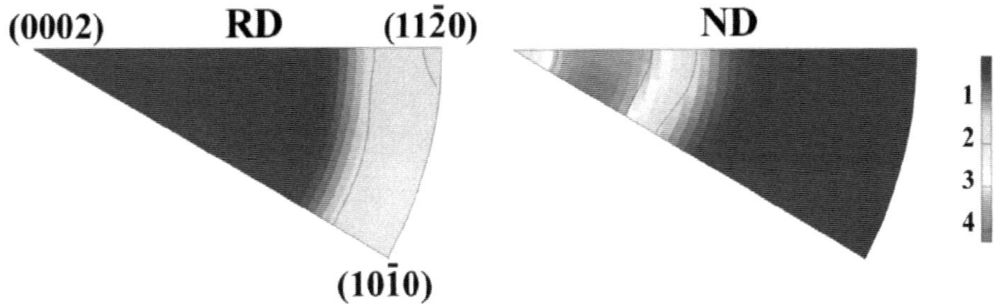

Figure 4. EBSD inverse pole figures for the experimental sample.

3.3. Corrosion behaviour

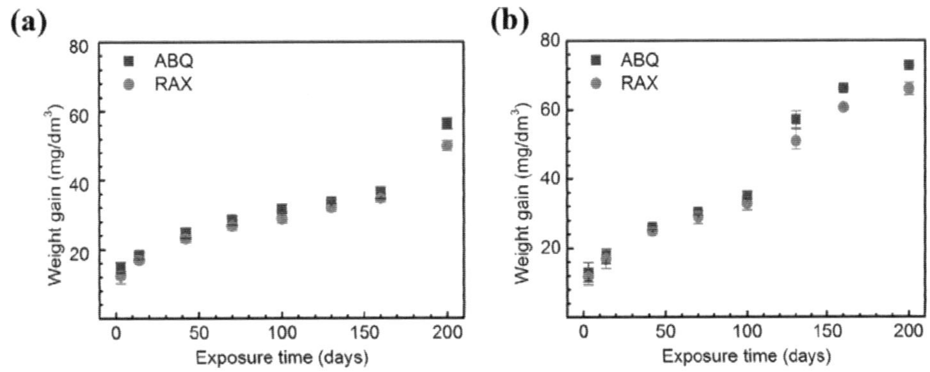

Figure 5. Autoclave corrosion weight gain profile of the experimental alloy (ABQ) and the regularly cold rolled and annealed sample (RAX) after 200 days at 360 °C/18.6 MPa in (a) deionized water and (b) 0.01 mol/L LiOH aqueous solution.

The results of corrosion tests are obtained as shown in figure 5, which are examined in deionized water and 0.01 mol/L LiOH aqueous solution at 360 °C under a saturation pressure of 18.6 MPa, respectively. The weight gains of the regularly cold rolled and annealed sample (denoted as RAX) at same experimental conditions are used to compare with the corrosion result of ABQ, which is published in our previous work [18]. As seen from figure 5, both samples have comparative weight gains before corrosion kinetics transition. This phenomenon means both samples have almost same corrosion rate at the initial stage of corrosion experiments. Once transition occurs, an obvious gap could be found between the weight gains of ABQ and RAX in both conditions. The corrosion resistance depends significantly on the transition point. Higher corrosion resistance is related to a longer pre-transition period [18, 31]. Hence, it indicates that the transition of ABQ takes place earlier than that of RAX. The corrosion resistance of ABQ is worse than that of RAX under the conditions of this work. Since the corrosion kinetics transition phenomenon is closely related to the oxide formed on the matrix [20], the cross-sectional microstructure of oxide of ABQ is thus investigated.

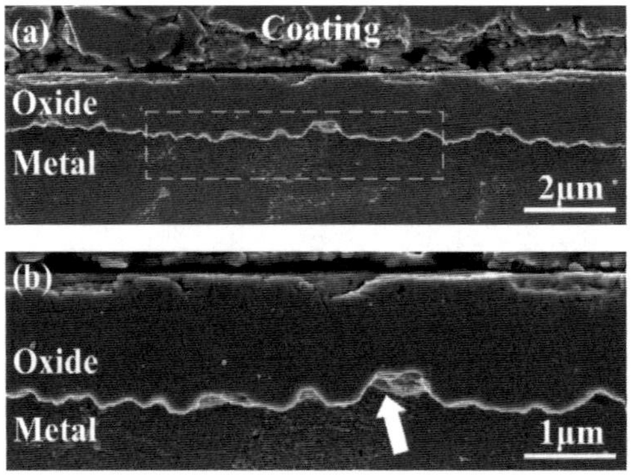

Figure 6. (a) SEM micrograph representative of the experimental alloy of oxide (cross-section) in deionized water, (b) is the magnified micrograph for the square regions marked in (a).

Figure 6 represents the cross-sectional microstructure of oxide of ABQ sample at 100 days in deionized water. The sample is still in pre-transition and a continuous oxide layer can be found. Beneath the oxide, the microstructure of metal matrix can be clearly observed in figure 6(a), which is as same as the description in figure 1. The undulated interface of oxide and metal matrix is produced by stress in oxide and matrix or diffusion of oxygen ions [18, 31-34], which is also observed in figure 6. Figure 6(b) is the magnified image of the dash line rectangle in figure 6(a). As shown by white solid arrow, the oxide is much thinner in this area. Below this area, densely located SPPs area is presented. It manifests that the corrosion of matrix is much slower in such area. Hence, discrepant corrosion rate of metal matrix and SPPs leads to the formation of protrusions of matrix. Meanwhile, such protrusions on matrix can lead to the redistribution of planar stress of oxide and trigger cracks in the oxide [20, 29, 31-33]. Therefore, the slower corrosion of matrix in densely located SPPs area may be the main reason accounting for the earlier corrosion kinetics transition of ABQ.

3.4. The possible corrosion mechanism

The precondition to obtain excellent corrosion resistance of zirconium alloys is uniform corrosion which means the corrosion rate is equivalent in the everywhere of metal surface [22, 36-38]. During inward diffusion of oxygen ions, the matrix gets oxidized. The corrosion rate has significant difference between metal matrix and precipitates [39], which depends on the nature of type of SPPs. Although SPPs can improve the corrosion resistance of zirconium alloys, they also play an important role in triggering cracks in the oxide due to their slower corrosion rate than that of matrix. More energy is needed for Fe and Nb to form oxide than Zr [40, 41]. C15 type precipitate is found in this work (figure 2), which was reported that has a slower corrosion rate than matrix [20, 39-41]. In the triangular grains area, SPPs are highly aggregated. These SPPs postpone the corrosion of alloy in such areas. Due to discrepant corrosion rate of metal matrix and SPPs, the thickness of oxide is uneven during corrosion. Therefore, protrusions form as shown in figure 6. Once protrusions form, stresses are apt to concentrate in these segments of oxide and then cracks are generated. It was reported that the cracks on the protrusions are the key factor influencing the occurrence of transition [20, 42]. Compared to the RAX sample the SPPs are evenly distributed [18], ABQ sample is liable to degrade earlier during corrosion tests. Hence, the corrosion resistance of ABQ sample is slightly worse than that of RAX sample. This work advances the understanding of applying appropriate processing procedure to keep evenly distribution of SPPs is an important prerequisite to ensure uniform corrosion of zirconium alloy.

4. Conclusions

In this paper, the microstructure of a Zr-Sn-Nb-Fe-Cu-O alloy prepared by vacuum arc re-melting and fabricated by α+β quenching processing is investigated by the combined utilization of SEM, EBSD and TEM. The corrosion behavior and morphology of cross-sectional oxide are also studied. Some key conclusions can be drawn as follows:

β-Zr is produced at triple junctions of polygonal grains when sample is annealed at α+β region. Most alloying elements concentrate at β-Zr grains due to their larger solid solubility. SPPs precipitate out from these areas in the afterward processing. Therefore, areal distribution of SPPs forms.

Strongly $<0002>$ directions are tilted approximately 25° from the normal direction of sample and the rolling direction tends to spread along the $<11\bar{2}0>$ - $<10\bar{1}0>$ symmetry boundary of the stereographic triangle. This kind of texture is always found in cold rolled or at a low temperature annealing (below 600 °C) after cold rolled zirconium alloys. It indicates that adding α+β quenching into processing procedure has no significant effect on the texture of sample.

The ABQ has an earlier corrosion kinetics transition compared to RAX. The reason is that the corrosion rate of SPPs is slower than that of matrix. Therefore, the corrosion rate is much slower in the densely SPPs located area. Result from non-uniform corrosion, protrusions which concentrating stress and generate cracks in the oxide forms on the matrix. Consequently, the corrosion resistance of ABQ is worse than that of RAX.

Acknowledgments

The authors would like to acknowledge financial support provided by National Science Foundation under Grant No: 51601075.

References

[1] H L Yang, Y Matsukawa, S Kano, Z G Duan, K Murakami and H Abe 2016 *J. Nucl. Mater.* **481** 117–124

[2] H Yang, J Shen, Y Matsukawa, Y Satoh, S Kano, Z Zhao, Y Li, F Li and H Abe 2015 *J. Nucl. Sci. Tech.* **52** 1162–1173

[3] Z N Yang, Y Y Xiao, F C Zhang and Z G Yan 2012 Effect of cold rolling on microstructure and mechanical properties of pure Zr *Mater. Sci. Eng. A.* **556** 728–733

[4] Z N Yang, F C Liu, F C Zhang, Z G Yan and Y Y Xiao 2012 Microstructural evolution and mechanical properties in Zr705 during the rolling process *Mater. Sci. Eng. A.* **544** 54–58

[5] L. Chai, B. Luan, K.L. Murty, Q. Liu, Effect of predeformation on microstructural evolution of a Zr alloy during 550–700° C aging after β quenching *Acta Mater.* 61 (2013) 3099–3109.

[6] L Chai, B Chen, Z Zhou, K L Murty, Y Ma and W Huang 2015 A special twin relationship or a common Burgers misorientation between α plates after β quenching in Zr alloy *Mater. Charact.* **104** 61–65

[7] L J Chai, B F Luan, S S Gao, J W Chen and Q Liu 2012 *J. Nucl. Mater.* **427** 274–281

[8] F Garzarolli, E Steinberg and H G Weidinger 1989 *ASTM STP* **1023** 202–212

[9] H J Beie, A Mitwalsky, F Garzarolli, H Ruhmann and H J Sell 1994 Zirconium in the nuclear industry *ASTM STP* **1245** 615–626

[10] A Yilmazbayhan, E Breval, A T Motta and R J Comstock 2006 *J. Nucl. Mater.* **349** 265–281

[11] L Chen, J Li, Y Zhang, L C Zhang, W Lu, L Zhang, L Wang and D Zhang 2015 *Corros. Sci.* **100** 651–662

[12] X Xie, J Zhang, M Yao, B Zhou, J Peng and X Liang 2014 *J. Nucl. Mater.* **451** 255–263

[13] A T Motta, A Yilmazbayhan, R J Comstock, J Partezana, G P Sabol, B Lai and Z Cai 2006 *ASTM STP* **1467** 205–216

[14] H G Kim, J H Baek, S D Kim and Y H Jeong 2008 *J. Nucl. Mater.* **372** 304–311

[15] A Ly, A Ambard, M Blat-Yrieix, L Legras, P Frankel, M Preuss, C Curfs, G Parry and Y Bréchet 2011 *J. ASTM STP* **1529** 1–29
[16] Y Z Liu, J Y Park, H G Kim and Y H Jeong 2010 *Mater. Chem. Phys.* **122** 408–416
[17] J Y Park, B K Choi, S J Yoo and Y H Jeong 2008 Corrosion and oxide properties of HANA alloys *J. ASTM STP* **1505** 471–485
[18] L Chen, J Li, Y Zhang, L C Zhang, W Lu, L Wang, L Zhang and D Zhang 2015 *Corros. Sci.* **100** 332–340
[19] L Chen, Q Zeng, J Li, J Lu, Y Zhang, L Zhang, X Qin, W Lu, L Zhang and L Wang 2016 *Mater. Des.* **92** 888–896
[20] L Chen, J Li, Y Zhang, L C Zhang, W Lu, L Zhang, L Wang and D Zhang 2015 *Corros. Sci.* **100** 651–662
[21] G P Sabol 2005 ZIRLO™—an alloy development success *J. ASTM STP* **1467** 3–24
[22] L Chen, J Li, Y Zhang, W Lu, L Zhang, L Wang and D Zhang 2016 *J. Nucl. Sci. Tech.* **53** 496–507
[23] Y Liu, W J Zhao, Q Peng, H Jiang and X T Zu 2008 *Mater. Chem. Phys.* **107** 524–527
[24] H Okamoto 1990 The Si-Zr (silicon-zirconium) system *Bulletin of Alloy Phase Diagrams* **11** 513–519
[25] H Kim, I Kim, B Choi, J Park, Y Jeong and K Kim 2010 *Corros. Sci.* **52** 3162–3167
[26] L Jiang, M T Pérez-Prado, P A Gruber, E Arzt, O A Ruano and M E Kassner 2008 *Acta Mater.* **56** 1228–1242
[27] L Jiang, O A Ruano, M E Kassner and M T Perez-Prado 2007 The fabrication of bulk ultrafine-grained zirconium by accumulative roll bonding *JOM.* **59** 42–45
[28] K Linga Murty and I Charit 2006 Texture development and anisotropic deformation of zircaloys *Prog. Nucl. Energy.* **48** 325–359
[29] M Parise, O Sicardy and G Cailletaud 1998 *J. Nucl. Mater.* **256** 5–46
[30] N Ni *et al.* 2012 How the crystallography and nanoscale chemistry of the metal/oxide interface develops during the aqueous corrosion of zirconium cladding alloys *Acta Mater.* **60** 7132–7149
[31] V V Likhanskii, T N Aliev, M Y Kolesnik, I A Evdokimov and V G Zborovskii 2012 *Corros. Sci.* **61** 143–147
[32] V Likhanskii and M Kolesnik 2014 On the evolution of wave structure at the metal/oxide interface during corrosion of Zr alloys *Corros. Sci.* **87** 416–420
[33] N Vermaak, G Parry, R Estevez and Y Bréchet 2013 New insight into crack formation during corrosion of zirconium-based metal-oxide systems *Acta Mater.* **61** 4374–4383
[34] P Tejland and H O Andrén 2012 *J. Nucl. Mater.* **430** 64–71
[35] D S Balint and J W Hutchinson 2003 Undulation instability of a compressed elastic film on a nonlinear creeping substrate *Acta Mater.* **51** 3965–3983
[36] C Li, R Zuo, Z Li, S Ying and B Shen 2004 *Thin Solid Films* **461** 272–276
[37] Y Hatano, K Isobe, R Hitaka and M Sugisaki 1996 Role of intermetallic precipitates in hydrogen uptake of Zircaloy-2 *J. Nucl. Sci. Tech.* **33** 944–949
[38] N Ni, S Lozano-Perez, J M Sykes, G D W Smith and C R M Grovenor 2011 *Corros. Sci.* **53** 4073–4083
[39] C Proff, S Abolhassani and C Lemaignan 2013 Corrosion behaviour of zirconium alloys and their precipitates – A mechanistic study *J. Nucl. Mater.* **432** 222–238
[40] E Werner and D R Gaskell 2008 *Introduction to the Thermodynamics of Materials (5th ed.)* (Taylor and Francis, New York and London)
[41] K Ohwada 1984 On the pauling electronegativity scales—II *Polyhedron* **3** 853–859
[42] J Wei, P Frankel, E Polatidis, M Blat, A Ambard, R J Comstock, L Hallstadius, D Hudson, G D W Smith and C R M Grovenor 2013 *Acta Mater.* **61** 4200–4214

Effect of microstructure and alloying elements distribution on welding quality of Zircaloy-4

Pengbo Ji, Yanhua Du*, Chunxia Chi and Hongbo Wang

China North Nuclear Fuel Co. Ltd. Inner Mongolia Baotou 014035 China

E-mail: *Corolin12345@163.com

Abstract. In order to understanding the appearance of the abnormal welding line during the CANDU fuel element end plug welding process, the analysis on the microstructure and the alloying elements distribution of three different batches of Zircaloy-4 rods, named as A753, B592 and C744, was carried out in the present study. The microstructure and localized composition of the alloys were studied by optical microscopy, X-ray diffractometer (XRD), high-resolution transmission electron microscopy (HRTEM) and energy disperse spectroscopy (EDS).The results showed that there was no obvious difference among the three specimens. However, it was found that there appeared etch pits in the middle area of A753 batch samples. And the grain size of welding qualified C744 batch sample was greater than the un-qualified samples of A753 and B592. According to HRTEM and EDS results, the second phase particles in the Zircaloy-4 were $Zr(Fe, Cr)_2$ phase with the Hexagonal Close-Packed (HCP) structure. The matrix belonged to α-Zr phase with the Hexagonal Close-Packed (HCP) structure. The second phase particles distributed inside the grains and at the grain boundaries of the three batches of the alloys. The size of the second phase particles in qualified sample C744 was around 200nm with the elements of Zr, Fe and Cr. However the size of the second phase particles in the un-qualified samples B592 and A753 was about 100nm, in which the elements of Fe and Cr could not be detected or only trace amount of Fe could be detected for part of the second phase particles. Therefore it may be concluded that the grain size, the second phase size and composition are the main factors to affect the welding quality of the alloy.

Keywords: Zircaloy-4, second phase particles, welding quality

1. Introduction

In the manufacture process of nuclear fuel elements, the Zirconium based alloys such as Zircaloy-2 and Zircaloy-4 are widely used as the structural material and fuel cladding material for the excellent performance in corrosion resistance and mechanical properties [1,2].As it is known that the preferable comprehensive performance has a direct relationship with the microstructure. In order to provide the reference for the improvement of technology, many literatures [3-10] study the microstructure of Zirconium based alloy prepared under different conditions. The microstructure analysis carried out with transmission electron microscope (TEM), scanning electron microscope (SEM), X-ray diffractometer (XRD), neutron diffraction and the electron backscatter diffraction(EBSD). Among these characterizing techniques, the TEM and XRD is mature and widely used in the microstructure analysis. For example, reference [3] studies the relationship between microstructure and corrosion behaviour of Zr-Nb alloy using the TEM and XRD. Reference [4] studies the microstructure of Zircaloy-4 tube under the condition of hot extrusion, cold rolling and annealing with TEM. At present the Zircaloy-4 is widely used as the end caps in Pressurized Heavy Water Reactor (PHWR) fuel elements.The abnormal welding line appeared in the detection of the end plug welding products of individual batches, the product will be judged to be substandard. In order to find out whether there are some differences between the qualified batch products and the unqualified batch products, the microstructure and composition of Zircaloy-4 are analysed by OM, HRTEM and EDS. It can provide the reference for the better application of Zircaloy-4 in the production.

2. Experimental

The material used in this study is Zircaloy-4 made in Canada. Three batches of Zircaloy-4 were selected and numbered C744, B592 and A753 respectively. The welding products were qualified of C744 batch. The welding products were unqualified of A753 and B592 batches. The chemical composition of the alloy is given in table 1. In order to find out the differences between three batches Zircaloy-4, Firstly, the macro-structure and the grain size were observed and measured by OM. The phase was tested by the XRD. The microstructure was studied by the F30 TecnaiG2 high resolution transmission electron microscopy. And EDS is chosen to analyse the composition of the second phase particles and matrix.

Table 1. Chemical composition of Zircaloy-4 (wt.%).

Element	Sn	Fe	Cr	Fe+Cr	Zr
Content	1.20~1.70	0.18~0.24	0.13	0.18~0.38	Balance

3. Results and Analysis

3.1. Optical metallographic study of Zircaloy-4

The macroscopic metallographic were carried out on the A753, B592 and C744 batches samples. Five samples for each batch were detected. There was no difference in the metallographic structure of the three batches samples, which was shown in figure 1. When the samples were observed at high magnification it was found that the samples of A753 batch was different from C744 and B592, as shown in figure 2. There were many pits on the surface of the sample, three of the five samples appeared this phenomenon, and the reason remains to be further studied.

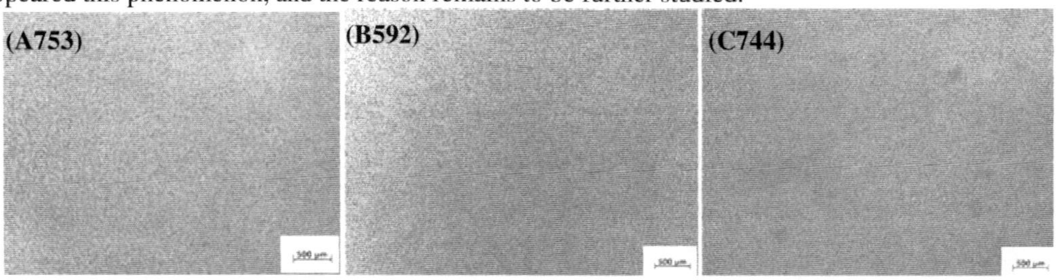

Figure 1. Microscopic metallographic photos of three batches of Zircaloy-4.

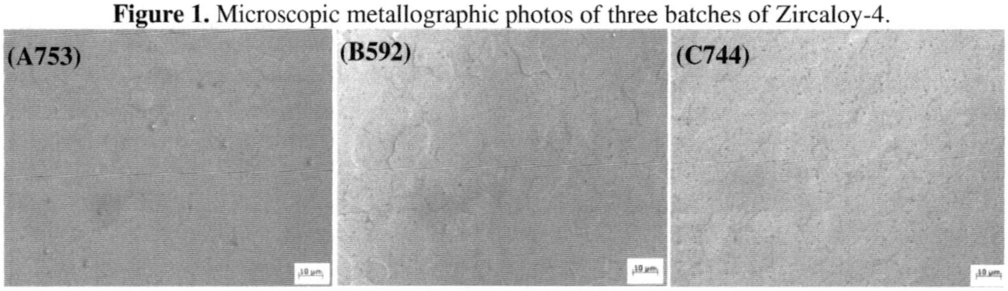

Figure 2. High magnification metallographic photos of three batches of Zircaloy-4.

IOP Conf. Series: Materials Science and Engineering 182(2017) 012020 doi:10.1088/1757-899X/182/1/012020

The grain size of the three batches samples was analyzed by the cut-off point method. Statistical analysis of the data of grain size was shown in table 2. It can be seen that the grain size of C744 batch samples was larger than that of B592 and A753 batch samples. The grain is shown in figure 3.

Table 2. Statistical data of grain size of three batches samples.

Sample number	Grain size (μ m)	average(μ m)
A753-1	9.48	
A753-2	9.51	9.24
A753-3	8.72	
B592-1	9.33	
B592-2	8.64	8.95
B592-3	8.90	
C744-1	9.20	
C744-2	10.01	9.94
C744-3	10.63	

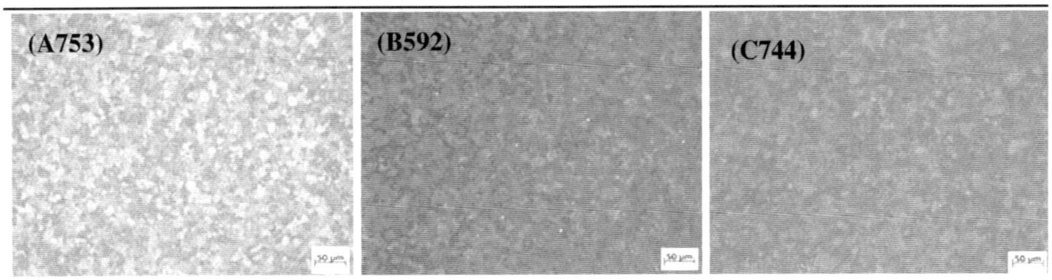

Figure 3. Grain size photographs of three batches samples.

3.2. *Phase analysis of Zircaloy-4*

Two samples of each batch were carried out on phase by XRD, as shown in figure 4. According to the query results of the PDF card the phase is α -Zr, The phase of three batches of Zircaloy-4 is the same.

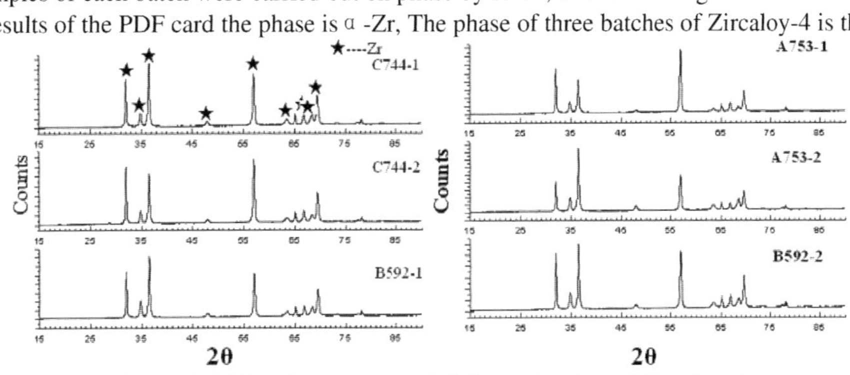

Figure 4. Diffraction patterns of different batches of Zircaloy-4.

3.3. *Analysis of microstructure and composition of Zircaloy-4*

The thin film samples were examined and analysed with high-resolution transmission electron microscopy. Figure 5 shows the TEM micrographs and EDS spectrum of Zircaloy-4 samples. From the results, it can be found that the second phase particles were distributed into the matrix and the main composition was Zr, Fe and Cr. The main composition of the matrix was Zr. Figure 6 presents the HRTEM and Fourier transform picture of the second phase particles and matrix.

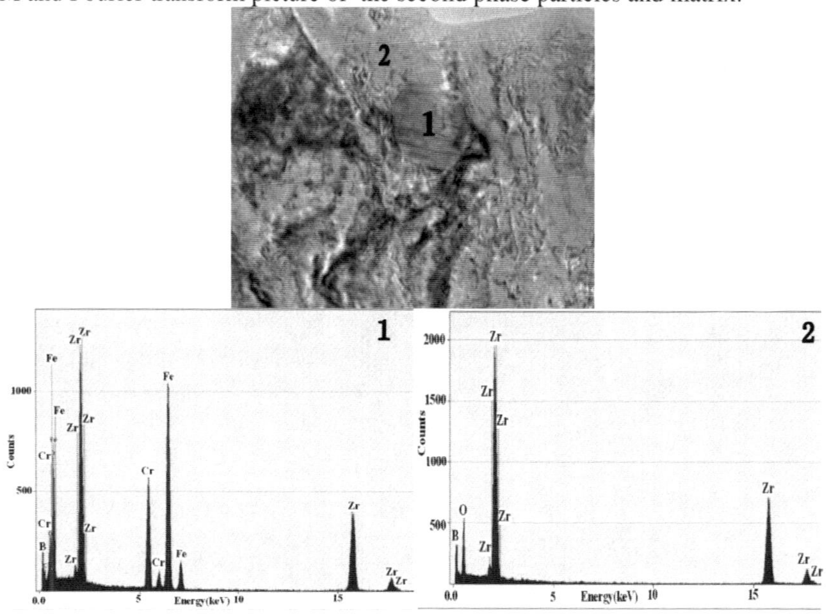

Figure 5. TEM micrographs and EDS spectrum of B592 batchZircaloy-4 samples.

Table 3. Statistics table for EDS data of the second phase particles and matrix in figure 5.

Position	Element	wt.%	at.%	Position	Element	wt.%	at.%
	Cr	14.79	19.76		Cr	0.00	0.00
TEM5-1	Fe	31.90	39.66	TEM5-2	Fe	0.23	0.38
	Zr	53.28	40.56		Zr	99.75	99.6

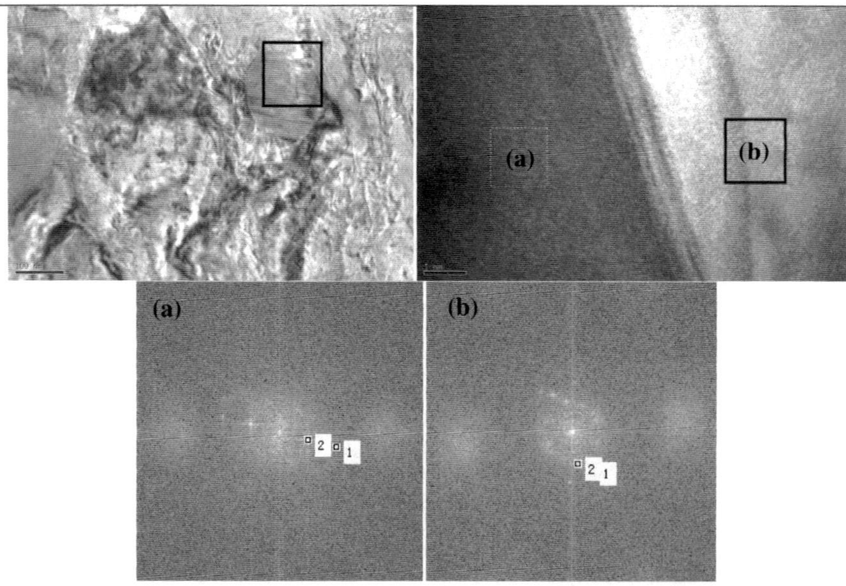

Figure 6. TEM micrographs, HRTEM and Fourier transform results of Zircaloy-4 samples.

Transform the HRTEM picture of the second phase particles and matrix with the Fourier transform, the statistics of the second phase particles were obtained from figure 6(a) (d_1=0.1279nm, d_2=0.2569 nm). The results were in accordance with the PDF #42-1289 results(d=0.1279 nm, d=0.2504 nm). According to the EDS results from table 3, the second phase particles were composed of Zr, Fe and Cr. And the reference[11]shows that the phase between n(Fe)/n(Cr)=1.55～3.45 is $Zr(Fe,Cr)_2$. According to the comprehensive analysis, the second phase particles in figure 6 are $ZrFe_{1.5}Cr_{0.5}$ with the Hexagonal Close-Packed (HCP) structure. And the lattice constant is a=5.007, c=8.193. The computation results from figure 6(b) (d_1=0.1859 nm, d_2=0.2462 nm)were identical to the PDF #65-3366 statistics (d=0.1894 nm, d=0.2459 nm). And the main composition of the matrix was Zr, the matrix in the figure 6 is α-Zr with the Hexagonal Close-Packed (HCP) structure.

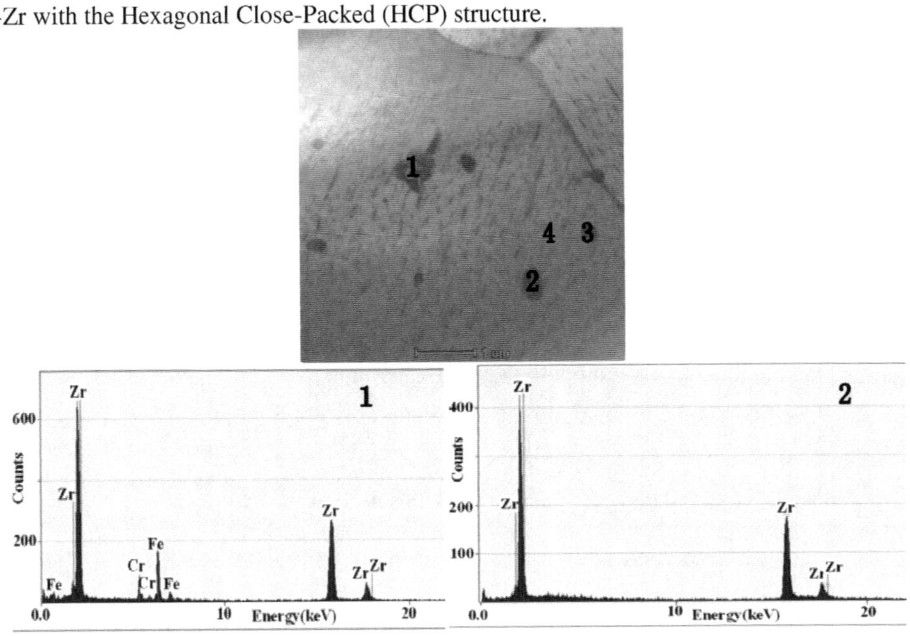

Figure 7. STEM micrographs and EDS spectrum of B592 batch Zircaloy-4.

Table 4. Statistics table for EDS data of the second particles and matrix in figure 7.

Position	Element	wt.%	at.%	Position	Element	wt.%	at.%
STEM 7-1	C	0.01	0.08	STEM 7-2	C	0.00	0.00
	O	0.23	1.21		O	0.04	0.27
	Cr	4.39	6.94		Cr	0.00	0.00
	Fe	10.26	15.11		Fe	0.00	0.00
	Zr	85.07	76.62		Zr	99.94	99.71
STEM 7-3	C	0.13	1.03	STEM 7-4 (matrix)	C	0.21	1.59
	Cr	0.00	0.00		Cr	0.00	0.00
	Fe	0.00	0.00		Fe	0.00	0.00
	Zr	99.86	98.96		Zr	99.78	98.40

Figure 7 shows the STEM and EDS results of the TEM samples of Zircaloy-4. From the STEM and EDS results listed in table 4, the second phase particles distributed at the grain boundaries and inside the grains of B592 batch. The second phase particles were round, ellipse and square with the size in the range of 100 nm. The main composition of the second phase particles was Zr, Fe and Cr, the contents of Fe and Cr varies a lot: n(Fe)/n(Cr)=1.55～3.45. According to the research [12-14], the Fourier transform spot and EDS statistical analysis indicate that the second phase particles were the

$Zr(Fe, Cr)_2$ with the Hexagonal Close-Packed (HCP) structure. The composition of the matrix was mainly Zr and there existed micro-scale of Fe in some location. The matrix is α-Zr phase with the Hexagonal Close-Packed (HCP) structure. The content of Fe and Cr of the point 2 and point 3 was zero in the scanning transmission photographs.

Figure 8 is TEM micrographs of the C744 batch Zircaloy-4. The composition of the second phase particles was Zr, Fe and Cr.

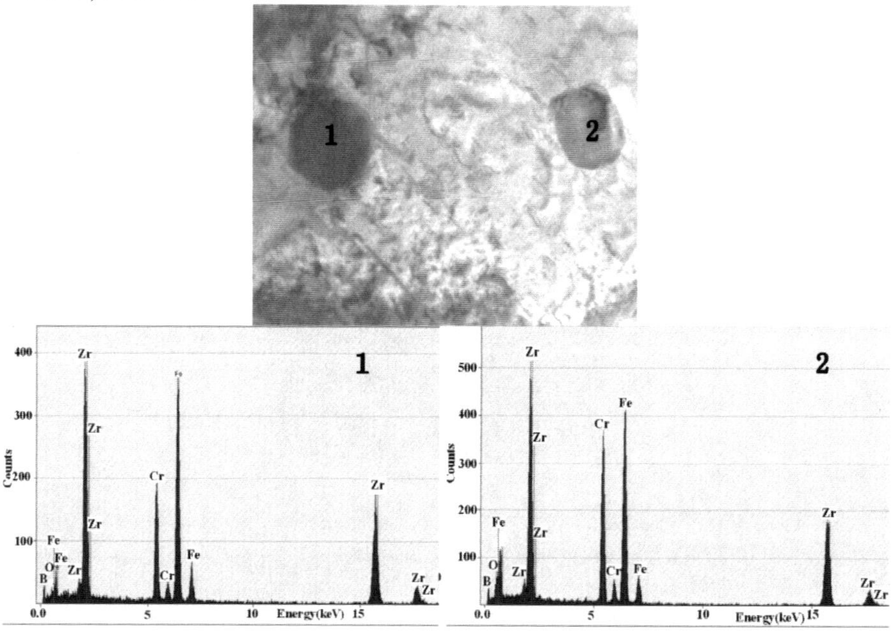

Figure 8. TEM micrographs and EDS spectrum of C744 batch Zircaloy-4.

Table 5. Statistics table for EDS data of the second particles in figure 8.

Position	Element	wt.%	at.%	Position	Element	wt.%	at.%
TEM8-1	O	0.31	1.37	TEM8-2	O	1.36	5.52
	Cr	15.17	19.95		Cr	22.05	27.34
	Fe	32.38	39.62		Fe	29.07	33.56
	Zr	52.11	39.04		Zr	47.48	33.55

Transform the HRTEM picture of the second phase particles and matrix with the Fourier transform, the statistics of the second phase particles were obtained from figure 9(a) (d_1=0.2479nm, d_2=0.1349nm,d_3=0.1553nm). The results were in accordance with the PDF #42-1289 results (d=0.2504nm, d=0.1364nm, d=0.1553nm). According to the EDS results from table5, the second phase particles were composed of Zr, Fe and Cr. And the reference[11]shows that the phase between n(Fe)/n(Cr)=1.22～2.37 is $Zr(Fe, Cr)_2$. According to the comprehensive analysis, the second phase particles in figure 9 were $ZrFe_{1.5}Cr_{0.5}$ with the Hexagonal Close-Packed (HCP) structure. And the lattice constant is a=5.007, c=8.193. The computation results from figure 9(b) (d_1=0.1610nm, d_2=0.2756nm) were identical to the PDF #65-3366 statistics(d=0.1615nm, d=0.2798nm). And the main composition of the matrix was Zr, the matrix in the figure 9 was α-Zr with the Hexagonal Close-Packed (HCP) structure.

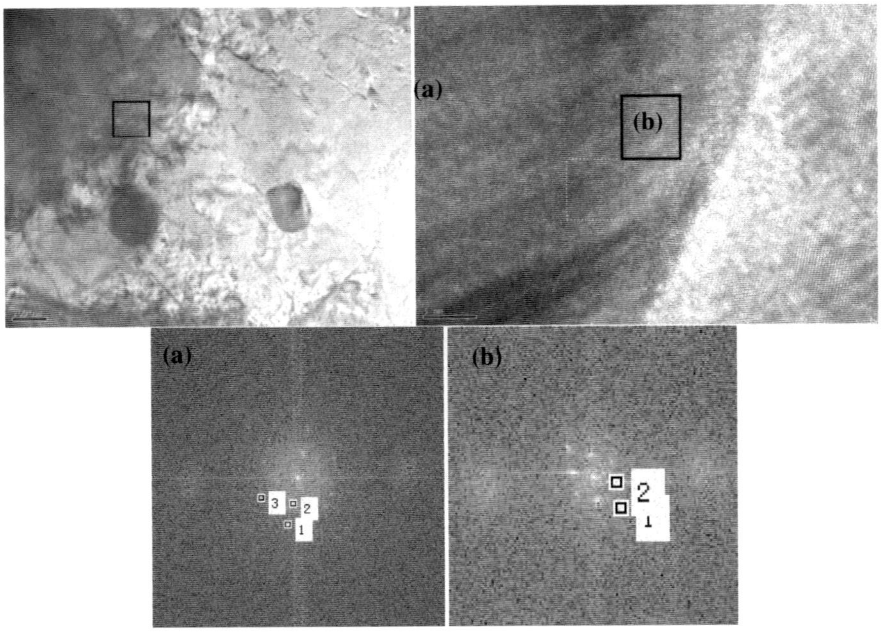

Figure 9. TEM micrographs, HRTEM and Fourier transform results of C744 batch Zircaloy-4.

Figure 10 is STEM micrographs of the C744 batch Zircaloy-4. The composition of the second phase particles was Zr, Fe and Cr.

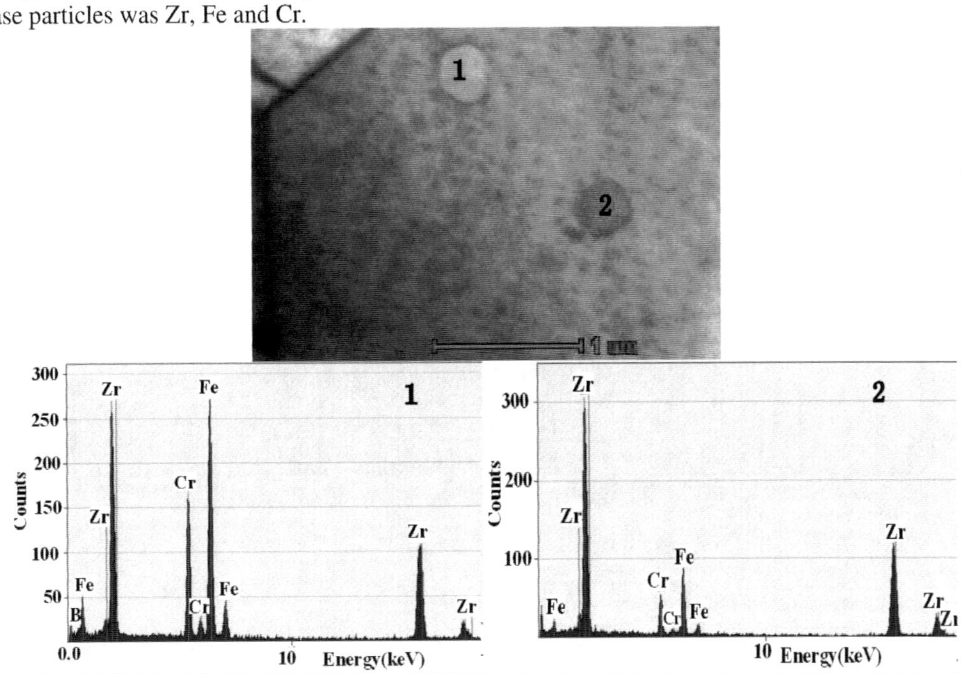

Figure 10. STEM micrographs and EDS spectrum of C744 batch Zircaloy-4.

Table 6. Statistics table for EDS data of the second particles in figure 10.

Position	Element	wt.%	at.%	Position	Element	wt.%	at.%
STEM 10-1	O	0.38	1.63	STEM 10-2	O	0.00	0.00
	Cr	16.14	20.97		Cr	5.58	8.72
	Fe	33.17	40.12		Fe	12.85	18.69
	Zr	50.29	37.25		Zr	81.54	72.56

From the above photos and EDS results it can be found the second phase particles distributed at the grain boundaries and inside the grain in the Zircaloy-4 of C744 batch. The second phase particles were round, ellipse and square with the size around 200nm.The main composition of the second phase particles was Zr, Fe and Cr. The contents of Fe and Cr varies a lot: $n(\mathrm{Fe})/n(\mathrm{Cr})=1.22\sim2.37$. According to the reference, the Fourier transform spot and EDS statistical analysis indicate that the second phase particles were the Zr (Fe, Cr)$_2$ with the Hexagonal Close-Packed (HCP) structure. The composition of the matrix is mainly Zr.

Figure 11 and figure 12 is TEM and STEM micrographs of the A753 batch Zircaloy-4. It can be seen the second phase particles distributed at the grain boundaries and inside the grain. The size of the second phase particles was around 100nm with the main composition of Zr, Fe and Cr.Individual second phase particles contained only Fe. The content of the Cr was zero, which is showed in table 8.

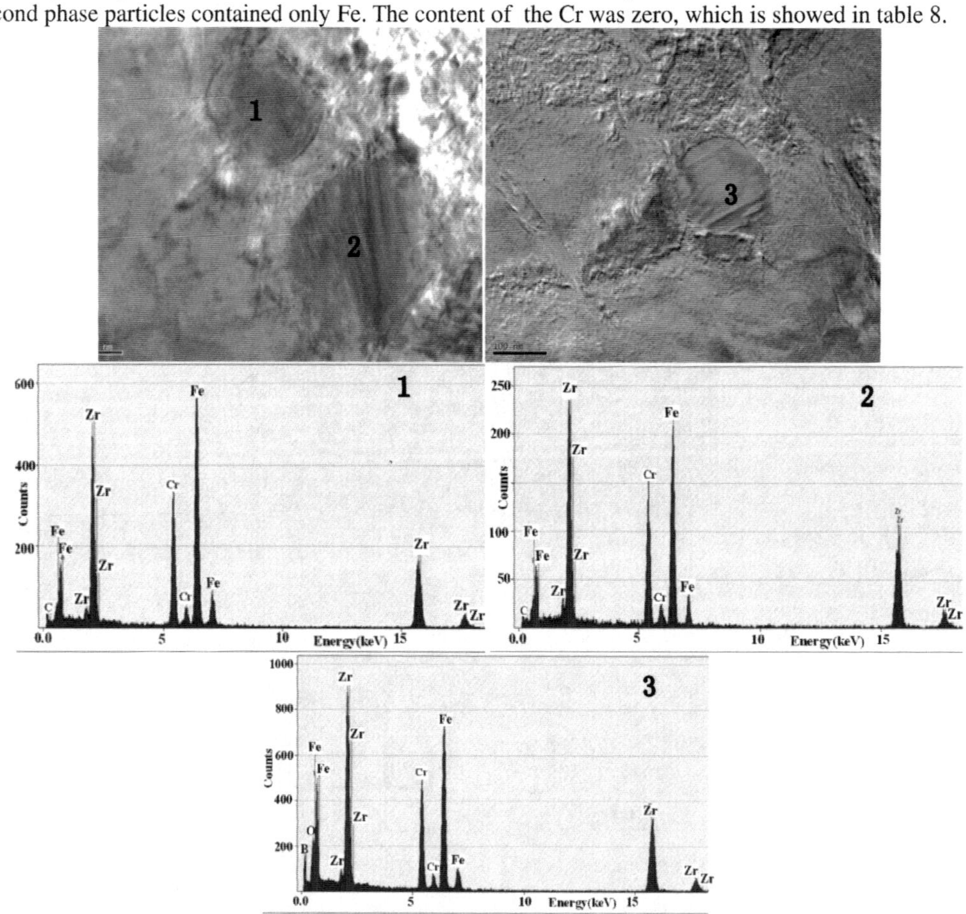

Figure 11. TEM micrographs, HRTEM and Fourier transform results of A753 batch Zircaloy-4.

Table 7. Statistics table for EDS data of the second particles in figure 11.

Position	Element	wt.%	at.%	Position	Element	wt.%	at.%
TEM 11-1	O	0.00	0.00	TEM 11-1	O	1.42	5.87
	Cr	17.06	21.73		Cr	16.75	21.30
	Fe	36.26	46.56		Fe	29.43	34.84
	Zr	43.66	31.69		Zr	52.39	37.97
TEM 11-3	O	3.66	10.43	--	--	--	--
	Cr	14.41	12.63				
	Fe	26.19	21.38				
	Zr	48.28	24.13				

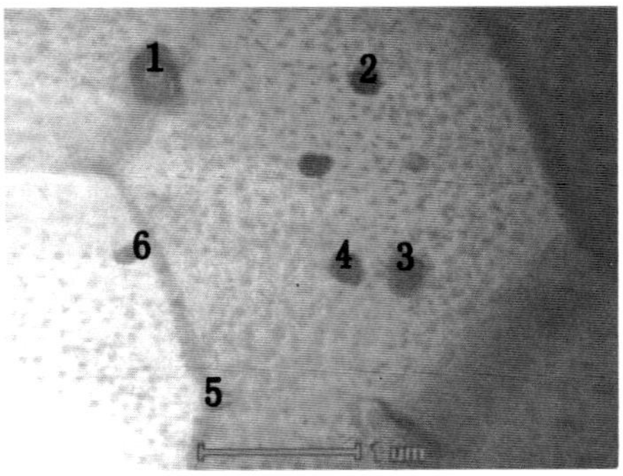

Figure 12. STEM micrographs of A753 batch Zircaloy-4.

Table 8. Statistics table for EDS data of the second particles in figure 12.

Position	Element	wt.%	at.%	Position	Element	wt.%	at.%
STEM 12-1	C	0.00	0.00	STEM 12-2	C	0.88	6.32
	Cr	12.62	17.75		Cr	0.16	0.26
	Fe	24.00	31.43		Fe	0.20	0.32
	Zr	63.36	50.80		Zr	98.74	93.08
STEM 12-3	C	0.65	4.67	STEM 12-4	C	0.93	6.68
	Cr	0.82	1.35		Cr	0.00	0.00
	Fe	2.69	4.13		Fe	0.01	0.02
	Zr	95.82	89.23		Zr	99.04	93.28
STEM 12-5	C	0.46	3.20	STEM 12-6	C	0.00	0.00
	Cr	0.04	0.06		Cr	12.08	17.02
	Fe	11.17	16.57		Fe	24.24	31.81
	Zr	88.31	80.15		Zr	63.67	51.15

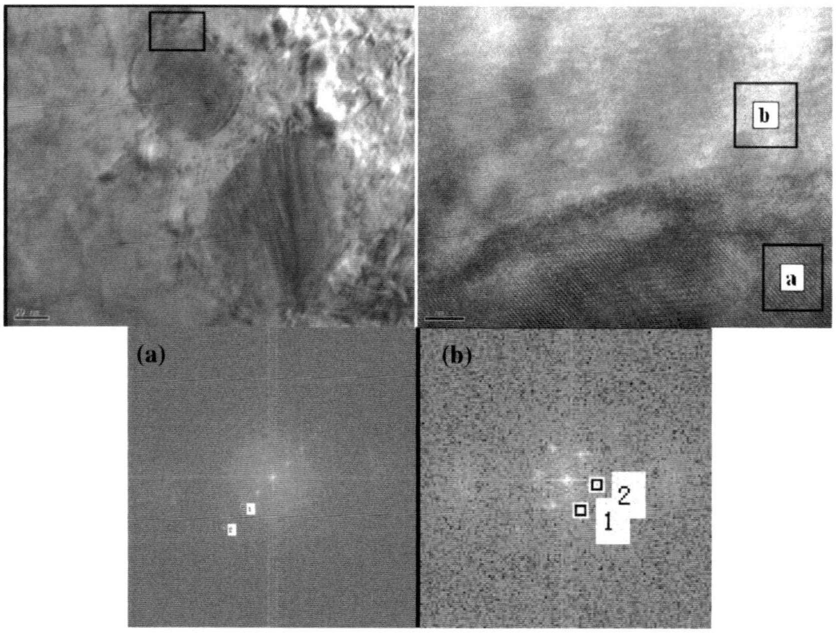

Figure 13. TEM micrographs, HRTEM and Fourier transform results of A753 batch Zircaloy-4.

Transform the HRTEM picture of the second phase particles and matrix with the Fourier transform, the statistics of the second phase particles were obtained from figure 13(a) (d_1=0.2105nm, d_2=0.1252nm). The results were in accordance with the PDF #42-1289 results (d=0.2136nm, d=0.1252nm). According to the EDS results from table7, the second phase particles were composed of Zr, Fe and Cr. And the reference shows that the phase between n(Fe)/n(Cr)=1.22～3.45 is $Zr(Fe, Cr)_2$. According to the comprehensive analysis, the second phase particles in figure 13 were $ZrFe_{1.5}Cr_{0.5}$ with the Hexagonal Close-Packed (HCP) structure. And the lattice constant is a=5.007, c=8.193. The computation results from figure 13(b) (d_1=0.2553nm, d_2=0.2826nm) were identical to the PDF #65-3366 statistics (d=0.2553nm, d=0.2798nm). And the main composition of the matrix was Zr, the matrix in the figure 13 was α-Zr with the Hexagonal Close-Packed (HCP) structure.

The microstructure and composition were studied of three batches Zircaloy-4. It can be found that the second phase particles distributed at the grain boundaries and inside the grain in the Zircaloy-4.The size of the second phase particles was around100-200nm with the main composition of Zr, Fe and Cr. The contents of Fe and Cr varies a lot: n(Fe)/n(Cr)=1.22～3.45. According to the reference, the Fourier transform spot and EDS statistical analysis indicate that the second phase particles were the $Zr(Fe, Cr)_2$ with the Hexagonal Close-Packed (HCP) structure. The composition of the matrix was mainly Zr.

In the above analysis it can be also found that individual second phase particles contained only Fe. The content of the Cr was zero or the content of Fe and Cr were zero according to EDS in the figure 7 and figure 12.There is no phenomenon in the second phase particles of the figure 10. The composition of the second phase particles contained Fe and Cr. This was the different between the three batches Zircaloy-4.

4. Conclusion

There is no obvious difference observed by the macro metallographic analysis among the three batches Zircaloy-4. However the etch pits in the middle area of the A753 batch samples have been observed. The grain size of C744 batch samples is greater than the samples A753 and B592 batches. The phase of three batches Zircaloy-4 is α-Zr by XRD.

The Zircaloy-4 matrix was α-Zr phase with the HCP structure. The second phase particles was Zr(Fe, Cr)$_2$ with the HCP structure. The second phase particles distribute at grain boundaries and inside the grains.

The size of the second phase particles in qualified sample C744 is about 200nm with the elements of Zr, Fe and Cr; and the sizes of the second phase particles in the un-qualified samples B592 and A753 are about 100nm, in which the elements of Fe and Cr could not be detected or only trace amount of Fe could be detected for part of the second phase particles.

Therefore it may be concluded that the grain size, the size and composition of the second phase are the main factors to affect the welding quality of the alloy.

References

[1] Liu Jianzhang *et al.* 2007 *Nuclear Structural Material* (Beijing: Chemical Industry Press) 5–6
[2] Li Xianjun 2011*Tianium Ind. Pro.* **28(1)** 38–38
[3] Yong H J, Kyoung O L and Hyun H G 2002 *J. Nucl. Mate.* **302(1)** 9–19
[4] Krishna KV M *et al. 2008 J. Nucl. Mater.* **383(1-2)** 78–85
[5] Hiwarkar VD *et al.* 2009 *J. Nucl. Mater.* **384 (1)** 30–37
[6] Zhou Bangxin, Li Qiang and Miao Zhi 2000 *Nucl. Power. Eng.* **21(4)** 339–343
[7] Zhou Bangxin, Li Qiang, Miao Zhi, Pan Shu fang, Li Cong and Jian Yourong 1995 *Chinese J. Nucl. Sci. Eng.* **15(3)** 242–249
[8] Daum R S, Chu Y S and Motta A T 2009 *J. Nucl. Mater* **392 (3)** 453–463
[9] Yan Qingsong, LiuWenqing, Lei Ming, Li Qiang, Yao Meiyi and Zhou Bangxin 2007 *Rare. Metal. Mat. Eng.* **36(1)** 104–107
[10] Bordona R A and Olmedo A M 1981 *J. Mater. Science* **16(6)** 1527–1532
[11] Chai Linjiang, Luan BaiFeng, Zhou Yu, Qiu Risheng and Chen Jianwei 2012 *Chinese J. Nonferrous. Met.* **22(6)** 1594–1604
[12] Liu Wenqing, Zhong liuming, Peng Jianchao, Liu Renduo, Yao Meiyi and Zhou Bangxin 2011 *Rare .Metal. Mat. Eng.* **40(7)** 1216–1219
[13] Zhao Wenjin and Zhou Bangxin 1991 *Nucl. Power. Eng.* **12(5)** 67–72
[14] Yao Meiyi, Zhou Bangxin, Li Qiang, Liu Wenqing, Wang Shu'an and Huang Xinshu 2007 *Rare. Metal. Mat. Eng.* **36(11)** 1916–1919

Influence of Mo content on the γ-phase stability and properties of U-Mo alloy

Xuewei Zhang*, Zhiyuan Yang, Ying Meng, Shidong Kang and Jinming He

China North Nuclear Fuel Co., Ltd Inner Mongolia Baotou 014035 China

E-mail: *969647786@qq.com

Abstract. U-Mo alloys have been considered as the potential new advanced nuclear fuel, due to its favourable irradiation stability associated with the presence of the metastable γ-phase in the microstructure. Thus its γ-phase stability and performance are of great importance. In this work, three kinds of U-Mo alloys with the content of 6, 8 and 10wt.% Mo were prepared by non consumable arc melting technology, the homogenization annealing and heat treatment of isothermal decomposition. Through metallographic microscope, scanning electron microscope (SEM), X-ray diffraction, Brinell hardness tester, the microstructure and mechanical properties of U-Mo alloys were investigated. The results indicate that the stability of γ-phase in the U-Mo alloy is directly related to the content of Mo. With the increase of Mo, the stability of γ-phase is enhanced and the decomposition of γ-phase is delayed during the isothermal decomposition at 565□. In addition, the mechanical properties of this alloy are closely associated with its phase rather than directly related to the content of Mo.

Keywords: U- Mo alloy, γ-phase, stability, properties

1. Introduction

In order to achieve the conversion of research reactors from high enriched uranium (HEU) fuel to low enriched uranium (LEU), high density uranium compounds or alloys are widely used [1-6]. For pure metallic uranium, the corrosion resistance and irradiation stability are very poor, which is related to the presence of α-phase at room temperature. Compared to the α-phase, the γ-phase of metallic uranium has a better radiation resistance and appears to possess the isotropic thermal expansion property, which makes it more suitable for the nuclear fuel [7]. Unfortunately, the γ-phase of metal uranium is not stabilized at the lower temperature range, and thus makes it a significant technical barrier for uranium using as the fuel. One solution to this problem is to stabilize the γ-phase at room temperature by the addition of Mo. While γ-phase is not thermodynamically stable, but a metastable phase. The high-temperature γ-phase will decompose into the α-phase and U_2Mo during the cooling process theoretically [8]. Since the diffusion velocity of molybdenum atoms is slow, the eutectoid reaction maintains sufficient thermal activation energy and soaking time, which keeps the γ-phase exist at room temperature at the normal cooling rate in the U-Mo alloy with the high content of Mo. Meanwhile, in the U-Mo alloys with low content of Mo, the γ-phase can also be maintained at room temperature by rapid cooling [9-11]. And the research shows that the phase structure has an important effect on the mechanical properties of the alloy [12].

In this paper, U-Mo alloy was prepared by non-consumable arc melting. The microstructure and phase composition of the U-Mo alloy were analyzed. Meanwhile, the mechanical properties of the alloys were studied, which provides the guidance for the research of U-Mo alloy fuels.

2. Materials and Experimental

2.1. Materials

The raw material was depleted uranium ingots (the content of carbon is less than 100μg/g) and molybdenum particles. In order to avoid the introduction of excessive inclusions, the depleted uranium ingots were refined and purified, and the chemical composition is shown in table 1. The metallographic photograph of the collected pure uranium is shown in figure 1. The size of molybdenum particle is Φ3mm× (5~10) mm. And the chemical composition is shown in table 2.

Table 1. Chemical composition of depleted uranium.

Items（μg/g）												
Fe	C	N	O	Ni	Cu	Mn	Al	Ca	Si	P	S	H
34	86	51	90	＜20	＜20	＜20	＜20	＜20	＜10	＜10	＜10	1.2

Figure 1. Microstructure of the depleted uranium ingots.

Table 2. The chemical composition of Mo.

Items（μg/g）								
C	O	Fe	Ni	Si	Al	Mg	Sn	Ca
70	30	20	20	20	20	20	5	20

From metallographic photograph of the pure uranium ingot, it can be seen that the refining and purification effect of ingot is good, and there is no obvious impurity and reunion phenomenon of inclusions.

2.2. Preparation of Molybdenum Uranium Alloy

U-6wt. % Mo, U-8wt. % Mo, U-10wt. % Mo alloy (hereafter called U-6Mo, U-8Mo, U-10Mo) were prepared by non-consumable arc melting technology. The melting output current was 500A; stirring current was 8A. In order to accomplish homogenization, the sample were turned over and melted 6 times.

2.3. Heat Treatment of U-Mo Alloys

The homogenization annealing heat treatment of U-Mo alloys was carried out at the temperature of 1000 °C with the cooling holding time of 24 hours in furnace. And the isothermal decomposition heat treatment of the alloys was conducted at 565℃for 20 hours.

2.4. Metallographic examination

The metallographic examination of polished and corroded states of U-Mo alloys under different heat treatments was carried out. The sample is observed by universal research-level automatic metallographic microscope (Axio Observer Z1m). At the same time, the alloy composition was analyzed by scanning electron microscopy (SEM, VEGA 3XMU).

The phase composition of U-Mo alloy was analyzed by X-ray diffract meter (D8 DAVINCI). The scanning mode is 2θ mode. The scanning range is 10°~90°. The step is 0.02°, and the acquisition time is 0.02s. The source of radiation was standard Cu Kα with a wavelength of 0.154 nm. The operating voltage of the X-ray tube was 40 kV and the current was 40 mA.

The hardness test of U-Mo alloy prepared with different heat treatment was carried out by using electronic hardness tester. The model was HBE-3000A, and the standard hardness block was HB2.5/187.5.

3. Results and Discussion

3.1. Microstructure Analysis of U-Mo Alloy

The metallographic polishing state and corrosive state photos of U-Mo alloy are shown in figure 2 and figure 3. The scanning test results are shown in figure 4 and figure 5.

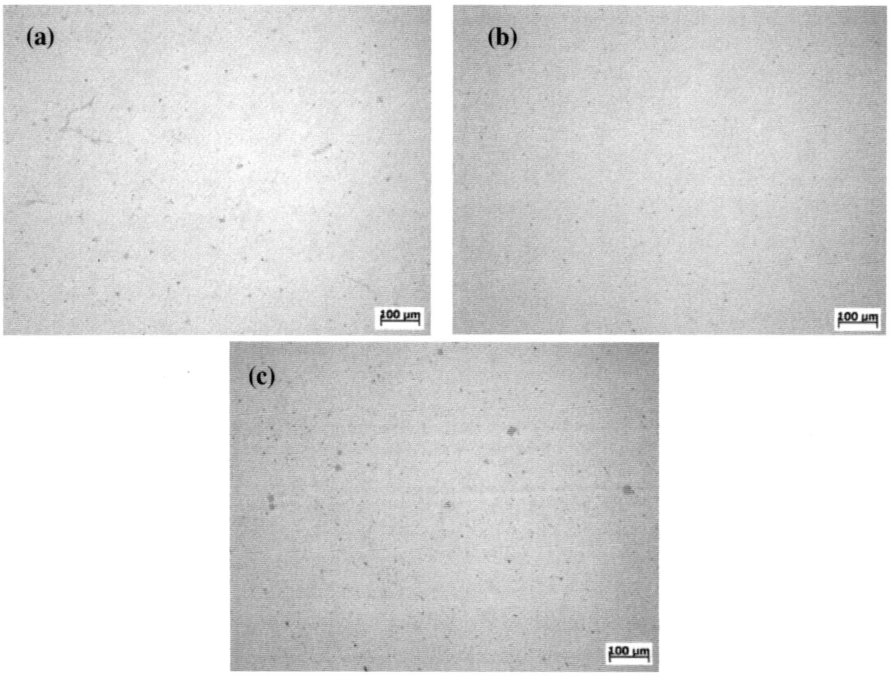

Figure 2. Metallographic photos of as-polished. (a) U-6Mo, (b) U-8Mo, (c) U-10Mo.

From figure 2, it can be seen that the prepared sample contains little inclusions and no obvious impurity agglomeration. Figure 3 shows that there are different contrasts caused by corrosion between the crystals and the adjacent crystal boundaries, which indicates that the presence of composition segregation within the crystals [13-14]. From the results of the scanning detection in figure 3 and figure 4, the matrix element is uranium and molybdenum, and there is no un-melted molybdenum with the inclusions mainly constituting of carbides or nitrides of uranium.

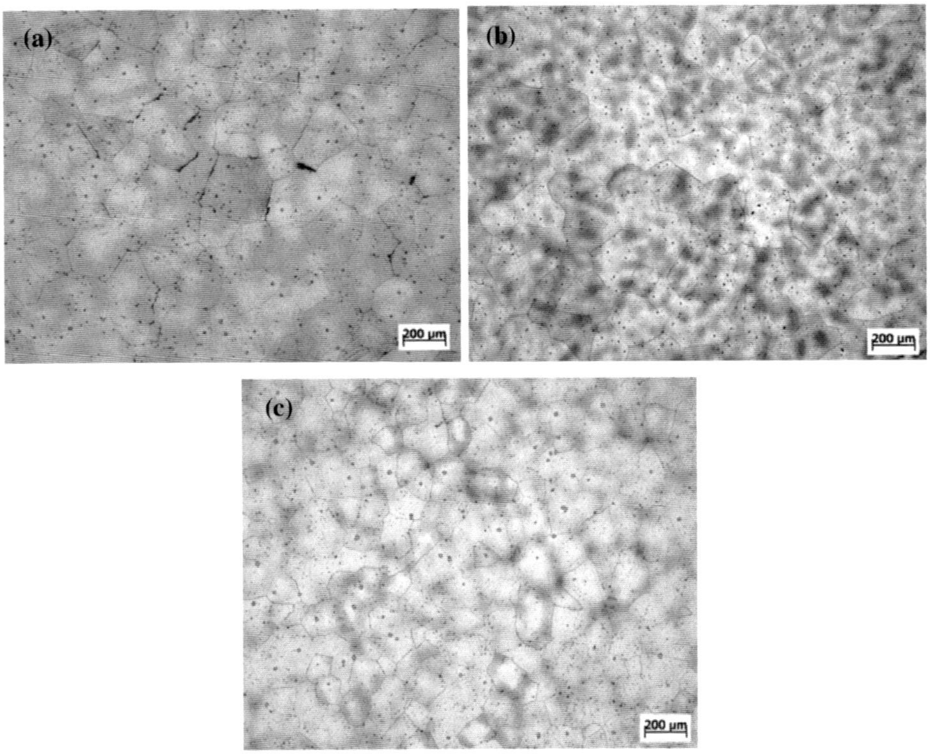

Figure 3. Metallographic photos after corrosion. (a) U-6Mo, (b) U-8Mo, (c) U-10Mo.

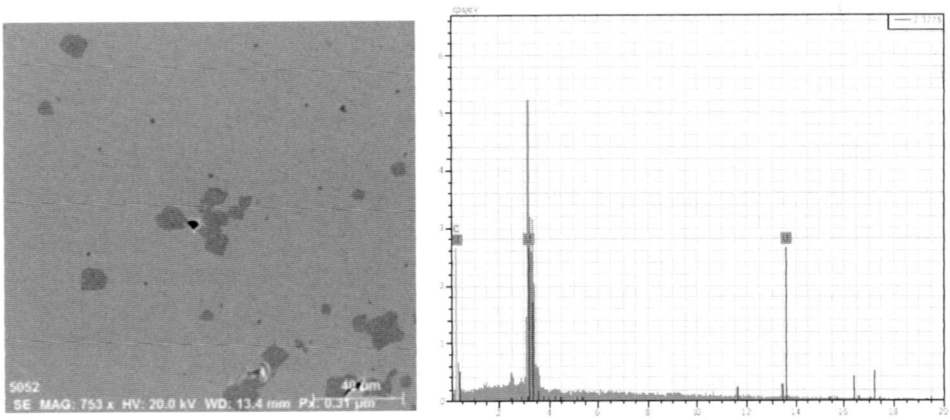

Figure 4. SEM images of inclusions.

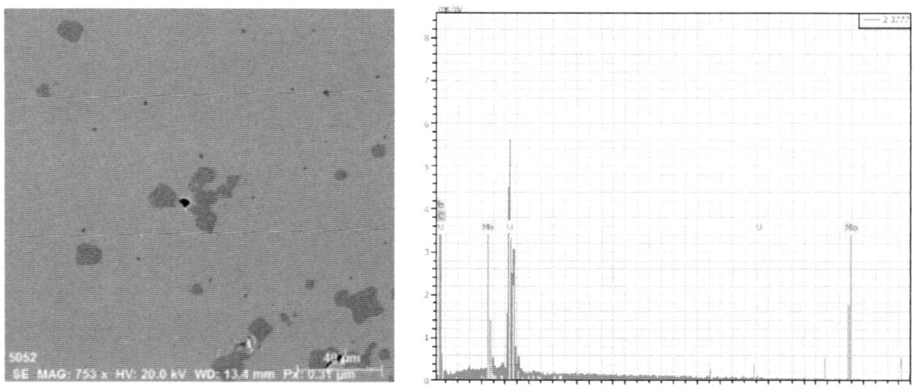

Figure 5. SEM images of matrix.

Brinell hardness test of three kinds of alloys were carried out, and the test results are shown in table 3.The results show that the hardness of U-Mo alloy enhances with the increase of Mo.

Table 3. Hardness of U-Mo alloy prepared by non consumable arc melting.

Samples	U-6Mo	U-8Mo	U-10Mo
1	217	242	252
2	218	249	258
3	212	245	253
Average	216	245	254

3.2. Phase Structure and Property Analysis of U-Mo Alloy after Homogenization Heat Treatment

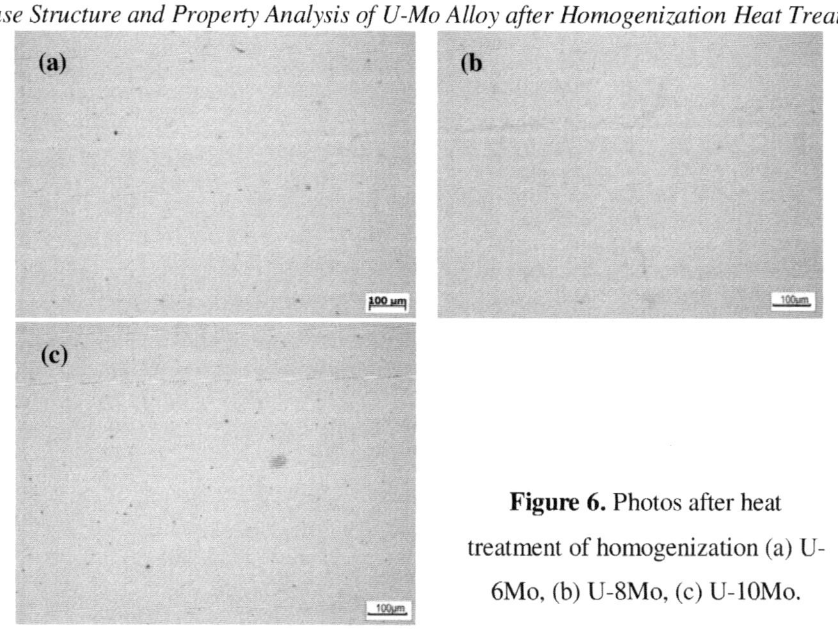

Figure 6. Photos after heat treatment of homogenization (a) U-6Mo, (b) U-8Mo, (c) U-10Mo.

Figure 6 shows the metallographic as-polished photos of the alloy after the homogenization. It can be seen that the homogenization reduces the agglomeration of the inclusions, and the composition segregation of the samples is obviously improved compared with that before the heat treatment, with sufficient alloying and uniform components.

Phase analysis of the samples showed that the U-6Mo alloy contains a small amount of α phase, mainly due to the low content of Mo. And the metastable γ-phase has changed during the cooling process in the furnace [9]. γ-phase is the main phase in the U-8Mo and U-10Mo. When the content of Mo is around 8%, it can be enough to guarantee the high temperature phase of uranium, that is, the γ-phase is stable at room temperature and does not decompose in the process of cooling in the furnace. (figure 7)

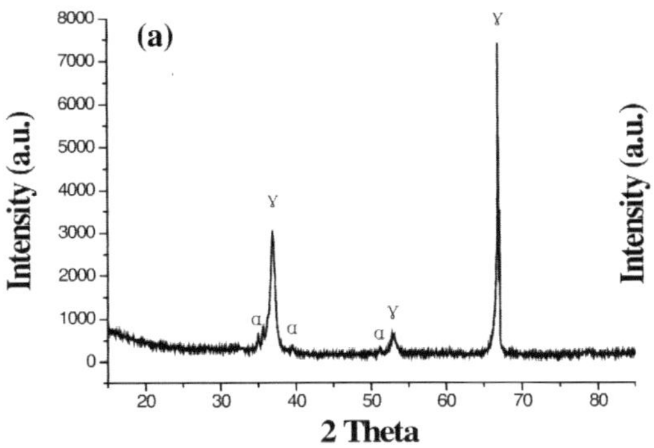

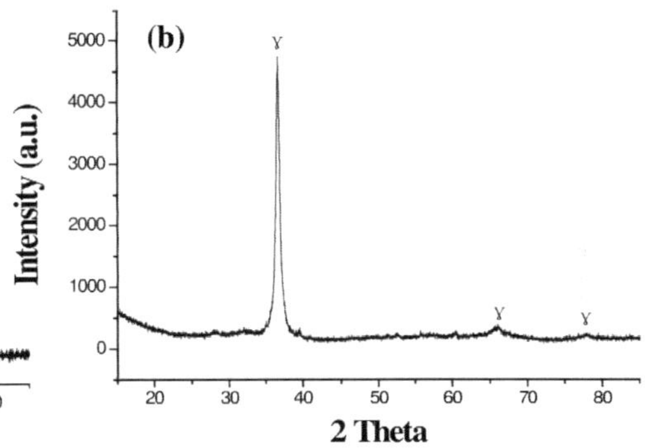

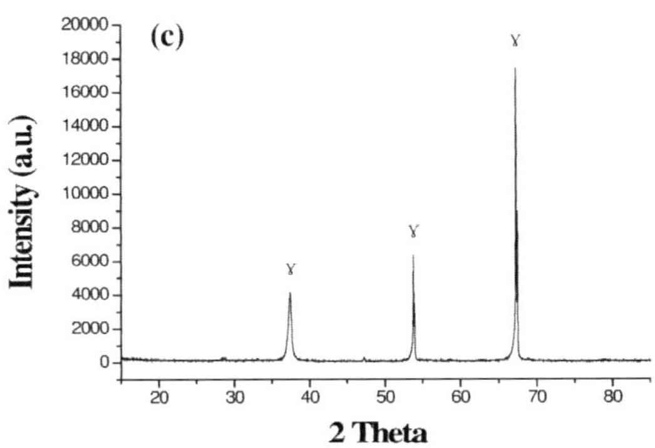

Figure 7. XRD pattern after heat treatment of homogenization. (a) U-6Mo, (b) U-8Mo, (c) U-10Mo.

Table 4. Hardness of U-Mo alloy after heat treatment of homogenization.

Samples	U-6Mo	U-8Mo	U-10Mo
1	223	245	253
2	228	247	256
3	225	245	251
Average	225	246	253

Brinell hardness testing of three kinds of alloys is carried out. The test results are shown in table 4. It was found that the hardness value of U-6Mo alloy after homogenization is higher than that of before, which may be related to the phase transformation and the change of the structure of U-6Mo alloy. This is consistent with the XRD test results. Compared with the hardness of as-cast, U-8Mo and U-10Mo alloy have no obvious change, as is shown in table 3. The hardness of U-Mo alloy increases with the increase of Mo at as-cast condition, which mainly due to the fact that γ-phase is stable at room temperature under rapid cooling [10] Compared with the U-6Mo, U-8Mo and U-10Mo alloys, the increase of Mo content leads to the increase of hardness.

3.3. Phase Structure and Property Analysis of U-Mo Alloy after Heat Treatment of Isothermal Decomposition.

Figure 8 is the metallographic photograph of the U-Mo alloy after homogenization annealing and heat treatment of isothermal decomposition. As it can be seen that for the (a) U-6Mo, γ-phase decomposition takes place at the crystal boundary and in the inner part of the crystal [15]. For the (b) U-8Mo sample, there is a significant phase transition (black region) at the crystal boundary. The contrast exhibited by corrosion shows the shape of crystal boundary, which is smaller than that of U-6Mo. (c) There is no obvious phase transition at the crystal boundary of U-10Mo samples, which may be related to the high content of Mo and the good stability of γ phase [10-11].

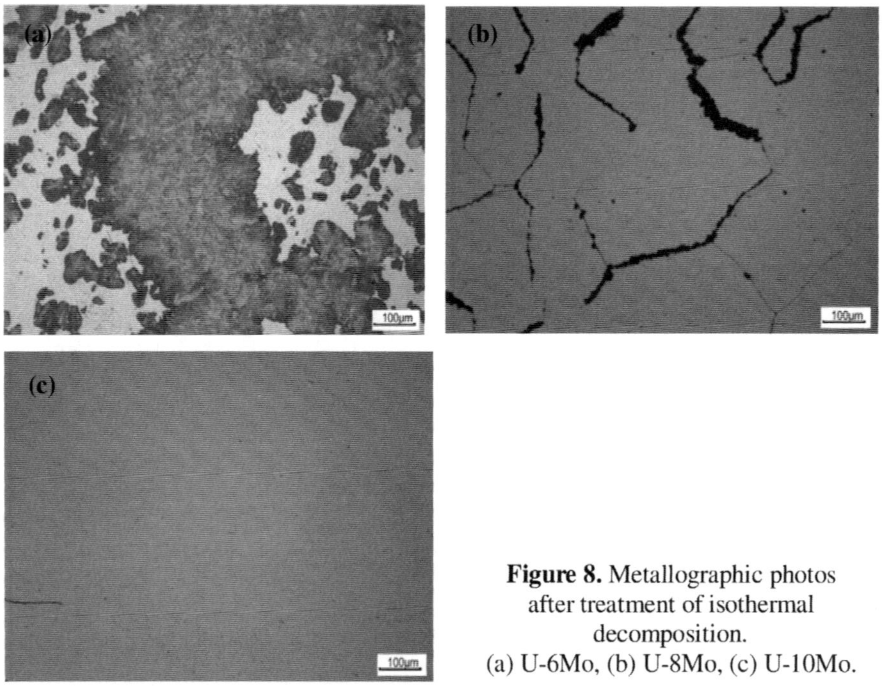

Figure 8. Metallographic photos after treatment of isothermal decomposition.
(a) U-6Mo, (b) U-8Mo, (c) U-10Mo.

XRD shows that (as shown in figure 9) in addition to the peak of the γ phase uranium molybdenum alloy, the α-phase U peak of (a) U-6Mo and (b) the U-8Mo also appeared, which indicates that γ⟶ α phase transformation occurred during the heat treatment of isothermal decomposition. The results of U-10Mo show that the phase structure of alloy is still γ-phase, indicating that U-10Mo does not undergo the phase transition during the isothermal decomposition. This is in accordance with the metallographic results. The Brinell hardness test of three kinds of alloys were carried out and the results are shown in table 5.

It can be seen from table 5 that the hardness values of U-6Mo and U-8Mo alloy are higher than that after homogenization heat treatment, which may be closely related to its structure phase decomposition. And the decomposition leads to Mo enrichment in the γ-phase [10], resulting to the hardness increase. While the U-10Mo alloy with γ-phase hardly changes its hardness value before and after isothermal decomposition.

Table 5. Hardness of U-Mo alloy after treatment of isothermal decomposition.

Samples	U-6Mo	U-8Mo	U-10Mo
1	232	265	258
2	239	262	263
3	237	269	254
Average	236	265	258

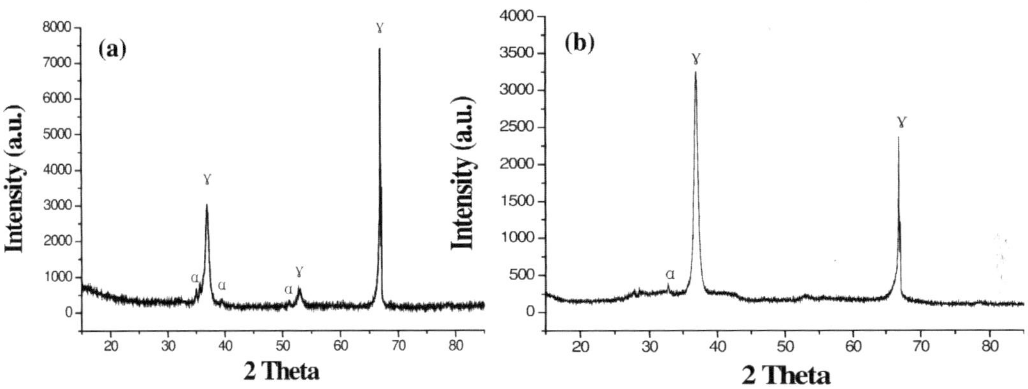

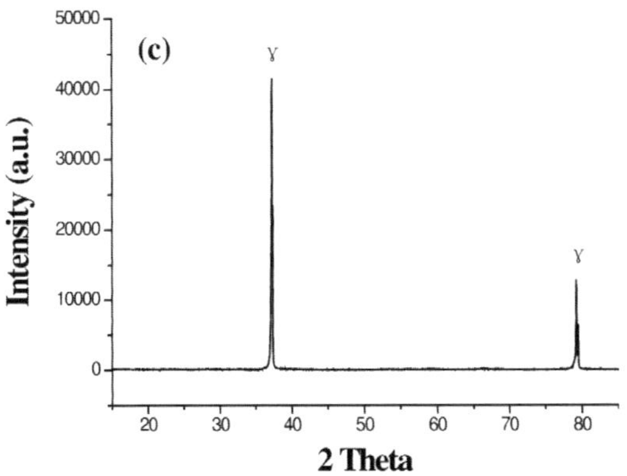

Figure 9. XRD pattern after isothermal decomposition treatment.
(a) U-6Mo, (b) U-8Mo, (c) U-10Mo.

4. Conclusions

The U-6Mo, U-8Mo and U-10Mo alloys prepared by non-consumable arc melting lead to the retention of the γ-phase at room temperature and there hardness increase with the increase of Mo content.

The stability of γ-phase of U-Mo alloy is directly related to the content of Mo. And the increasing addition of Mo causes a delay in the γ-phase decomposition during isothermal heat treatment.

The hardness of U-Mo alloys after the isothermal heat treatment is measured, it is found that the hardness of U-Mo alloys are closely associated with its phase rather than directly related to the content of Mo.

Reference

[1] Dmitriev V D, Ibragimov S S and Karmilov A G 1967 *At. Energy. (USSR)*, **22** 459–465
[2] Xiao H X, Long C S, Tian X F and Li S J 2015 *Mater. Des.* **74** 55–60
[3] Jue J F, Trowbridge T L, Breckenridge C R, Moore G A, Meyer M K and Keiser D D 2015 *J. Nucl. Mater.* **460** 153–159

[4] Miao Y B *et al.* 2016 *Scripta Mater.* **114** 146–150
[5] Ye B *et al.* 2015 *J. Nucl. Mater.* **464** 236–244
[6] Liu Y M, Yin C G and Sun C L 2011 *Nuclear Power Engineering* **32(4)** 19–23
[7] Leal Neto R M, Rocha C J, Carvalho E Urano de, Riella H G and Durazzo M 2014 *J.Nucl. Mater.* **445** 218–223
[8] Burkes D E, Papesch C A, Maddison A P, Hartmann T and Rice F J 2010 *J.Nucl. Mater.* **403** 160–166
[9] Yin B Y and Qu Z H 2012 *Atomic Energy Science And Technology* **46(05)** 596–602
[10] Pedrosa T A, Dos Santos A M M, Lameiras F S, Cetlin P R and Ferraz W B 2015 *J. Nucl. Mater.* **457** 100–117
[11] Steiner M A, Calhoun C A, Klein R W, An K, Garlea E and Agnew S R 2016 *J.Nucl. Mater.* **477** 149–156
[12] Liu C H, Jiang M Z and Yin C G 2006 *Nuclear Power Engineering* **27(2)** 50–53
[13] Burkes D E, Hartmann T, Prabhakaran R and Jue J F 2009 *J. Alloy Compd.* **479** 140–147
[14] Jeong G Y, Kim Y S and Sohn D S 2015 *J.Nucl. Mater.* **466** 509–521
[15] Parida S C, Dash S, Singh Z, Prasad R and Venugopal V 2001 *J. Phys. Chem. Solids* **62** 585–597

Welded-Ag-nanowires/FTO conducting film with high transmittance and its application in transparent supercapacitors

Zhensong Qiao, Xiaopeng Yang[1,*], Feng Liu, Guangbin Duan and Bingqiang Cao[2,*]

Materials Research Centre for Energy and Photoelectrochemical Conversion, School of Material Science and Engineering University of Jinan, Jinan 250022, Shandong, China.

E-mail: [1,*] mse_yangxp@ujn.edu.cn, [2,*] mse_caobq@ujn.edu.cn.

Abstract. Silver nanowires (AgNW)with a small diameter weresynthesized by a facile and novel polyol reduction method. Ag nanowires ink was then spun on the surface of F-doped SnO_2 (FTO) to form the AgNW/FTO conducting film. Welding treatment of the AgNW/FTO conducting film not only increased the optical transmittance from 71.9 % to 79.3 % at 550 nm and decreased the sheet resistance from 11.4 ohm sq^{-1} to 9.8 ohm sq^{-1}, but also improved the adhesivity of AgNW network on FTO substrate. Furthermore, MnO_2 nanosheets were directly deposited on welded-AgNW/FTO (wAF) substrate to prepare a transparent MnO_2/weled-AgNW/FTO (MwAF) composite electrode. The MwAF electrode displayed excellent electrochemical performance, including high specific capacitance (375 F g^{-1} at 5 mV s^{-1}) and superior cycle stability (173.3 % of the initial capacitance after 20000 GCD cycles).

Keyword: silver nanowires, welding treatment, transparent, MnO_2, supercapacitors

1. Introduction

Transparent conductive electrodes (TCEs) due to high conductivity and high transparency have been widely used as the critical components of many optoelectronic devices, like solar cells, light-emitting diodes (LEDs), touch screen panels, and plasma displays [1-5]. In general, tin doped indium oxide (ITO) is the most widely used TCE material in these applications because of its excellent electrical conductivity and optical transmittance. However, ITO is subject to some drawbacks such as scarcity and high cost of indium, inherent brittleness, and the high temperature processing to restrict its practical applications. Therefore, various transparent electrodes, like carbon nanotubers (CNT), graphene, conducting polymers, and metal nanowires mesh, have been investigated to replace ITO [3,6].

Ag nanowires (AgNW) network has attracted considerable attention as one of the most promising candidates due to high transparency, excellent conductivity, good mechanical flexibility, high surface area, and solution process-ability [7]. Moreover, large-scale and high-yield synthesis of AgNW has been achieved using polyol reduction methods. However, for practical application of AgNW film, there are several issues that need to be addressed. Firstly, the AgNW-AgNW contact resistance is too large because of the insulating coating of polyvinylpyrrolidone (PVP) on the surface of AgNW and the loose contact at the crossed section of AgNWs [8,9]. Several post-processes such as mechanical pressing, fast light sintering, and thermal annealing, have been required to decrease the resistance of AgNW network [10]. Secondly, AgNW network is easy to fall off owing to poor adhesion between AgNW and substrates [11]. For the initial AgNW network, the connections at the crossed section are mainly via physical contacts, such as van der Waals forces and capillary force form solvent evaporation.[12] Therefore, the formation of junction contacts at the crossing points of AgNWs and between AgNW and substrate on surface of AgNW can effectively improve the mechanical adhesion[13]. Brongersma and co-workers demonstrated a light-induced plasmonic nano-welding technique to achieve welding together AgNW selectively at junction points [5]. Recently, our group also reported that a fluorine doped ZnO_2 (FZO) layer coating on surface of AgNW, which not only decreased sheet resistance of AgNW network but also provided strong adhesion and weld [14].

In the paper, the optical transmittance and conductivity of AgNW/F-doped SnO_2 (FTO) conducting film were improved by a post-process method. For this purpose, we developed a facile polyol method to prepare AgNW with high aspect ratio, which were spin-coated on surface of FTO to form AgNW/FTO conducting film. Then, the AgNW/FTO conducting film was welded to form junctions using a rapid annealing process, which increased the optical transmittance from 71.9 % to 79.3 % at 550 nm and reduced the sheet resistance from 11.4 ohm sq^{-1} to 9.8 ohm sq^{-1}. Moreover, we adopted welded-AgNW/FTO (wAF) conducting film to prepare transparent MnO_2/weled-AgNW/FTO (MwAF) composite electrodes, which exhibited excellent capacitive performance with high specific capacitance (375 F g^{-1} at 5 mV s^{-1}) and superior cycling stability (~173.3 % after 20000 cycles).

2. Experimental

2.1. Synthesis of AgNW and preparation of Ag NW network

The silver nanowires with high aspect ratio were synthesized using a new polyol process. During the synthesis, 0.34 g of polyvinylpyrrolidone (PVP) with high molecular weight ($Mw \approx 1300000$) was dissolved into a 15 mL ethylene glycol (EG) solution of 0.1mM $FeCl_3$ and heated to 170°C, and was named as solution A. Next, 0.16 g of $AgNO_3$ was added to 5 mL EG solution and stirred to uniformity as Solution B. Then, 30 mg of $AgNO_3$ was added into the Solution A stirred for initial nucleation of the silver seeds. After 5 min, the Solution B was injected into the Solution A drop by drop using a syringe within 10 min. The mixed solution was allowed to react for 30 min. The cooled solution at room temperature was centrifuged five times at 9000 rpm. Then, the AgNW/ethanol solution was spun on the surface of FTO substrates at a 3000 rpm for several times to prepare the AgNW/FTO conducting films. Finally, the AgNW/FTO conducting films were welded by a rapid thermal annealing process for 30 min by the rate of 40 °C/s and then cooled down naturally under Ar atmosphere.

2.2. Preparation of transparent MwAF composite electrodes

The transparent MwAF composite electrodes were fabricated by an electrochemical deposition technology. The carbon, saturated calomel and welded-AgNW/FTO conducting films were as the counter electrode, reference electrode and working electrode, respectively. The electrolyte was the mixture of 0.1 M $Mn(CH_3COO)_2 \cdot 4H_2O$ and 0.1 M Na_2SO_4, and its pH was set to 6 by adding dilute H_2SO_4. The MnO_2 nanosheets were deposited onto the welded-AgNW/FTO (wAF) conducting films in potentiostatic mode under 0.45 V for 400s to fabricate the transparent MwAF composite electrodes. Finally, the composite electrodes were washed in distilled water.

2.3. Characterization

Morphology and crystal structure studies of the samples were performed with a field emission scanning electron microscope (SEM, Quanta 250, FEI), and X-ray diffraction (XRD, D8-Advance, Bruker), respectively. The transmittance spectra were measured using a UV-vis-NIR spectrophotometer (UV-3600, Shimadzu) equipped with an integrating sphere. The sheet resistances were measured by a four-point probe method with a source/meter (Keithley 2162A) and probe station

(ECOPIA, ESP300). Cyclic voltammetry (CV), Galvanostatic charge/discharge (GCD) and electrochemical impedance spectroscopy (EIS) were carried out in a three-electrode system using an electrochemical workstation (Zahner/Zennium E 6.0). The electrolyte was a 0.5 M Na_2SO_4 aqueous solution.

3. Results and discussion

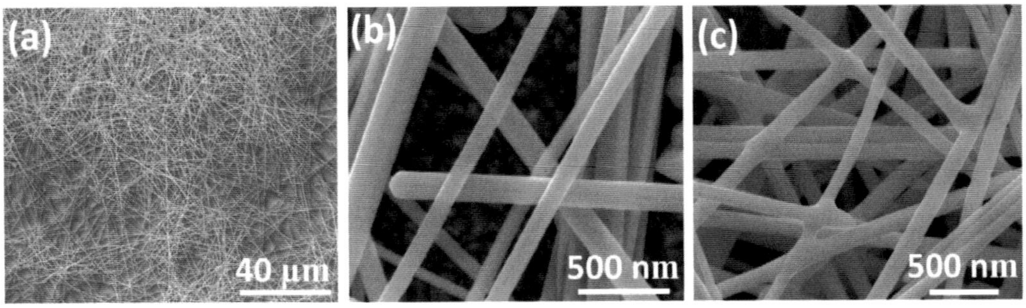

Figure 1. (a) (b) SEM images of AgNW networkand, (c) the weldled-AgNW/FTO conducting film.

Figure 1(a) and 1(b) show the typical SEM images of the AgNW/FTO conducting film before annealing treatment. The diameters of these as-grown AgNW are about 50 to 200 nm and the lengths are about 40 to100 μm. However, the initial AgNW network is easy to loose or peel off due to the connections at the crossed section by physical contacts, like van der Waals forces and capillary force form solvent evaporation[14].Then, the AgNW/FTO conducting film is welded by annealing process, and the welding points at the crossing points of AgNWs are clearly observed, as shown in figure 1(c). Importantly, away from each junction the nanowires are unaffected by the annealing treatment and no further morphological changes were observed.

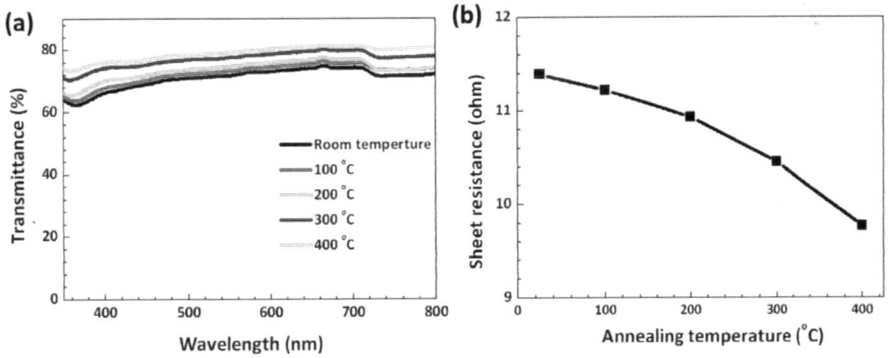

Figure 2. (a) the optical transmittance spectra of the same AgNW/FTO conducting film and (b) the corresponding sheet resistances at different annealing temperature.

To investigate the photoelectric properties of AgNW/FTO conducting film, the same AgNW/FTO conducting film is annealed at various temperatures, respectively. In figure 2(a) and 2(b), the optical transmittance of the as-prepared AgNW/FTO conducting film is only about 71.9 % at 550 nm, and the corresponding sheet resistance is 11.4 ohm sq^{-1}. Then, the AgNW/FTO conducting film is welded by a rapid thermal annealing process. With increasing annealing temperature, its sheet resistances gradually decrease and the corresponding optical transmittances increase. When the annealing temperature is up to 400 °C, the optical transmittance increases to 79.3 % at 550 nm and the sheet resistance reduces to

9.8 ohm sq^{-1}.On the one hand, the more welding points of AgNWare formed to decrease the sheet resistance of AgNW network with increase the annealing temperature; on the other hand, the whole thickness of AgNW network can become thinner to increase the optical transmittance due to the top Ag nanowires into the bottom Ag nanowires by welding process.

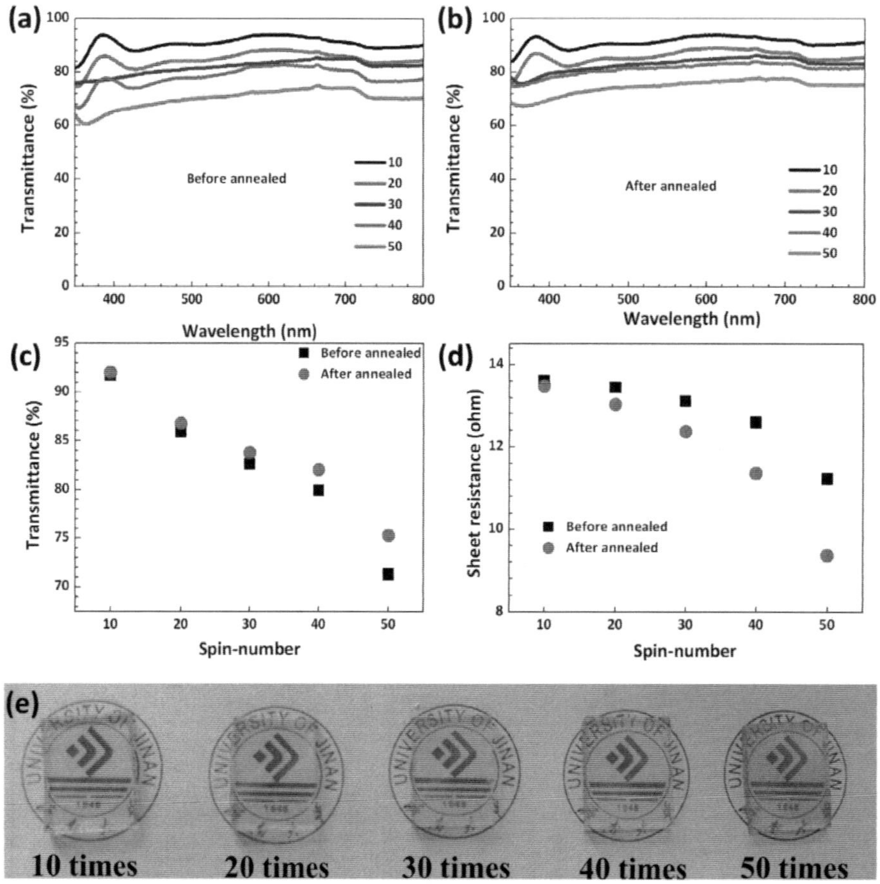

Figure 3. (a) optical transmittance spectra of AgNW/FTO conducting films with different spin-number before and (b) after annealing process, (c) transmittance versus spin-number for AgNW/FTO conducting films, (d) sheet resistances versus spin-number for AgNW/FTO conducting films, (e) A photograph of the weld-AgNW/FTO conducting films with different spin-coating numbers of AgNW ink .

Figure 3 shows the effect of annealing treatment on the optical transmittance and sheet resistance of the AgNW/FTO conducting films with different spin-number of AgNW ink. As shown in figure 3(a) and 3(b), the transmittances of AgNW/FTO conducting films decrease with the increase of AgNW spin-coating times before and after annealing process. Figure 3(c) shows the optical transmittance of the AgNW/FTO conducting film before and after annealing treatment. The transmittance of the as-prepared AgNW/FTO conducting film at 550 nm reaches 91.8 % when the spin-coating number is 10 times; 85.9 % at 20 times, 82.6 % at 30 times, 79.9 % at 40 times, and 71.3 % at 50 times. After annealing treatment, their transmittance increase to 92 %, 86.8 %, 83.8 %, 82.1 %, and 75.3 %, respectively. In figure 3(d), the sheet resistance of the AgNW/FTO conducting film is about 13.6 ohmsq^{-1} at 10 spin-coating times before annealing. With increasing to 50 times, the resistance sheet reduces to 11.2 ohm sq^{-1}. Then, the AgNW/FTO conducting films are welded at 400°C. Compared with as-prepared AgNW/FTO conducting films, the decreases of sheet resistance for the wAF conducting films are even more obvious with increasing the number of spin-coating. For instance, the sheet resistance decreases from 11.2 to 9.4 ohm sq^{-1} at spin-coating times of 50 after annealing process while the sheet resistance only decreases by 0.1 ohm sq^{-1} at 10 times. Figure 3(e) shows the photograph of the wAF conducting films with different spin-coating times of AgNW/FTO conducting films.

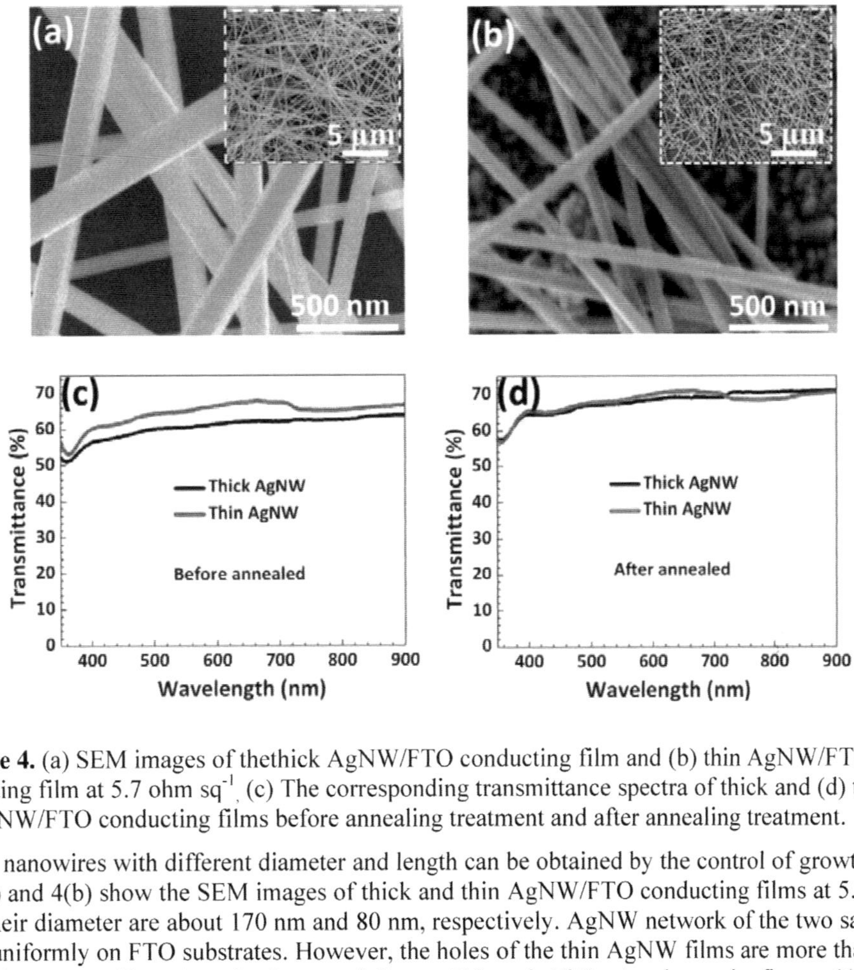

Figure 4. (a) SEM images of thethick AgNW/FTO conducting film and (b) thin AgNW/FTO conducting film at 5.7 ohm sq^{-1}. (c) The corresponding transmittance spectra of thick and (d) thin AgNW/FTO conducting films before annealing treatment and after annealing treatment.

The Ag nanowires with different diameter and length can be obtained by the control of growth time. Figure 4(a) and 4(b) show the SEM images of thick and thin AgNW/FTO conducting films at 5.7 ohm aq^{-1}, and their diameter are about 170 nm and 80 nm, respectively. AgNW network of the two samples distribute uniformly on FTO substrates. However, the holes of the thin AgNW films are more than that of the thick AgNW films (see the insets of figure 4(a) and 4(b)). As shown in figure 4(c), the transmittances of as-prepared thick AgNW/FTO conducting film are 60.6 % at 550 nm while that of the as-prepared thin AgNW/FTO conducting film are 65.1 %, indicating the greater optical transmittance of as-prepared thin AgNW/FTO conducting film. Then, the samples were carried out annealing treatment, and their optical transmittance spectra are displayed in figure 4(d). The thick AgNW/FTO conducting film shows a transmittance of 67.5 % after annealing, giving 6.9 % difference from the initial value, while the thin AgNW/FTO conducting film with initial transmittance of 65.1 % becomes only 3.3 % more transparent after annealing treatment.

AgNW/FTO conducting film with unique properties like good electrical conductivity, high specific surface area, and high transmittance, can be applied to transparent supercapacitors as an ideal current collector material. In our paper, MnO_2 is directly grown on wAF conducting film using electrochemical deposition method to prepared flower-like transparent MwAF composite electrodes.

Figure 5(a) and 5(b) clearly show that MnO_2 nanosheets are uniformly coated on the wAF conducting film. The MwAF composite electrodes have large surface area, which can give sufficient the electrochemical active sites to enhance the capacitive performance [15]. Figure 5(c) shows the XRD pattern of the transparent MwAF composite electrode. The diffraction peaks at about 38.1°, 44.3°, 64.4° and 77.5° are observed, which can be assigned to the (111), (200), (220), and (311) planes of metallic Ag (JCPDFS 04-0783). The peak at 32.3° is the characteristic peaks of AgO (JCPDFS 43-1038). The peaks at 26.5°, 33.8°, 37.8°, 51.6°, 61.7°, and 65.7° can demonstrate the presence of cassiterite SnO_2 (JCPDFS 77-0451) from the FTO film. The diffraction peaks at 26.6°, 34.0°, 37.9°, and 54.8° are observed, which correspond to the (201), (301), (111), and (402) reflections ramsdellite MnO_2 (JCPDFS 42-1316), respectively, supporting the successful preparation of MwAF composite in our samples. Figure 5(d) shows the MwAF composite electrode, and its transmittance is about 23.8 % at 550 nm. According to the inset of figure 5(d), the brown electrode material is a transparent film, and can make the words in the background clearly visible.

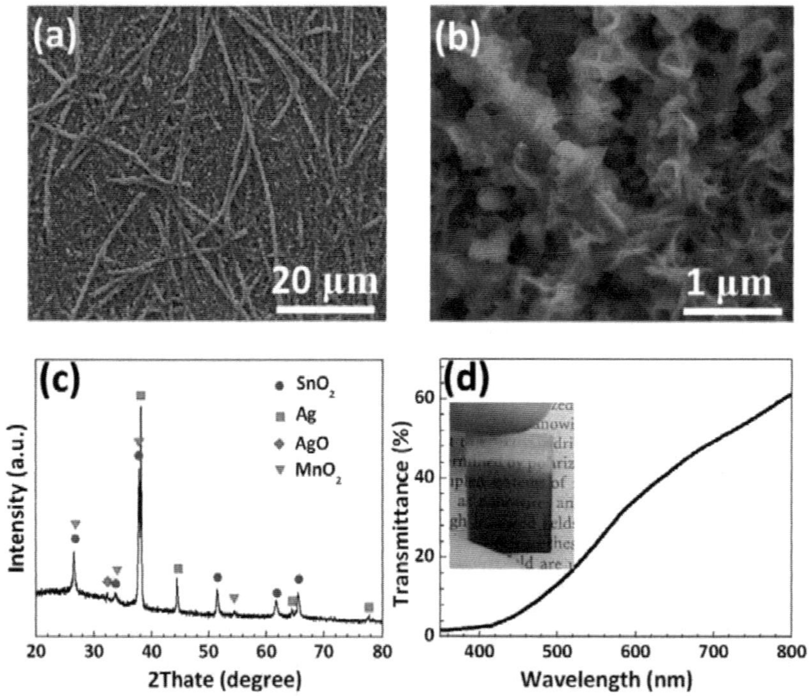

Figure 5. (a) and (b) SEM image of the MwAF electrode, (c) XRD pattern of the MwAF electrode, (d) optical transmittance spectra of the MwAF electrode, the inset is a photograph of the MwAF composite electrode.

To further evaluate the capacitive properties of the transparent MwAF composite electrodes, electrochemical studies are conducted in a three-electrode cell in 0.5 M Na_2SO_4 electrolytes. Figure 6(a) shows CV curves of the MwAF composite electrode at different scan rates. The shapes of the CV curves are roughly rectangular mirror images characteristic, which indicates ideal capacitive behaviour and good high-rate capability [16, 17]. Figure 6(b) displays the variation in the specific capacitance of the MwAF electrode as a function of the scan rate. The specific capacitance of the MwAF electrode are up to 375, 286, 218, 177, 142, and 110 F g^{-1} at scan rates of 5, 10, 25, 50, 100, and 200 mVs^{-1}, respectively. The good capacitive performance of the MwAF electrode is mainly due to both the larger surface area of flower-like architectures MnO_2 layer and excellent conductivity of AgNW network. Figure 6(c) shows the cycling stability of the MwAF electrode, which is achieved via conducting GCD cycling at a current density of 7 Ag^{-1}.The specific capacitance of the MwAF electrode continuously increases in the initial cycles, which is due to the electrochemical active process of electrodes [18, 19]. During the first 5000 cycles, its specific capacitance increases from 191.1 F g^{-1} to 312 F g^{-1}, showing an 170 % retention. Until 10000 cycles, the specific capacitance of the MwAF electrode reaches the maximum value, 181 % of its initial capacitance. After 20000 cycles, the MwAF electrode still remains at 173.3 % of its initial capacitance, demonstrating its excellent long-term cycling stability. The inset of figure 6(c) is the GCD curves at different current densities from 3.5 to 28 A g^{-1}, which are linear and symmetrical charge and discharge profiles, indicating good super-capacitive behaviour of the MwAF electrode. The electrochemical performance of the MwAF electrode is further studied by electrical impedance spectroscopy (EIS) measurement, and the Nyquist plot is shown in figure 6(d). A small charge transfer resistance of the MwAF electrode is confirmed by the small diameter of the semicircleof 34.1 ohm, which is lower than that of the pure MnO_2, indicating that the MwAF composite electrode has a good capacitance behaviour [20].

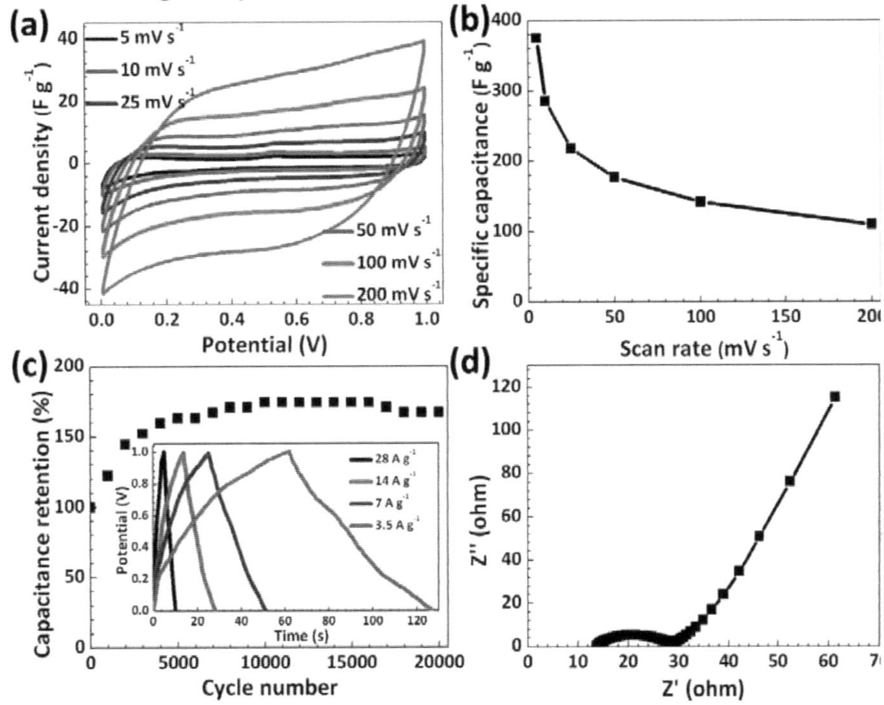

Figure 6. (a) CV curves of the MwAF electrode at different scan rates, (b) Specific capacitance of the MwAF electrode at different scan rates, (c) Cycling performance of the MwAF electrode at current density of 7 A g^{-1}, the inset is the GCD curves at different current densities, (d) EIS curves of theMwAF electrode.

4. Conclusion

We have demonstrated a facile and novel polyol method to prepare AgNW with high aspect ratio. The AgNW/FTO conducting film is then obtained by spin-coating, which presents good optical transmittance with 71.9 % at 550 nm and low sheet resistance with 11.4 ohm sq^{-1}. Then, welding treatment using a rapid annealing process for AgNW/FTO conducting film form welding junctions at AgNW crossing points, which canimprove the transmittance to 79.3 % and decrease the sheet resistance to 9.8 ohm sq^{-1}. Furthermore, the wAF film shows good adhesivity of AgNW network and high surface area. Based on the unique properities, we also prepared transparent MwAF composite as a supercapacitor electrode. The transparent MwAF composite electrode not only has large surface area, but also shows high specific capacitance with 375 F g^{-1} at 5 mV s^{-1}, and good cycle stability with 173.3 % retention after 20000 cycles. This study provides a new route for the application and development of AgNW in future optoelectronic and energy storage devices.

Acknowledgement
This work is supported by the National Natural ScienceFoundation of China (No. 11404138, 51401239, 21477047),Shandong Provincial Natural Science Foundation of China(No. BS2014CL019, ZR2013BL002).

References
[1] Sukanta D, Thomas M, Philip E, Evelyn M, Peter N, Werner J, John J and Jonathan N 2009 *ACS nano* **3**17
[2] Gaynor W, Burkhard G, McGehee M and Peumans P 2011 *Adv.Mater.* **23** 2905
[3] Kim T, KimY, LeeH, Kim H, Yang W and Suh K 2013 *Adv.Funct. Mater.* **23** 1250
[4] Selzer F, Floresca C, Kneppe D, Bormann L, Sachse C, Weiß N, Eychmüller A, Amassian A, Müller-Meskamp L and Leo K 2016 *Appl. Phys. Lett.* **108** 163302
[5] Garnett E, Cai W, Cha J, Mahmood F, Connor S, Greyson C, Cui Y, McGehee M and Brongersma M 2012 *Nat.Mater.* **11** 241
[6] Im H, Jin J, Ko J, Lee J, Lee J and Bae B 2014 *Nanoscale* **6** 711
[7] Oh H and Lee M 2016 *Mater. Lett.* **176** 110
[8] Lee J, Lee I, KimT and Lee J 2013 *Small* **9** 2887
[9] Jin Y, Deng D, Cheng Y, Kong L and Xiao F 2014 *Nanoscale* **6** 4812
[10] Weiß N, Müller-Meskamp L, Selzer F, Bormann L, Eychmüller A, Leo K and Gaponik N 2015 *RSC Adv.* **5** 19659
[11] Zhu, Chung C, Cha K, Yang W, Zheng Y, Zhou H, Song T, Chen C, Weiss P, Li G and Yang Y 2011 *ACS nano* **5** 9877
[12] Qiao Z, Yang X, Yang S, Zhang L and Cao B 2016 *Chem. Commun.* **52** 7998
[13] Nian Q, Saei M, Xu Y, Sabyasachi G, Deng B, Chen Y and Cheng G 2015 *ACS nano* **9** 10018
[14] Han J, Yuan S, Liu L, Qiu X, Gong H, Yang X, Li C, Hao Y and Cao B 2015 *J. Mater. Chem. A* **3** 5375
[15] Chi H, LiY, Xin Y and Qin H *Chem. Commun.* **50** 13349
[16] Lang X, Hirata A, Fujita T and Chen M 2011 *Nat.Nanotechnol.* **6** 232
[17] Lu X, Zheng D, Zhai T, Liu Z, Huang Y, Xie S and Tong Y 2011 *Energy Environ. Sci.* **4** 2915
[18] Liu Z, Xu K, Sun H and Yin S 2015 *Small* **11** 2182
[19] Yu M, Zeng Y, Han Y, Cheng X, Zhao W, Liang C, Tong Y, Tang H and Lu X 2015 *Adv. Funct. Mater.* **25** 3534
[20] Xia H, Hong C, Shi X, Li B, Yuan G, Yao Q and Xie J 2015 *J. Mater. Chem. A* **3** 1216

IOP Conf. Series: Materials Science and Engineering **182** (2017) 012023 doi:10.1088/1757-899X/182/1/012023

Tribological performances of multilayer-MoS_2 nanoparticles in water-based lubricating fluid

B M Zhang and J L Sun*

School of Material Science and Engineering, University of Science & Technology Beijing, Beijing 100083, China

E-mail: *sjl@ustb.edu.cn

Abstract. Multilayer-MoS_2 nanoparticles were dispersed in water-based lubricating fluid in a convenient and economical way. Oleic acid and triethanolamine were used as the main surfactants in this solution. Tribological performances of this fluid were tested under high pressure and high rotating speed. FEI Tecnai G20 TEM, Malvern Zetasizer Nano ZS were used to study the particle size and Zeta potential of the lubricants. Tribological performances of this water system with multilayer-MoS_2 nanoparticles and the 3D-surfaces of wear scars after the experiments were analyzed by means of four-ball wear test machine and Olympus laser confocal microscope. According to Hertz theory and experimental data, it is practical significance to combine MoS_2 nanoparticles with water-based lubricating fluid. Under high loading and high rotating speed the mixed suspension can provide good lubricating properties. On the basis of calculation the binding energy of layers in MoS_2 nanoparticles is less than the energy of shearing in friction pairs. The layers in MoS_2 nanoparticles slide in the process of friction. The best lubrication effects can be displayed when friction process lasts about 500 s.

Keywords: Water-based lubricating fluid, Multilayer-MoS_2 nanoparticles, Four-ball tribological test

1. Introduction

With the improvement of nanotechnology and the much deeper research of nano-tribology, scholars all over the world have made a lot of researches on nanoparticles as a new type of lubricant additives [1-2]. Water-based lubricating fluid has become a hot research topic with the advantages of good cleaning property, energy saving, environmental friendly, large heat capacity and so on [3]. At present, it has been proved that the nanoparticles can bring in good lubrication and cooling effects [4]. Yaohua Xu used water-soluble nano-copper as an additive, which improved the performance of wear and friction reducing, and thus enhanced the thermal conductivity of the solution [5]. Liang et al. has confirmed that the ultrathin graphite sheet had a good lubrication effect as the additives of the water-based lubricant. MoS_2 is similar to the structure of graphite and widely used in the process of lubrication. The fracture strength of single-layer MoS_2 is 16 GPa-30 GPa, which has been determined by S Bertolazzi et al. Showing the single-layer MoS_2 have good mechanical properties [7]. When the solid MoS_2 reach the nanometer scale, the physicochemical properties are better than the ordinary solid MoS_2 [8-9]. Qingjuan Meng carried out the researches on water dispersible nanometer MoS_2 [10]. But there are still not enough researches on the tribological properties of MoS_2 nanoparticles as water-based lubricating fluid additives [11].

IOP Conf. Series: Materials Science and Engineering **182** (2017) 012023 doi:10.1088/1757-899X/182/1/012023

Currently, most of the researches on the nanoparticle lubrication mechanism are proceeded within low pressure and low rotating speed, which is far from the actual production processes [12-13]. Furthermore, the ultra-thin graphite, fullerene like MoS_2 and single-layer nano-MoS_2 as the research objects of water-based lubricant additives are higher production costs, less application in actual production, in that it is necessary to carry researches about the application and mechanism with the combination of multilayer-MoS_2 nanoparticles and water-based lubricant. Because of oleic acid and triethanolamine are lamellar liquid crystal structure, not affected by pH, low requirement of working environment [14], and the reactant of oleic acid triethanolamine ester under high temperature, which has good lubricity and corrosion resistance [15]. All of the three reagents are conducive to degradation and smaller affection of pollution. In this experiment, we use oleic acid / triethanolamine as a water-based lubricant system. So in this paper, multilayer-MoS_2 was dispersed in the acid / triethanolamine water system to study the tribological properties under high pressure and high rotating speed.

2. Experiment

2.1. Preparation of water-based lubricating fluid with nanoparticles

The use of sodium dodecyl benzene sulfonate (SDBS) as surfactant, sodium polyacrylate as stabilizer, aqueous boric acid esters as both extreme pressure agent and anti-rust agent, sodium hydrogen diacetate as preservatives, 2, 2, 4-Trimethyl-1, 3-pentanediol monoisobutyrate (cs-12) as film-forming agent disperse the nanoparticles in the nominal particle size 100 nm in water-based lubricant. Figure 1 is the flow chart for the preparation of the quality score of 0.1% water-based lubricant with nanoparticles:

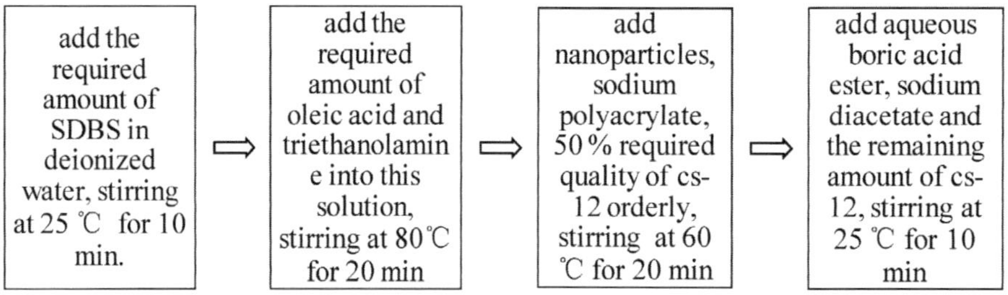

Figure 1. Flow chart of preparing water-based lubricating fluid with MoS_2 nanoparticles.

2.2. Tribological test

Based on the American standard ASTM D2783-2003, the tribological properties of the water-based nano-lubricant were tested by four-ball wear test machine, determination of the last non-seizure value (P_B) of the lubricant and the friction coefficient under different load. In the condition of 588 N load, the friction coefficient at different speeds was measured. Three-dimensional morphologies of the steel balls after long milling were analyzed by Olympus laser confocal microscope. Special steel balls were used for four ball testing machine. The steel ball material is GCr15A, the diameter is 12.7 mm, and the hardness is HRC 64 - 66.

3. Results and discussion

3.1. Stability of water-based lubricating fluid with nanoparticles

First of all, before dispersion, XRD was carried to test the nanoparticles. The result was shown in figure 2.

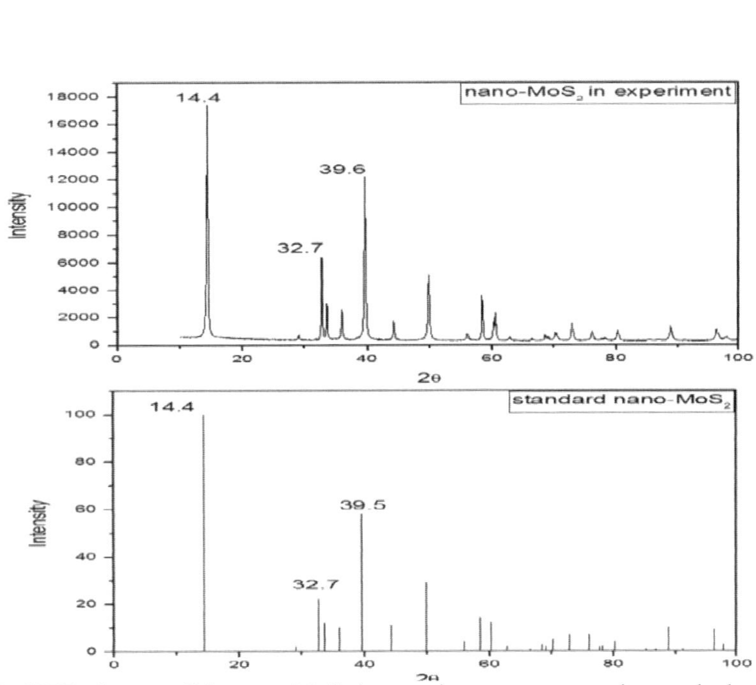

Figure 2. The XDR pictures of the nano-MoS_2 in experiment contrast to the standard nano-MoS_2.

The main diffraction peak of the experimental nano-MoS_2 is almost the same as the standard nano-MoS_2. It is demonstrated that the nanoparticles used in the experiments are nano-MoS_2. The Tecnai G20 FEI was used to measure the dispersion of nanoparticles before and after dispersion. The TEM diagrams were shown in figure 3. The particle size of MoS_2 particles is above 500 nm before dispersion, which shows that the particles aggregate largely at room temperature.

Figure 3. TEM images of undispersed nanoparticles (a) and dispersed nanoparticles (b) in room temperature.

From the figure 3(b), the sizes of the nanoparticles are mostly concentrated in the 100 nm-300 nm. The results show that the nanoparticles dispersion effect is ideal, and there is no large-scale agglomeration.

Then the Zeta potential of the solution was determined by using the Malvern Zetasizer Nano ZS. The three measurements were -44.7 mv, -45.4 mv and -46.4 mv, respectively. Since the absolute value of Zeta potential is greater than 30, the dispersion system is stability [16].

3.2. The last non-seizure value(P_B) of lubricating fluid

In the four-ball experiment, a point contact between the balls can be calculated accordingly to Hertz theory [17]. In the Hertz theory, the theoretical contact radius between two spheres is:

$$\alpha = \sqrt{\frac{3WR_1R_2}{4(R_1+R_2)}\left(\frac{1-v_1^2}{E_1}+\frac{1-v_2^2}{E_2}\right)} \quad (1)$$

In the formula: ν is the Poisson coefficient of GCr15, E is GCr15 elastic modulus. Based on force analysis to top ball and bottom ball, contact force is 0.408 W, W represents the force, R=6.35 mm.

Table 1. Actual diameters of wear scars under different pressure after 10s testing and academic diameters.

Experimental force (N)	Actual wear spot diameter (mm)	Theoretical contact diameter (mm)
392	0.313	0.299
490	0.343	0.323
588	0.367	0.343
696	0.408	0.363
784	0.437	0.378
883	0.456	0.393
991	0.469	0.408

The theoretical results obtained by the Hertz theory are in the presence of no friction. As seen in the table 1, when the load is increased to 991 N, the difference between the actual grinding spot diameter and the theoretical contact diameter is only 13%. If the load is low, the actual grinding spot diameter will be equal to the theoretical contact diameter, which can lead to a conclusion that multilayer-MoS_2 nanoparticles have a good effect on wear and friction reduction. In case of a larger load, it can still maintain good wear and abrasion resistance. PB value of the 0.1% aqueous solution was determined under 559 N. However, compared with the solution without nanoparticles, the P_B value of the nano-lubricant is increased by 43%.

3.3. Effects of nanoparticles on friction coefficient and surfaces of friction pairs

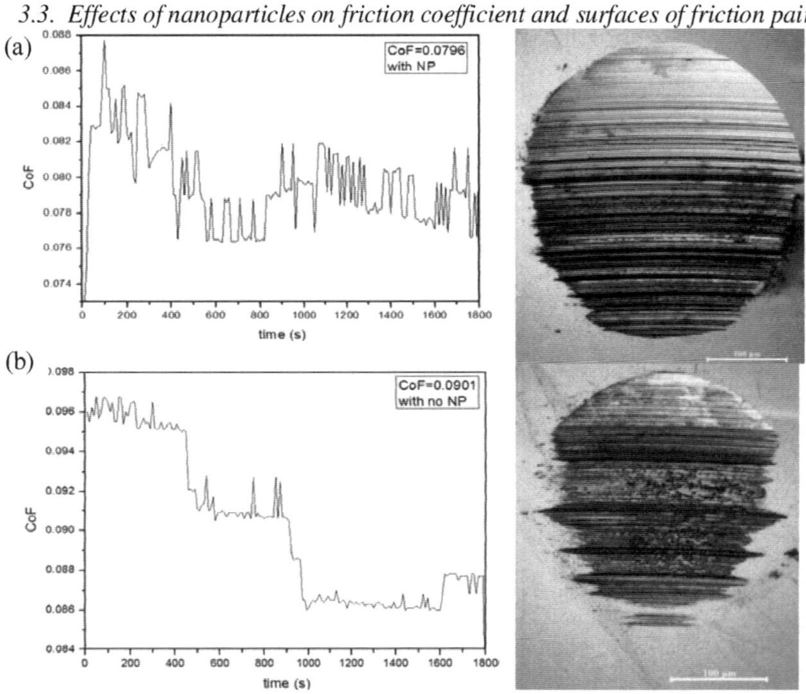

Figure 4. Time variation curves of friction coefficients and the corresponding surfaces of wear scars in water-based lubricating fluid with nanoparticles (a) and without nanoparticles (b) under 392N.

The determinations of friction coefficients of the water-based lubricants with nanoparticles and without nanoparticles were done under 392 N. Friction surfaces were observed after these tests. Time variation curve of friction coefficient and wear patterns are shown in figure 4.

After adding nanoparticles, the friction coefficients decrease significantly from 0.0973 to 0.0796. However, due to the friction in the process, the energy of the lubricants is increased, and the aggregation of nanoparticles occurs. If large particles attach to the surface of the friction pair, the friction coefficients will fluctuate. From the pictures of surfaces, it is obvious that the nanoparticles play an important role in reducing the grinding and greatly reducing wear and tear. In addition, friction surface is relatively smooth and no scratches appear.

3.4. Relationship between pressure and friction coefficients

Four-ball wear test machine under different experimental conditions with 1200 r/min rotating speed was used, and the friction coefficients of the lubricants standing after 1 day and 30 days were determined. The experimental data are as follows:

Table 2. Data of friction coefficient under different pressures.

Experimental force (N)	friction coefficient after 1day	friction coefficient after 30 days
392	0.079	0.090
490	0.078	0.084
588	0.066	0.067
696	0.067	0.069
784	0.069	0.071
883	0.061	0.064
991	0.064	0.067

From the table 2, the friction coefficients decreased with the increase of experimental force. Because the MoS_2 film layers are held together mainly by week Van der Waals force connection, they will slide under a smaller force. When the experimental force is increased, the sliding occurs between the slice layers, so the friction coefficients decrease [18].

The calculation results show that there is a sliding between the layers of MoS_2 nanoparticles. Each wear experiment requires 10g lubricant, which contains 0.01g nanoparticles. When the experimental force is 991 N, the average friction force is 7 N, and the line speed is 0.798 m/s, so the friction work is 5.686 J per second. The friction coefficients chart under 588 N is as follows:

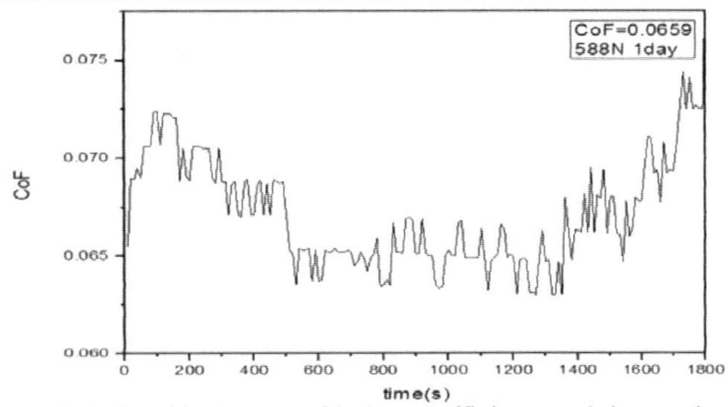

Figure 5. Relationship between friction coefficients and time under 588N.

From the pictures of friction coefficients under 392 N and 588 N, friction coefficients are larger at the beginning, because the beginning of friction produces a large number of new surfaces. The friction coefficients of the 500 s are suddenly reduced because MoS_2 particles begin to slide between the layers. Supposedly, there are only two MoS_2 layers in the nanoparticles. With the energy 76 KJ of the 1 mol nanoparticles, which greater than the maximum Van der Waals energy-hydrogen bond energy 40 KJ, the nanoparticles could slide at the beginning of friction. But the multilayer MoS_2nanoparticles are used in experiment. Therefore, the accumulation of energy needs to make all the MoS_2 layers sliding. 1mol MoS_2 particles work up to 45000 KJ energy for 500 s. Therefore, in the experiment, multilayer MoS_2 nanoparticles can overcome Van der Waals force of the layers to slide.

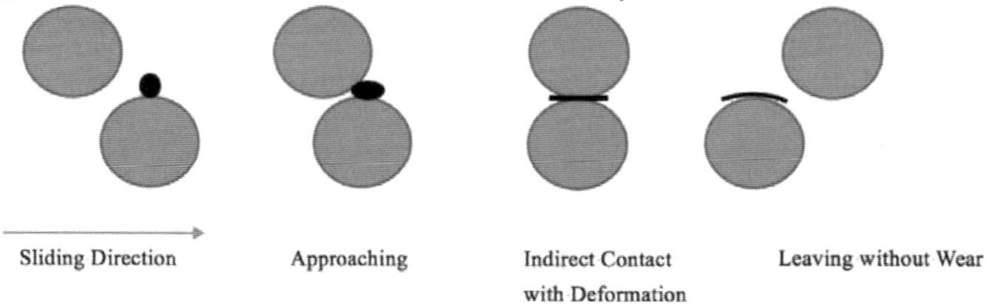

Figure 6. Diagram of the lubricating mechanism withMoS_2 nanoparticles.

With the friction time going on, the wear of the friction pair surfaces increase, and the friction coefficients increase with the accumulation of friction particles. With the friction process, the chances of contact between nanoparticles are increased so that agglomeration of nanoparticles is easy to occur to raise the friction coefficient. The following figures are three-dimensional morphologies of wear surface under the experimental force 883 N and 991 N after 30 min testing:

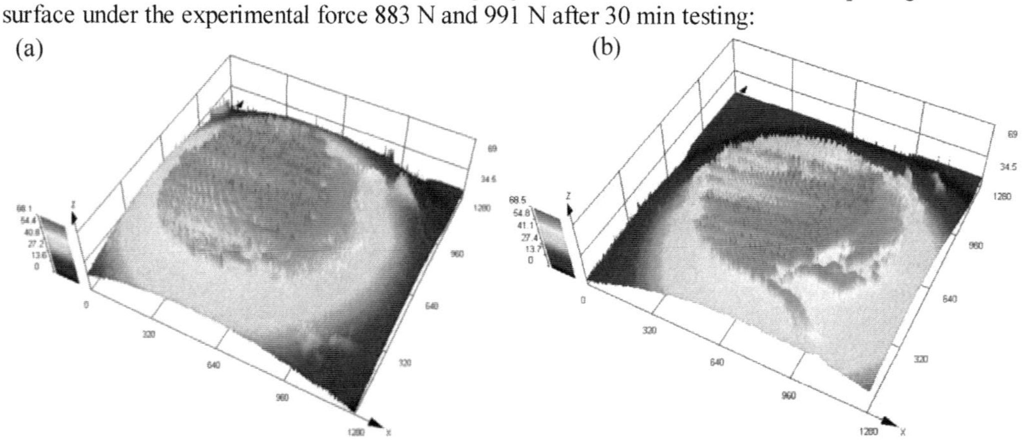

Figure 7. 3D-surface topography picture of wear scars under different pressure (a) 883 N (b) 991 N after 30 min testing.

From the morphologies, with the higher of pressure, the micro convex peaks on the surfaces of the friction pairs and the wear rate increase at the same time. Moreover, the distribution of oil film is no longer uniform under the high pressure of 991 N and grinding spot is not a complete round. The results confirm that the unstable and aggregation number of nanoparticles in the lubricating fluid of high pressure are increased, the oil film is partially damaged, resulting in poor lubrication effect. Under low pressure, the lubricating effect is ideal, the nanoparticles are not in a large scale, and the friction surfaces are smooth and the wear amount is small.

Under the experimental conditions of 588 N, the friction coefficients of the lubricating fluid after 1 day were measured at different speed. The changes of friction coefficients are shown in figure 8.

From the figure 8, with the faster of rotational speed, the friction coefficient has a tendency to decrease. High speed improves the thermal effect between the friction pairs, so the lubricant viscosity drop, lubricating effect is enhanced. In addition, when wear becomes slightly increase, a small amount of grinding particles play a role in lubrication, the friction coefficient decrease. But when the speed exceeds a certain value, the friction will destroy the integrity of the lubricating layer, so that the friction coefficients increase [20]. In the pace of the further aggravation of wear, wear particles accumulate, resulting in the third body friction, so that the friction coefficients increase. With the higher of rotational speed, the stable time of friction coefficients of lubricants is reduced. This is due to the decrease of the time with the increase of the rotational speed, which the lubricants are attached to the surface of the friction pairs. However, the high speed movement is not conducive to the adhesion of nanoparticles and the wear rate of both top ball and bottom ball is higher with more wear particles accumulation, so the friction coefficients increase slightly.

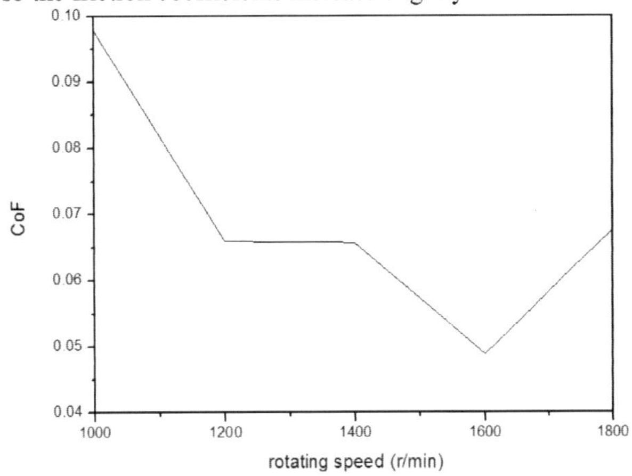

Figure 8. Graph of friction coefficients under different rotating speeds.

4. Conclusion

With the use of sodium dodecyl benzene sulfonate, 2, 2, 4-Trimethyl-1,3-pentanediol monoisobutyrate (cs-12)and sodium polyacrylate as dispersants, multilayer-MoS_2 nanoparticles are dispersed into oleic acid / triethanolamine water-based lubricants. Through the determination of MoS_2 particles' size in this lubricating liquid and Zeta potential, MoS_2 nanoparticles with smaller size can be stably dispersed in the solution, with no large-scale reunion, and dispersion effect is ideal.

After adding nanoparticles, the bearing capacity of water-based lubricating liquid is obvious improved. Under 883 N the lubricants still have good lubrication effect, and the solution can withstand the pressure of 5.9 GPa. The friction coefficients of water-based nano-MoS_2 lubricating fluid decrease with the increase of rotational speed. The yield strength of multilayer-MoS_2 nanoparticles is lower than that of singlelayer-MoS_2, but higher than the yield strength of the general steels, so it can provide good lubrication effects in the actual rolling process. The calculation results show that the low friction coefficients under high load are caused by the sliding between the layers of MoS_2 nanoparticles. The water-based lubricants can be used under the condition of high load and high speed in the actual production process. In the process of friction to 500 s or so, the lubricating fluid shows the best lubrication effect.

Nanoparticles as additives in the large particle size will have a scratch on the friction surface, causing fluctuations in the coefficient of friction, the short storage time and other issues arise. It is still a problem remains to be solved that how to stabilize the nanoparticles in water with smaller particle size in aneconomic, convenient and efficient way.

Acknowledgements

The present study is financially supported by the National Natural Science Foundation of China(51274037)

References

[1] Gulzar M *et al.* 2016 Tribological performance of nanoparticles as lubricating oil additives *Journal of Nanoparticle Research* **18(8)**

[2] Hong Mei X *et al.* 2016 An Investigation on the Tribological Performances of the SiO_2 /MoS_2, Hybrid Nanofluids for Magnesium Alloy-Steel Contacts *Nanoscale Research Letters* **11(1)**
[3] Tomala A *et al.* 2010 Tribological properties of additives for water-based lubricants *Wear* **269(11)** 12804–12810
[4] J L Sun, Z X Zhu, and P F Xu 2015 Study on the Lubricating Performance of Nano-TiO_2 in Water-Based Cold Rolling Fluid *Materials Science Forum* 817219–817224
[5] Xu Y H *et al.* 2012 Preparation of water-soluble copper nanoparticles and evaluation of their tribological properties *Mocaxue Xuebao/tribology* **32(2)** 165–170
[6] Liang S *et al.* 2015 In-situ exfoliated graphene for high-performance water-based lubricants *Carbon* **96** 1181–1190
[7] Bertolazzi, Simone, J Brivio and A Kis 2011 Stretching and Breaking of Ultrathin MoS2 *Acs Nano* **5(12)** 9703–9709
[8] X H Cui, H Chen and T Yang 2016 Research Progress on the Preparation and Application of Nano-sized Molybdenum Disulfide *Acta Chimica Sinica -Chinese Edition* **74(5)**
[9] L L Zhang *et al.* 2015 The preparations of Graphene-Like Molybdenum Disulfide and Research in Tribological Properties in High Vacuum *Tribology* **6** 746–753
[10] Q J Meng *et al.* 2011 Preparation of the Nanoscale Water-Dispersible Molybdenum Disulfide and Evaluation of Its Tribological Behavior *Tribology* **02** 144–149
[11] S M Zhang *et al.* 2014 Investigation on Wear-resistance and Anti-friction Performance of Water-based Nano Hydraulic Fluid *Tribology* **2** 211–216
[12] Spear, Jessica C, B W Ewers and J D Batteas 2015 2D-nanomaterials for controlling friction and wear at interfaces *Nano Today* **10(3)** 301–314
[13] J L Qiao *et al.* 2014 Tribological Properties of Graphene-based Fe_3O_4 Nanocomposite Materials *Journal of Inorganic Materials* **30(1)** 41–46
[14] H M Yang *et al.* 2001 Oleic acid / triethanolamine / liquid paraffin / water lubrication performance of the lamellar liquid crystal system *Chinese Journal of Applied Chemistry* 03176–03179
[15] A M Al-Sabagh *et al.* 2014 Synthesis of Some Surfactants Based on Polytriethanolamine and Investigation of Their Surface Active Properties *Ui Sahak* **23** 2239–2268
[16] J S Pang *et al.* 2009 Research on Dispersion of ZnO Nanoparticles in Aqueous Coating System *Bulletin of Chinese Ceramic Society* 01108-01112+116
[17] Antonyuk S *et al.* 2005 Breakage behaviour of spherical granulates by compression *Chemical Engineering Science* **60(14)** 4031–4044
[18] Vattikuti *et al.* 2015 Synthesis and structural characterization of MoS_2, nanospheres and nanosheets using solvothermal method *Journal of Materials Science* **50(14)** 5024–5038
[19] Hussain *et al.* 2016 Water driven flow of carbon nanotubes in a rotating channel *Journal of Molecular Liquids* **214** 136–144
[20] Y F Fu *et al.* 2016 The preparation and the friction and wear behaviours of TiO/CNT/PI composite film *Journal of Experimental Nanoscience* **49** 21–11

IOP Conf. Series: Materials Science and Engineering **182** (2017) 012024 doi:10.1088/1757-899X/182/1/012024

Size-dependent photoluminescent property of hybrid nanoparticlesconsisted with $YVO_4:Eu^{3+}$andgold

W Zhang[1], Y L Yan[1,*], F Q Guo[1], L S Zhong [1], J Wang[1], Y Wang[1] and Y H Xu[1]

[1] School of Materials Science and Engineering, Xi'an University of Technology, Xi'an 710048, China

E-mail: *yyl3550@xaut.edu.cn

Abstract. The hydrothermally synthesized$YVO_4:Eu^{3+}$nanophosphors were mixed with gold nanoparticles with different sizes to form hybrid nanoparticles. The microstructure and luminescent properties of all hybrid nanoparticles were charaterized by X-ray diffraction (XRD), transmission electron microscopy (TEM), UV-Visible absorption spectrum(UV-Vis),and photoluminescence (PL) spectroscopy. The growth process of gold nanoparticles prepared by seeding growth method was discussed firstly. Furthermore, the effects of grain size of gold nanoparticles on the fluorescent property of hybrid nanoparticles were investigated in detail. All the emission spectra presented the characteristic red emission of Eu^{3+}. The emission intensity of hybrid nanoparticles was increasing with the growing of gold nanoparticles. The influential mechanism was discussed as well.

Keywords: $YVO_4:Eu^{3+}$, gold nanoparticles, size-dependent, photoluminescent properties

1. Introduction

YVO_4 crystal has been considered as an excellent host material because of its outstanding chemical, thermal, mechanical and optical properties as well as easy to be doped with trivalent rare earth ions [1,2]. Eu^{3+} ions are excellent luminescent centers with the parity forbidden characteristics exhibiting strong red line emissions with high color purity [3]. As a kind of excellent commercial phosphor, $YVO_4:Eu^{3+}$ has been widely used in many fields, such as luminescent lamps, solid-state lasers, optical fiber communication systems, flat displays, and photonic devices [4-7].

Recent years, higher requirements of fluorescence materials about stability, brightness, color purity and so on are put forward due to its expansive application in drug delivery, biolabeling and solar cell. Therefore, $YVO_4:Eu^{3+}$ phosphor has become a hot research topic because of its high purity and high quantum yield. Currently, the methods to improve the luminescent property of $YVO_4:Eu^{3+}$ phosphors are including ions doping [8-11], core-shell structure [12-17], metal enhancing [18-20], and so on.

Metal enhanced fluorescence is a promising method to enhance the luminescent property of lanthanide nanophosphors [21,22]. The highly localized electromagnetic (EM) fields generated around metal nanoparticles (MNPs) are responsible for the enhancement. The EM fields are generated by resonant excitation of conduction electrons in MNPs also known as surface plasmon, upon excitation by incident light [23]. It is suggested that the plasmon absorption band and scattered light of MNPs both contribute to the enhancement process [24].The MNPs can modify the free-space conditions in ways that increase the incident EM and decrease the radiative decay rate, namely, the MNPs provide more radiative decay pathways and larger radiative decay rates [25].

In this paper, we present a study of a hybrid nanoparticles consisting of $YVO_4:Eu^{3+}$ nanocrystals and gold nanoparticles (Au NPs). The fluorescent property of the hybrid nanoparticles with different Au NPs sizes was investigated. The excitation and emission intensity of $YVO_4:Eu^{3+}$ nanocrystals were influenced by Au NPs due to the local plasma resonant effect. The details were discussed below.

2. Experimental

All the chemical reagents purchased from Sinopharm Chemical Reagent Beijing Co., Ltd. (China) were of analytical grade and used as received without further purification.

2.1. $YVO_4:Eu^{3+}$ nanoparticles preparation

The $YVO_4:Eu^{3+}$ nanoparticles were synthesized by typical hydrothermal method. Firstly, all of rare earth chloride stock solutions of 0.1 M were prepared by dissolving the corresponding metal oxide in hydrochloric acid at elevated temperature. In a typical procedure, the Eu^{3+} concentration in the YVO_4 host was 5 mol.%, a certain amount of YCl_3 was added into aqueous solution containing 2 mmol of trisodium citrate. Then, 1 mmol Na_3VO_4 was introduced into the solution. After additional agitation for 1h, the as-obtained mixing solution was transferred into a Teflon bottle held in a stainless steel autoclave, sealed, and maintained at 180℃ for 24h. As the autoclave cooled to room temperature naturally, the precipitates were separated by centrifugation, washed with ethanol and deionized water in sequence, and finally dried in air at 80℃ for 12h.

2.2. Au NPs preparation

Au NPs with various diameters were obtained by using classic seeding growth method [26].

2.2.1. Preparation of Gold Seed. A 20 mL aqueous solution containing 2.5×10^{-4}M $HAuCl_4$ and trisodium citrate were prepared in a conical flask. Then,0.3mL ice-cold, freshly solution of 0.1M $NaBH_4$ was added to the above-mentioned solution while stirring. The formatted particles in this solution were used as seeds after the preparationof 2 h. (Citrate serves only as a capping agent since it cannot reduce the gold salt at room temperature in this section).

2.2.2. Preparation of Growth Solution. A certain amout of cetyltrimethylammonium bromide (CTAB) and $HAuCl_4$ aqueous solution were stirring under heated condition. And the solution was cooled to room temperature. The detailed process was described in Ref 25.

2.2.3. Seeding Growth. Four groups of 50 mL conical flasks were labeled A, B, C, and D. The appropriate volume of growth solution was mixed with 0.05 mL of freshly prepared 0.1M ascorbic acid solution. Next, the appropriate volume of particle seeds was added while stirring. The volumes of growth solution and seeds solution added in the conical flasks were listed in table 1.

Table 1. Reaction parameters of seeding growth method for Au nanoparticles.

Sample name	Ascorbic acid solution (0.1M)	Growth solution	Seed Solution		
			Original	B sample	C sample
A	0.05 mL	7.5 mL	2.5 mL		
B	0.05 mL	9.0 mL	1.0 mL		
C	0.05 mL	9.0 mL		1.0 mL	
D	0.05 mL	9.0 mL			1.0 mL

2.3. Synthesis of $YVO_4:Eu^{3+}$–Au hybrid nanoparticles

The as prepared citrate-coated YVO_4 nanoparticles were dissolved in deionizedwater to form suspension, and then four different CTAB-coated Au nanoparticles were added to the suspension. Kept stirring for 2h.The YVO_4:Eu^{3+}-Au hybrid nanoparticles were prepared after excess Au colloid was wiped out by centrifugation.

2.4. Characterization methods

The X-ray diffraction(XRD) analysis (XRD-7000 X-ray diffractometer, Shimadzu, Japan) was used to identify the crystalline phases and operated at 40 kV and 40 mA, Cu Ka radiation was used and the goniometer was run from 15° to 80°. Transmission electron microscopy (TEM) and selected area electron diffraction (SAED) were performed on a JEOL 3010 transmission electron microscope, using an accelerating voltage of 300 kV. UV-Vis spectra were recorded from 400 nm to 600 nm using a UV-3010 UV-Vis spectrophotometer. The photoluminescence (PL) excitation and emission spectra were recorded with 970CRT fluorescence spectrophotometer equipped with a 150 W xenon lamp as the excitation source.

3. Results and discussion

3.1. Phase, structure and morphology of YVO_4:Eu^{3+} nanoparticles

Figure 1(a) shows the XRD patterns of YVO_4:Eu^{3+} nanoparticles. According to JCPDS standard card No. 72-0861, all peaks are assigned to tetragonal crystalline phase YVO_4 with lattice contents a = b = 7.12 Å and c = 6.29 Å. No other impurity peaks are detected indicating pure YVO_4powder was prepared and Eu^{3+} doping don't affect its crystal structure. The morphologies of YVO_4:Eu^{3+} nanocrystals characterized by TEM are presented in figure 1(b). The TEM image indicates that as-prepared YVO_4:Eu^{3+}nanocarystals possess rod-like shape with a length of ~50 nm and a width of ~20 nm. Furthermore, the SAED image (inset as figure 1(c)) reveals the polycrystallinity of YVO_4:Eu^{3+} nanocrystals with spotty diffusion rings indexing to the tetragonal phase of YVO_4.

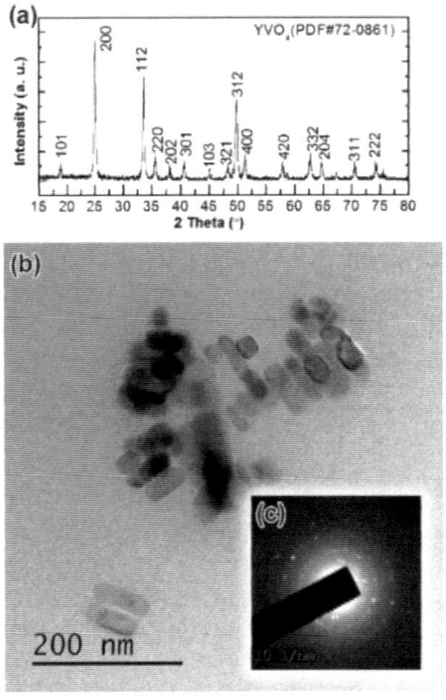

Figure 1. XRD pattern(a), TEM image(b), and SAED image(c) of YVO_4:Eu^{3+} nanoparticles.

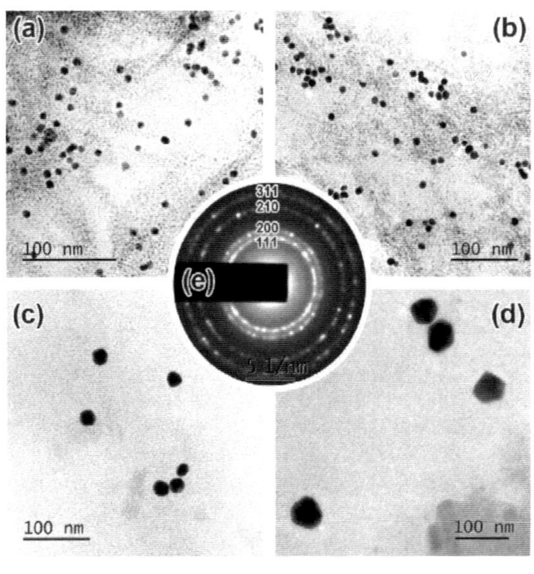

Figure 2. TEM images of Au NPs with four different diameters (a) 8.5±0.6nm, (b) 10.0±0.8 nm, (c) 24.5±0.8 nm, and (d) 55±0.5 nm. The SAED image of Au NPs.

IOP Conf. Series: Materials Science and Engineering **182** (2017) 012024 doi:10.1088/1757-899X/182/1/012024

3.2. Size-dependent Au NPs
The TEM images of Au NPs obtained from A, B, C and D groups were exhibited in figure 2. As can be seen that the products' diameter is increasing from 8.5±0.6nm, 10.0±0.8 nm, 24.5±0.8 nmto55±0.5nm corresponding to A, B, C and Dgroups, respectively. Otherwise, all the NPs distribute homogeneously and independently, which can be attributed to the feature of CTAB and the negative electrical property of Au NP. The Au NPs with sizes of 8.5 nm and 10 nm (in figure 2(a-b)) present good size distribution and uniform spherical morphology. But the size distribution of the Au NPs with sizes of 24.5 nm and 55 nm (in figure 2(c-d)) become wider and present polygonmorphology, which results from the non-uniform adsorption of chlorauric acid on the crystal surface during the step by step reduction process. Furthermore, the SAED image (inset as figure 2(e)) presents bright spots rings, which verify the polycrystallinity of Au NPs. All four products exhibit similar SAED images those are not repeated here.

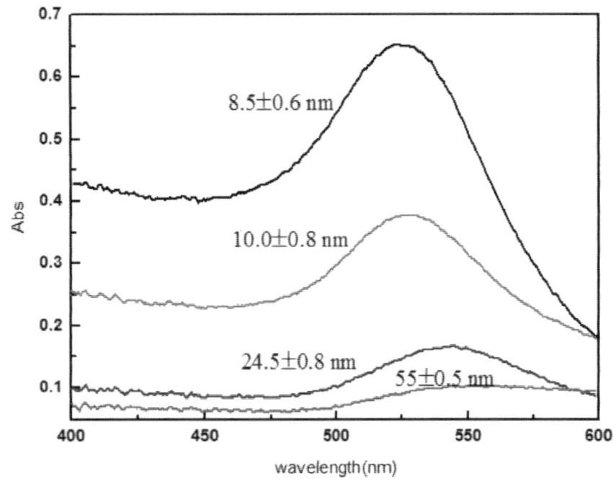

Figure 3. UV-Vis absorption spectrum of Au NPs with four different particle size.

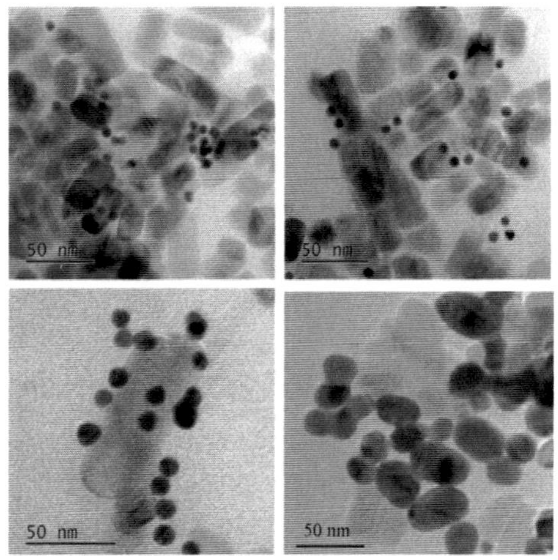

Figure 4. The TEM image of four different diameters Au NPs and $YVO_4:Eu^{3+}$ hybrid nanocomposite.

For the UV-Vis absorption spectrum of nano dimension, the electron cloud of the particle surface can generate oscillation through the electromagnetic radiation of specific energy absorption. It could produce resonance when the frequency of electromagnetic wave equals to the plasma oscillation. The resonance macroscopically shows the absorption of light by metal nanoparticles, which is seriously affected by the particle size. Figure 3 shows the corresponding UV-Vis absorption spectra of Au NPs with different particle sizes. It can be seen that the plasma absorption band is broadened and the strength is weaken with the increasing of the Au NPs size from 8.5nm to 55nm, which means the coupling of plasmon mode is weaken. Meanwhile, the plasma absorption band location is red shifted from 525 to 540 nm due to size shrinkage.

3.3. Morphology of the hybrid nanoparticles
Figure 4 is the TEM image of four kinds of hybrid nanoparticles with different Au NPs' diameters. As can be seen, the dark particles represent Au NPs, while the light particles represent $YVO_4:Eu^{3+}$ nanoparticles. It also can be seen clearly from the figure 4 (a-c) that the spherical Au NPs with four different sizes and $YVO_4:Eu^{3+}$ nanoparticles distribute alternatively. Some of the Au NPs are attached on the surface of $YVO_4:Eu^{3+}$ nanoparticles due to the Van der Waals' force. In figure 4 (d), the Au NPs possess the similar size with $YVO_4:Eu^{3+}$ nanoparticles.

3.4. Size-dependent photoluminescent properties

Figure 5 is the excitation and emission spectra of the hybrid nanoparticles consisted with $YVO_4:Eu^{3+}$nanoparticles and Au NPs at room temperature. All the excitation spectra are monitored at the emission wavelength of 618nm.It is obviously that the excitation intensity is higher than pure $YVO_4:Eu^{3+}$ nanoparticles, and the intensity increase with the grown of the grain size of Au NPs. The excitation band at 200 ~ 300 nm is caused by energy transition (ET) between VO_4^{3-} and Eu^{3+}. And the violent fluctuation of the excitation curve is attributed to the strong reflection of local plasma resonant field around the Au NPs. The emission spectra are obtained under excitation of 281 nm. Two main identifiable peaks located at 593 nm and 618 nm are connected with $^5D_0 \rightarrow {}^7F_1$ and $^5D_0 \rightarrow {}^7F_2$ characteristic transition of Eu^{3+}. Luminescent properties of Eu^{3+} in the crystal lattice are related to the position, when in the non-spatial inversion symmetry, Eu^{3+} exhibit electric dipole transitions ($^5D_0 \rightarrow {}^7F_2$), however, at the center of the inversion sites, Eu^{3+} exhibits magnetic dipole transitions ($^5D_0 \rightarrow {}^7F_1$). So in the crystal of as-prepared $YVO_4:Eu^{3+}$ nanoparticles, Eu^{3+} ions occupied the non-spatial inversion sites. The intensity of the main emission peaks is enhanced with the increasing of theAu NPs'grain size. It can be seen in the figure 3 that with the size reduction of Au NPs, the local plasma resonant field is stronger, but the fluorescent property decreased. According to Lakowicz report [25], the metal enhanced fluorescence is due to at least three known mechanisms: (1) energy transfer quenching, (2) the increasing of the local incident field on the fluorophore, (3) the increasing of the intrinsic radiative decay rate of the fluorophore. In the hybrid nanoparticles, the absorption band of Au NPs is not overlapped with the emission peaks of $YVO_4:Eu^{3+}$nanoparticles. So the energy transfer quenching is not presented in the $YVO_4:Eu^{3+}$nanoparticles mixed with smaller Au NPs (8~10 nm). But with the increasing of Au NPs' size, the local incident field is enhanced. Thus the $YVO_4:Eu^{3+}$ nanoparticles mixed with larger Au NPs (25~55 nm) exhibit enhanced emission intensities. The radioactive decay rate is not detected in this experiment. The schematic diagram of fluorescence enhancement is shown in figure 6.

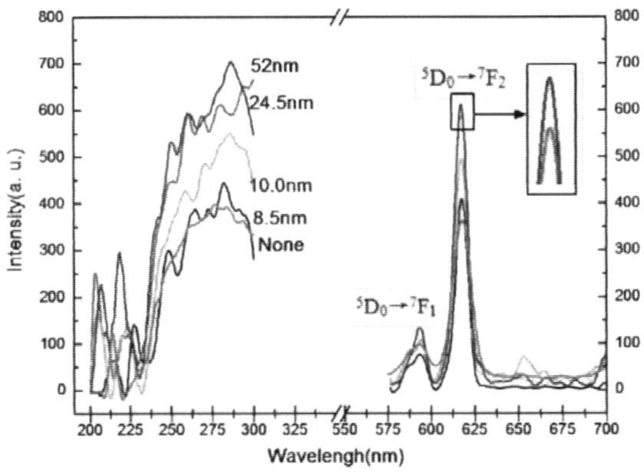

Figure 5. Excitation and emission spectrum of $YVO_4:Eu^{3+}$ hybrid four sizes Au NPs.

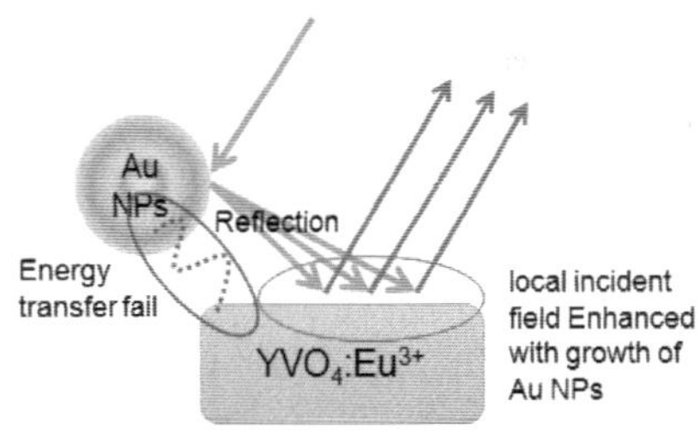

Figure 6. The schematic diagram of fluorescence enhancement.

4. Conclusions

In summary, a series of hybrid nanoparticles consisting of $YVO_4:Eu^{3+}$ nanophosphors and Au NPs were prepared and examined for fluorescent property. The excitation and emission intensity were increased with the size growing of Au NPs. This is a result of enhanced reflection of plasma resonant effect of Au NPs.

Acknowledgment

This work was supported by the International Science and Technology Cooperation Program of China (Nos.2015DFR50350) and Scientific Research Program Funded by Shaanxi Provincial Education

Department (Nos. 15JK1536, 14JS047), Innovation project of Xi'an University of technology（Nos. 2015CX002, 2014CX024).

References

[1] Wiglusz R J, Bednarkiewicz A and Strek W 2012 *Inorg. Chem.* **51** 1180–1186
[2] Wang E, Li G and Zhang J 2014 *J. Phys. Chem.* **118** 3820–3827
[3] Ma Q and Wang Q 2015 *Displays* **39** 6–10
[4] Liu Y, Tu D , Zhu H and Chen X 2013 *Chem. Soc. Rev.* **42** 6924–6958
[5] Hu Y B, Qiu J B and Jiao Q 2013 *Adv. Mater. Res.* **815** 171–178
[6] Xu W, Song H and Yan D 2011 *J. Mater. Chem.* **21** 12331–12336
[7] Ai Y, Tu D and Zheng W 2013 *Nanoscale* **5** 6430–6438
[8] Zhao M, Li G and Zheng J 2012 *Crystengcomm* **4** 2062–2070
[9] Wang G, Qin W and Zhang D 2008 *J. Phys. Chem.C* **112** 17042–17045
[10] Tian L and Mho S 2007 *J. Lumin.* **122** 99–103
[11] Yu M., Lin J and Fang J 2005 *Chem. Mater.* **17**
[12] Pan G, Song H and Bai X 2007 *J. Phys. Chem. C* **111** 12472–12477
[13] Chang M and Tie S 2008 *Nanotechnology* **19** 267–274
[14] Zhou J C, Sun L D and Shen J 2011 *Nanoscale* **3** 1977–1983
[15] Eichelbaum M and Rademann K 2009 *Adv. Funct. Mater.* **19** 2045–2052
[16] Lee S M, Choi K C and Kim D H 2011 *Opt. Express* **19** 13209–13217
[17] Xu W, Bai X and Xu S 2012 *Rsc Adv.* **2** 2047–2054
[18] Iso Y, Takeshita S and Isobe T 2014 *Langmuir* **30** 1465–1471
[19] Luwang M N, Ningthoujam R S and Srivastava S K 2011 *J. Mater. Chem.* **21** 5326–5337
[20] Sun Z, Tong L and Liu D 2012 *J. Mater. Chem.* **22** 6280–6284
[21] Feng W, Sun L D and Yan C H 2009 *Chem. Commun.* **29** 4393–4395
[22] Lakowicz J R 2005 *Anal. Biochem.* **337** 171–194
[23] Buch Z, Kumar V and Mamgain H 2013 *Chem. Commun.* **49** 9485–9487
[24] Mendes M J, Hernández E and López E 2013 *Nanotechnology* **24** 2589–2593
[25] Jana N R, Gearheart L and Murphy C J 2001 *Langmuir* **17** 6782–6786
[26] Lu Z H, Lockwood D J and Baribeau J M. 1995 *Nature* **378** 258–260
[27] Sharma P K, Nass R and Schmidt H 1998 *Opt. Mater.* **10** 161–169
[28] Singh N S, Ningthoujam R S and Singh S D 2010 *J. Lumin.* **130** 2452–2459
[29] Mishra K C, Berkowitz J K and Johnson K H 1992 *Phy. ReV. B* **45** 10902–10906

IOP Conf. Series: Materials Science and Engineering **182**(2017) 012025 doi:10.1088/1757-899X/182/1/012025

Tribological properties and lubricating mechanism of SiO_2 nanoparticles in water-based fluid

Y Y Bao, J L Sun* and L H Kong

School of Material Science and Engineering, University of Science & Technology Beijing, 30 Xue Yuan Road, Beijing

E-mail: sjl@ustb.edu.cn

Abstract. The tribological properties of surface modified SiO_2 nanoparticles suspension in water-based lubricant have been studied. SiO_2 (30 nm) nanoparticles were dispersed through surface modification with polyethylene glycol-200. Transmission electron microscope (TEM) and infrared (IR) spectroscopy show that SiO_2 nanoparticles disperse well and stably in the water-based lubricant. The diameter of the nanoparticles is about 60 nm. Tribological properties of the water-based lubricant were evaluated using four-ball wear test machine and pin-on-disk tester under different loads and different concentrations of SiO_2 nanoparticles. Wear surface morphology, element chemistry configuration of steel balls and steel rings were studied by means of X-ray photoelectron spectroscopy (XPS) and scanning electron microscopy (SEM). Results show that the tribological properties of the water-based fluid have been improved by addition of nanoparticles ranging from 0.1% to 0.3% concentrations. SiO_2 nanoparticles deposited onto the wear surface during the sliding, which helped to reduce the friction coefficients and increase the anti-wear properties due to the miniature ball bearing effect and self-repairing performance of nanoparticles between the friction pairs. With the increase of test load, the friction coefficients decrease but the wear of the surface increase.

Keywords: SiO_2 nanoparticles, Tribological properties, Lubrication mechanism

1. Introduction

With the development of nanomaterials in lubrication technique and the deepening understanding of the particularity of nano-functional materials, nanomaterials have a broad application prospect in the field of metal processing lubrication with its unique physical and chemical properties. Using different additives to improve the tribological performance of lubricants is a widely used method [1-2]. With the researches of the nano-tribology gradual deepen, metal and oxide in nanoscale as functional additives in lubricants have been reported have good anti-wear, antifriction and extreme pressure performance, also have a certain of self-repairing function, as well as solving environmental problems of oil lubricants [3].

At present, there are many researches of water-based nano-fluid using in the lubrication, but the application of SiO_2 nanoparticles in water-based fluid are rarely reported. In this paper, a stably dispersed SiO_2 nanoparticles water-based fluid was obtained, with polyethylene glycol-200 as dispersants. Tribological performances of the water-based lubricants were evaluated using Four-ball friction and wear tester and pin-on-disk tester under different loads and different concentrations of SiO_2 nanoparticles. Wear surface morphology and element chemistry configuration of steel rings were analyzed by the means of optical microscope, X-ray photoelectron spectroscopy (XPS) and scanning electron microscopy (SEM).

2. Experimental

2.1. The preparation of water-soluble SiO_2 nanoparticles

The surface modification process is described as follows: SiO_2 nanoparticles（30 nm） were stirred for 5 minutes in 15% ammonia solution at room temperature and then dried in the oven at room temperature. After drying, nanoparticles were putted in the aqueous solution of polyethylene glycol－200, stirred for 30 minutes under 60 ℃, the quality ratio of nanoparticles and polyethylene glycol－200 was 1:1, then the mixture was placed in the ultrasonic disperser with frequency of 20 KHZ and scattered for 5 minutes, ultimately, a stably dispersed water-soluble SiO_2 nanoparticles were obtained.

2.2. Tribological performance test

Two groups of water-based fluid were used to study the tribological performance of nanoparticles, one group without nanoparticles (base fluid), and the other was added with nanoparticles with the concentration varied from 0.1 wt.% to 0.5 wt.%. The last nonseizure load (PB) of water-based fluid has been evaluated with four-ball wear test machine. According to the ASTM D2783, the spindle speed was 1760 r/min, and experiments were lasted for 10s at room temperature, GCr15 steel balls with the hardness of HRC 61 ~ 65 and the diameter of 12.7 mm were used. The friction and wear studies of water-based fluid were performed using pin-on-disk tester, the spindle speed was 300r/min and experimenting for 30 minutes at room temperature. The ring was made from 45 # steel, and the inside diameter was 16 mm, outside diameter was 31.7 mm, and the surface hardness was HB162. Before experiment, the surfaces were grinded on 2000 # sandpaper to ensure that the initial surfaces were all at the same. The chemical composition of 45 # steel was shown in Table 1. After the test, the samples were used for surface analysis after rinsing with petroleum ether. The morphology and element analysis of the samples were acquired by scanning electron microscope (SEM), and elements combining state and binding energy of the sample surface were evaluated by X-ray photoelectron spectrometer (XPS).

Table 1. Chemical composition of 45 # steel.

Element	C	Si	Mn	Cr	Ni	Cu
Content/%	0.42 ~ 0.50	0.17~0.37	0.50~0.80	<0.25%	<0.30	<0.25

3. Results and discussion

3.1. Characterization of water-soluble SiO_2 nanoparticles

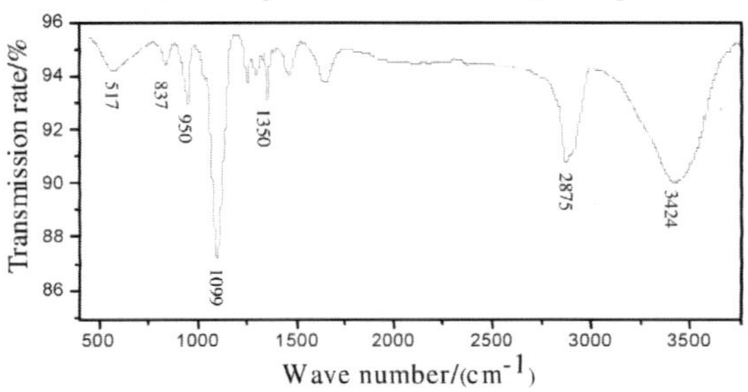

Figure 1. Infrared spectra of surface modified SiO_2 nanoparticles.

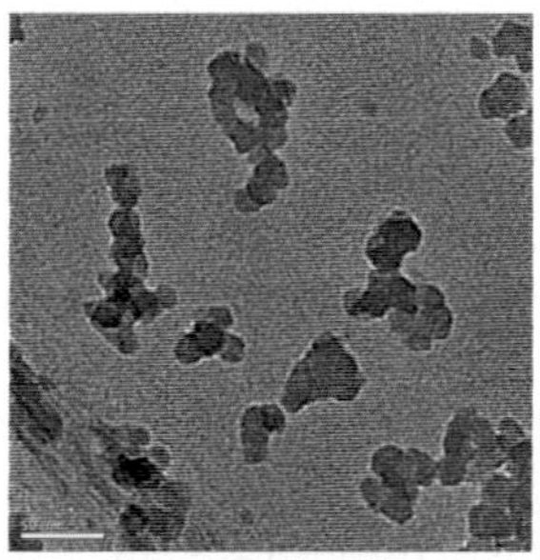

Figure 2. Transmission morphology of SiO_2 nanoparticles.

Figure1 was the infrared spectra of SiO_2 nanoparticles. The bands close to 3424 cm^{-1} correspond to the stretching vibration peak of O–H, the peaks of stretching vibration in the plane and antisymmetric stretching vibration at 2875 cm^{-1} and 1350 cm^{-1} show that CH_2 is present in the molecular structure, and the antisymmetric stretching vibration, symmetric stretching vibration and bending vibration of Si–O–Si are close to 1099 cm^{-1}, 837 cm^{-1} and 571 cm^{-1}. The weak band at 950 cm^{-1} corresponds to the characteristic stretching vibration of Si–O–C [4]. According to the infrared spectrum of SiO_2 nanoparticles [5,6], there were stretching vibration peak of Si-OH at 2300 cm^{-1}, which didn't seen in figure 1. Thus the infrared spectrum of the samples indicates that the Polyethylene glycol-200 adsorbed on the surface of the SiO_2 nanoparticles by condensation reaction, which result in homodisperse of SiO_2 nanoparticles. The transmission morphology of surface modified SiO_2 nanoparticles (0.1 wt.%) is presented in figure 2. It can be seen that SiO_2 nanoparticles is amorphous and the average diameter is about 50 nm, and also the SiO_2 nanoparticles dispersed evenly in water, which confirms that nanoparticles were dispersed successfully.

3.2. Effect of the concentration of SiO_2 nanoparticles

Figure 3 shows the P_B and friction coefficients of water-based fluid with varying concentration of SiO_2 nanoparticles. According to figure 3, it can be seen that the P_B increased first but then decreased with the increasing concentration of SiO_2 nanoparticles, and reached the top of 883 N at 0.3 wt.% SiO_2 nanoparticles. In addition, the P_B of water-based fluid which contains SiO_2 nanoparticles were improved obviously than that with base fluid, and the minimum increase is 13%. This is mainly because that the size of nanoparticles is small and well dispersed in the base fluid. In the process of sliding, nanoparticles easily entered into the friction pair, which prevents direct contact of friction pair, and the bearing capacity of water-based fluid is highly improved because of the high hardness of nanoparticles [7]. When the concentration of nanoparticles is more than 0.3 wt.%, SiO_2 nanoparticles would produce irreversible agglomeration which will result in the increases of particle size, reducing the amount of the nanoparticles entering into the friction pair, thus reduce the bearing capacity of water-based fluid.

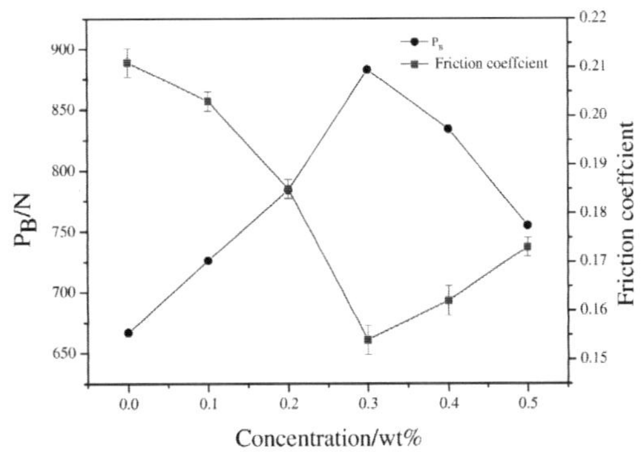

Figure 3. P_B and friction coefficients of water-based fluid with varying nanoparticle concentration.

The variation of friction coefficients with the increasing concentration of SiO_2 nanoparticles was plots in figure3, which illustrates that the friction coefficients decrease when concentrations of SiO_2 nanoparticles are less than 0.3 wt.%, while increase when concentrations exceed 0.3 wt.%, the minimum value of friction coefficients is 0.154, reduced by 28% compared with the base fluid. Besides, the friction coefficients of water-based fluid with addition of nanoparticle were all less than that of base fluid. Accordingly, SiO_2 nanoparticles have good anti-friction performance, it's contributes to the miniature ball bearing effect of SiO_2 nanoparticles, which changed the friction conditions from sliding friction to rolling friction, thereby reduce the friction coefficients [8,9]. When the amount of SiO_2 nanoparticles was exceed 0.3 wt.%, the probability of nanoparticles forming agglomeration will increase, which made it difficult to get into the friction pair, and the larger particles lead to the surface scratched, leading to the friction coefficients increase [10].

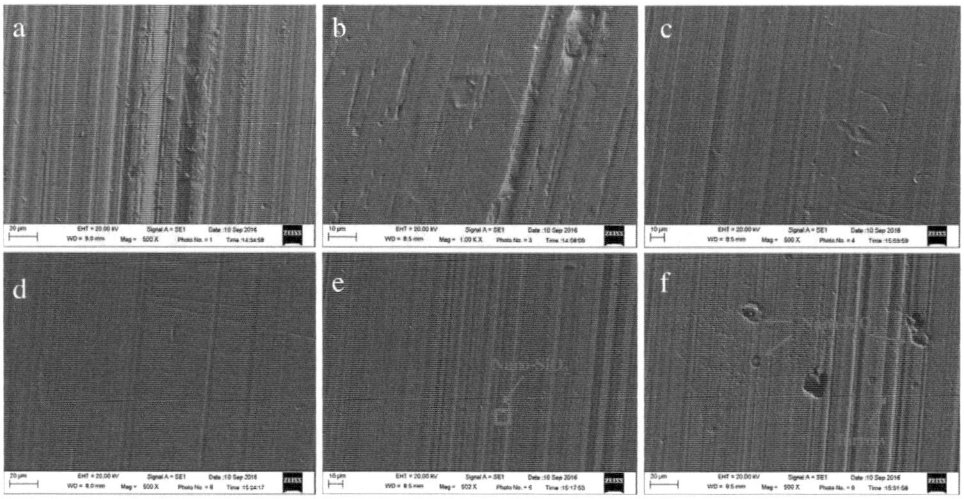

Figure 4. Wear surface morphology of steel rings lubricated under water-based fluid with (a)base fluid, (b)0.1 wt.% SiO_2, (c)0.2 wt.% SiO_2, (d)0.3 wt.% SiO_2, (e)0.4 wt.% SiO_2, (f)0.5 wt.% SiO_2.

Figure 4 presents the wear surface morphology of steel rings lubricated under water-based fluid with different concentrations of nanoparticles. In figure 4(a), there were many scratches, and the scratches were intensive and deep. What's more, two furrows were observed, indicating that the surface was severely weared when lubricated with base fluid. While when adding nanoparticles, the wear surfaces were more smooth and had less furrows compared with that of base fluid. When the nanoparticles content was more than 0.3 wt%, the number of scratches became more, more pits were also observed compared with figure 4(c), (d), nonetheless, the wear surfaces were better than that lubricated with base fluid. The reasons will be discussed in the mechanism of lubrication.

3.3. *Influence of the test load*

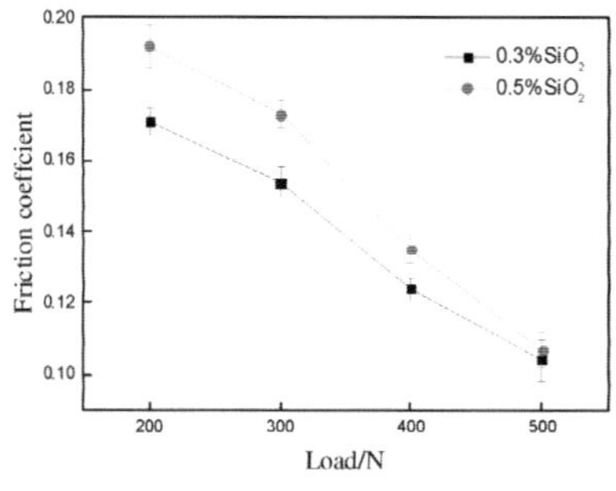

Figure 5. Friction coefficients of water-based.

The friction coefficients of water-based fluid with 0.3%wt and 0.5 wt.% SiO_2 nanoparticles under different test load were shown in figure 5. We can see that with the increase of test load, the friction coefficients were decreased, and the difference of friction coefficients between two kinds of water-based fluid were becoming smaller. This phenomenon will be discussed in detail in the following section. The wear surface morphology lubricated with 0.5 wt.% SiO_2 nanoparticles water-based fluid under different pressures were shown in figure 6. Scratches became wider and the pits on the wear surface were disappeared with the increase of test load. In addition, there were adhesive wear and plough wear in figure 6(b), (c) and (d).

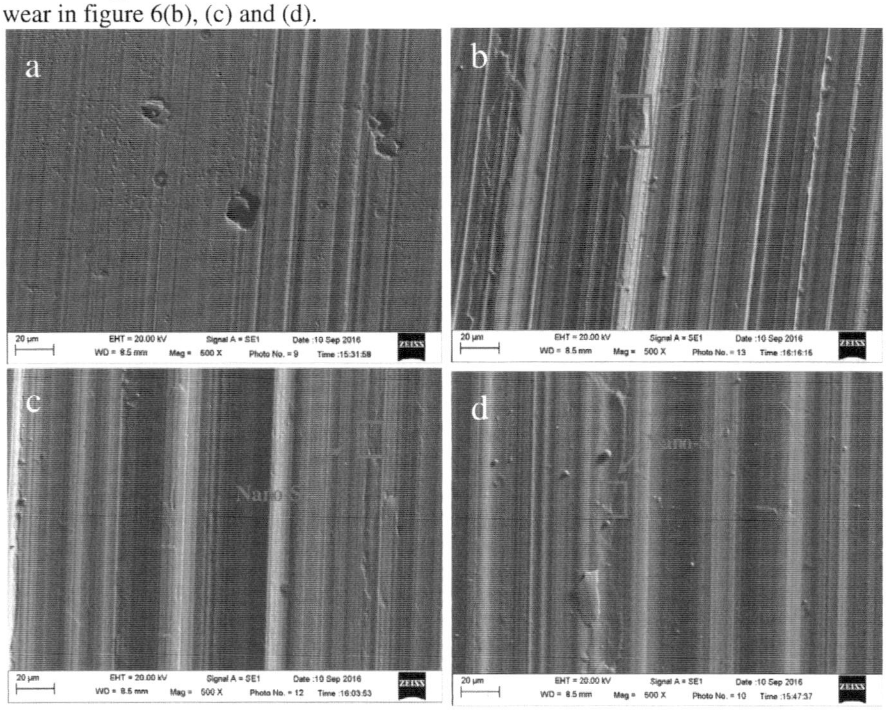

Figure 6. Wear surface morphology under Nano-rolling fluid under different test load. (a) 200N, (b) 300N, (c) 400N, (d) 500N.

3.4. *Inquiry of lubrication mechanism*

Wettability of lubricants reflects the adsorption and spreading performance of lubricants on the work piece and roll surface, therefore, it is necessary to discuss the effect and mechanism of nanoparticles on wettability. Figure 7 plots the influence of nanoparticles concentration on the wettability of water-based fluid, which are corresponding to 0, 0.1 wt.%, 0.3 wt.% and 0.5 wt.% nanoparticles respectively. The figures show that with the increase of nanoparticle concentration, the contact angle of water-based fluid was decreased. The higher concentration of nanoparticles, the easier water-based fluid adsorbed and spread on the metal surface, which is mainly because nanoparticles have high surface energy. With the addition of nanoparticles, the surface energy of the base fluid is reduced, the same with surface tension, therefore make the contact angle decrease. In addition, there are a large number of hydroxyl groups in the surface of nanoparticles, which easily adsorbed on the metal surface, making the adsorption increase and the contact angle decrease.

In order to study the mechanism of SiO_2 nanoparticles as water-based additives, the EDS and XPS of the wear surface under 0.5 wt.% SiO_2 nanoparticles were analyzed. The elemental composition and typical chemical state of element analysis results were shown in figure 8 and figure 9 respectively. Seen from the figure 8, the element of silicon was detected on wear surface, which illustrated that nano-SiO_2 were filled in the defects. The figure 9 showed there were two peaks at the binding energy of 102.7 eV for silica and 532 eV for oxygen, which were belong to the chemical state of SiO_2 [11,12]. It showed that there was no reaction of SiO_2 nanoparticles, which were just deposited on the surface of the samples.

There are many lubricating mechanisms of nanoparticles as additives of lubricant have been proposed, such as the rolling friction mechanism, the formation of protective film, third-body effect and so on [13].

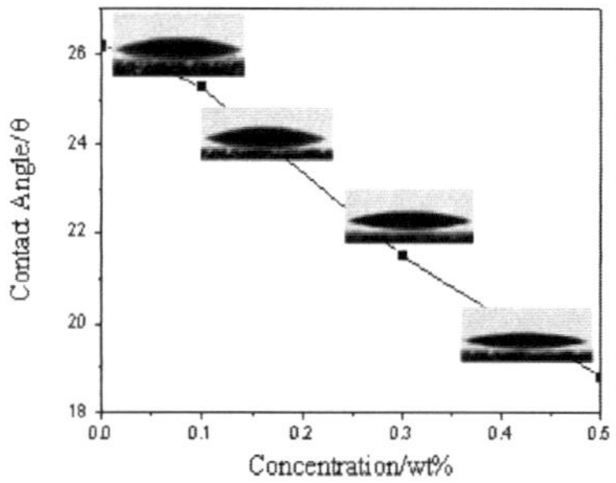

Figure 7. Influence of nanoparticle concentration on the wettability of water-based fluid.

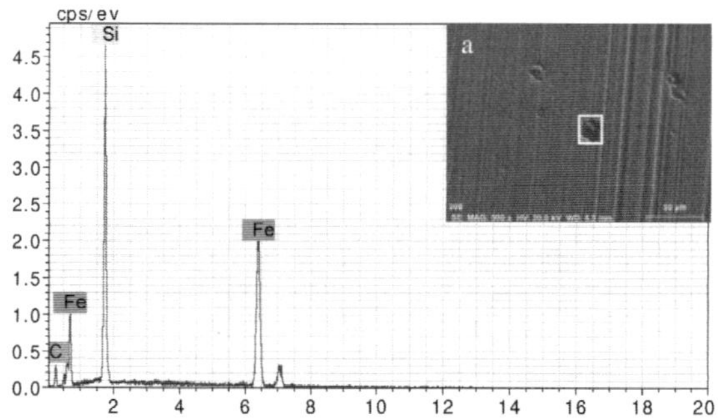

Figure 8. EDS spectra of worn surface lubricated with 0.5 wt.% SiO_2 nanoparticles. (a) 200N, (b) 300N.

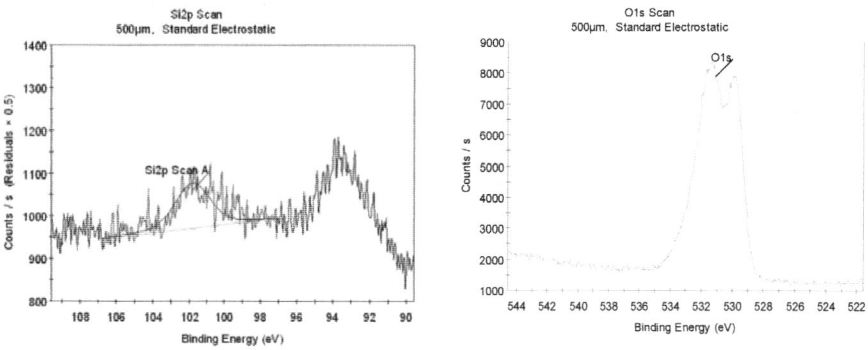

Figure 9. XPS spectra of the wear surface lubricated with0.5 wt% SiO_2 nanoparticles with concerning element. (a) Si2p, (b) O1s (200N)

The lubrication principle is proposed as follows, under the high contact load, the based fluid and nanoparticles were penetrated into the friction pair and thus forming a thin physical lubricating film [14]. This lubricating film possesses a lot of important functions, such as avoiding direct contact by separating the friction pair, and bearing the load [15]. Besides, spherical nanoparticles are most likely to roll between the friction pair and reduce the friction coefficients [16,17]. Figure4 showed that, with the increase concentration of nanoparticles, the scratches became less and shallow, and the wear of surface was decreased. The reasons can be summed up in the following two points: Firstly, the "miniature ball bearing" effect of SiO_2 nanoparticles turn a part of sliding friction into rolling friction, thus reduce the wear of surface; Secondly, there are nanoparticles in the defects of figure 4(e), (f), proving that nanoparticles can deposit in the defects and repair the surface, thereby reducing the wear of surface. Interestingly, the wear of surface become serious when the concentration is more than 0.3 wt%, due to the agglomeration of nanoparticles, which causing surface scratched and particle deposition. As shown in Figure6, scratches became wider and the pits on the wear surface were disappeared with the increase of test load. What is more, there were adhesive wear and plough wear on the surface, but the scratches become wider and wear surface roughness was decreased. However, the effect of nanoparticles during sliding depends on the thickness of the lubricant film. Assumed that the shape of nanoparticles is preserved, when the film thickness is close to the size of nanoparticles, rolling friction is considered to be the leading mechanism [18].

As shown in figure 6, with the increase of test load, there were obvious furrows in wear surface, indicating that nanoparticles did not melt, and the shape of nanoparticles was preserved. With the increase of test load, the thickness of lubricant film decreases. When test load is small, the lubrication film is denser, which plays an important role in lubrication; while with the increase of test load, the thickness of lubrication film would decreased. When the thickness was approach to the size of nanoparticles, the nanoparticles would plays a major role in lubricant, making rolling friction become the main mechanism, thus reducing the friction coefficient, on the other hand, the increaseing of test load will reduce agglomeration of nanoparticles, thus making the effect of nanoparticle concentration on the friction coefficients decreases. But because of the increasing load, nanoparticles embedded into the sample surface, making the surface wear heavier.

4. Conclusion

SiO_2 nanoparticles modified by polyethylene glycol-200 were well dispersed in water-based fluid through condensation reaction.

Carrying capacity and antifriction performance of the base fluid can be improved by adding nanoparticles when the concentration of nanoparticle is less than 0.3 wt%. The optimal concentration of SiO_2 nanoparticles is 0.3 wt%.If excess nanoparticles are added into the base fluid, SiO_2 nanoparticles would tend to form irreversible agglomeration, which will result in worse friction.

With the increase of test load, antifriction performance of SiO_2 nanoparticles has been improved, while the antiwear performance is decreased.

Tribological performance of the base fluid are improved mainly by the "miniature ball bearing" effect and self-mending effect of SiO_2 nanoparticles,when the test load increase, the thickness of lubrication film decreases and nanoparticles plays a main role in lubrication, nanoparticles embedded into the sample surface which increasing the wear of surface.

Acknowledgements

The present study is financially supported by the National Natural Science Foundation of China (51274037)

References

[1] Hongxin D, Yanbao Z and Zhijun Z 2004 Tribological performance study of Bi nanoparticles additives in liquid paraffin *Tribology* **24** 185–187

[2] Zhiguang G, Kali G and Jiansheng X 2005 Friction and wear properties of organic molybdenum and its composites with nanometer lubricating additive *Tribology* **25** 317–318

[3] Patrick F, Valerie A and Jens R 2004 Nano-emulsion formation by emulsion phase inversion *Colloids and Surface* **8** 53–54

[4] Xiaohong L, Qinghua L and Zhijun Z 2005 Synthesis, Characterization and tribological properties of a reactable nano-silica *Tribology* **25** 499–503

[5] C P Tripp 1997 Low f requency infrared studies of silica at the solidgas and solid-fluid interface *Journal of Molecular Structure* **408 /409** 133–137

[6] Michaela K, And re A and Christi an G 1999 Silica coating on colloidal maghemite particles *Collid Interface Science* **20** 357–361

[7] D X Peng, Y Kang and R M Hwang 2009 Tribological properties of diamond and SiO_2 nanoparticles added in paraffin *Tribology International* **42** 911–917

[8] Tao X, Jiazheng Z and Kang X 1996 The ball-bearing effect of diamond nanoparticles as an oil additive *Applied Physics* **29** 2925–2932

[9] Shen M, Luo J and Wen S 2001 The tribological properties of oils added with diamond nanoparticles *Tribology Transaction* **44** 486–494

IOP Conf. Series: Materials Science and Engineering **182** (2017) 012025 doi:10.1088/1757-899X/182/1/012025

[10] S Y Sia, E Z Bassyony and A A D Sarhan 2014 Development of SiO_2 nanolubrication system to be used in sliding bearings *Advanced Manufacturing Technology* **71** 1277–1284
[11] Jianhua W, Min S and Jinlong L 2011 The preparation and tribological properties of water-soluble nano-silicaparticales *Tribology* **31** 118–123
[12] Zhi C, Xiaohong L and Zhijun Z 2005 Effect of SiO_2 nanoparticles as additive on antiwear and extreme pressure properties of lithium grease *Tribology* **5** 38–40
[13] Hongmei X, Bin J and Junjie H 2016 Lubrication performance of MoS_2 and SiO_2 nanoparticles as lubricant additives in magnesium alloy-steel contacts *Tribology International* **93** 63–70
[14] Dexing P, Cheng Chen, Yuan K, Yeonpun Ch and Shiyan C 2010 Size effects of SiO_2 nanoparticles as oil additives on tribology of lubricant Industrial *Lubrication and Tribology* **62** 111–120
[15] H Kato, K and Komai 2007 Tribofilm formation and mild wear by tribo-sintering of nanometer sized oxide particles on rubbing steel surfaces *Wear* **262** 36–41
[16] M Sayuti, A A D Sarhan and M Hamdi 2012 An investigation of optimum SiO_2 nanolubrication parameters in end milling of aerospace Al6061-T6 alloy *Advanced Manufacturing Technology* **67** 833–849
[17] Mohd S, Sarhan Ahmed A D and Sarhan F 2014 Novel uses of SiO_2 nanolubrication system in hard turning process of hardened steel AISI4140 for less tool wear, surface roughness and oil consumption *Cleaner Production* **67** 265–276
[18] L Rapoport, V Leshchinsky, LapskerI, Y Volovik, O Nepomnyashchy and Lvovsky M 2003 Tribological properties of WS_2 nanoparticles under mixed lubrication *Wear* **255** 785–793

Synthesis of core-shell structured magnetic nanoparticles with a carbide shell

Shushan Hou , Yue Chi* and Zhankui Zhao*

College of Material Science and Engineering, Key Laboratory of Advanced Structural Materials, Ministry of Education, Changchun University of Technology, Changchun, 130012, China

E-mail: * yuechi@ccut.edu.cn, * zhaozk@ccut.edu.cn

Abstract. Core-shell structured materials combining the functionalities of the core and shell have great application potential in many fields. In this work, by combining solvothermal, polymerization and the high temperature carbonization, we have successfully developed a facile method to generate core-shell structured nanoparticles which possess an internal magnetic nanoparticle with a carbide shell. The thickness of resorcinol formaldehyde resin as intermediate transition shell could be easily adjusted by changing the concentration of the RF precursor. The resulting nanoparticles possess well-defined structure, uniform size and high magnetization. The unique nanostructure of the magnetic core-shell structured nanoparticles could lead to many promising applications in areas ranging from drug delivery to the purifyication of sewage.

Keywords: solvothermal method, core-shell structure, carbide, magnetic materials

1. Introduction

Multishell composite nanostructure materials, since their discovery [1, 2], have gained global attention because their potential applications in separation [3], catalysis [4], magnetic resonance imaging (MRI) [5], bioseparation [6], drug delivery [7, 8] and sensor [9]. In the other hand, magnetic nanomaterials have recently attracted numerous researchers' interest in catalysis, adsorption and enrichment due to their convenient manipulation, such as magnetic separation and recycling [10-14]. However, magnetic nanoparticles (Fe_3O_4 nanoparticles) are easy to rust corrosion in the water, especially they have a large tendency to aggregate driven by vander Waals forces and high interface energy because of their ultrasmall particle size. Therefore, there is an impending need in developing methods to create a protective shell on the surface to improve its performance and practicability, such as polymer nanoshell, silica nanoshell and so on.

Taking the above considerations into account, there are some experimental reports about preparing magnetic nanoparticles (Fe_3O_4 nanoparticles) with multi-shell structures [15], such as polymer/gold nanoshell [16], Pd/silica [17] and C/silica [18]. Thus it can be seen that it is also very interest and valuable to develop a new method to generate a carbide shell on magnetic nanoparticles.

Herein, we reported a controllable and facile route to fabricate magnetic core-shell structured nanoparticles that comprise Fe_3O_4 core and carbide shell. The introduction of carbide shell and the construction of core-shell structure are considered to improve the potential applications of magnetic nanomaterials, such as separation, drug delivery, bioseparation, sensor and so on.

2. Experimental section

2.1. Chemicals
All chemicals were purchased from Beijing Chemical Corp. Other chemicals were used as received without any further purification. Deionized water was used for all experiments.

2.2. Materials synthesis

2.2.1. Synthesis of magnetic Fe_3O_4 nanoparticles. The magnetic Fe_3O_4 nanoparticles were synthesized through a solvothermal method according to the method reported previously with some modification [19]. Briefly, 1.08 g of $FeCl_3 \cdot 6H_2O$ was dissolved in 20 mL of ethylene glycol with magnetic stirring until the solution became clear. Then 1.8 g of NaAc was added to this solution and stirred for another 1 h. Afterwards, 0.25 g trisodium citrate was added. When the mixed solution was stirred for 10 h to form a homogeneous dispersion, The obtained black solution was then transferred and sealed into a Teflon-lined stainless-steel autoclave (50 mL in capacity), The autoclave was heated at 200 °C for 10 h, and then allowed to cool down to room temperature. The black sample was washing with ethanol and deionized water several times, and then dried under vacuum at 60 °C overnight.

2.2.2. Synthesis of magnetic Fe_3O_4@C nanoparticles. 0.1 g of magnetic Fe_3O_4 nanoparticles was dispersed in a mixture solution of ethanol (20 mL) and deionized water (10 mL) by ultrasonication. Into the resulting dispersion, ammonium hydroxide aqueous solution (0.5 g, 28 wt %), resorcinol (0.1 g, 0.09 mM), and formaldehyde (0.1 g, 37 wt %) were added consecutively, and the mixed dispersion was mechanically stirred for polymerization for 2 h at 30 °C, forming a shell of RF resin. After polymerization, the obtained Fe_3O_4@RF nanoparticles were collected by a magnet and washed with deionized water and alcohol several times, respectively. In order to try to adjust the thickness of the RF shell, the concentration of the RF precursor will be appropriate to increase [20]. Finally, the obtained sample was calcined at 500 °C in argon atmosphere for 3 h to form a carbide shell.

2.3. Characterization
Scanning electron microscope (SEM) were measured on JSM-5600. Transmission electron microscopy (TEM) was performed on JEM 2000. Crystal identification was carried out using a Bruker D8 Advance X-ray diffractometer with a Cu Kα X-ray source operating at 40 kV and 100 mA. Magnetic measurement of the samples was investigated with a Quantum Design vibrating sample magnetometer (VSM) at room temperature in an applied magnetic field sweeping from - 15 to 15 kOe. Raman data was recorded on a LabRAM HR Evolution using an Ar+ laser at ambient temperature. An excitation source with a wavelength of 532 nm was used.

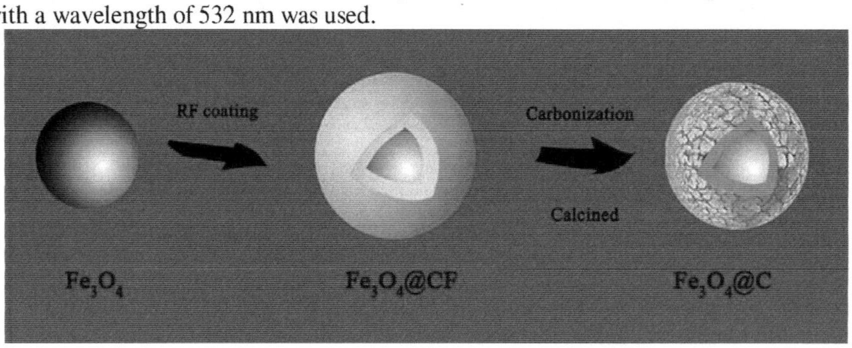

Scheme 1. Schematic representation of synthesis of Fe_3O_4@C.

3. Results and discussion

The fabrication of magnetic core-shell structured nanoparticles is illustrated in Scheme 1. Firstly, uniform magnetic Fe_3O_4 nanoparticles were synthesized by the solvothermal method. Secondly, through polymerization of resorcinol and formaldehyde, a controllabe shell (50 ~ 150 nm) can be deposited on the surface of Fe_3O_4 nanoparticles. Thirdly, after calcination in argon atmosphere, Fe_3O_4@C could be obtained. The synthesized nanoparticle is composed of Fe_3O_4 core, and carbide shell.

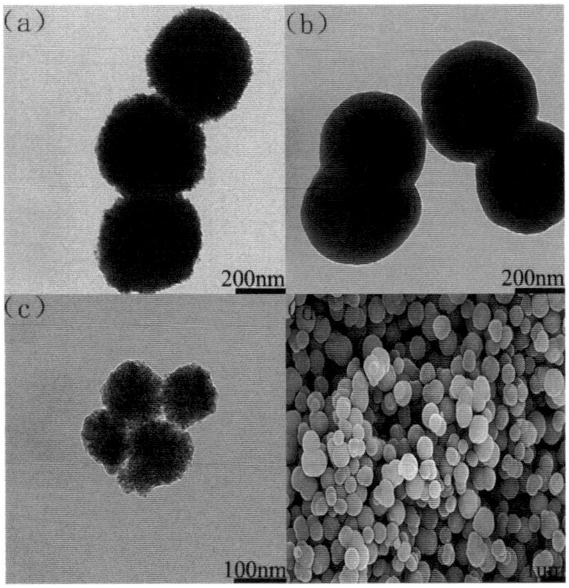

Figure 1. TEM images of (a) Fe_3O_4, (b) Fe_3O_4@RF, (c) Fe_3O_4@C, (d) SEM image of Fe_3O_4@RF.

As shown in figure 1(a), the obtained magnetic Fe_3O_4 nanoparticles have a mean diameter of ~ 250 nm. After polymerization, a shell of RF of ~ 50 nm in thickness was individually deposited on magnetic Fe_3O_4 nanoparticles, resulting in core-shell structured Fe_3O_4@RF with spherical morphology (figure 1(b) and 1(d)). This RF shell plays some important roles in this process. First, it can prevent the magnetic Fe_3O_4 nanoparticles from being oxidized under some conditions. Second, it can also prevent the magnetic Fe_3O_4 nanoparticles from losing magnetic properties in the final carbonization process. Last, it provides the carbonyl and hydrogen ions[20], thus providing conditions for the subsequent reaction. After high temperature carbonization in argon atmosphere, magnetic Fe_3O_4 nanoparticles was coated with a carbide shell (figure 1(c)), then core-shell structured nanoparticles with a size of ~ 400 nm were obtained.

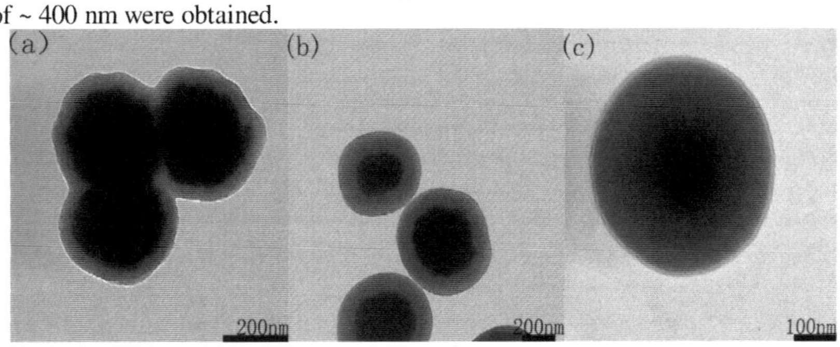

Figure 2. TEM images of Fe_3O_4@RF nanoparticles with different thickness of RF shell.

Under the same conditions except for the concentration of RF precursor, the core-shell structured Fe_3O_4@RF, nanoparticles with different thickness of RF shell can be easily adjusted. By increasing the concentration of RF precursor, the thickness of the RF can increase from 50 to 150 nm. As can be seen from TEM images (figure 2(a) and 2(b)), when the concentration of RF precursor is 1:2, the thickness of the RF shell of the resultant core-shell nanoparticles can be easily adjusted from 50 to 100 nm. Further increasing the concentration of RF precursor to 1 : 3 leads to a much thicker shell at ~ 150 nm (figure 2(c)).

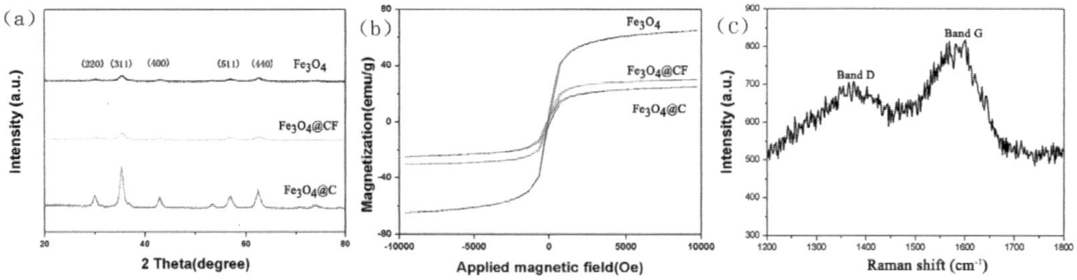

Figure 3. (a) XRD patterns of Fe_3O_4, Fe_3O_4@RF, Fe_3O_4@C, and (b) Magnetization curves of Fe_3O_4, Fe_3O_4@RF, and Fe_3O_4@C at room temperature. (c) Raman spectrum of Fe_3O_4@C.

The phase structures of Fe_3O_4, Fe_3O_4@RF, and Fe_3O_4@C were analyzed via XRD (figure 3(a)). For Fe_3O_4, the characteristic diffractions at 30.0°, 35.5°, 43.1°, 57.0°, and 62.6° of 2θ suggest a face-centered cubic phase structure with the space group of Fd3m according to JCPDS card no. 19-0629. In the case of Fe_3O_4@RF and Fe_3O_4@C composites, the phase structure of Fe_3O_4 core was intact, but the intensity of XRD peaks gradually stronger, when the thickness of shell was increased. This result suggests that the Fe_3O_4 crystal was successfully used as a magnetic center of the Fe_3O_4@C. The magnetization curves measured for Fe_3O_4, Fe_3O_4@RF and Fe_3O_4@C are compared (figure 3(b)). There was no hysteresis in the magnetization for the three tested nanoparticles. Compared with Fe_3O_4 (64.99 emu/g), the saturation magnetization of Fe_3O_4@RF and the Fe_3O_4@C decreased to 29.90 and 24.81 emu/g, respectively. It should be noted that although the magnetization saturation value of Fe_3O_4@RF and the Fe_3O_4@C are lower than that of Fe_3O_4 due to the increase in the mass caused by the surface depositing of Fe_3O_4, it still shows strong magnetization, which suggests its suitability for magnetic separation and targeting. Raman spectroscopy was used (figure 3(c)). The spectrum consists of two characteristic bands, namely D band at ~ 1370 cm^{-1} and G band at ~ 1580 cm^{-1}, respectively. The D band corresponds to the disordered structure of a carbonaceous solid from an A1g mode, which is attributed to the presence of amorphous or disordered carbon. The G band originates from graphitic lattice vibration mode with E2g symmetry. The disordered D band is weaker than the ordered G band, indicating a substantial coating of definite carbon on the surface of Fe_3O_4.

4. Conclusions

In conclusion, we have successfully synthesized the Fe_3O_4@C nanoparticles by combining solvothermal, polymerization and the high temperature carbonization. The magnetic Fe_3O_4 nanoparticles with a diameter of ~ 250 nm have been synthesized by solvothermal method. Through adjusting the concentration of RF precursor in the reaction solution, the thickness of RF shell can be adjusted in the range of 50 ~ 150 nm. The transformation of RF shell to carbide protective shell could be implemented by the high temperature carbonization. Considering the facile synthesis method and the useful superparamagnetic properties and unique microstructure, Fe_3O_4@C has promising application in many fields.

Acknowledgements

This work was supported by the Natural Science Foundation of Jilin Province of China (20150520020JH), National Natural Science Foundation of China (grant nos. 51608050 and 51601018) and Jilin Provinical Science & Technology Department (Grant nos: 20160519020JH).

References

[1] Kresge C T *et al.* 1992 Ordered mesoporous molecular sieves synthesized by a liquid-crystal template mechanism *Nature* **359(6397)** 710–712
[2] Che S *et al.* 2004 Synthesis and characterization of chiral mesoporous silica *Nature* **429(6989)** 281–284
[3] Vinu A *et al.* 2004 Adsorption of cytochrome c on mesoporous molecular sieves: Influence of pH, pore diameter, and aluminum incorporation *Chemistry of materials* **16(16)** 3056–3065
[4] Alayoglu S *et al.* 2008 Ru-Pt core-shell nanoparticles for preferential oxidation of carbon monoxide in hydrogen *Nature materials* **7(4)** 333–338
[5] Shin J *et al.* 2009 Hollow manganese oxide nanoparticles as multifunctional agents for magnetic resonance imaging and drug delivery *Angewandte Chemie* **121(2)** 327–330
[6] Chen O *et al.* 2013 Compact high-quality CdSe-CdS core-shell nanocrystals with narrow emission linewidths and suppressed blinking *Nature materials* **12(5)** 445–451
[7] Melancon M P, Zhou MIN and Li C 2011 Cancer theranostics with near-infrared light-activatable multimodal nanoparticles *Accounts of chemical research* **44(10)** 947–956
[8] Kam N W S *et al.* 2005 Carbon nanotubes as multifunctional biological transporters and near-infrared agents for selective cancer cell destruction *Proceedings of the National Academy of Sciences of the United States of America* **102(33)** 11600–11605
[9] Reiss P, Protiere M and Li L 2009 Core/shell semiconductor nanocrystals *Small* **5(2)** 154–168
[10] O'handley R C 2000 *Modern magnetic materials* (Wiley)
[11] Miller J S and Epstein A J 1994 Organic and organometallic molecular magnetic materials-designer magnets *Angewandte Chemie International Edition in English* **33(4)** 385–415
[12] Standley K J 1972 *Oxide magnetic materials* (Oxford University Press)
[13] Spaldin N A 2010 *Magnetic materials: fundamentals and applications* (Cambridge University Press)
[14] Hu M *et al.* 2012 Tailored design of multiple nanoarchitectures in metal-cyanide hybrid coordination polymers *Journal of the American Chemical Society* **135(1)** 384–391
[15] Park H *et al.* 2009 Multifunctional nanoparticles for combined doxorubicin and photothermal treatments *ACS nano* **3(10)** 2919–2926
[16] Wan Y *et al.* 2009 Ordered mesoporous Pd/silica-carbon as a highly active heterogeneous catalyst for coupling reaction of chlorobenzene in aqueous media *Journal of the American Chemical Society* **131(12)** 4541–4550
[17] Han S and Hyeon T 1999 Simple silica-particle template synthesis of mesoporous carbons *Chemical Communications* **19** 1955–1956
[18] Fuertes A B, Valle-Vigón P and Sevilla M 2012 One-step synthesis of silica@resorcinol-formaldehyde spheres and their application for the fabrication of polymer and carbon capsules *Chemical Communications* **48(49)** 6124–6126
[19] Chi Y *et al.* 2012 Synthesis of Fe_3O_4@SiO_2-Ag magnetic nanocomposite based on small-sized and highly dispersed silver nanoparticles for catalytic reduction of 4-nitrophenol *Journal of colloid and interface science* **383(1)** 96–102
[20] Yue Q *et al.* 2015 Magnetic yolk-shell mesoporous silica microspheres with supported Au nanoparticles as recyclable high-performance nanocatalysts *Journal of Materials Chemistry A* **3(8)** 4586–4594

Getting graphite nano-sheets with different sizes by choosing parent graphite: ultrasonication assisted preparation

Chunyu Wang*, Haiping Liu, Chunlin Qin and Sifu Bi

School of Materials Science and Engineering, Harbin Institute of Technology at Weihai, Weihai 264209, China

E-mail: *wcyadam@126.com

Abstract. In this work, a simple and effective method for preparing graphite nano-sheet (GNS) was investigated. The size choosing of the parent graphite is important to obtain smaller size of GNSs diameter or thickness with the ultrasonication treatment. The ultrasonication treating parameter was confirmed under the optimal preparing condition,. The microstructure was characterized with scanning electron microscopy (SEM), high resolution transmission electron microscopy (HTEM). It is The experimental results show that the GNS size decreases decreased sharply at the high power factor stage of ultrasonication and then the decrement trend was slow down with the decreasing power factor of ultrasonication treatment. It is employed that choosing smaller size of parent graphite (>200-mesh) can obtain smaller size GNSs with the diameter value of (<10μm) and thickness value of (~4nm).

Keywords: GNS, Parent graphite, Sizes, Ultrasonication

1. Introduction

Natural graphite flakes are composed of graphene layers that are stacked in a nearly perfect manner. By intercalation and exfoliation, the natural graphite flakes can become expanded graphite (EG), which is composed of partially connected graphite nano-sheets (GNSs) [1-3]. Because the graphite nano-sheets are only weakly connected, they can be separated by using dispersion techniques, such as high speed mixing, sonication, and high shear-strain rate [4]. Both GNSs and CNTs are known for possessing exceptionally high thermal conductivity, with measured values higher than 1200 and 3000 W/m·K, respectively [5]. They are also able to carry an extremely high current density (109 A/cm2) [6]. New methods for graphite nano-sheet (GNS) production have recently emerged. NG turns into a graphite intercalation compound after acid intercalation, easily producing expanded graphite (EG). GNS can be exfoliated from EG via ultrasonic powdering. The thermal conductivity properties of different forms of graphite have been studied with various polymer resins such as polypropylene, polystyrene, polyoxymethylene, high density polythelene, nylon and polyaniline as well as silicone rubber, and metal hydride [7-10]. However, the effect of GNS size on parent graphite with ultrasonication assisted preparation has seldom been studied.

It is known that the size of graphene sheets plays an important role in controlling their properties and applications. It was also reported that the thermal conductivity of graphene mainly depends on its sheet size [11]. Therefore it is highly demanded to develop simple and efficient methods for the preparation of GNSs with specified size and large quantities for their broad range of technological applications. So far, much effort has been made to study various factors which would affect the size of

GO sheets, such as post-treatment of parent graphite, oxidation conditions, controlled centrifugation, and pH value [12-15]. However, the large scale preparation of high-quality GNSs with specified size still remains a great challenge. The mechanism of ultrasonication relies on the acoustic phenomenon, of which ultrasonic cavitation is the most important, involving the formation, growth, and implosive collapse of bubbles in liquids [16-17]. During the ultrasound process, extremely high temperature about 5000 K, rapid cooling rate about 1010 Ks^{-1} and high pressure about 20 MPa are driven by high intensity ultrasound, thus enabling some chemical reactions generally not accessible in the conventional methods [18].

It is worth noting that ultrasonication is an important experimental step in the preparation method to produce GNSs. In order to decrease the high consumption during the preparing GNS, and further simplify the technological process, we proposed a new method, which was choosing parent graphite sizes. So far, however, little attention has been paid to analyze in detail about the effects of ultrasonication to the structure of GNSs. On the basis of the fact that reactions can be accessible by using ultrasonication in preparing nano-materials due to its cavitation and vibration effects, it is motivated to investigate the effects of ultrasonication to the size and structure of GNSs.

2. Experimental

2.1. Materials

Nature graphite power with the average sizes of 80-mesh, 120-mesh, 150-mesh, 200-mesh, was purchased from Qingdao Tai Xing graphite Co., Ltd. Potassium permanganate ($KMnO_4$), sulfuric acid (H_2SO_4 98%), hydrogen peroxide (H_2O_2), and glacial acetic acid (CH_3COOH), perchloric acid ($HClO_4$) were purchased from Kermel Chemical reagent plant (Tianjin, China). All these raw materials were directly used without further purification.

2.2. Preparation of EG

Preparations of graphite sheets from large-size to small-size were synthesized by ultrasonication. The largest-size EG was using 80-mesh expanded graphite powder, and the smallest-size EG was relatively using 200-mesh.

Graphite (10g) was mixed with CH_3COOH (10mL) in $HClO_4$ (50mL) at 35℃ (water bath), then $KMnO_4$ (2g) was gradually added in 10min while keeping the temperature at 35℃. After that, the reaction mixture was stirred at 35℃ for 1.5 h. Next, deionized water (500mL) was rapidly dropped into the reaction system. Then 2L of deionized water was added, to dilute the mixture. In the final procedure, mixture particles washed with deionized water till pH5, leaching remove excess water, then the expandable graphite was produced with drying in oven at 60℃ for 6h. The dried graphite intercalation compound was subjected to a thermal shock at 950℃ for 50s. The thermal shock vaporized intercalates in the intercalation compound, providing the necessary force to overcome the van der Waals interactions that bind the graphite layers. The exfoliation of the graphite flakes resulted in EG, as shown in Fig.1. The EG is a highly porous material with an extremely low density, occupying a volume about 270-500 cc for 1g of different EG size.

2.3. Preparation of GNSs

The EG is composed of GNSs that are partially and weakly connected. The GNSs can be separated from the EG by using dispersion techniques. The thickness of the GNSs depends on the intercalation and exfoliation processes. In the present study, most GNSs range from 20 to 50 nm in thickness. The mechanism of ultrasonication relies on the acoustic phenomenon, of which ultrasonic cavitation is the most important, involving the formation, growth, and implosive collapse of bubbles in liquids. Different power factor of ultrasonication are employed on producing GNSs. Effect of GNSs sizes relies on parent graphite size and parameters of ultrasonication. The sizes of parent graphite choose on four kinds of 80-mesh, 120-mesh, 150-mesh and 200-mesh. The power factor of ultrasonication is used on 70% (U-70), 80% (U-80), 90% (U-90) and 99% (U-99), corresponding to work 40min.

IOP Conf. Series: Materials Science and Engineering 182(2017) 012027 doi:10.1088/1757-899X/182/1/01202

2.4. Characterization

The morphologies of GNSs were obtained by scanning electron microscopy (SEM), and energy disperse spectroscopy (EDS) was used to confirm elements composition. The Inspect F (FEI) scanning electron microscopy (SEM) with an acceleration voltage of 5 kV was used to measure the size of GNSs. The structure and morphology of the samples obtained were extensively characterized by X-ray powder diffraction (XRD, Rigaku D/max-gB X-ray diffractometer with Cu K radiation (λ=0.154178nm). Specimens for transmission electron microscopy (TEM) observation were prepared by ultrasonicating the as-synthesized products in alcohol for about 20 min, then dipping holey-carbon TEM sample grids into the dispersion. Specific surface area measurements were conducted on a Micrometrics BET 3H-2000PS1 static volume method apparatus with liquid nitrogen at 77K.

3. Results and discussion

3.1. Expanded graphite and GNSs

Figure 1. SEM images of EG (a) 80-mesh (b) 120-mesh (c) 150-mesh (d) 200-mesh.

Preparation of EG is using nature graphite power of 80-mesh, 120-mesh, 150mesh, and 200-mesh to produce expanded graphite. Fig.1 shows typical SEM images of EG obtained by the similar exfoliation. It is worth noting that most of the EG are similarity on microstructure, which is similar to this reported recently [2], and expanded graphite with different sizes are vermicular or wormlike in shape. As is apparent from figure 1(a) ~ figure 1(d) reveals that EG has many pores of different sizes ranging from 10 nm to 10μm. There was no difference for all specimens microstructure; vermicular and wormlike in shape are not affected in each group. The Expandable graphite also possesses some closed pores along with a significant amount of open pores. The conductivity of the bulk material is certainly influenced by the expansion of the EG due to the spacing of the graphite layers being increased. In addition, a large number of polar groups, such as hydroxyl, ether and carboxylate groups, on the surface of the

IOP Conf. Series: Materials Science and Engineering 182 (2017) 012027 doi:10.1088/1757-899X/182/1/012

EG layers were easily processed. So multi-pores with some functional groups such as C–O–C, C–OH and–COOH on the surface of EG layers greatly promote the ability of adsorption of other particles. In order to investigate the effects of sizes for EG, different sizes from natural graphite were suitable for grading granulated screening out powder. The expansion volumes are checked for it. Table 1 shows the results of that effect. Measurement of expansion volume shows that expansion volume of different sizes from natural graphite increase with the increase of particle sizes. The best expansion volume of acid treatment is 497ml/g. It is ordered to contrast on imparity of four EG, results of the specific surface area shows in table 1, too. It is shows that specific surface of different sizes from EG increase with the increase of expansion volume. The best specific surface of EG is 220.6 m^2/g.

Figure 2 panels a, b, c, and d show SEM images of GNSs obtained from different ultrasonication power factor. In order to obtain size change regularity from ultrasonication power factor, it employs 80-mesh EG for it. Based on the analysis of the SEM images for ultrasonication power factor U-70, U-80, U-90 and U-99, it is found that the size of the GNSs decreases regularly with increasing the power factor of ultrasonication.

Table 1. Expanding volume and BET specific surface of EG with different parent graphite sizes.

	Particle size (μm)	Expanding volume (mL/g)	Specific surface BET (m^2/g)
80-mesh	180	497	220.6
120-mesh	120	421	108.2
150-mesh	106	391	89.3
200-mesh	74	277	38.2

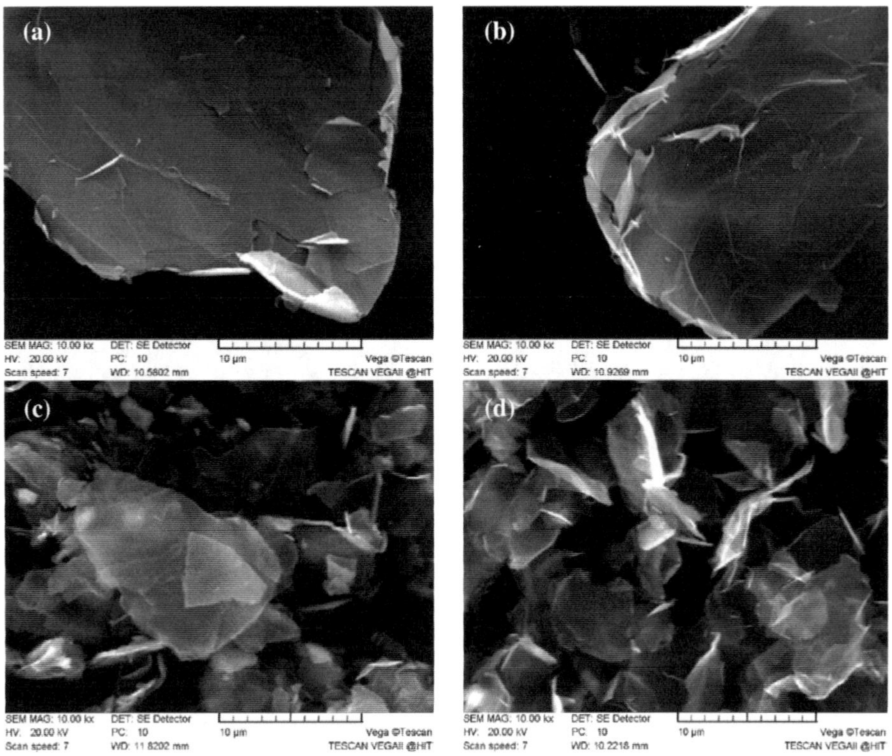

Figure 2. SEM image of GNS (a) U-70 (b) U-80 (c) U-90 (d) U-99.

The mean size of U-70, U-80, U-90 and U-99 sheets, obtained from the SEM diameter size statistical analysis above, is >30μm, ~20μm, ~10μm, and <10μm, respectively. Moreover, the different diameter size of U-70, U-80, U-90 and U-99 sheets all exhibit an about similar thickness correspondingly. The thickness are about 10~30nm, the SEM images cannot confirm thickness for each sample of U-70, U-80, U-90 and U-99 sheets, TEM images will be employed to measurement it. The results show that the sheet size decreases sharply at the high power factor stage of ultrasonication and then the decrease slow down with the decrease of ultrasonication power factor. Ultrasonication belongs to high-energy sound waves. Long time of ultrasonication could release large localized amounts of energy and result in the extreme reaction conditions such as high temperature, pressure, rapid cooling times that provide the energy for the peeling off GNSs. Here, it is found that ultrasonication not only can exfoliate and break GNSs, but also increase the shockwaves created from the collapse of cavitating bubbles.

3.2. *Layers of GNSs*

It is employed that U-99 is to obtain GNSs which come from different parent graphite 80-mesh, 120mesh, 150-mesh, and 200-mesh. Fig.3 and Fig.4 show the low-magnification and high-magnification TEM images of the GNSs. As can be seen from figure 3, the GNSs exhibit a wrinkled paper-like multilayer structure, for example Fig.3 panels a, b, c, and d. The edge thickness of the graphite nano-sheets measured in Fig.4 panels a, b, c, and d, is about 5~20nm. The thickness of U-70, U-80, U-90 and U-99 sheets analysis above, is ~14nm, ~8nm, ~5nm, and ~4nm, respectively. The interlayer spacing can be calculated to be about 0.47 nm which is larger than that of graphite (0.334 nm). It is worth noting that, different from graphite, the adjacent graphene layers of the multilayer grapheme derived from chemical exfoliation are not parallel. Figure 4 shows the high-resolution TEM images of synthetic GNSs. The graphite shows twenty or thirty of graphene sheets stacked as shown in figure 4a. Figure 4b shows about ten layers of stacked grapheme sheets. Fig.4c and Fig.4d shows about five layers of stacked grapheme sheets. All of these GNSs images suggest the formation of exfoliated graphite nano-sheets varying from 5 to ~30 layers similar to that obtained by other chemical methods [19-21].

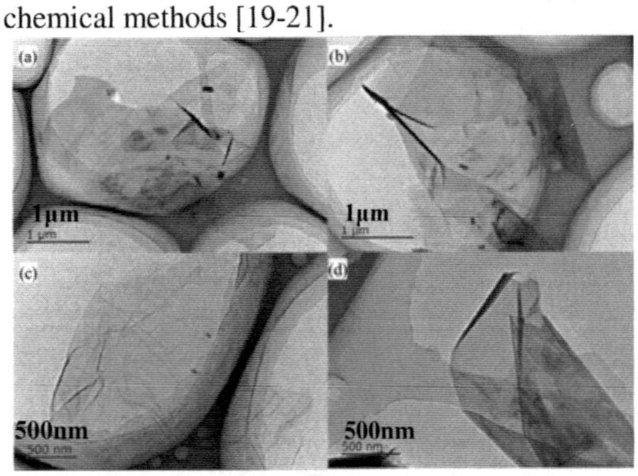

Figure 3. TEM image of the GNS with U-99(a)80-mesh (b) 120-mesh(c) 150-mesh (d) 200-mesh.

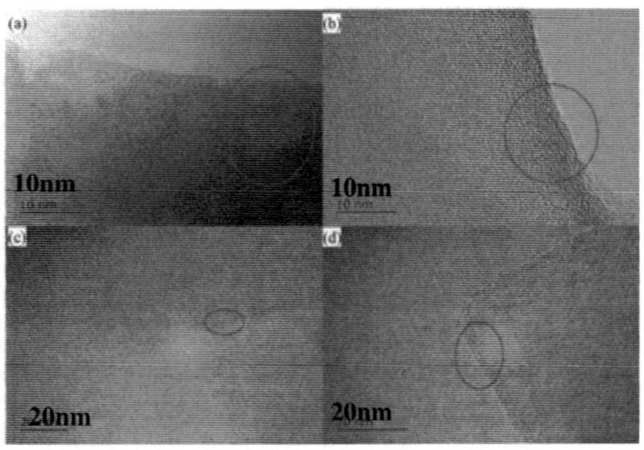

Figure 4. High-magnification TEM image of the GNS with U-99 (a) 80-mesh (b) 120-mesh (c) 150-mesh (d) 200-mesh 3.

3.3. *Diffraction peaks of GNSs*

The XRD patterns of four GNSs in figure 5. As we know, graphite contained a very sharp and high intensity peak at about $2\theta = 26°$ with the layer-to-layer distance (d-spacing) of 0.334 nm. The feature diffraction peak of 200-mesh GNS appears at about $2\theta= 21°$ after the ultrasonication treatment of 200-mesh EG, moreover other GNSs have no peak at this station. It is indicated that the feature diffraction peak of 200-mesh GNS structure has different character contrast on other three peaks of GNSs. XRD pattern of 200-mesh GNS has not peak of $2\theta= 44°$, but which appears at $2\theta= 45°$. The graphite feature diffraction peaks have this characteristic on 80-mesh GNS, 120-mesh GNS and 150-mesh GNS ($2\theta=26°$, 44°, 54°), but 200-mesh GNS has not all peaks for it ($2\theta=21°$, 26°, 54°). This phenomenon is complying with results of SEM and TEM for smaller size of diameter (<10μm) and thickness (~4nm) of 200-mesh GNS after the ultrasonication treatment.

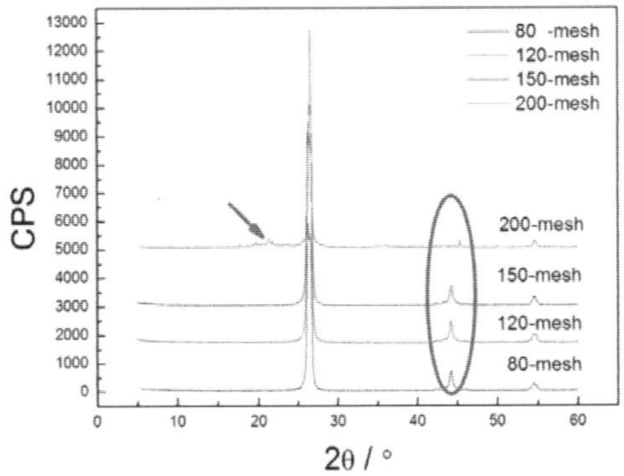

Figure 5. XRD of the GNS with U-99 (a)80-mesh (b) 120-mesh (c) 150-mesh (d) 200-mesh.

4. Conclusions

A simple and effective method for preparing GNS is presented. GNSs with the diameter of (<10μm) and thickness of (~4nm) have been prepared by the ultrasonication treatment. Under the optimal preparing condition, ultrasonication treating parameter is 40min for 99% power factor. The results show that the GNS size decreases sharply at the high power factor stage of ultrasonication and then the decrease trend is slow down with the decrease of ultrasonication power factor. It is concluded that the choosing smaller size of parent graphite (>200-mesh) can obtain smaller size GNSs with diameter of (<10μm) and thickness of (~4nm).

Acknowledgment

This work was financially supported by National Science Foundation of China, No. NSFC51302050 and NSFC 51302051.

References

[1] Tsung-Lin Wu, Tzu-Sen Lo, Wen-Shyong Kuo 2010 *J. Polymer Composites* **31** 292–298
[2] Chang Yanhong, Wang Bin, Luo Hui 2011 *Advanced Materials Research* **168–170** 2571–2574
[3] Leilei Wanga, Liqun Zhanga, Ming Tiana 2012 *J. Wear* **276–277** 85– 93
[4] Xiaoqing Su, Gang Wang, Weilong Li 2013 *Advanced Powder Technology* **24** 317–323
[5] D. Li, M.B. Müller, S. Gilje. 2008 *Nature Nanotechnology* **3** 101–105
[6] Y. Hernandez. 2008 *Nature Nanotechnology* **3** 563–568
[7] G.H. Zheng, J.S. Wu, W.P. Wang 2004 *Carbon* **42** 2839–2847
[8] Bin Li, Wei-Hong Zhong 2011 *Materials Science* **46** 5595–5614
[9] Leilei Wang, Liqun Zhang, Ming Tian 2012 *Polymer Advanced Technology* **23** 652–659
[10] Minwei Xu, Huiting Sun, Cai Shen. 1998 *Nano Research* **124** 1–7
[11] Arnab Mukherjee, JungHo Kang, Oleksandr Kuznetsov. 2011 *Chemical Material* **23** 9–13
[12] Botas C, A ´ lvarez P, Mene´ndez R. 2012 *Carbon* **50** 275–282
[13] Chang H, Wang G, Zheng ZJ. 2010 *Advanced Function Materials* **20** 2893–2902
[14] Khan U, O'Neill A, Coleman JN. 2012 *Carbon* **50** 470–475
[15] McAllister MJ, Li JL, Aksay IA. 2007 *Chemical Materials* **19** 4044–4396
[16] Xiaodong Qi, Tiannan Zhou, Sha Deng, et al 2014 *Material Science* **49** 1785–1793
[17] Tung V. C., Allen M. J., Yang Y. 2009 *Nature Nanotechnology* **4** 25–29
[18] AvourisP. , Dimitrakopoulos C. 2012 *Materials Today* **15** 86–97
[19] Geim, A. K.; Novoselov, K. S. 2007 *Nature Materials* **6** 183–191
[20] Viculis L. M., Mack J. J., Mayer O. M. 2005 *Materials Chemical* **15** 974–978
[21] Zhang L, Liang J, Chen YS. 2009 *Carbon* **47** 3365–3368

New practical method of homogeneous dispersion of multi-walled carbon nanotubes (MWCNTs) into Mg matrix composites

Teng Zhang, Shuangming Du, Wanchang Sun, Jumei Zhang, Libin Niu and Xiaohu Hua

School of Material Science and Engineering, Xi'an University of Science and Technology, Xi'an 710054, China

E-mail: Zhangteng517@163.com

Abstract. It is difficult to disperse multi-walled carbon nanotubes (MWCNTs) into Mg matrix composites homogeneously due to the high specific surface energy between MWCNTs. This would affect the mechanical properties of magnesium matrix composites reinforced by MWCNTs tremendously. The key is to overcome MWCNTs aggregate together, and gain the homogeneous dispersion of MWCNTs. Moreover, the density difference of Mg matrix and MWCNTs is another problem which prevents MWCNTs from dispersing homogeneously into Mg matrix composites. In order to solve these problems, a new practical method of homogeneous dispersion of carbon nanotubes into Mg matrix composites was presented basing on combination the tow step ball milling and ultrasonication dispersion methods. First, a certain amount of MWCNTs was ball-milled in ethanol solution contain 5 wt.% sodium stearate at various rotation speeds (100 rpm-300 rpm) and ball-milled times (0.5 H-2 H) to break MWCNTs hard aggregation. Second, MWCNTs which were subjected to the ball-milling were dispersed further in ethanol solution containing (SDS, DBS) using ultrasonication method. Third, the Mg powers were mixed with good dispersion of carbon nanotubes solution and ball-milled for 1 H to make MWCNTs coated uniformly on the surface of the magnesium substrate. The results were mainly examined by field emission scanning electron microscopy (SEM). It was found that MWCNTs were dispersed well and some of these MWCNTs had directional arrangement. Furthermore, MWCNTs were homogeneously dispersed into Mg matrix composites. These results suggest that this method has good application for magnesium matrix composites reinforced by carbon nanotubes.

Keywords: MWCNTs, Mg matrix composites, dispersion, tow step ball milling

1. Introduction

The investigations for Mg matrix composites reinforced by MWCNTs have been carried out extensively on the account of its remarkable mechanical properties [1-3]. Although, the mechanical properties of composite materials are improved, there is great difference between experiment and expected results. The main reasons for this phenomenon are the aggregation of MWCNTs and non-uniform distribution of MWCNTs into Mg matrix. In order to overcome these problems, some researchers try to apply powder metallurgy method to disperse MWCNTs into metal matrix which is a more conventional method to prepare desired ratio materials [4-7], and other researchers used the solution containing zwitterionic surfactant isopropyl alcohol to transfer CNTs onto the powders surface [8]. However, single process couldn't meet the demand of dispersion MWCNTs uniformly into Mg matrix. Thus, multi-step methods such as mechanical stirring and high-intensity ultrasonic dispersion methods were taken into consideration by some researchers [9-10]. Unfortunately, a little achievement has been obtained on the uniform dispersion of MWCNTs into Mg matrix because of higher specific surface energy between MWCNTs clusters and the density difference between Mg matrix and MWCNTs. Consequently, the homogeneous dispersion of high volume fractions of MWCNTs into Mg matrix has not been achieved yet.

Ball milling is effective way to break the hard agglomeration of MWCNTs with the aid of grinding agent in semi-wet environment [11-12]. In this paper, a practical method was developed which combined semi-wet ball milling and ultrasonic dispersion methods with the aids of dispersing agent (SDS, SDBS) in absolute ethyl alcohol environment, then the dispersed MWCNTs were adsorbed onto the surface of Mg powers uniformly by secondary ball milling process in absolute ethyl alcohol environment. Moreover, the effect of dispersing agent (SDS, SDBS) and ball milling condition(ball-milled time and rotation speed) on the dispersion of MWCNTs were investigated and discussed basing on the analysis of dispersion characteristics of MWCNTs by field emission scanning electron microscopy (SEM).

2. Experiment

Multiwall carbon nanotubes (MWCNTs) with 40-60 nm in diameter, 97% in purity and 5>μm in length were supplied by Shenzhen Nanotech Port Co Ltd. The surfactant agents of sodium dodecyl sulfate sodium salt (SDS), sodium dodecyl benzene sulfonate(SDBS) and the grinding-aids of sodium stearate which are analytical purity were used to prevent MWCNTs from aggregating, magnesium powders is also anlytical purity (99.5%, 100μm).

2.1. MWCNTs dispersion procedure

0.5g MWCNTs mixed with 5 wt.% sodium stearate were placed in 200 ml nylon mixing jars which contained 20 zirconia balls and 20 ml absolute ethyl alcohol. The diameter of balls is 4, 8, 10 mm respectively, and the number of different diameter balls is 4, 6, 10 accordingly. The mixtures were ball milled by planetary ball mill (BXQM-2L, Nanjing, China) at 100-300 rpm for milling time changing from 0.5 H to 2 H. the samples were ball-milled for 2 min at 2 min intervals in order to prevent absolute ethyl alcohol from volatilizing during ball milling process. Subsequently, further dispersion treatment of ball-milled MWCNTs was performed using ultrasonic dispersion method at a frequency of 40 KHz for 1.0 H in absolute ethyl alcohol containing certain mount of SDS (0.5 g/L) and SDBS (0.5 g/L) respectively.

2.2. MWCNTs dispersion in Magnesium matrix procedure

For the purpose of dispersion MWCNTs homogeneously into magnesium matrix, magnesium powder was ball-milled with the dispersed MWCNTs in anhydrous ethanol environment on condition of the ball milling rotary speed of 100 rpm and milling time of 1 H. Then, the samples were compacted into 20 mm diameters and 10 mm height under the pressure of 20 MPa by a cold isostatic press. Moreover, the homogeneous dispersion of MWCNTs and MWCNTs into magnesium matrix were mainly characterized by field emission scanning electron microscopy (SEM) analysis.

3. Results and discussion

The original morphology of MWCNTs which were used in this paper is shown in figure 1(a). It can be observed that smaller and larger diameters of MWCNTs are tangled around each other. In order to break the MWCNTs hard aggregation, carbon nanotubes were subjected to medium energy level ball milling with the ball milling rotary speed of 100-300 rpm and milling time of 0.5-2 H in absolute ethyl alcohol environment containing 5 wt.% sodium stearate. It can be seen that MWCNTs were dispersed further. In addition, the length of MWCNTs was shortened when the ball milling rotation speed was

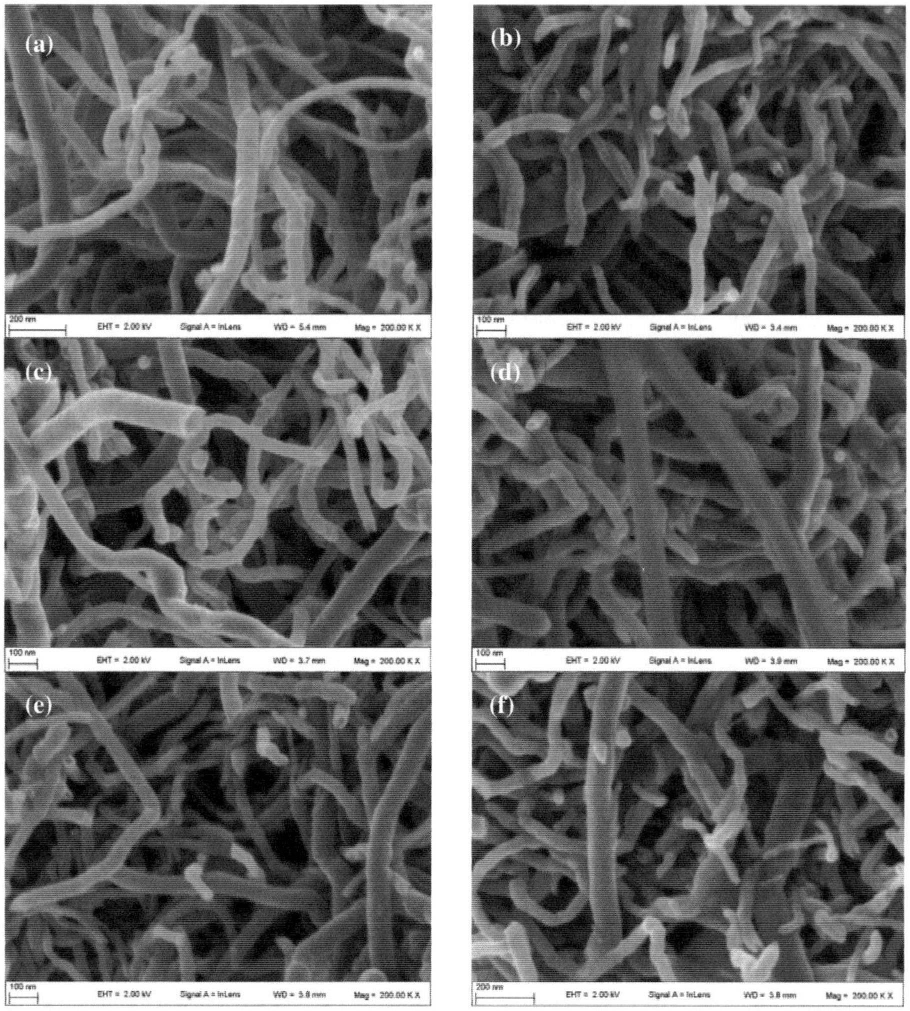

Figure 1. (a) , original carbon nanotubes; (b), (c), (d) are the dispersion of MWCNTs with the ball-milled condition of 100, 200, 300 rpm, and the milling time of 0.5 H in absolute ethyl alcohol containing 5 wt% sodium stearate respectively; (e), (f) are the dispersion of MWCNTs with he ball-milled condition of 200 rpm, and the milling time of 1.0, 2.0 H in absolute ethyl alcohol containing 5 wt% sodium stearate respectively.

up to 200 rpm which are shown in figure 1(b)-1(c), the diameter of MWCNTs become larger, and aggregated together again as the ball milling rotation speed up to 300rpm which is shown in figure1(d). This suggested that moderate ball milling rotation speed is required for ball milling process of MWCNTs dispersion. Moreover, it is interesting to observe that some MWCNTs have directional alignment which is shown in figure 1(b). Figure 1(e) and figure 1(f) show the morphology of MWCNTs variation along with ball milling time, longer ball milling time could increase the diameter of carbon nanotubes and lead to MWCNTs aggregation again. Therefore, short ball milling time process is more effective on carbon nanotubes dispersion in absolute ethyl alcohol environment containing 5wt% sodium stearate.

Figure 2 shows the effect of ball milling on the dispersion of MWCNTs in liquid environment containing 5% sodium stearate + absolute ethyl alcohol, SDS (0.5g/l) + absolute ethyl alcohol and

SDS (0.5g/l) + deionized water respectively. It can be seen that some MWCNTs show oriented distribution, and it is evidently observed that MWCNTs were dispersed well. This means that ball milling could break the hard agglomeration of MWCNTs with the aid of 5% sodium stearate+absolute ethyl alcohol. The reason for this could be that the specific surface energy of the carbon nanotubes was reduced due to the adsorption of sodium stearate in absolute ethyl alcohol environment [13]. Obviously, single ball milling process couldn't meet the complete dispersion of carbon nanotubes.

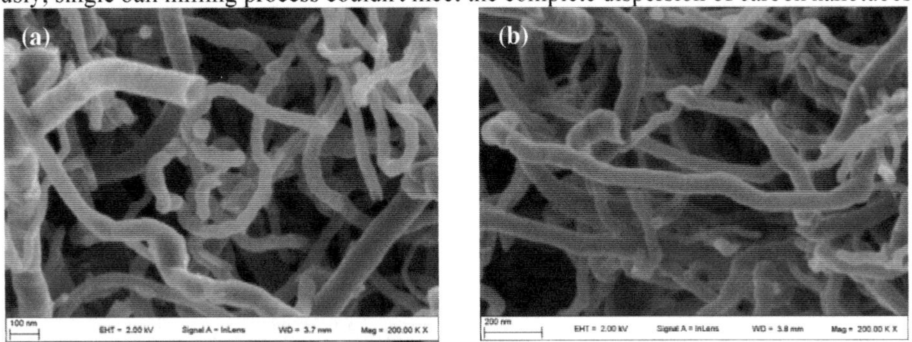

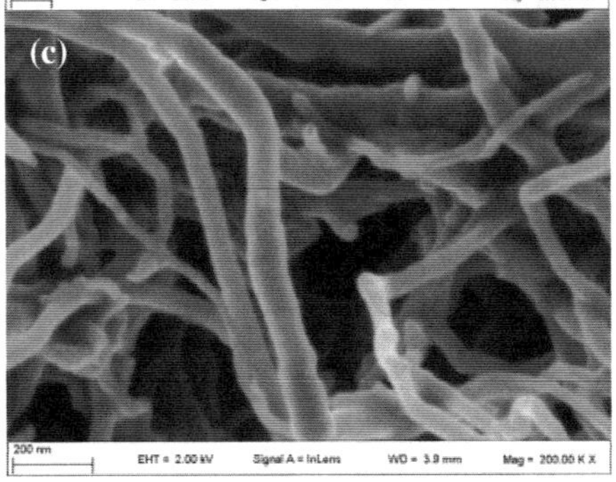

Figure 2. (a), (b), (c) are the dispersion of MWCNTs with the ball-milled condition of 200 rpm, and the milling time of 0.5 H in absolute ethyl alcohol containing 5 wt% sodium stearate, absolute ethyl alcohol containing SDS (0.5g/l), aqueous solution containing SDS (0.5g/l) respectively.

In order to obtain homogeneous dispersion of MWCNTs which were ball-milled by the above process were dispersed further using ultrasonic dispersion method. As figure 3 shown, ball milling rotary speed and disperse agents have different effects on MWCNTs dispersion. Compared with the rotation speed of 200 rpm and 300 rpm, there was no aggregation phenomena when the rotation speed was 100 rpm during ultrasonic dispersion process in absolute ethyl alcohol solution which are shown in figure 3(a), figure 3(b), and figure 3(c) respectively. Similar results were observed during ultrasonic dispersion process in absolute ethyl alcohol + SDS (0.5 g/l) solution, which are presented in figure 3(d), figure 3(e) and figure 3(f) respectively. However, the difference is that MWCNTs were distributed more uniformly when the rotation speed was 100 rpm in absolute ethyl alcohol+SDS (0.5g/l) solution as figure 3(a) and figure 3(d) shown. The main reason for this may be that the specific surface energy of MWCNTs was decreased because of the absorption of SDS on the surface of MWCNTs [15]. It was found that some of the MWCNTs were shortened and intertwined together once more obviously on the account of higher rotation speed of 300 rpm which are shown in figure 3(c) and figure 3(f).

SDBS was used for better dispersion of MWCNTs on the condition of 300 rpm. Figure 4 shows the effect of SDBS on the dispersion of MWCNTs which were subjected to ball milling process on the ball-milled condition of 300 rpm, and 0.5 H-2 H in 5% sodium stearate + absolute ethyl alcohol solution and then were ultrasonic - dispersed in absolute ethyl alcohol + SDBS (0.5g/l) solution. It suggested the re-aggregation of MWCNTs were completely broken up with the aids of SDBS. Furthermore, some MWCNTs present oriented distribution which is shown in figure 4(c).

Based on the above dispersion treatment of MWCNTs, homogeneous dispersion MWCNTs have been obtained, secondary ball milling process was carried out for the mixture of MWCNTs into magnesium powder with the ball milling rotation speed of 100 rpm and ball mill time of 1 H in absolute ethyl alcohol environment. It is evidently found that MWCNTs were uniformly distributed into Mg matrix composites with the content of MWCNTs up to 2 wt.% which are shown in figure 5(a) and figure 5(b). The present results would provide a promising way to prepare Mg matrix composites reinforced by MWCNTs with higher content of MWCNTs.

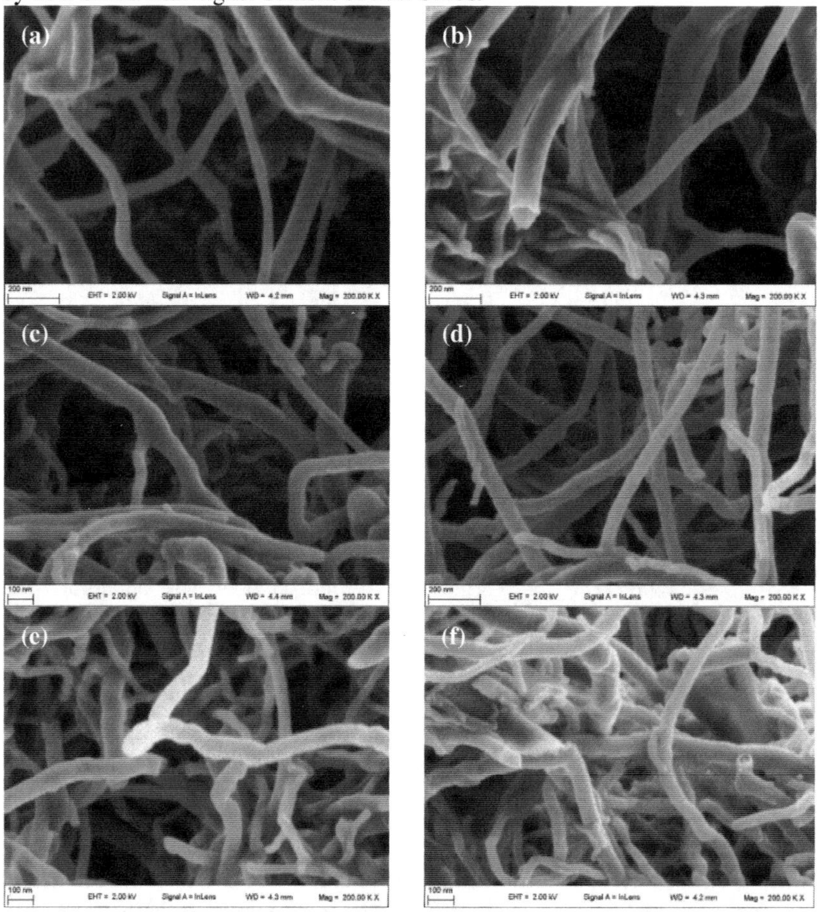

Figure 3. The dispersion of MWCNTs with the ball-milled condition of 100、200、300 rpm, and the milling time of 0.5 H in absolute ethyl alcohol containing 5 wt% sodium stearate, ultrasonic dispersion time is 1.0 H in absolute ethyl alcohol in (a), (b), (c); and ultrasonic dispersion time is 1.0 H in absolute ethyl alcohol containing SDS (0.5g/l) in (d), (e), (f) respectively.

The dispersion mechanism of carbon nanotubes is discussed as follows: first, sodium stearate was completely dissolved in alcohol and adsorbed onto MWCNTs which may lead to reduce the zeta potential of MWCNTs [13]. Then, the surfactant agent SDS or SDBS were adsorbed onto MWCNTs to overcome van der Waals force between carbon nanotubes for MWCNTs dispersion furtherly [14]. Due to the adsorption of surfactants onto the surface of MWCNTs, the bind energy between MWCNTs was also decreased with the respect to the adsorption density of SDS increasing [15]. This may result in the orientation arrangement of MWCNTs which are shown in figure 2(b), figure 3(a), figure 3(d) and figure 4(c). Therefore, It is reasonable to concluded that the dispersion of MWCNTs was greatly improved by the adsorption of sodium stearate and surfactants (SDS or SDBS), which are evidently observed in figure 3(d) and figure 4.

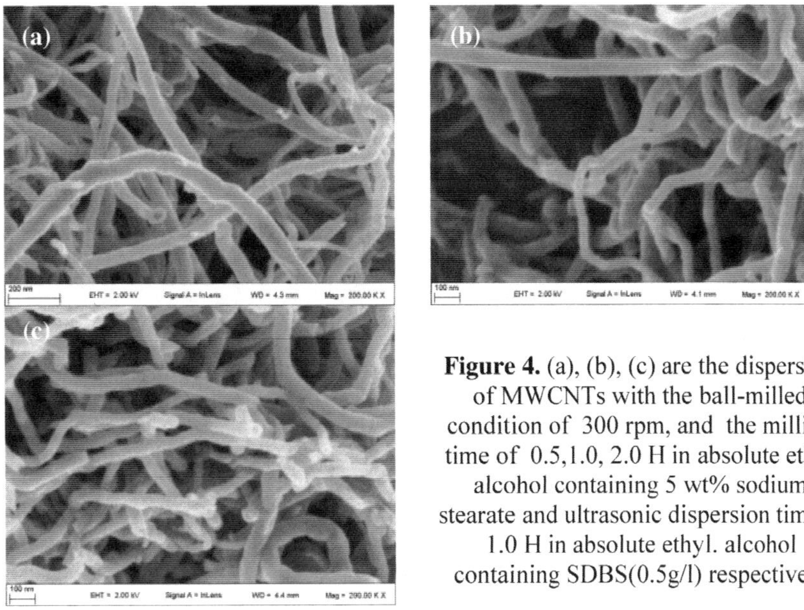

Figure 4. (a), (b), (c) are the dispersion of MWCNTs with the ball-milled condition of 300 rpm, and the milling time of 0.5,1.0, 2.0 H in absolute ethyl alcohol containing 5 wt% sodium stearate and ultrasonic dispersion time is 1.0 H in absolute ethyl. alcohol containing SDBS(0.5g/l) respectively

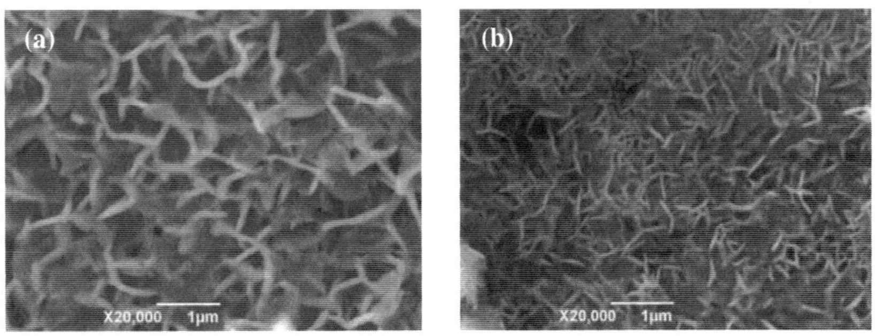

Figure 5. The dispersion of MWCNTs into magnesium matrix which were dispersed by the present method. (a), MWCNTs (wt1%); (b), MWCNTs (wt 2%).

4. Conclusions

A practical method have been developed to disperse MWCNTs into Mg matrix composites homogeneously with higher content of MWCNTs by combining two step high-energy ball milling and ultrasonic dispersion processes in wet environment. The following conclusions were obtained:

The aggregation of MWCNTs was completely broken up by wet-ball milling in 5% sodium stearate + absolute ethyl alcohol environment on the ball-milled condition of 100 -200 rpm and the milling time of 0.5 H. The dispersing agents (SDS, SDBS) have great effect on the further dispersion of MWCNTs in absolute ethyl alcohol using ultrasonic dispersion method. Furthermore, some of MWCNTs present oriented arrangement. The secondary wet ball milling process was contributed to obtain Mg matrix composites containing higher mass fraction of MWCNTs.

Acknowledgements

Thanks are given to the Dr. Start-up fund of Xi'an University of Science and Technology (2014QDJ067), and the Scientific research cultivated fund of Xi'an University of Science and Technology (201602).

References

[1] Peng B *et al.* 2008 Measurements of near-ultimate strength for multiwalled carbon nanotubes and irradiation-induced crosslinking improvements *Nature Nanotechnology* **3(10)** 626–631
[2] Demczyk B G *et al.* 2002 Direct mechanical measurement of the tensile strength and elastic modulus of multiwalled carbon nanotubes *Mater Sci Eng A* **334(1-2)** 173–178
[3] Ruoff R S, Qian D and Liu W K 2003 Mechanical properties of carbon nanotubes: theoretical predictions and experimental measurements *Comptes Rendus Physique* **4(9)** 993–1008
[4] Thakur S K, Srivatsan T S and Gupta M 2007 Synthesis and mechanical behavior of carbon nanotube-magnesium composites hybridized with nanoparticles of alumina *Materials Science and Engineering A* **466(1-2)** 32–37
[5] Fukuda H *et al.* 2011 Fabrication of magnesium basedcomposites reinforced with carbon nanotubes having superior mechanicalproperties *Mater. Chem. Phys* **127(3)** 451–458
[6] Sun F, Shi C, Rhee K Y and Zhao N 2013 In situ synthesis of CNTs in Mg powder at lowtemperature for fabricating reinforced Mg composites *J. Alloy. Compd* **551** 496–501
[7] Mindivan H *et al.* 2014 Fabrication and characterization of carbon nanotube reinforced magnesium matrix composites *Applied Surface Science* **318** 234–243
[8] Fukuda H *et al.* 2011 Fabrication of magnesium based composites reinforced with carbon nanotubes having superior mechanical properties *Materials Chemistry and Physics* **127(3)** 451–458
[9] Liu S Y *et al.* 2010 Fabrication of carbon nanotubes reinforced AZ91D composites by ultrasonic processing *Trans. Nonferrous Met. Soc. China* **20(7)** 1222–1227
[10] Li Q *et al.* 2009 Improved processing of carbon nanotube/ magnesium alloy composites *Composites Science and Technology* **69(7-8)** 1193–1199
[11] Munkhbayar B *et al.* 2013 Influence of dry and wet ball milling on dispersion characteristics of the multi-walled carbon nanotubes in aqueous solution with and without surfactant *Powder Technology* **234(234)** 132–140
[12] Simões S *et al.* 2016 Microstructural Characterization of Aluminum-Carbon Nanotube Nanocomposites Produced Using Different Dispersion Methods *Microscopy and Microanalysis* **22(3)** 1–8
[13] Song M G, Kim J Y and Kim J D 2003 Effect of sodium stearate and calcium ion on dispersion properties of precipitated calcium carbonate suspensions *Colloids and Surfaces A: Physicochem. Eng. Aspects* **229(1-3)** 75–83
[14] Clar J G *et al.* 2013 Interactive Forces between Sodium Dodecyl Sulfate-Suspended Single-Walled Carbon Nanotubes and Agarose Gels *Journal of the American Chemical Society* **135(47)** 17758–17767
[15] Duan W H, Wang Q and Collins F 2011 Dispersion of carbon nanotubes with SDS surfactants: a study from a binding energy perspective *Chemical Science* **2(7)** 1407–1413

Influence of chemical admixtures on the dispersion of carbon nanotubes in water and cement pastes

Hui Liu[1], Suping Cui[1,*], Jiachen Wang[1] and Jianfeng Wang[1]

[1] Department of Materials Science and Engineering, Beijing University of Technology, China

E-mail: cuisuping@bjut.edu.cn

Abstract. The influence of ultrasonic and three types of chemical surfactants (including cationic surfactants: CTAB, anionic surfactants: SDS, and nonionic surfactants: TX-405) on the dispersion of CNTs was investigated. The techniques include UV-Vis-NIR spectrophotometer, laser particle size analyser and scanning electrical microscope (SEM). The results show that: 1) Ultrasonic leads to a dispersive effect on CNTs in water, and the optimal ultrasonic time is 120 s; 2) Three types of surfactants have positive effects on the dispersion of CNTs in water, among which cationic surfactant (CATB) leads to the best dispersibility; 3) CNTs with more carboxyl groups show better dispersion in water indicated from UV-vis-NIR spectra and particle size measurement; 4) The optimum concentration of surfactants is 5:1 (the mass ratio of dispersant to CNTs); 5) Three types of surfactants can improve the dispersion of CNTs in cement pastes indicated from SEM images at the optimum dosage.

Keywords: Carbon nanotubes, dispersion, surfactants, cement-based materials

1. Introduction

The application of carbon nanotubes (CNTs) has generated tremendous interests due to their excellent physical and chemical properties [1]. Nowadays, CNTs are considered an attractive candidate for cement reinforcement [2]. The Young's modulus and tensile strength of CNTs are tens and hundreds of times that of the strongest steel respectively [3,4].

Over the last few decades, extensive research endeavors have been made to development of composited cementitious materials with CNTs [5-8]. Researchers have carried out in studying the effects of CNTs (0.02% to 2.0% by weight of cement) on mechanical properties of various cementitious systems incorporated with CNTs, and found that the addition of CNTs can increase strength and toughness of cementitious materials [7,9-11]. However, it is noticeable that the reported investigations focused on the reinforced cement based materials with CNTs are always divergent and conflicting, which is related to the dispersion of CNTs within the cement paste matrix [12,13]. CNTs have tendency to aggregate and form bundle structures duo to the extremely strong Van der Waals forces, thus making it difficult to disperse and separate CNTs [14]. Poor dispersion of CNTs results in the formation of many microcracks in the composited matrix and limits the efficiency of the CNTs in the cement based materials [15].

Ultrasonic technology is one effectively mechanical method to disperse the CNTs. However, the mechanically dispersion method through ultrasound have potential to damage the structure of CNTs. [16] Therefore, it is necessary to determine the optimal ultrasound time. Chemical methods include pre-treatment of the surface of CNTs chemically by using covalent or non-covalent modification approaches. A mixture of H_2SO_4 and HNO_3 was commonly used to modify the CNTs to form carboxyl functional groups in their structure, thus leading to steric repulsion between CNTs [5]. The non-covalent modification approaches include employing cationic, anionic and neutral surfactants to improve the dispersibility of CNTs [17-19]. A lot of researching studies indicated that some chemical surfactants such as CATB (1-hexadecylpyridine bromide); SDS (sodium dodecyl sulfate); and TX-405 (Triton X-405) were effectively in dispersibility of CNTs [18,20,21]. Results from dispersing CNTs with non-covalent modification approaches also suggest the concentration of surfactants have significance influence the dispersibility of CNTs in water [19]. However, there is limited literature focused on the effect of surfactants on the dispersion of carboxylic CNTs.

In this study, the influences of ultrasound time, surfactant type and concentration of chemical surfactants (cationic, anionic and nonionic surfactants) on the dispersion of CNTs in water were systematically investigated.

2. Experimental methods

2.1. Raw materials

Multi Walled Carbon Nanotubes (MWCNTs) with COOH functional groups, provided by Beijing DK nano S&T Ltd, were used. The MWCNTs have outside diameter from 8 nm to 30 nm, inside diameter from 3 nm to 10 nm and length ranging from 10 μm to 30 μm, according to the manufacturer, as presented in table 1. Chemical surfactants include CATB (1-hexadecylpyridine bromide; cationic surfactants); SDS (sodium dodecyl sulfate; anionic surfactants); and TX-405 (Triton X-405; nonionic surfactants) were supplied by J&K Scientific Ltd. Deionized water was used for all experiments.

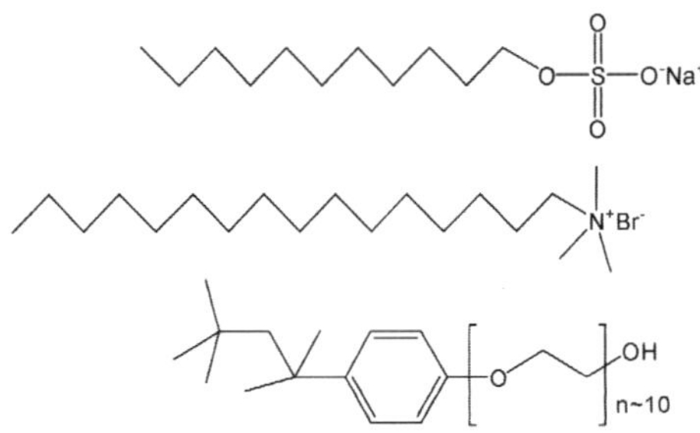

Figure 1. Chemical structure of surfactants used in this study.

Cement pastes were prepared from ordinary Portland cement, type CEM I 42.5 R, MWCNTs, deionized water and Chemical surfactants. Chemical composition of CEM I 42.5 R is presented in table 2.

Table 1. Physical properties of MWCNTs.

Type	Inside diameter	Outside diameter	Length	Density	Thermal conductivity	Special surface area	Carboxyl content
CNT-1	3~10 nm	8~30 nm	10~50 μm	2.1 g/cm^3	>100 s/cm	>110 cm^2/g	1.23%
CNT-2	3~10 nm	8~30 nm	10~50 μm	2.1 g/cm^3	>100 s/cm	>110 cm^2/g	2.00%

2.2. CNTs suspensions preparation and characterization

For preparation of MWCNT dispersions, three surfactants and a sonicator were used. In a typical procedure, suspensions were prepared by mixing the MWCNTs in a concentration of 16 mg/ml in an aqueous solution with or without different amounts of surfactant and the resulting dispersions were sonicated at room temperature. Constant energy was applied to the samples using a 300 W cup-horn high intensity ultrasonic processor with a 6 mm cylindrical tip and temperature controller. Throughout sonication the beaker with the suspended MWCNTs was cooled in a water bath.

Table 2. Chemical composition and mineralogical composition of cement (wt/%).

Chemical composition						M6ineralogical composition			
CaO	SiO_2	SO_3	MgO	Al_2O_3	Fe_2O_3	C_3S	C_2S	C_3A	C_4AF
61.91	22.10	2.87	2.66	4.04	3.38	57.34	18.9	6.47	11.25

2.2.1. Determination of ultrasounds time. The carbon nanotubes (MWCNT-1) were dispersed in a concentration of 0.16 g/ml in an aqueous solution, and the ultrasounds time was set at 60s, 120s and 180s. At first, the amount of CNT-1 was added into a 150 ml breaker with water previously prepared. Then, the suspension in the beaker (water + CNT-1) was subjected to ultrasounds for different time. The last stage corresponds to the assessment of the dispersion of the suspension through conductivity test and zeta potential analysis to determination of ultrasounds time (t). Before measurements, each sample was diluted by distilled water to the concentration of one-twentieth of its initial concentration.

2.2.2. Effect of surfactant type and concentration on dispersion of CNTs suspensions. The content of MWCNTs was kept constant in an aqueous solution at an amount of 0.16 g/ml, and the surfactant to carbon nanotubes weight ratio is 3:1, 4:1 and 5:1 respectively. The solutions were subjected to ultrasounds for a regular time (t) to make MWCNTs suspensions.

The dispersion of CNTs suspensions was evaluated by UV-vis-NIR spectrophotometer (U-4100, Hitachi, Japan), with a wavelength range from 230 nm to 800 nm. Before measurements, each sample was diluted by distilled water to the concentration of one-twentieth of its initial concentration. The corresponding surfactant aqueous solution was used as the blank.

2.2.3. Effect of carboxyl content of CNTs on dispersion of CNTs suspensions. At the presence of cationic surfactants (1-hexadecylpyridine bromide), CNT suspensions prepared from CNTs with different carboxyl content were characterized by UV-vis-NIR spectrophotometer and particle size distribution.

3. Results and discussion

3.1. Effect of ultrasounds time on dispersion of CNTs

Conductivity test and zeta potential are significant parameters determining the dispersion and stability of nanoparticles in solution. In order to better understand the effect of different ultrasounds time on dispersion of MWCNTs, conductivity test and zeta potential analysis of MWCNTs suspensions were performed.

Table 3. Conductivity and zeta potential of MWCNTs suspensions measured for three times.

Ultrasounds time	Conductivity（μS/cm）			Zeta potential（mV）		
0s	0.0034	0.0038	0.0031	-10.4	-9.9	-10.6
60s	0.0243	0.0244	0.0245	-20.3	-19.8	-19.5
120s	0.0303	0.0310	0.0352	-19.3	-19.3	-19.5
180s	0.0246	0.0247	0.0247	-18.3	-17.9	-17.3

The results presented in table 3 show that ultrasounds time have effects on dispersion and stability of MWCNTs in water. It is evident that MWCNTs suspensions sonicated for different time show better Conductivity than that without ultrasounds treatment, among which suspension sonicated for

120 s shows the best Conductivity. This indicated that ultrasounds treatment can effectively disperse the MWCNTs in water. From the results obtained by zeta potential analysis, it can be found that CNT-2 are negative charged as the presence of functional groups (-COOH). In addition, ultrasounds treatment leads to a positive effect on stability of CNTs suspensions, indicated by increased zeta potential value. According to literature [22], the higher the zeta potential is, the more probable the preservation of suspensions stability. Different ultrasounds time (60s~180s) have no obviously effect on stability of CNTs suspensions. Therefore, the optimal ultrasounds time (t) was 120s and performed in the following experiments.

3.2. Effect of surfactant type and concentration on dispersion of MWCNTs

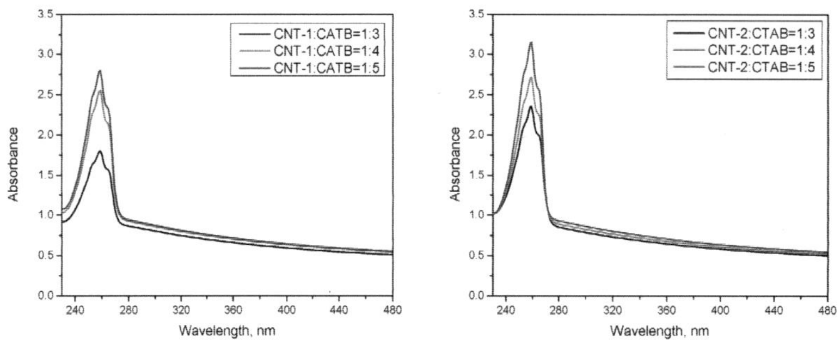

(a) (b)

Figure 2. UV-vis-NIR absorption spectra of MWCNT dispersed in CATB surfactant solutions: (a) CNT-1 and (b) CNT-2

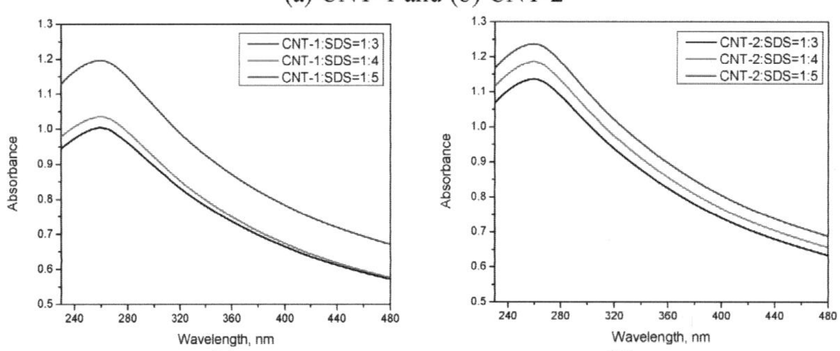

(a) (b)

Figure 3. UV-vis-NIR absorption spectra of MWCNTs dispersed in SDS surfactant solutions: (a) CNT-1 and (b) CNT-2

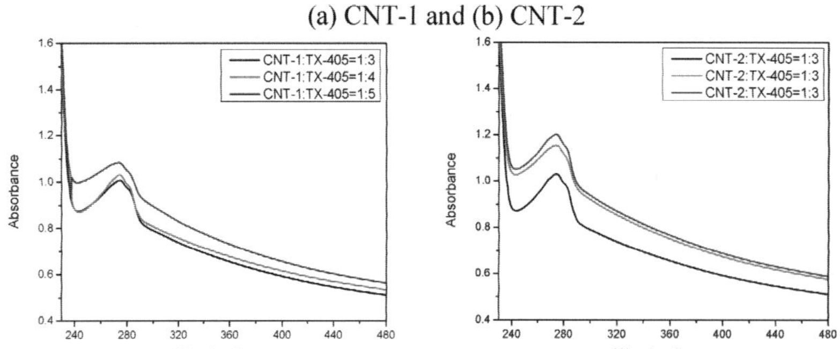

(a) (b)

Figure 4. UV-vis-NIR absorption spectra of CNT dispersed in TX-405 surfactant solutions: (a) CNT-1 and (b) CNT-2

IOP Conf. Series: Materials Science and Engineering **182** (2017) 012029 doi:10.1088/1757-899X/182/1/012029

The effect of surfactant type and concentration on dispersion of MWCNTs suspensions has been investigated by the UV-vis-NIR spectroscopy. Figure 2-4 show the UV-vis-NIR absorption spectra of dispersions in which MWCNTs were dispersed in CTAB, SDS and TX-405 solutions, respectively. The absorbance of all solutions show a maximum between 240 nm and 280 nm which agree with the report from literature [23], and the intensity of absorbance is related to the concentration of CNTs in suspensions according to the Lambert-Beer' law. From figure 4(a) and 4(b), it can be seen that the maximum intensity in the absorbance curves increases continuously with increasing the concentration of CATB, indicating CATB leads a dispersion effect on MWCNTs. In addition, similar results obtained from MWCNTs prepared from the same surfactant for CNT-1 and CNT-2. However, the absorbance intensity beyond 280nm did not increase when more CATB was added in suspensions. For SDS and TX-405 samples, the absorbance intensity and peak values increase with increasing concentration of surfactants. Therefore, it is obviously that the three surfactants at high concentration have positive dispersion effect on MWCNTs in water. Among different surfactants samples at dosage of 1:5 (mass ratio of CNTs to surfactant), MWCNTs suspensions prepared by CATB have the highest absorbance value, which indicated that CATB (cationic surfactants) has a better ability to disperse CNTs.

The measurement of the mean particle size of MWCNTs is important to explain the effect of three surfactants on the dispersion of MWCNTs particles in aqueous suspensions. The mean cluster size of MWCNTs suspension with CTAB, SDS or TX-405 is depicted in figure 5(a-c). From the data, it can be found that the cluster size of MWCNTs is distributed between 240 nm and 300 nm, which indicated that MWCNTs still formed agglomerations even in surfactant suspensions. However, it is obvious that the mean particle size decreases with increasing the concentration of surfactants.

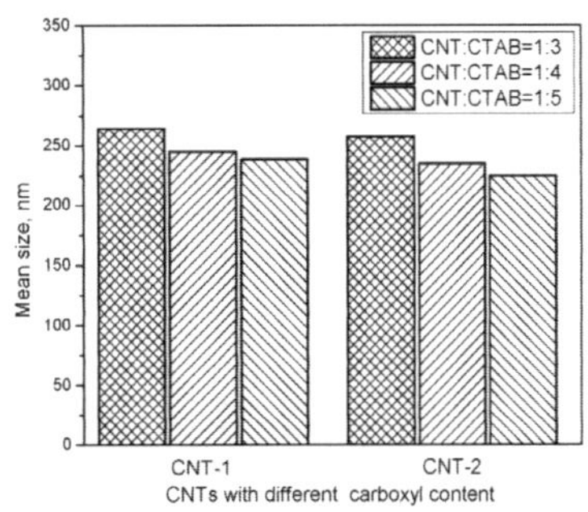

(a)

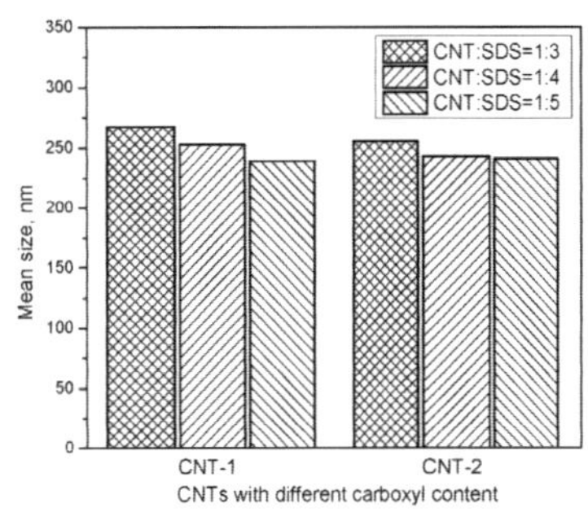

(b)

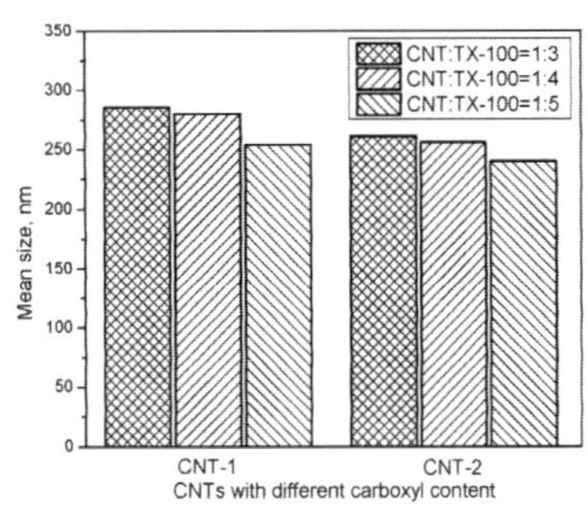

(c)

Figure 5. Mean particle size of MWCNT agglomerates in suspensions prepared from (a) CTAB; (b) SDS and (c) TX-405.

3.3. The effect of carboxyl content of MWCNTs on dispersion of suspensions

The maximum absorbance values and the mean particle size are two important parameters to understand the dispersion of MWCNTs with different carboxyl content in aqueous solutions. Table 4 and table 5 depict the variation of the maximum absorbance values and the mean particle sizes of two types of MWCNT with the different type and concentration of surfactants. When MWCNTs were added into suspensions with the same type and concentration of surfactant, the maximum absorbance values of CNT-2 suspensions are larger than that of CNT-1 suspensions, and the mean particle sizes of CNT-2 become smaller. Therefore, it is obvious that the dispersing ability of CNT-2 with more carboxyl content is better than that of CNT-1 at the same type and concentration of surfactant.

Table 4 The maximum absorbance values of CNTs suspensions with different surfactants.

Type	CTAB			SDS			TX-405		
	1:3	1:4	1:5	1:3	1:4	1:5	1:3	1:4	1:5
CNT-1	1.802	2.539	2.788	1.005	1.035	1.196	1.007	1.031	1.083
CNT-2	2.346	2.700	3.150	1.136	1.186	1.237	1.031	1.155	1.202

Table 5. Mean particle size of CNTs in the suspensions prepared from different surfactants.

Type	CTAB			SDS			TX-405		
	1:3	1:4	1:5	1:3	1:4	1:5	1:3	1:4	1:5
CNT-1	264.5	245.1	238.7	267.5	253.1	239.1	285.5	279.8	254.0
CNT-2	257.4	235.2	225.0	255.6	243.0	240.7	260.9	255.8	240.0

3.4. Dispersion of MWCNTs in cement pastes

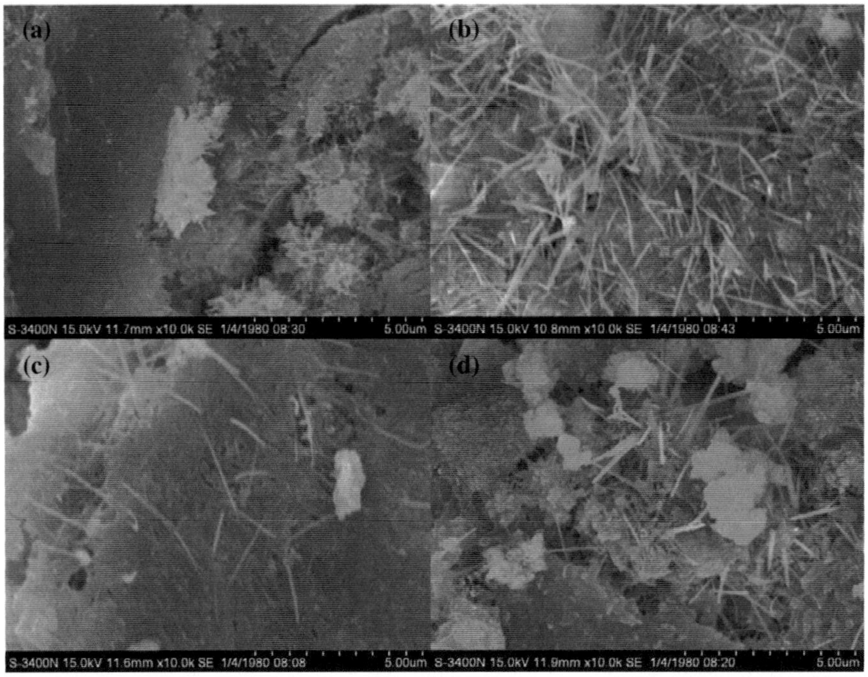

Figure 6. Effect of different types of surfactants on the dispersion of CNTs in hardened cement pastes curing for 7 days: (a) no surfactant; (b) CTAB; (c) SDS; and (d) TX-405.

The effects of the different surfactants on the dispersion of MWCNTs in hardened cement pastes (the surfactant to CNT weight ratio is 5:1) are shown in figure 6. All pastes mixtures were composed with water/cement ratio of 0.5, CNTs at concentration of 0.16 g/ml. Figure 6(a-d) describes the microstructures of cement pastes prepared from CNT-2 suspensions within the following mixtures: (a) no surfactant; (b) cationic surfactant: CTAB; (c) anionic surfactant: SDS and (d) nonionic surfactant: TX-405. It can be observed that in case of surfactant, CNTs seem well dispersed in cement pastes and only individual MWCNTs can be identified on the fracture surface. In contrast, MWCNTs in cement paste without any surfactant were hardly to find out in the micrographs, which may result from the aggregation of CNTs to form bundle structures and totally coated by hydration products.

4. Conclusions

This paper reports the research results of the dispersion of two types MWCNTs with different carboxyl in the aqueous solutions prepared from three different surfactants including CATB, SDS and TX-405. The following conclusions can be drawn:

(1) Ultrasonic leads a dispersive effect on CNTs in water and the best ultrasonic time is 120 s;

(2) Three types of surfactants have positive effect on the dispersion of MWCNTs in water, among which cationic surfactant (CATB) leads to the best dispersability;

(3) MWCNTs with more carboxyl groups show better dispersion in water indicated from UV-vis-NIR spectra and particle size measurement;

(4) The optimum concentration of surfactants is 5:1 (the mass ratio of dispersant to MWCNTs) among the three concentrations;

(5) Three types of surfactants can improve the dispersion of MWCNTs in cement pastes indicated from SEM images at the optimum dosage.

References

[1] Parveen S, Rana S and Fangueiro R 2013 A review on nanomaterial dispersion, microstructure, and mechanical properties of carbon nanotube and nanofiber reinforced cementitious composites *J. Nanomater.* **7** 3271–3283

[2] Mukhopadhyay AK 2011 *Nanotechnology in Civil Infrastructure* (Berlin Heidelberg Springer) p 207–223

[3] Peng B, Locascio M, Zapol P, Li S, Mielke S L and Schatz G C 2008 Measurements of near-ultimate strength for multiwalled carbon nanotubes and irradiation induced crosslinking improvements *Nat. Nanotechnol.* **10** 626–31

[4] Belytschko T, Xiao S P and Schatz G C 2002 Atomistic simulations of nanotube fracture *Physical Review B* **23** 121–121

[5] Geng Y L, Pei M W and Zhao X 2005 Mechanical behavior and microstructure of cement composites incorporating surface-treated multi-walled carbon nanotubes *Carbon* **6** 1239–1245

[6] Collins F, Lambert J and Duan W H 2012 The influences of admixtures on the dispersion, workability, and strength of carbon nanotube–OPC paste mixtures *Cem. Concr. Compos.* **2** 201–207

[7] Konsta-Gdoutos M S, Metaxa Z S and Shah S P 2016 Highly dispersed carbon nanotube reinforced cement based materials *Cem. Concr. Res.* **7** 1052–1059

[8] Parveen S, Rana S and Fangueiro R 2015 Microstructure and mechanical properties of carbon nanotube reinforced cementitious composites developed using a novel dispersion technique *Cem. Concr. Res.* **73** 215–227

[9] Kerienė J, Kligys M and Laukaitis A 2013 The influence of multi-walled carbon nanotubes additive on properties of non-autoclaved and autoclaved aerated concretes *Constr. Build. Mater.* **49** 527–535

[10] Musso S, Tulliani J M and Ferro G 2009 Influence of carbon nanotubes structure on the mechanical behavior of cement composites *Composites Science & Technology* **69** 1985–1990

[11] Chaipanich A, Nochaiya T and Wongkeo W 2010 Compressive strength and microstructure of carbon nanotubes–fly ash cement composites *Materials Science & Engineering A* **527** 1063–1067

[12] Al-Rub R K A, Ashour A I and Tyson B M 2012 On the aspect ratio effect of multi-walled carbon nanotube reinforcements on the mechanical properties of cementitious nanocomposites *Constr. Build. Mater.* **35** 647–655
[13] Luo J, Duan Z and Li H 2009 The influence of surfactants on the processing of multi-walled carbon nanotubes in reinforced cement matrix composite *Physica Status Solidi Applications & Materials* **206** 2783–2790
[14] Grobert N 2007 Carbon nanotubes – becoming clean *Materials Today* **10** 28–35
[15] Xie X L, Mai Y W and Zhou X P 2005 Dispersion and alignment of carbon nanotubes in polymer matrix: A review *Materials Science & Engineering R Reports* **49** 89–112
[16] Krause B, Mende M and Pötschke P 2010 Dispersability and particle size distribution of CNTs in an aqueous surfactant dispersion as a function of ultrasonic treatment time *Carbon* **48** 2746–2754
[17] Rausch J, Zhuang R C and Mäder E 2010 Surfactant assisted dispersion of functionalized multi-walled carbon nanotubes in aqueous media *Composites Part A Applied Science & Manufacturing* **41** 1038–1046
[18] Kumar P and Bohidar H B 2010 Aqueous dispersion stability of multi-carbon nanoparticles in anionic, cationic, neutral, bile salt and pulmonary surfactant solutions *Colloids & Surfaces A: Physicochemical & Engineering Aspects* **96** 376–385
[19] Figueiredo D T R, Correia A A S and Hunkeler D 2014 Surfactants for dispersion of carbon nanotubes applied in soil stabilization *Colloids & Surfaces A: Physicochemical & Engineering Aspects* **480** 405–412
[20] Bai Y, Lin D and Wu F 2010 Adsorption of Triton X-series surfactants and its role in stabilizing multi-walled carbon nanotube suspensions *Chemosphere* **79** 362–367
[21] Datsyuk V, Landois P and Fitremann J 2009 Double-walled carbon nanotube dispersion viasurfactant substitution *Journal of Materials Chemistry A* **19** 2729–2736
[22] Goscianska J, Olejnik A and Nowak I 2016 Stability analysis of functionalized mesoporous carbon materials in aqueous solution *Chemical Engineering Journal* **290** 209–219
[23] Lu F, Zhang S and Zheng L 2012 Dispersion of multi-walled carbon nanotubes (MWCNTs) by ionic liquid-based phosphonium surfactants in aqueous solution *Journal of Molecular Liquids* **173** 42–46

Effect of oxygen plasma etching on graphene's mechanical and electrical properties

Pengfei Jia*, Fengming Pan* and Tianhang Chen

College of Science, Nanjing University of Aeronautics and Astronautics, Nanjing 211106 China

E-mail: * jiapengfei@vip.163.com, fmpan@nuaa.edu.cn

Abstract. This paper reports a study of the effect of oxygen plasma etching on the mechanical and electrical properties of chemical vapor deposited (CVD) multilayer graphene. By means of scan probe microscopy (SPM), it was found that the defects were initially induced to the top graphene layer by oxygen plasma etching, which plays an incentive role in further etching. Oxygen plasma bombarding on surface of graphene enhanced the surface roughness, as well changed the tribological properties. The results of electronic transport measurements show a decrease in mobility with the increase of etching duration. These findings are valuable for studying the effects of plasma etching on graphene, and modifying the physical properties of graphene through artificially generated defects.

Keywords: Graphene, Oxygen plasma etching

1. Introduction

Due to its excellent physical properties [1-5], graphene has drawn significant attention in electronic, chemical, biochemical and mechanical applications [6], such as high-speed transistors, DNA sequencing [7,8], and biochemical sensors. In order to expand its applications in technology, efforts have been tried to modify the band structure (semimetal) of graphene. Oxygen plasma etching on multilayer graphene has been taken to study the effects on the mechanical and electrical properties through changing the energy gap between conduction and valence bands [9]. Oxygen plasma etching has become a topic of intense research in the past years [10-14].This process is isotropic that particles attack the wafer from all angles. As well, plasma treatment, especially oxygen plasma treatment with high chemical reactivity to carbon has become an effective way for the modification of carbon based on materials to alter their nano-structures and properties. In this study, we explore the effect of oxygen plasma etching of chemical vapor deposited (CVD) graphene on Si/SiO_2 substrate over different durations.

Graphenes were initially produced with chemical vapor deposition (CVD) using copper as a substrate. Then they were transferred onto a Si substrate with a 300nm SiO_2 buffer layer. Optical microscope was used to confirm and identify that graphene has been transferred successfully. By carefully tuning the time of the oxygen plasma etching, we studied the effect of oxygen plasma etching on graphene. The durations of oxygen plasma etching were selected as 10 seconds, 30 seconds, 60 seconds and 180 seconds, individually. Atomic force microscopy (AFM, Veeco multiMode) was used to explore the morphology and defects on multilayer graphene. Influence of oxygen plasma etching on mechanical properties of graphene was discussed. By comparing four cases of the etching duration, we analyzed the etching effect on electrical properties of graphene taking advantage of the electrostatic force microscopy module of AFM and ferroelectric analyzer (TF Analyzer 2000). Based on the experimental results, we discussed how oxygen plasma interacted with graphene and how the electrical properties were affected. A conclusion was drawn in the end.

2. Experimental

Silicon with a top layer of oxide is generally used as substrate for various graphene devices due to easy optical identification of the sample. In this study, multilayer graphene produced by chemical vapor deposition (CVD) on copper were transferred to silicon substrate with thermally grown oxide layers (300 nm). Optical microscope was used to locate and identify graphene samples. Before oxygen plasma etching, all samples were cleaned with acetone rinse for 5 minutes in order to remove tape residues and other organic contaminations. Plasma cleaner (PDC- M) with oxygen plasma source was used to etch multilayer graphene under the following parameters: oxygen flow rate of 400 sccm (standard cubic centimeter per minute), the RF power of 150W and pressure of 0.37 Torr (50pa). Etching process durations include 10 seconds, 30 seconds, 60 seconds, 180 seconds.

After the treatments, atomic force microscopy was taken to present the morphology and thicknesses of the samples. Taking advantage of the force imaging mode of AFM, the tribological features of graphene were investigated. A combination of electrostatic force microscopy (EFM) and ferroelectric analyzer module (TF Analyzer 2000 FE-Module) was used to characterize the electrical properties of graphene treated by oxygen plasma etching.

3. Results and Discussion

By analyzing the surface scan images, the average thickness of multilayer graphene is obtained as shown in figure 1. It is found that the thickness of graphene decreases versus etching duration, from initial thickness of 8.4nm approximately to 0.6nm of graphene treated by 180s oxygen plasma etching. Etching rate seems to decreasing with etching duration.

We have realized that the morphology of graphene changes with the oxygen plasma etching as well as roughness and thickness of graphene. While what interests us more is how the oxygen plasma interacts with graphene. To explain the question, we must further understand the mechanisms of oxygen plasma interacting with the remaining graphene sheets after the transient etching process. There are two possible etching mechanisms including isotropic etching and anisotropic etching [15]. If single sheet of graphene removed by oxygen plasma is isotropic, the degree of etching should be identical on a graphene sheet regardless of graphene defects and edges. When a single atomic layer is removed from the top, a single or a couple of chains of atoms at the edge will also be removed. The reason is removing one carbon atom from the top sheet of graphene will need enough energy to break three bonds, but at the edges or defects, only one or two bonds need to be broken due to the existence of dangling bonds. So the energy needed to break C-C bonds at the edges or defect is lower than other parts on graphene obviously [16]. And etching mechanisms must choose the lower energy-requiring method. That means the isotropic process is not likely to happen in this study. For an anisotropic etching, there are usually two methods, anisotropic vertical etching (layer-by-layer) and anisotropic horizontal etching (direct etching from edges or defect to inside) [15].

Unfortunately, we cannot distinguish the removal situation at the edge or defect by using optical microscopy or atomic force microscopy (AFM) due to limitation of the resolution of these instruments. Nevertheless, the modification behavior of surface morphology versus etching time provides useful information helping us with understanding the detailed etching mechanism. For the anisotropically horizontal etching mode, the etching rates at the graphene edges and defective area are anticipated to be higher than those at other intact regions due to the fact that oxygen plasma tends more likely to interact with carbon atom located at the edges/defects of graphene. In addition, this oxygen plasma will react with atoms at exposed edges/defects, regardless of their layer position. This results in the defective regions of the top graphene sheet enlarging, and further generates larger defective areas on the lower layers. While in anisotropic vertical etching, oxygen plasma will interact with graphene layer by layer. The defective regions of the top sheet will not expand. The surface morphology and the defective region size are not expected to change much during the etching in the case of anisotropically vertical etching. In the experiments, roughness of graphene decrease versus etching process, implying an un-expanded defect region. These results suggest that the removal of graphene by oxygen plasma is most likely anisotropically vertical etching, i.e., layer by layer etching mechanism is preferred. So it is possible to use oxygen plasma etching to change the property of top-layer graphene without affecting interior structure.

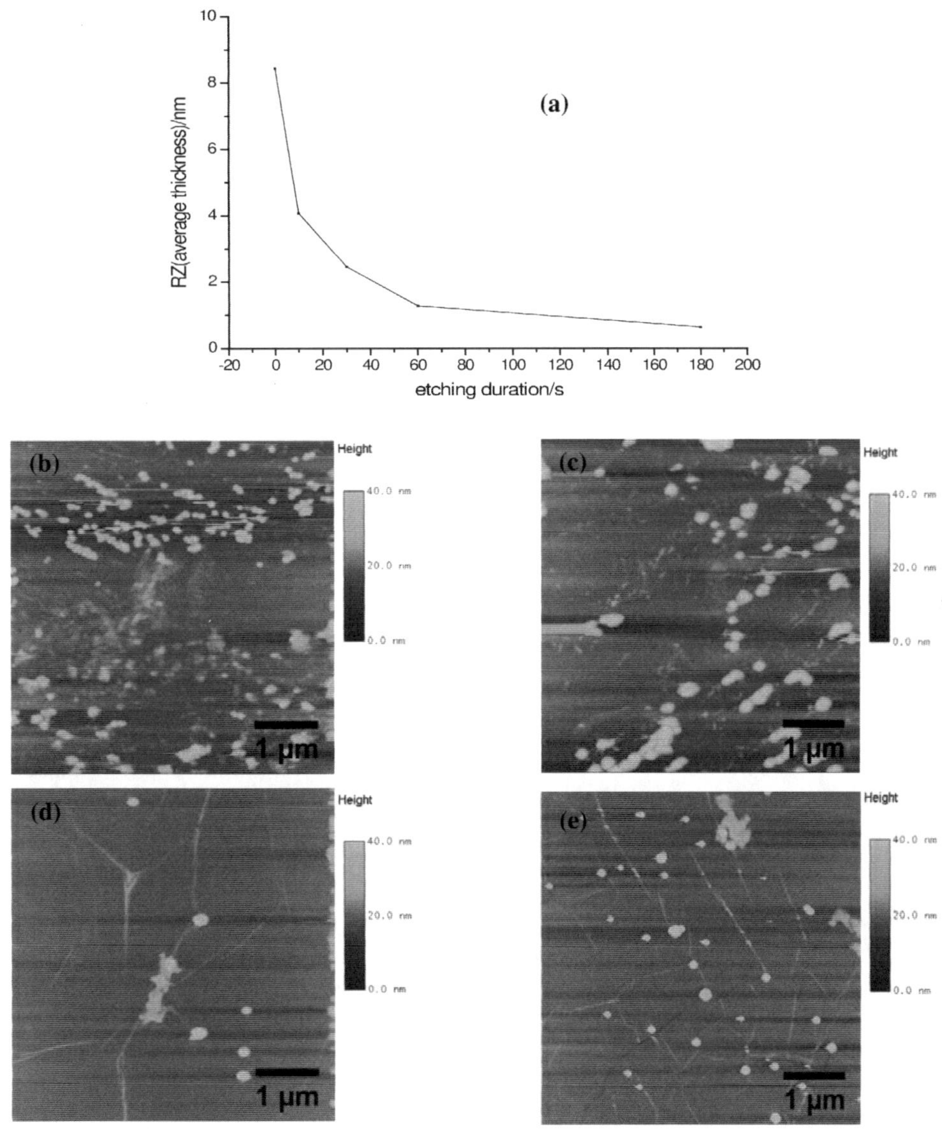

Figure 1. (a) Variation of the average thickness of graphene with the decrease of the cumulative etching time. (b) ,(c) ,(d) and (e) AFM images of graphene being etched with oxygen plasma after (b) 10 seconds ,(c) 30 seconds ,(d) 60 seconds and (e) 180 seconds.

IOP Conf. Series: Materials Science and Engineering **182** (2017) 012030 doi:10.1088/1757-899X/182/1/012030

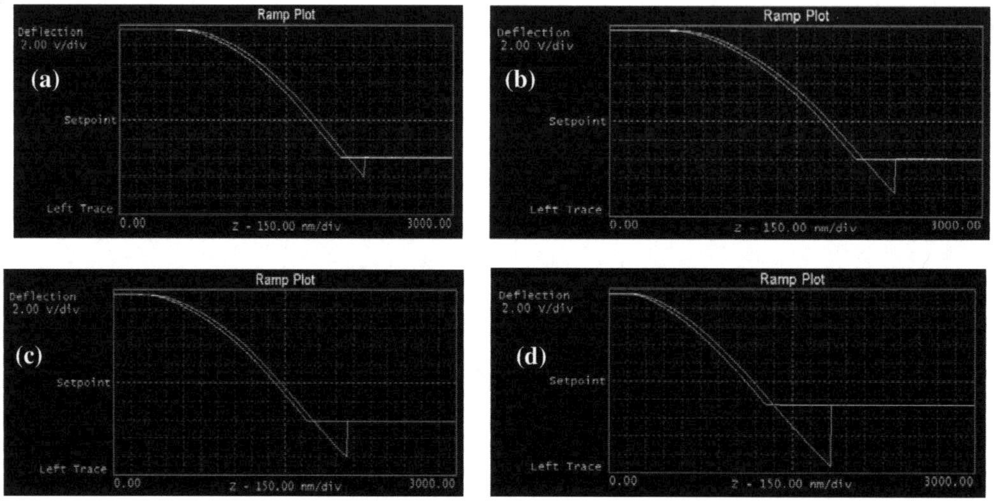

Figure 2. Images show the tracks of pinpoint on etched graphene surfaces with different etching durations (a)etching duration is 10 seconds (b) etching duration is 30 seconds (c) etching duration is 60 seconds .(d) etching duration is 180 seconds. Pinpoint distance of lateral is 3μm.

For purpose of measurement of mechanical properties influenced by oxygen plasma etching, force calibration of AFM has been performed. Figure 2 shows tracks of pinpoint on etched graphene surfaces with different etching durations. The friction force can be extracted from these data (diMutlimode SPM Instruction Manual 004-210-000) [17], as shown in figure 2.

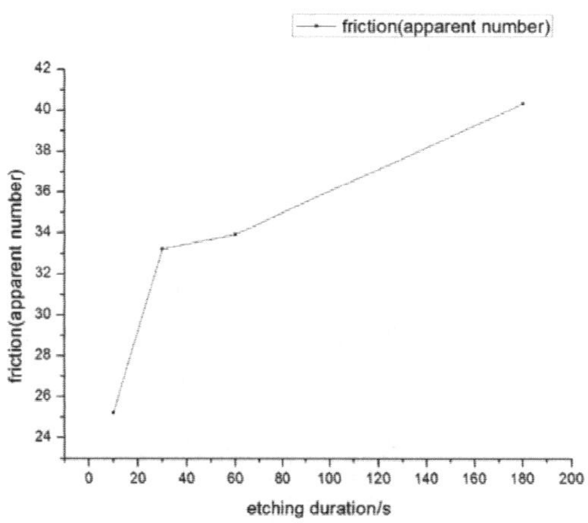

Figure 3. Friction force on graphene being extracted from SPM image cycles as a function of duration of oxygen plasma etching.

It is found that the surface friction decreases with the etching duration. We contribute this phenomenon to puckering effect, taking into account of bending stiffness and in-plane rigidity. The degree of the puckering on graphene and the resulting contact area will depend on the bending stiffness of graphene, as a thinner sample has lower bending stiffness and more thereby puckers [18], leading to a larger contact area and higher frictional force. When graphene layers increases, both the bending stiffness and the in-plane rigidity will be increased, causing a lower puckering degree. Due to this reason, the friction on graphene intends to increase with the decrease of numbers of layers of graphene. This may explain the increase of friction on graphene with the oxygen plasma etching duration, as shown in figure 3. It should be noticed that when number of graphene layers exceeds five layers, friction on graphene will approach friction on graphite [17, 19] where friction is much less related to puckering effect, or the friction is rarely sensitive to graphene layer numbers. According to the variation behavior of friction versus etching duration, friction does not change much within 60s etching, probably due to the lower decrease in the numbers of layers of multilayer graphene. With increase of etching time friction increases sharply, indicating a thinner graphene where puckering effect plays an important role on friction.

To further explore the electrical property of graphene treated by oxygen plasma etching , electrostatic force microscopy (EFM) has been used to present electric domains of graphene, as shown in figure 4.

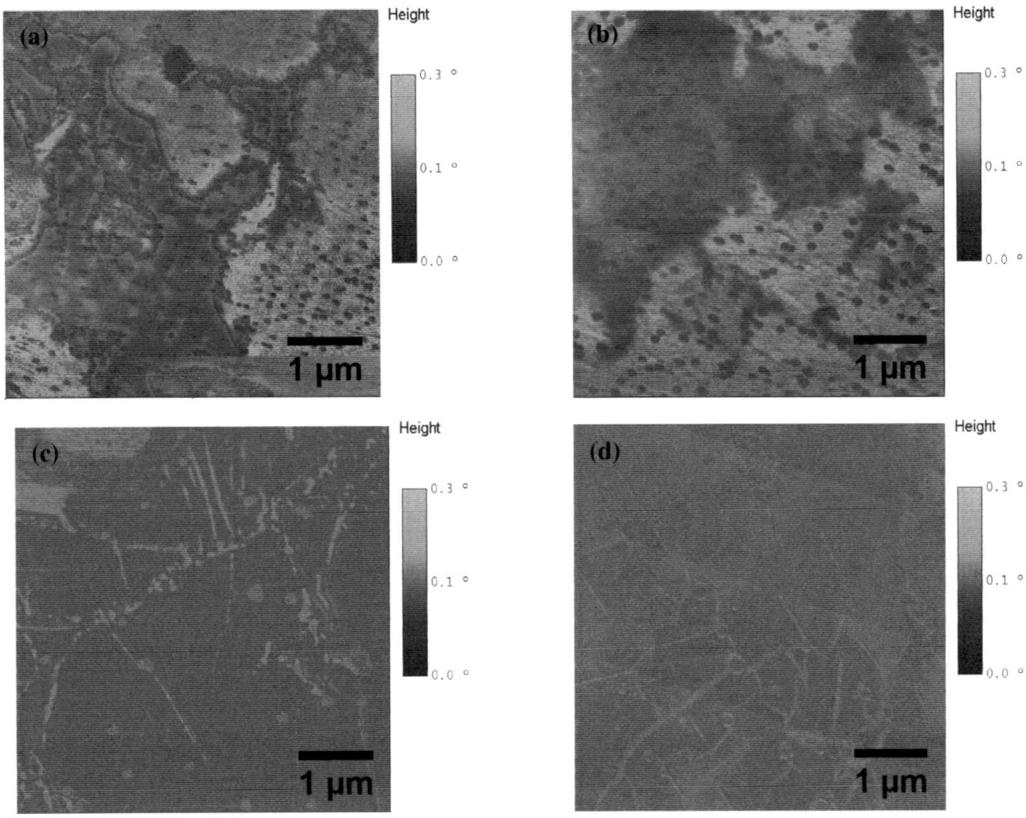

Figure 4. EFM images show the distribution of electric domains for graphene samples treated by different etching time (a) 10 seconds, (b) 30 seconds, (c) seconds and (d) seconds. Height of EFM images represent the surface electric potential when applied a voltage as the brightness of images.

Figure 4 shows the distribution of electric domains for samples treated by different etching time. The brightness represented the surface electric potential when applied a voltage. It can be seen that both the brightness and bright areas decrease with the etching time, indicating a decrease of surface electric domain size with oxygen plasma etching duration, which might be attributed to the fact that oxygen plasma etching reduces the quantity of free electrons or holes on the graphene surface. Oxygen plasma bombardment on the surface of graphene brings more defects on the top layer. Defect expanding reduces movement of the free electrons or holes resulting in a decrease of electric domains [20]. Nevertheless, the electric domain distribution does not appear to be much different when etching time is over 60s.

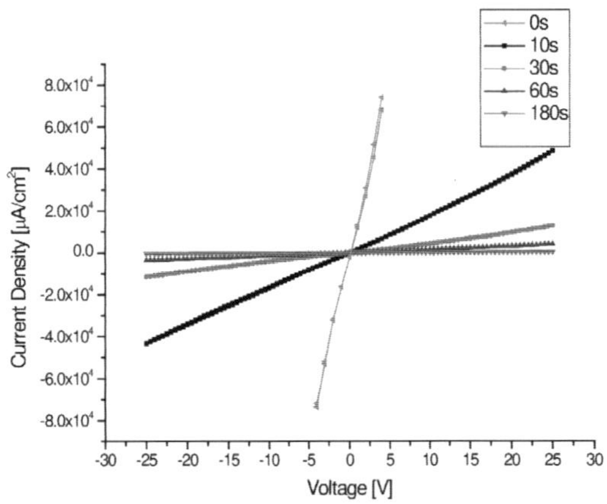

Figure 5. The I-V characteristics for different cumulative plasma exposure, different color of the lines show the different I-V characteristics for different cumulative plasma exposure.

In order to further characterize electrical properties of graphene treated by oxygen plasma etching, we measured the electrical conduction of multilayer graphene using ferroelectric analyzer (TF Analyzer 2000). The I-V characteristic curves are presented in figure 5 for samples with etching duration of 10s, 30s, 60s and 180s, individually. For all samples, a linear relationship is well established between the electric current and applied voltage, besides that the conductivity decreases with the etching duration. As a matter of fact, one of the main purposes to perform oxygen plasma etching on graphene is a try to induce semiconductor characteristic. In this study, the ohmic contact characteristic of graphene does not seem to be broken by oxygen plasma etching. The decrease of conductivity with increase of etching time is due to the reduction in numbers of layers of graphene.

4. Conclusions

This paper reports a study of the effect of oxygen plasma etching on the chemical vapor deposited (CVD) graphene. It is suggested that an anisotropically vertical etching mechanism happens for the oxygen plasma etching on graphene. The oxygen plasma etching increases the graphene surface friction based on puckering effect. The results provide a post-synthesis approach to change the properties of graphene and the possibility of producing the graphene thin films carrying expected properties including monolayer graphene.

Acknowledgement

Great thanks should be provided for the support of National natural science foundation of China (A020102, A040215). We are also appreciated much for the help with sample preparations from prof. Tai guoan' group in the State Key Laboratory of Mechanics and Control of Mechanical Structures, College of Aerospace Engineering, Nanjing University of Aeronautics and Astronautics.

References

[1] Y Yang and J Hrbek 1995 Oxidation of cesium-modified graphite supported on a Ru (001) surface *The Journal of Physical Chemistry* **99** 3229–3234

[2] L Liu, S Ryu, M R Tomasik, E Stolyarova, N Jung, M S Hybertsen, M L Steigerwald, L E. Brus and G W Flynn 2008 Graphene oxidation: thickness-dependent etching and strong chemical doping *Nano letters* **8** 1965–1970

[3] J M Carlsson, F Hanke, S Linic and M Scheffler 2009 Two-step mechanism for low-temperature oxidation of vacancies in graphene *Physical review letters* **102** 166104

[4] H Zhang, Q Fu, Y Cui, D Tan and X Bao 2009 Growth mechanism of graphene on Ru (0001) and O_2 adsorption on the graphene/Ru (0001) surface *The Journal of Physical Chemistry C* **113** 8296–8301

[5] Y Cui, Q Fu, H Zhang, D Tan and X Bao 2009 Dynamic characterization of graphene growth

and etching by oxygen on Ru (0001) by photoemission electron microscopy *The Journal of Physical Chemistry C* **113** 20365–20370

[6] E Starodub, N C Bartelt and K F McCarty 2010 Oxidation of graphene on metals *The Journal of Physical Chemistry C* **114** 5134–5140

[7] P Sutter, J T Sadowski and E A Sutter 2010 Chemistry under cover: tuning metal– graphene interaction by reactive intercalation *Journal of the American Chemical Society* **132** 8175–8179

[8] L Jin, Q Fu, R Mu, D Tan and X Bao 2011 Pb intercalation underneath a graphene layer on Ru(0001) and its effect on graphene oxidation *Phys Chem Chem Phys* **13** 16655–16660

[9] R Larciprete *et al.* 2012 Oxygen Switching of the Epitaxial Graphene-Metal Interaction *Acs Nano* **6** 9551–9558

[10] D L Duong *et al.* 2012 Probing graphene grain boundaries with optical microscopy *Nature* **490** 235–239

[11] K Gotterbarm, W Zhao, O Hofert, C Gleichweit, C Papp and H P Steinruck 2013 Growth and oxidation of graphene on Rh(111) *Phys Chem Chem Phys* **15** 19625–19631

[12] Y Zhang, Q Fu, Y Cui, R Mu, L Jin and X Bao 2013 Enhanced reactivity of graphene wrinkles and their function as nanosized gas inlets for reactions under graphene *Phys Chem Chem Phys* **15** 19042–19048

[13] I Vlassiouk, P Fulvio, H Meyer, N Lavrik, S Dai, P Datskos and S Smirnov 2013 Large scale atmospheric pressure chemical vapor deposition of graphene *Carbon* **54** 58–67

[14] R Blume *et al.* 2014 The influence of intercalated oxygen on the properties of graphene on polycrystalline Cu under various environmental conditions *Phys Chem Chem Phys* **16** 25989–26003

[15] H Al-Mumen, F Rao, W Li and L Dong 2014 Singular sheet etching of graphene with oxygen plasma *Nano-Micro Letters* **6** 116–124

[16] U A Schröder, E Grånäs, T Gerber, M A Arman, A J Martínez-Galera, K Schulte, J N Andersen, J Knudsen and T Michely 2016 Etching of graphene on Ir(111) with molecular oxygen *Carbon* **96** 320–331

[17] Q Li, C Lee, R W Carpick and J Hone 2010 Substrate effect on thickness-dependent friction on graphene *Physica Status Solidi* **247** 2909–2914

[18] C Lee, X Wei, Q Li, R Carpick, J W Kysar and J Hone 2009 Elastic and frictional properties of graphene *Physica Status Solidi* **246** 2562–2567

[19] B I Yakobson, C J Brabec and J Berhnolc 1996 Nanomechanics of carbon tubes: Instabilities beyond linear response *Physical Review Letters* **76** 2511–2514

[20] I Childres, L A Jauregui, J Tian, and Y P Chen 2011 Effect of oxygen plasma etching on graphene studied using Raman spectroscopy and electronic transport measurements *New Journal of Physics* **13** 104–109

Study on voids of epoxy matrix composites sandwich structure parts

Simin He,Youyi Wen, WenjunYu, Hong Liu, Cheng Yue and Jing Bao

AVIC Chengdu Aircraft Industry (Group) Co. Ltd., Chengdu 610091 China

E-mail: he.simin@ccac.cc

Abstract. Void is the most common tiny defect of composite materials. Porosity is closely related to composite structure property. The voids forming behaviour in the composites sandwich structural parts with the carbon fiber reinforced epoxy resin skins was researched by adjusting the manufacturing process parameters. The composites laminate with different porosities were prepared with the different process parameter. The ultrasonic non-destructive measurement method for the porosity was developed and verified through microscopic examination. The analysis results show that compaction pressure during the manufacturing process had influence on the porosity in the laminate area. Increasing the compaction pressure and compaction time will reduce the porosity of the laminates. The bond-line between honeycomb core and carbon fiber reinforced epoxy resin skins were also analyzed through microscopic examination. The mechanical properties of sandwich structure composites were studied. The optimization process parameters and porosity ultrasonic measurement method for composites sandwich structure have been applied to the production of the composite parts.

Keywords: Composites processing, Void, Microscopic analysis, Composites mechanical properties, Ultrasonic inspection

1. Introduction

Carbon fiber reinforced resin matrix composites (CFRP) is widely used in the aircraft because of its excellent specific strength, specific modulus and comprehensive properties, they can effectively reduce the weight of the aircraft, improve the load capacity of aircraft [1]. However, the void defects produced during the manufacturing in composites are receiving greater attention. Porosity is one of the most common defects in composite structures [2]. Voids are the critical problem for the polymer composites and are harmful to the mechanical properties of composites [3-4]. Therefore, it is significant to study the influence of process parameters on the microstructure and mechanical properties to analysis the void content and distribution for the application of composites. A lot of research have been done on improving the voids content by adjust the autoclave process [5]. Many researches have been focused on improving the voids content for composite laminate structure [6-7]. For the epoxy matrix composites honeycomb core sandwich structure can greatly reduce the weight of the composite structure and have good mechanical properties compare to the laminate, honeycomb core structure has already been widely used in civil aircraft structure [8]. This study by adjusting the lay-up manufacturing process parameters, analysis the voids formation and distribution in epoxy matrix composites honeycomb core sandwich structure. An optimization process parameters and porosity ultrasonic measurement method for composites sandwich structure have been established and applied to the production.

2. Experiment procedure

2.1. Materials and Composite preparation
HexPly® M21E/IMA carbon fiber reinforced epoxy prepreg, ECK4.0-96 Honeycomb core and adhesive film FM73 were used in the research. M21E/IMA was unidirectional tape.

The composite parts were fabricated by sandwich structure parts. The sandwich part studied in this work was co-cured with upper skin, adhesive film, honeycomb core, adhesive film and lower skin. The Upper skin stacking sequences studied in this work was:[45 135 (0)4 (45) 4 90 (135) 4 0 (45) 3 90 (135) 3 90 (45) 4 (0) 4 135 45]. The Lower skin stacking sequences studied in this work was:[45135(0)3 (45)3 90 (135) 3 (0)2 45(90) 3 (135) 3 0 (90) 3 (45) 2 (135) 2 90 0 (135) 2 (45) 2 (90) 2 0 135 (90) 3 (45) 3 (0) 3 (135) 3 90(45) 3 (0)3 13545]. The lower skin, upper skin and laminate area fiber volume content (FVC) tested in according to EN2564 type B should be around 59% ± 4%.

The lay-up process, the composite parts with different void contents were produced by using different compaction vacuum, the compaction vacuum were set as 30% and 100% of full vacuum respectively. The composite panels with different void contents were produced with a vacuum bag and autoclave using the same cure cycle. The extent of cure should be >95% determined by DSC. After machining, microscopic examination and metallographic microscopic analysis were done. Test samples were trimmed from each of those areas.

2.2. Microscopic analysis
For the microscopic analysis of CFRP, the micro-sections were previously embedded in a resin (cold mounting with an epoxy resin). Then embedded micro-sections were submitted to an adequate preparation by grinding & polishing. Grinding sequence begins with removal of material by disk grinding with SiC papers at high speed (300 rpm) using 180-grit silicon carbide paper and abundant running water. Subsequent grinding proceeds through 500, 1200 and 4000 grit papers. Polishing sequence begins with removal of scratches induced by grinding with clothes at low speeds (150 rpm), using 6μm diamond paste and abundant lubricant oil. Subsequent polishing proceeds through 3 μm and 1 μm diamond paste. Between different polishing clothes, all the samples have been cleaned using water, and observed under stereomicroscope(ZEISS "STEMI 2000-C") to check good progression of polishing. Microscopic analysis on each field is analyzed under optical microscope in order to detect and characterize possible defects over micro-cut section, using a low magnification. When it is found a cracked area, this will be focused at higher magnification.

The micro sections of CFRP were prepared in accordance with ASTM E3-11.The microscopic examination was carried out with small magnification, and the void content was determined using Automatic Image Analysis with a larger magnification.

The automatic image analysis system consisting of the following components must be used: A high quality reflected light microscope equipped with bright-field objectives. A programmable automatic stage control the movement of the specimen in x and y direction. An automatic focus system can control the z-direction of the stage. A camera with adequate resolution and sensitivity, and a computer with an automatic image analyzer were used to analyze and store the measurement data.

2.3. Mechanical Test and Ultrasonic Inspection
The samples corresponding to Filled Hole Compression (FHC) test results were also investigated. The layup for the FHC is $[135/45/90/135/0/45/135/0/45/90]_s$. The specimens were produced separating to the part but on the same tool and under the same vacuum bagging. FHC test specimens were manufactured with extra plate manufactured with the part in the same bagging as shown in figure 1. The specimen dimensions should be width = 36 ± 0.2 mm, length = 160 ± 1.0 mm and thickness =

2.54 mm. Bond10 glass fabric plies M21 / 120 at 45°/-45° for tabs (thickness=1.08 mm) onto the specimens with FM300. The length must be 60 mm on each side.

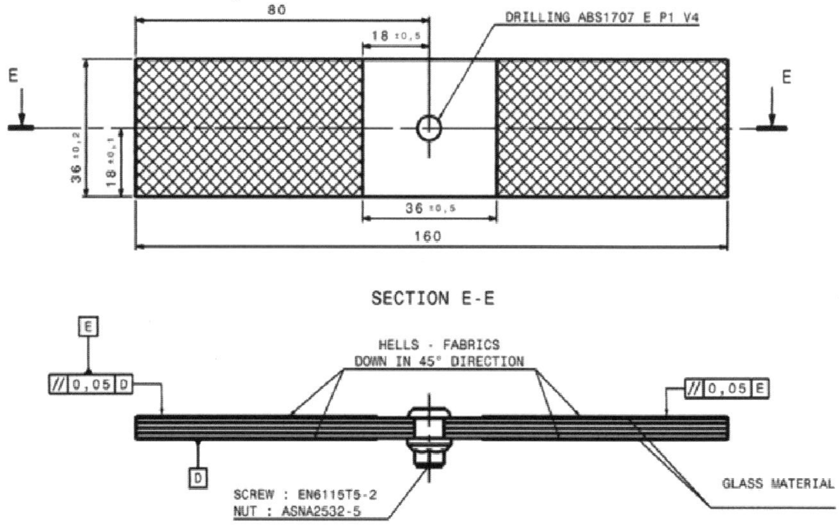

Figure 1. Filled Hole Compression specimens.

All panels were inspected with ultrasonic detector (Masterscan 380 and OminiScan MX2).

3. Results and discussion

3.1. Microscopic analysis

Figure 2 shows the photo of typical CFRP sandwich structure cross section manufactured with 30% vacuum compaction, we selected 2 area: MS-B-F-1 Verification of laminate bonding on trailing edge area, MS-B-F-2 and MS-B-F-3Verification of bonding between laminate and core ramp.

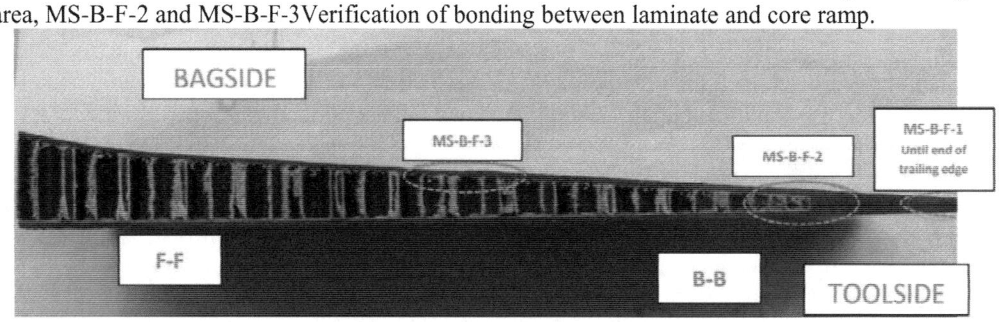

Figure 2. CFRP sandwich structure section photo.

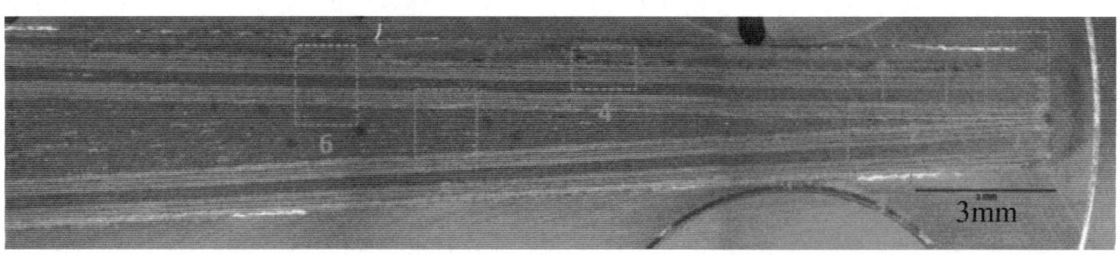

Figure 3. Low magnification picture of MS-B-F-1.

Figure 3 shows the laminate bonding on trailing edge area cross section MS-B-F-1. In the figure1, 2, 3, 4, 5, 6 were selected and marked for the higher magnification observation in figure 4.

Figure 4(a) shows the microstructure of the laminate voids in area 1, figure 4(b) shows the detail of figure 4(a) with higher magnification. Figure 4(c) shows the laminate voids of area 3, figure 4(d) shows the detail of figure 4(c) with higher magnification. Figure 4(e) shows the detailed morphology of laminate 5. Figure 4(f) shows the detailed morphology of laminate 6.We can see the laminate presents voids in all the length of the interested area.

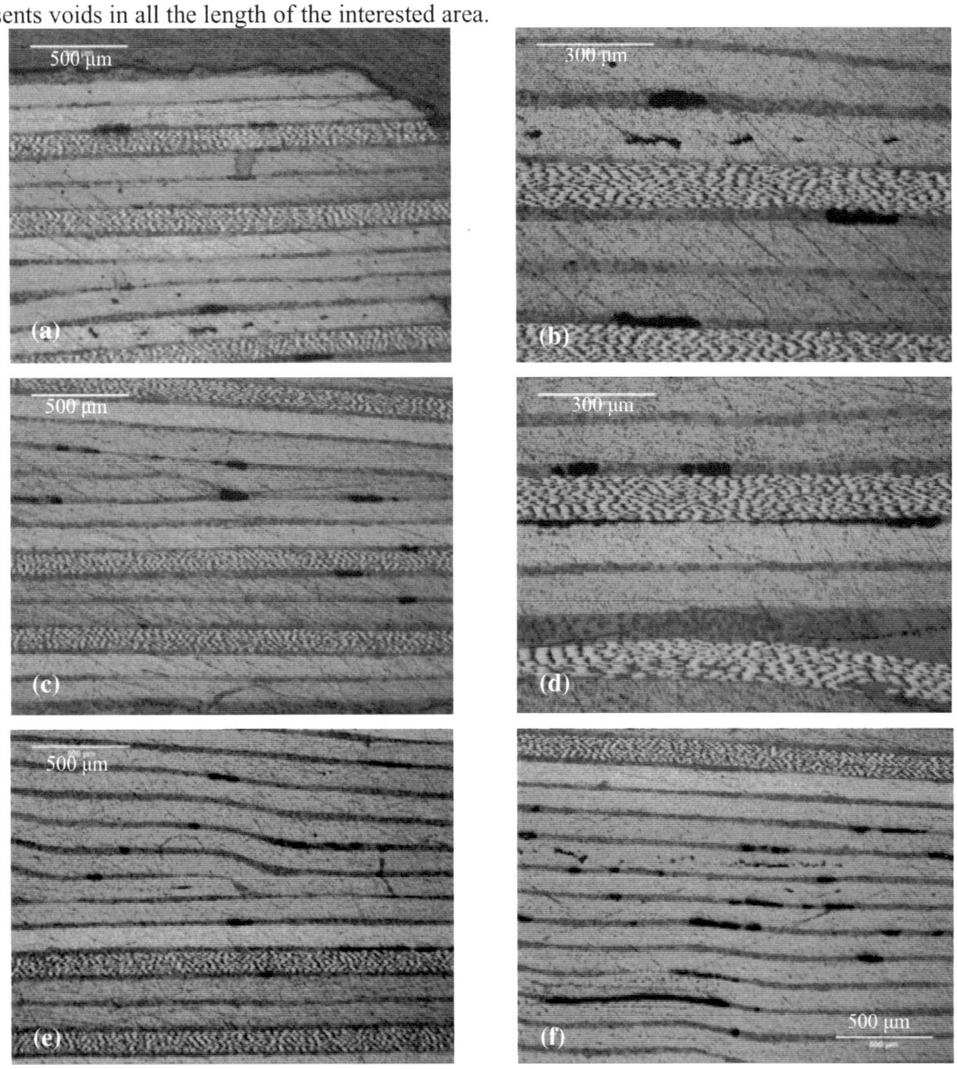

Figure 4. Microscopicphotos of MS-B-F-1.
(a)1, voids, (b)1, voids, (c)3, voids, (d)3, voids, (e)5, voids, (f)6, voids.

Figure 5 shows 12.5X of the bonding between laminate and core ramp cross section MS-B-F-3. We selected areas and marked as 1, 2, 3, 4for higher magnification observation.

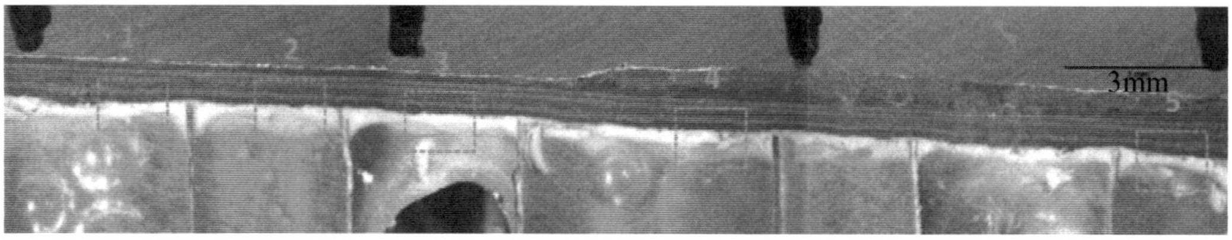

Figure 5. Low magnification picture of MS-B-F-3.

Figure 6(a) shows morphology of the bonding 1 area, the defects, voids, porosity in interface laminate and core can be observed. Figure 6(b), 6(c), 6(d) shows morphology of the bonding 2, 3, 4 area porosity presents in interface laminate and core. The existence of porosity promotes the formation and propagation of inter-laminar resin cracks.

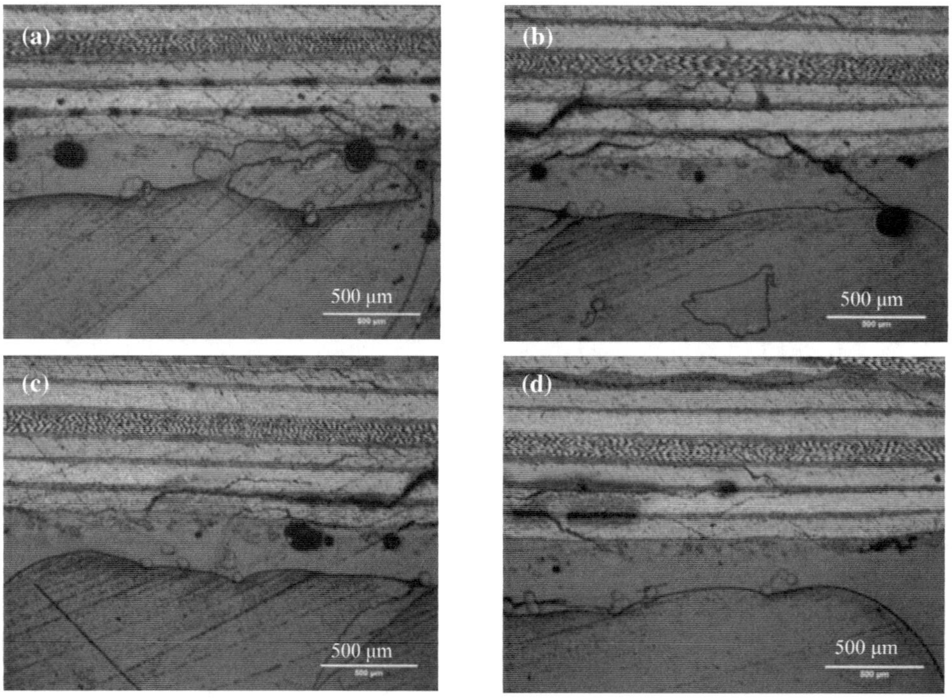

Figure 6. Microscopicphotos of MS-B-F-3.

Voids generally occur in the between layer and the junction resin/fiber. There are two main reasons for the voids formation: The air from the process of manual lay-up and the volatile component from the heat process. It is generally believed that there are three main reasons for the growth of void: (1) Diffusion of gas in the interface between the core and the resin. (2) The temperature or pressure change around the voids. (3) There is a temperature gradient in the resin that causes voids expansion. For the production and application, the most convenient method is to adjust the compaction parameters, but there is no relevant research report.

3.2. Voids Content Evaluation

In order to analyze the local porosity, the voids content was calculated by statistical method. Figure 7 and figure 8 show the typical voids content evaluation results through microscopic analysis in the samples of the 30% vacuum compaction and 100% vacuum compaction, respectively.

IOP Conf. Series: Materials Science and Engineering **182** (2017) 012031 doi:10.1088/1757-899X/182/1/012031

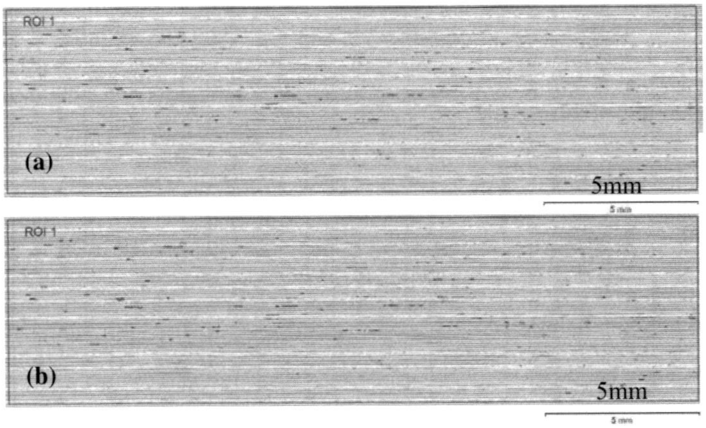

Figure 7. Voids content of 30% vacuum compaction part. (a)voids content: 0.4 %, (b)voids content: 0.4 %.

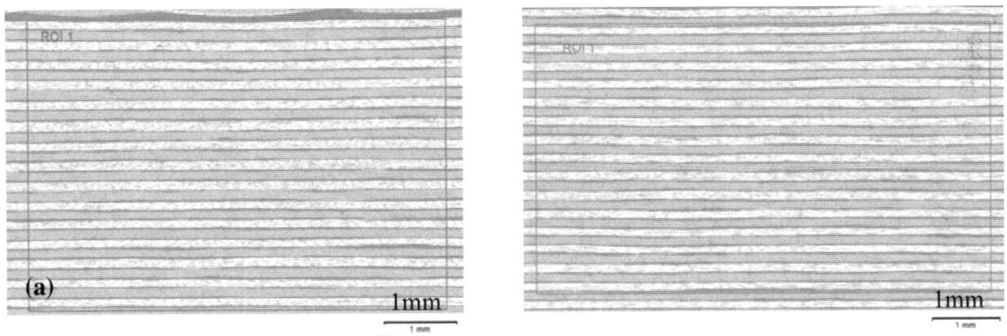

(b)

Figure 8. Voids content of 100% vacuum compaction part. (a)voids content: <0.1 %, (b)voids content: <0.1 %.

At the beginning, the air was laid up with the prepreg and the air from the volatile substance formed the air bubbles in the resin under the action of vacuum pressure. Under the action autoclave pressure, the air bubbles move upward from below, became more and more big when moving. The pressure inside the bubble is getting lower and lower. Finally, the bubbles burst discharge air and volatile content. When honeycomb core sandwich structure manufacturing in autoclave, the pressure should not as high as laminate manufacturing. The pressure inside the bubble is reduced slowly, before the resin gel, the bubbles and volatiles were not completely removed, then the voids formation. As can be seen from the figure 7, the voids are generally clustered in the middle and the upper layer of the composite structure. Increase the lay- up compaction pressure, the honeycomb core structure was not affected, but it can be as much as possible to reduce the trapped air. Finally reduce the porosity generation.

The voids content evaluated microscopic samples can be used as reference standards for ultrasonic inspection. The reference standards can be used to verify that the inspection technique for a specific configuration is consistent with the applicable classification for reporting, to check inspection program set-up, probes, accuracy and deterioration of equipment and to calibrate the whole system.

3.3. Mechanical Test

Table 1 and table 2 show FHC specimen dimensions and test results of the 30% vacuum compaction and 100% vacuum compaction respectively. The results show that the effects of compaction vacuum on FHC were small. The mechanical properties are related to the size of the voids.

Table 1. FHC specimen dimensions and test results of 30% vacuum compaction.

Specimen ID	Length (mm)	Thickness (mm)	Width (mm)	Free length (mm)	Tab length (mm)	Hole diameter (mm)	Pu (KN)	Strength(*) (MPa)
FHC-1	159.99	2.59	35.97	36.30	61.77	7.960	31.10	340.43
FHC-2	160.07	2.57	35.90	36.29	61.77	7.964	31.97	350.63
FHC-3	160.03	2.59	35.93	36.29	61.78	7.956	36.36	398.42
Mean	160.03	2.58	35.93	36.29	61.77	7.96	35.15	363.16
S.D	0.04	0.01	0.03	0.01	0.01	0.00	2.82	30.96
COV(%)	0.02	0.45	0.07	0.02	0.01	0.05	0.90	8.53

(*) Values based on nominal thickness (2.54 mm) and measured width

Table 2. FHC specimen dimensions and test results of 100% vacuum compaction.

Specimen ID	Length (mm)	Thickness (mm)	Width (mm)	Free length (mm)	Tab length (mm)	Hole diameter (mm)	Pu (KN)	Strength(*) (MPa)
FHC-1	160.05	2.59	35.98	36.39	61.81	7.936	36.17	395.74
FHC-2	160.07	2.57	36.03	36.43	61.76	7.962	36.10	394.48
FHC-3	159.97	2.58	35.99	36.45	61.73	7.959	35.58	389.13
Mean	160.03	2.58	36.00	36.42	61.77	7.95	35.95	389.19
S.D	0.05	0.01	0,03	0.03	0.04	0.01	0.32	3.48
COV(%)	0.03	0.39	0,07	0.08	0.07	0.18	0.90	0.88

(*) Values based on nominal thickness (2.54 mm) and measured width

Figure 9(a) and figure 9(b) show FHC test failure mode of the 30% vacuum compaction and 100% vacuum compaction, respectively. Form the failure mode figures of compressive strength test we can see that the failure occurred under the fastener or at the fastener edge. The matrix cracking and fiber breakage caused the failure of the samples. There was no big difference between specimens with different void content by adjusting the compaction vacuum.

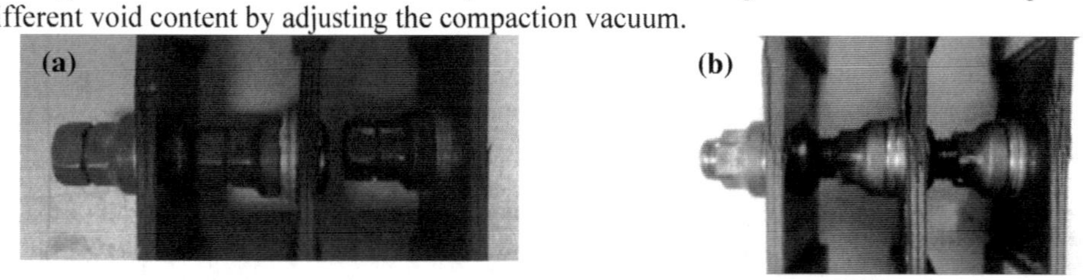

Figure 9. FHC test failure mode. (a)30% vacuum compaction, (b)100% vacuum compaction.

The research shows that the existence of the voids can cause local deformation of carbon fiber, and the local deformation of carbon fiber can reduce the compressive of the composites.

4. Conclusion

(1) The microscopic analysis showed that the voids exist in all the laminate area and the porosity presents in interface laminate and core.
(2) The porosity ultrasonic non-destructive measurement method are developed and verified through microscopic examination.
(3) Using the automatic image analysis system, the voids content was compared for different compaction pressure in the manufacturing process. Increasing the compaction pressure will reduce the voids content of the laminates.

(4) The results of the mechanical properties for sandwich structure composites indicate that the less voids, the higher composites mechanical properties, but the influence is not great for the case that the void content of the test panel are relatively low.

References

[1] Du Shanyi 2007 Advanced composite materials and aerospace engineering. *Acta Materiae Compositae Sinic* **24(1)** 1−12
[2] PARK S Y, CHOI W J and CHOI H S 2010 The effects of void contents on the long-term hygrothermal behaviors of glass /epoxy and GLARE laminates *CompositeStructures* **92** 18−24
[3] Liu zhizhen, LiHongyun and Yi Xiaosu 2005 Influence of void contet on mechanical properties of polyimide composite *Journal of Materials Engineering* **9** 56−58
[4] Judd and NCW 1978 Voids and their effects on the mechanical properties of composites *SAMPE Journal* **14(1)** 10−14
[5] F Y C Boey and S W Lye 1992 Void Reduction in Autoclave Processing of Thermoset Composites Part1: High Pressure Effects on Void Reduction *Composites* **23(4)** 261−265
[6] K J Bowles and S Frimponq 1992 Void Effects on the Interlaminar Shear Strength of Unidirectional Graphite-Fiber-Reinforced Composites *Journal of Composite Materials* **26(10)** 1487−1509
[7] L E Asp and F Brandt 1997 Effects of Pores and Voids on the Interlaminar Delamination Toughness of a Carbon/Epoxy Composite *Proceedings of ICCM-11* Australia
[8] H G Allen 1983 Sandwich constructions Today and Tomorrow, Sandwich construction *Res Mechanica* **8(1)** 29−38

Synergy effects of Cu and Sn onpittingcorrosion resistance of ultra-purified medium chromium ferritic stainless steel

XiangJun Zhang and ZhenYu Liu

The State Key Laboratory of Rolling and Automation, Northeastern University, Shenyang, China

E-mail: zh.xj.good@163.com

Abstract. Theinfluence of combination of Cu and Sn on pitting resistance of ultra-purified medium chromium ferritic stainless steel in 3.5 wt.%NaCl at 25°C was investigated by using electrochemical method. The results show that there is synergy effect between Cu and Sn, and the strong interaction between Cu and Sn in ferritic stainless steels clearly affects their pitting corrosion behaviour in 3.5% NaCl.A mechanism of the synergy of Cu and Sn was discussed.

1. Introduction

As is known to all, Cu and Snaredetrimental to surface quality, hot workability, and low temperature toughness forsteels [1-4]. It is very difficult to completely get rid of these tramp elements in steel making [5-7]. Thus,the researchers turn tostudy their effects on the performance of the steels, attempting to make the full use of their positive effects and toevade their negative ones.There is no doubt about the positive effect of Cu on the general corrosion behaviour of stainless steels in sulphuric acid solutions. The mechanism of beneficial effect of Cu is mainly based on the suppression of anodic dissolution by elemental Cu deposition and other Cu-containing particles on the steel surface [8-10]. Sn can effectively improve general corrosion resistance in H_2SO_4, mainly due to both highly improved hydrogen evolution overpotential, and inhibitive effect of dissolved Sn^{2+} on elementary anodic reactions [11-12]. The beneficial influence of Sn on the behaviour of stainless steels against pitting corrosionis mainly attributed to the improved firmness or stability of passive film for the presence of SnO_2, and decreased Fermi energy of steel by Sn addition [13-14].

Regarding the synergy effect of Cu and Sn on the corrosion behaviour of steel, a few researchers havepaid some attention to it. Osozawa [15] and Takizawa et al.[16] observed positive synergy of Cu and Sn in austenitic stainless steel in both diluted sulphuricand organic-chlorinated acid mediums. They concluded that the synergy is derived from the formation of a stable film of metalliccopper and tin oxides on the steel surface due to preferential dissolution of Cu and Sn.Pardo et al. studied the influence of Cu and Snon the corrosion behavior of AISI 304 and 316 stainless steels in 30% H_2SO_4 solution, revealed that the corrosion resistance of both steels was greatly improved by positive synergy of Cu and Sn, and that Sn addition reduced the corrosion rate in lower magnitude than Cu addition [17]. It was also found that Cu addition favours pit nucleation but inhibits its growth, whereas Sn exerts the opposite effect, favouring pit growth and inhibiting its nucleation in AISI 304 and 316 in 3.5%NaCl [18].

However, the unified conclusions about the synergyof Cu and Sn on corrosion resistance of ferritic stainless steels have not yet to be reached. This paper attempts to evaluate the effect of the combination of Cu and Sn on pitting corrosion behaviour of ultra-purified medium chromium ferritic stainless steels in chloride-containing media.

2. Experimental Procedure

2.1. Materials
Table 1 shows the chemical compositions of experimental steels. All specimens in this paper were cut from 2 mm thick cold rolled and annealed sheets.

Table 1. Chemical compositions of the tested steels (wt.%).

No.	Cu	Sn	Cr	Si	Mn	Nb	Ti	C	N	Fe
A	--	--	16.00	0.433	0.236	0.108	0.023	0.019	0.0032	Bal.
B	0.243	0.220	16.50	0.500	0.250	0.160	0.070	0.006	0.0043	Bal.
C	0.451	0.238	16.80	0.522	0.260	0.116	0.026	0.010	0.0021	Bal.
D	0.448	0.464	16.70	0.530	0.251	0.101	0.009	0.007	0.0030	Bal.

2.2. Electrochemical measurements
Specimens cut from cold rolled and annealed sheets were ground into 1×1 cm^2 with a series of SiC grit papers. After pre-passivation in 30 wt.% HNO_3 at 35°C for 30 min, specimens were cold mounted by using epoxy resin, leaving the surface area in the size of 1×1 cm^2. Before the experiments, all the mounted specimens were ground by using 400–1500 SiC grit papers and cleansed with acetone.

Electrochemical tests were carried out in 500 ml 3.5 wt.%NaCl solution using three electrodes electrochemical flat cell, and open to the air at 25 °C.Specimenswere used as the working electrode, and platinum foil was used as the counter electrode. The reference electrode was a saturated calomel electrode (SCE) connected via a Luggin's capillary, and all potentials in this paper refer to the SCE scale. Cathodic reduction of passive film on working electrodes was performed at -1.5 V for 10min to remove the possible existing surface film before electrochemical tests. Polarization measurements were carried out at a scan rate of 0.33 mV/s from an initial potential of -150 mV versus open-circuit potential (OCP) to pitting potential (E_{pit}) at which the current density continuously exceeded 10 μA/cm^2.In each electrochemical measurement test, five parallel specimens were used to guarantee the reliability and reproducibility of the results.

3. Results and Discussion

Figure 1 shows representative potentiodynamic polarization curves obtained in 3.5 wt.%NaCl solution.As can be seen, all specimens present similar cathodic curves due to the same predominant cathodic half-cell reaction, $O_2 + 2H_2O + 4e^- \rightarrow 4OH^-$ ($E^{\theta}\approx 0.157$ V), while all the anodic polarization curves exhibit direct passivation after slightly active dissolution, and finally reach different E_{pit} with no apparent active-passive transitions. The E_{pit} values of A, B, C and D specimens are 189 mV, 252 mV, 229 mV and 235 mV respectively.

Figure 2 shows the relation between the contents of Cu, Sn and the E_{pit} of test steels. As compared to the A specimen (without Cu and Sn), E_{pit} values of B, C, and D specimens alloyed with both Cu and Sn, are improved by about 33%, 21%, and 24% respectively, revealing positive effect of Cu + Sn addition on pitting resistance of stainless steels. B specimen presents the highest E_{pit} value indicating the best performance among tested steels. However, it can also be found that with increasing the Cu content nearly two times than that in B specimen,E_{pit} values of both C and D specimensare decreased by about 9.1% and 6.7% respectively.Furthermore, compared with C specimen, E_{pit} of D specimen is slightly improved by increasing Sn content to 0.464 wt%.These interrelatedresultsjust indicate that Cu addition reduces E_{pit}value of ferritic stainless steels, while Sn addition can increase E_{pit}value in chloride-containing medium.This paradoxical effects of Cu and Sn are in good agreement with A. Pardo group's research results [18]. Figure 2 also revealsthat there is a strong positive synergy between Cu and Sn in corrosion behaviour of ferritic stainless steel in chloride-containing medium.

Nevertheless, additions of Cu and Sn in appropriate proportion can further improve the corrosion resistance of experimental steels, just as B specimen shows in figure 2.

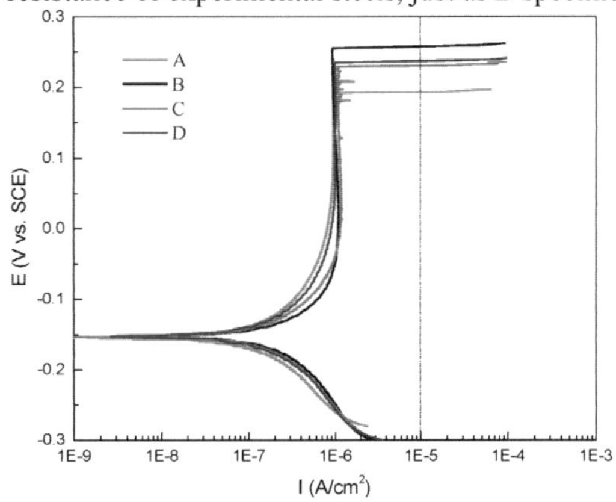

Figure 1. Potentiodynamic polarization curves of experimental ferritic stainless steels in 3.5 wt% NaCl solutions.

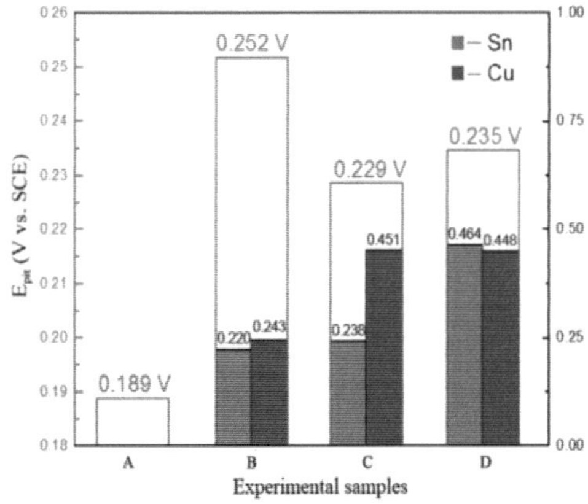

Figure 2. Variation of Epitwith Cu, Sn contents in test steels.

Our group has studied the effect of Sn aloneon the pitting behaviour of ultra-purified 17% Cr ferritic stainless steel [19]. We found that Sn-bearing 17% Cr ferritic stainless steels can form faster growing and more compact passive films than that in Sn-free steel, probably due to the presence of SnO_2 and Sn in passive film. Nam et al. also concluded that the presence of SnO_2 on the surface ofSn-containing low alloysteel decreasesthe corrosion rate due to a decrease in anodic reaction kinetics [20]. Furthermore, we also observed that with increasing Sn in 17% Cr ferritic stainless steels, the charge carrier densities (both N_d and N_a) in passive film decrease effectively, with respect to those of Sn-free steel [19]. This beneficial effect of Sn on reducing the charge carrier density of passive film, probably owes to a very wide band gap of SnO_2 (E_g = 3.6 eV, at 300 K) detected in passive film, and decreased Fermi energy of steel by Sn addition [13-14]. Thismechanism probably explains why E_{pit} values of B, C and D steels increase with increasing Sn, as shown in figure 2.

The beneficial effect of Cu on the corrosion behaviour of stainless steels was explained by the protective effect of Cu redeposition on anodic surface of steels during active dissolution, which improved the corrosion resistance by suppressing active dissolution even in acid chloride media [17]. However, Ujiro et al.[10], and some other authors [21-23] concluded that Cu has a harmful effect on the localized corrosion resistance in noble potential range for both ferritic and austenitic stainless steels. According to their research, Cu induced decreasing of E_{pit} is mainly attributed to the fact that ε-Cu phases tend to precipitate in grain boundaries and grains of ferritic stainless steels.The solubility of Cu in ferrite was low, about 0.2 wt.%.Thus, with the increasing of Cu content, ε-Cu phases might occur at grain boundaries and grains in ferrite stainless steels [24]. The dispersed inclusions of precipitates,such as Cu precipitates and (Nb, Ti)(C, N) precipitatescould negatively impair the stability of passive film of stainless steels, and therefore increase the susceptibility of ferrite to the pitting corrosion.This probably explains why E_{pit} values of both C and D steelsdecrease with further increasing Cu content up to about 0.45 wt.%, as shown in figure 2.

Besides, according to our previous study [19] effects of Sn on pitting corrosion resistance depends on the competitive relationship between beneficial effects and harmful effects of Sn. The excessive addition of Sn (about 0.33 wt%) in 17 wt% Cr ultra-purified ferritic stainless steel may deteriorate the corrosion resistance than good. Sn grain segregations [25-28] over increase of hydrogen evolution overpotential [29-30] and subsequent hydrolysis of Sn^{4+} ions ($Sn^{4+} + 4H_2O \rightarrow Sn(OH)_4 + 4H^+$) [31] may accelerate the corrosion of steels.These intensively adverse effects can no longer be counteracted by beneficial roles of Sn.However, as shown in figure 2, the E_{pit} values of experimental steels B, C and D are improved due to Cu addition,indicating that there isindeed a positive synergy between Cu and

Sn. The mechanismof beneficial synergy could beprobably attributed to the strong partitioning of large Sn atoms into Cu-rich precipitates. Sha et al.[32] reported that Sn segregates strongly to the precipitate/matrix interface of a large Cu precipitate, and particularly in the region where a dislocation appears to intersect the precipitate in a low-alloy pressure-vessel steel with0.43 at.% Cu and 0.007 at.% Sn, as shown in figure 3. Liu et al. reported that Sn promotes more and smaller sulfides precipitation at low temperature, and these precipitates bearing Cu and Sn are shown to reduce the micro-segregation degree of Sn in ultra-low carbon steel [33]. Sarafianos found that with the presence of Cu the diffusion range of Sn decreases [34]. Thus, it is reasonable to conclude thatSn diffusion to the precipitate/matrix interfacecould also reduce the segregation degree of Sn inexperimentalsteels herein.This probably explains why E_{pit} of specimen D (with the highest Sn content) is higher than that of C, and is not decreased for the excessive addition of Sn.

Furthermore, Liu et al. also concluded that the dissolved Sn in sulfide particles segregates to the sulfide/iron interface, and will restrict the growth of sulfide described by Ostwald ripening model [33, 35]. The Sn segregation at the interface can decrease the interfacial energy, and then stabilize small precipitates. This suppression effect of Sn on precipitate size is a much more common case in silicon

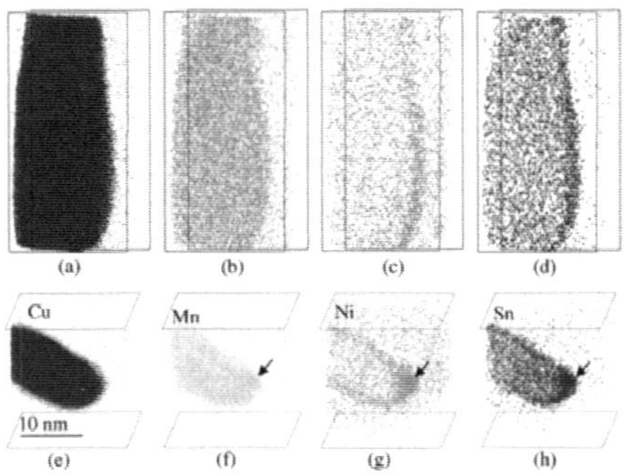

Figure 3. 3DAP maps of a large Cu-rich precipitate aged for 18,620 h at 405 °C. (a)–(d) Side view of the precipitate mapping Cu, Mn, Ni and Sn respectively; (e)–(h) plan view mapping Cu, Mn, Ni and Sn respectively.

The arrow points to a region of very high solute enrichment, probably associated with a dislocation steels [32]. Suziki et al.found that the size of MnS precipitates may keep small in Sn-added Fe-3 wt.% Si alloys [36]. Nakashima et al. also found the fine and uniform dispersion of MnS and AlN precipitates in Sn added grain-oriented electrical steel [37]. Liu et al. reported that the sizes of AlN precipitates in high magnetic induction grain-oriented silicon steel are significantly decreased due to the Sn segregation at precipitate/matrix interface [38]. Iwayama et al. also reported that the precipitates in 0.23mm thick high permeability grain-oriented silicon steel containing 0.1 wt.% Sn are much smaller in size than those without tin [39]. Thus, it can be concluded that the co-segregation of Cu and Sn may decrease the size and morphology of harmful precipitates in experimental steels. The much finer and high dispersion of the precipitates may have little negative impact on the pitting corrosion behaviour. This may explain why E_{pit} value of experimental steel D is further improved compared with that of steel C. Therefore, the pitting corrosion resistance of ultra-purified medium chromium ferritic stainless steel in 3.5% NaCl can be enhanced by the proper addition of Cu + Sn.

4. Conclusion

The result obtained from present work can be summarized in the following points.

(1) Appropriate addition of Cu + Sn is in favor of improving the pitting corrosion resistance. The pitting corrosion resistance of the steel containing an equivalent amount of Cu and Sn (appr. 0.22 wt%) shows the best performance due to the beneficial synergistic effect of Cu + Sn.

(2) The strong interaction between Cu and Sn in ferritic stainless steels clearly affects their pitting corrosion behaviour in 3.5% NaCl. A mechanism of Cu-Sn synergy was proposed: Cu can reduce the harmful micro-segregation of Sn in steel, and Sn restricts the growth of detrimental precipitates to the stability of passive film on the steel, such as ε-Cu, (Nb, Ti)(C, N) and other particles.

Acknowledgement
This work is financed by the National Science-technology Support Plan Project (2012BAE04B02), National Natural Science Found Project (51271050) of China.

References

[1] N Imai, N Komatsubara and K Kunishige 1997 *ISIJ Int.* **37** 217
[2] J Botella, M T Fernández and I Fernándezdel Castillo 1998 *Rev. Metal. Madrid* **34** 12
[3] C L Yan, F M Wang, L J Wei, Q Fu and C J Wu *J.* 2004*Univ. Sci. Technol. B.* **26** 277
[4] J Calvo, J M Cabrera, A Rezaeian and S. Yue 2007 *ISIJ Int.***47** 1518
[5] K Noro, M Takeuchi and Y Mizukami 1997 *ISIJ Int.* **37** 198
[6] L Savov, E Volkova and D Janke 2003 *RMZ- M & G* **50** 627
[7] K Kunishige and M Hatano 2007 *Mater. Sci. Forum* **539-543** 4113
[8] A Yamamoto, T Ashiura and E Kamisaka 1986 *BoshokuGijutsu* (Corros. Eng.) **35(8)** 448
[9] A A Hermas, K Ogura, S Takagi and T Adachi 1995 *Corros.***51(1)** 3
[10] T Ujiro, S Satoh, R W Staehle and W H Smyrl 2001 *Corros. Sci.* **43(11)** 2185
[11] A Pardo *et al.* 2003 *ZeitschriftfürMetallkunde* **94(12)** 1339
[12] X J Zhang, F Gao and Z Y Liu 2016 *J. Iron Steel Res. Int.* **23(10)**
[13] H Li *et al.* 2015 *Mater. Design* **84** 1
[14] Z W Chen *et al.* 2011 *Prog Mater. Sci.* **56(7)** 901
[15] K Osozawa 1971 *BoshokuGijutsu* **20(5)** 221
[16] K Takizawa *et al.* 1990 *BoshokuGijutsu* **39 (1)** 3
[17] A Pardo *et al.* 2006 *Corros. Sci.* **48(5)** 1075
[18] A Pardo *et al.* 2007 *Corros. Sci.* **49(2)** 510
[19] X J Zhang, F Gao and Z Y Liu 2016 *Steel Res. Int.* (DOI: 10.1002/srin.201600059)
[20] N D Nam *et al.* 2010 *Corros. Sci.* **52(1)** 14
[21] J Banas and A Mazurkiewicz 2000 *Mater. Sci. Eng. A* **277(1)** 183
[22] W Wu *et al.* 2015 *Int. J. Electrochem. Sci.* **10(12)** 10689
[23] J Shu *et al.* 2012 *Corros. Sci.* **57** 89
[24] 1973*Metals Handbook, Vol. 8,8th ed. Metal Park, Ohio: ASM*263
[25] T Watanabe, S Kitamura and S Karashima 1980 *Acta Metall.* **28(4)** 455
[26] W T Nachtrab and Y T Chou 1984 *J. Mater. Sci.* **19(7)** 2136
[27] Z X Yuan *et al.* 2003 *Scripta Mater.* **48(2)** 203
[28] W Yang *et al.* 2015 *Ultramicroscopy* **159** 432
[29] W F Wang 1999 *J. Mater. Eng. Perform.* **8(6)** 649
[30] H L Lin, J Q Wang and E H Han 2013 *Mater. Sci. Forum* **743–744** 607
[31] C I House and G H Kelsall 1984 *Electrochim. Acta* **29(10)** 1459
[32] G. Sha *et al.* 2007 *Philos. Mag. Lett.* **87(5)** 327
[33] Z Liu *et al.* 2009 *ISIJ Int.* **49(7)** 1087
[34] N Sarafianos 1986 *Mater. Sci. Eng.* **80(1)** 87
[35] T Gladman 2002 *The Physical Metallurgy of Microalloyed Steels* (London: Institute of Materials) p 204
[36] S Suzuki *et al.* 1992 *Mater. Trans. JIM* **33(11)** 1068
[37] S Nakashima, K Takashima and J. Harase 1992 *Tetsu-to-Hagane* **78(9)** 1495
[38] Z F Liu *et al.* 1991 *Acta Metall. Sin.* **27.4** 38
[39] K Iwayama *et al.* 1984 *J. Appl. Phys.* **55(6)** 2136

Synthesis and performances of polycarboxylate superplaticizer with clay-inerting and high slump retention capability

Shanshan Qian, Haidong Jiang, Bei Ding, Yi Wang, Chunyang Zheng and Zhaolai Guo

Jiangsu China railway ARIT new materials Co.Ltd., Jiangsu Nanjing 211505 China

E-mail: aritqss@163.com

Abstract. Macromolecules with pendant chlorine groups on their main chains were synthesized via free-radical copolymerization of 2-hydroxyethyl acrylate(HEA), TPEG and vinylbenzyl chloride(VBC) in the presence of intiator and chain transfer agent according to molecular structure design principle. The subsequent Arbuzov reaction between trimethoxyphosphine(TMP) and chlorine groups of macromolecules(PHVT) gave rise to new type of polycarboxylate superplaticizer with clay-inerting and high slump retention capability. The molecular structure of superplasticizer was determined by Fourier Transform Infrared Spectroscopy (FT-IR) and Gel Permeation Chromatography (GPC), respectively. The adaptability of clays,Zata potential, adsorption behavior and application performance in concrete were measured. The results shows that the polycarboxylate superplaticizer we prepared has good clay compatibility, excellent clay-inerting and high slump retention capability.

Keywords: clay-inerting, slump retention, Arbuzov reaction, trimethoxyphosphine, polycarboxylate superplaticizer

1. Introduction

Polycarboxylate superplaticizer, with its outstanding performance of low adding amount, high water reduction rate, well restraint of slump loss, high adjustability of molecular structure and environmental protection etc., has been used more and more widely in major projects like water conservancy, hydroelectricity, bridges, high-speed railways and motorways, and general civil construction [1]. However, with polycarboxylate superplaticizer, the expanding rang of using and largely applied in the practical engineering, a great amount of project cases and researches indicate that, comparing with traditional naphthalene, melamine and aminosulfonic acid water reducing agent, polycarboxylate superplaticizer, is more sensitive to clay content of aggregates, which absorbs the water reducing agent and with the increase of clay, the water reducing and dispersing ability and the effects of slump keeping decrease sharply, which results in the working performance of fresh mixed concrete decreases largely, all these problems disturbs cement manufacturer, concrete construction company and additive manufacturer [2-4].

As the fast development of construction industry, sandstone resources of high quality have gradually been depleted, and degraded in various places, especially in large and medium-sizes cities, which the sandstones used containing more clays [5,6].To solve these problems, it is usually to add a lot of slump agent, retarding agent and other agents which can improve the performance, while the methods can solve the problem of heavy loss of slump, but it also brings problems like high cost, delaying the setting time and compressive strength [7-9]. Therefore based on the molecular structure of polycarboxylate superplasticizer, introduce some functional groups which are able to decrease adsorption of clay, develop a polycarboxylate superplasticizer which resists clays and keeps slump to solve the construction problems in the sites, improve the application of PCE nationwide, which will be of vital importance to the development of construction industry.

In this paper, hydroxyethyl acrylate (HEA), 4-vinylbenzyl chloride (VBC) and terminal alkenyl polyoxyethylene ether(TPEG) are used as the main raw materials, which is first synthesized a halogen-containing copolymer (PHVT), and then polycarboxylate superplaticizer containing phosphate functional groups is synthesized by Arbuzov reaction with trimethyl phosphate (TMP). The polycarboxylate superplaticizer grafts the free phosphate ester into the polymer by chemical bonding. On the one hand, the phosphate group is chemically bonded to the polymer backbone. Compared with the free phosphate, it has strong steric hindrance and can solve the problem of large slump loss. On the other hand, adsorption ability of the phosphonic acid groups formed by hydrolysis of phosphate groups is stronger than carboxylic acid groups, preferentially adsorbs to the clay surface in the aggregate, which occupies the adsorption activity point of clay surface, and reduces the adsorption of carboxylic acid groups on the clay, so that reduces the sensitivity of the polycarboxylate superplaticizer to the mud-containing aggregate. This significantly improves the dispersion of superplaticizer and the ability to maintain slump, especially for the mud-containing gravel material has a significant slump retention effect.

2. Experiment

2.1. Materials

Macromonomer (TPEG) with molecular weight of 2400, commercial grade, Oxiranchem Yangzhou Company Ltd.. Vinylbenzyl chloride(VBC), trimethoxyphosphine(TMP), AR, available from Shandong Xiya Chemical Reagent Co., Ltd.. 2-hydroxyethyl acrylate(HEA), 3-Mercaptopropionic acid (MPC), hydrogen peroxide, AR, available from Sinopharm Chemical Reagent Co., Ltd. vitamin C, commercially available.

The chemical and mineral compositions of reference cement (P•O 42.5) are shown in table 1.

Table 1 Chemical and mineral compositions (wt.%)of reference cement.

SiO_2	A_2O_3	F_2O_3	CaO	MgO	SO_3	N_2Oeq	f-CaO	C_3S	C_2S	C_3A	C_4AF
22.91	4.29	2.89	66.23	1.93	0.34	0.71	0.63	58.68	21.48	6.49	8.79

Ca-MTT, illite and Kaolin were commercial grade provided by Hangzhou Chong Ke Chemical Co., Ltd., the chemical and mineral compositions of clay are shown in table 2.

Table 2. Chemical compositions (wt.%) of clay.

Clay	SiO_2	Al_2O_3	MgO	Fe_2O_3	CaO	Na_2O	K_2O	TiO_2	SO_3	Loss
Ca-MTT	77.07	10.78	2.80	0.96	4.29	0.29	1.53	0	0	15.69
illite	74.17	14.25	1.81	0.72	2.32	3.61	2.26	0.08	0	7.65
Kaolin	58.05	25.70	0.68	0.68	1.16	1.41	1.58	1.00	0.37	12.76

The macadam is 5~20 mm continuous grading macadam. the density is 2630kg/m^3 and the bulk density is 1540 kg/m^3.

The medium sand has a fineness modulus M=2.7, density of 2551kg/m^3 and bulk density of 1461kg/m^3.

The macadam and medium sand is washed and dried, Tap water is used.

2.2. *Synthesis of polycarboxylate superplaticizer*

2.2.1. Synthesis of polycarboxylate superplaticizer with clay-inerting and high slump retention capability. 1) Preparation of chlorine - containing copolymer (PHVT): Hydroxyethyl acrylate (HEA), terminal alkenyl polyoxyethylene ether (TPEG) and 4-vinylbenzyl chloride (VBC) by the action of hydrogen peroxide, Vc, 3-Mercaptopropionic acid (MPC), through the oxidation-reduction radical-polymerization reaction at room temperature obtains a chlorine-containing copolymerization product.

2) Preparation of polycarboxylate superplaticizer with clay-inerting and slump retention(PCE) : The chlorine-containing copolymerization product is reacted with a certain proportion of trimethyl phosphite (TMP) for Arbuzov reaction, and then dialysis purification is carried out for a certain period of time to obtain a containing phosphoric acid ester group polycarboxylate superplaticizer with clay-inerting and slump retention(20 wt%).

2.2.2. Synthesis of polycarboxylate superplaticizer with high slump retention capability. Preparation of polycarboxylate superplaticizer (PCE-0) with general slump retention: Hydroxyethyl acrylate (HEA) and terminal alkenyl polyoxyethylene ether (TPEG) by the action of hydrogen peroxide, Vc, 3-Mercaptopropionic acid(MPC),through the oxidation-reduction radical-polymerization reaction at room temperature obtains a copolymerization product, and then dialysis purification to obtain a polycarboxylate superplaticizer with general slump retention(20 wt%) .

2.3. *Structural characterization and property measurement*

Dry the sample to a constant weight and mix the sample with KBr powder to make tablets. Place the sample under VERTEX70-type Fourier infrared spectrometer of German Bruker to make a record of infrared spectrum. Waters 1515 gel permeation chromatograph is used to test the molecular weight and its distribution of the synthesized samples. The mobile phase is 0.1 mol.L-1 $NaNO_3$ solution, the flow rate is 1.0 mL/min, and the stationary phase is gelatinous porous filler. The fluidity and maintaining performance of cement paste are conducted according to GB 8077-2012 Methods for Testing Uniformity of Concrete Admixture. The water-cement ratio is 0.29 and the mixing amount of superplasticizer is 0.13%. TOC-II total organic carbon analyzer is used to measure concentration of original polycarboxylate superplasticizer solution and solution after absorption balance concentration difference of which is based to calculate adsorbing capacity of cement particle. Zetasizer Nano series of nanoparticle size analyzer produced by Malvern Instruments Ltd. could be used to measure Zeta potential on the surface of cement particle mixed with polycarboxylate superplasticizer. The fluidity and maintaining performance of cement paste are conducted according to GB 8077-2012 Methods for Testing Uniformity of Concrete Admixture. The water-cement ratio is 0.29 and the mixing amount of superplasticizer is 0.5%. Test the slump flow of concrete mixes and the variation of slump flow according to GB / T50080-2002 "ordinary concrete mixture performance test method standard". Test concrete 7d, 28d compressive strength, the content of water reducer is 0.2% of cement quality (converted into solid) according to GB / T50081-2002 "ordinary concrete mechanical properties test method standard".

3. Results and discussion

With HEA, 4-vinylbenzyl chloride (VBC) and terminal alkenyl polyoxyethylene ether (TPEG) as reactants, preparation of side-chain chlorine-containing copolymer products (PHVT) and then with trimethyl phosphite (TMP) for Arbuzov reaction, and forms a new containing phosphoric acid ester group of PCE. The reaction equation is shown in figure 1.

3.1. *The molecular weight and distribution of PCE and fluidity of cement paste*

The molecular structure of polycarboxylate superplasticizer is capable of being designed. Design the molecular structure by oxidation-reduction radical polymerization and Arbuzov reaction to obtain polycarboxylate superplasticizer with the expected molecular structure. This is mainly to analyze the dispersion and dispersion retention of the synthesized containing phosphate ester functional group polycarboxylate superplasticizer with clay-inerting and slump retention in the reference cement by analyzing the ratio of monomers, the molecular weight and its distribution. The results are shown in table 3.

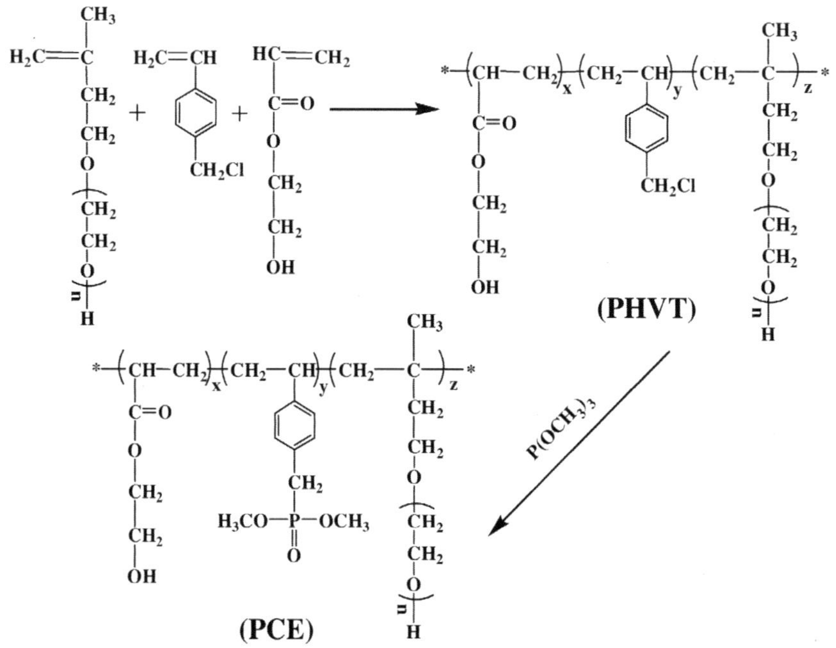

Figure 1. Reaction equation of PCE.

Table 3. Cement paste fluidity and GPC results of PCE.

Sample	PCE	Feed ratio ($n_{(HEA)}$:$n_{(VBC)}$:$n_{(TPEG)}$)	Dosage%	Fluidity/mm			GPC results		
				0.5h	1h	2h	M_n g.mol^{-1}	M_w g.mol^{-1}	PDI
A	PCE-0	$n_{(HEA)}$:$n_{(TPEG)}$ = 6:1		105	225	205	25500	34200	1.34
B	PCE-1	4:0.5:1		110	210	190	28000	39500	1.39
C	PCE-2	4:1:1		120	230	210	31000	43800	1.41
D	PCE-3	4:2:1		100	200	180	35200	50100	1.42
E	PCE-4	5:0.5:1		105	210	195	31300	45100	1.43
F	PCE-5	5:1:1		125	235	215	35100	49500	1.40
G	PCE-6	5:2:1		-	190	160	36200	53300	1.47
H	PCE-7	6:0.5:1	0.5	110	225	200	35400	52400	1.48
I	PCE-8	6:1:1		130	245	225	35700	53700	1.50
J	PCE-9	6:2:1		105	205	185	36100	56100	1.55
K	PCE-10	7:0.5:1		120	225	185	35900	56000	1.56
L	PCE-11	7:1:1		110	220	200	36400	56100	1.54
M	PCE-12	7:2:1		100	200	170	39200	60100	1.52

From table 3, the initial fluidity and fluidity retention composite propertie of the polycarboxylate superplasticizer with clay-inerting and slump retention are: PCE-8＞PCE-5＞PCE-2＞PCE-0＞PCE-11＞PCE-7＞PCE-4＞PCE-10＞PCE-1＞PCE-12＞PCE-9＞PCE-6＞PCE-3. This is due to the molecular structure of polycarboxylate superplasticizer with clay-inerting and slump retention contains hydrophobic groups, so that the surface tension becomes low, the lower the surface tension, the easier to reduce the gas-liquid interface surface energy. Superplasticizer would adsorb to the gas-liquid interface, and form a relatively solid liquid film, plus the ester-based in the molecular structure hydrolysis into a negative group adsorption, so that liquid film with negative electricity to form a stable, dense bubbles [10,11]. Air bubbles with negative and the gap between the air bubbles and cement particle are dispersed due to electrostatic repulsion, showing a significant excellent fluidity retention performance [12]. But with increasing the content of 4-vinylbenzyl chloride(VBC) ,when the hydrophobic group phenyl group increases beyond a certain range, the association in such a comb-like structure is enhanced. Aggregation occurs in an aqueous solution in a manner that the hydrophobic structure shrinks together, the hydrophilic structure is distributed inside the aggregates, resulting in water-soluble decline, and making the water-reducing agent macromolecule failed to anchor cement particle surface. The repulsion of the side chain is sufficiently realized, so the fluidity of the slurry is deteriorated [12].

3.2. Infrared spectroscopy analysis of PCE-8

Infrared spectrogram of the synthesized PCE-8 is shown as figure 2.

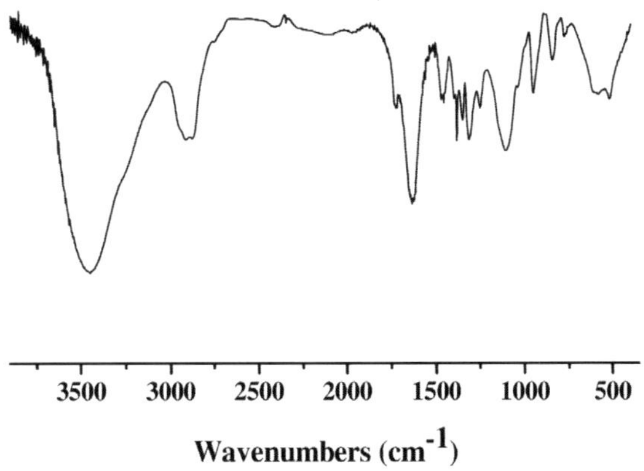

Figure 2. IR spectrum of PCE-8.

From literature [13~16] and IR spectrum in figure 2. In the IR spectrum of the synthesized PCE-8, 3024cm^{-1}, 3060cm^{-1}, 3082cm^{-1} is the stretching vibration peak of C-H on benzene ring, 2092cm^{-1}, 2849cm^{-1} is the stretching vibration peak of saturated CH and CH_2, 1600cm^{-1} is the framework vibration peak of benzene ring, 1494cm^{-1} is the in-plane bending vibration peak of C-H on the benzene ring, 676cm^{-1} is the stretching vibration peak of C-Cl, 1428cm^{-1} is the in-plane bending vibration peak of C-Cl, 1252cm^{-1} is the stretching vibration peak of P=O, 1093cm^{-1} is the absorption peak of C-O-C ether linkage on long-chain of polyoxyethylene of polyether side chain, its stretching vibration peak is at 2866cm^{-1}, 1380cm^{-1} is the vibration peak of C-H on the methyl and the main chain of methyl of the side chain phosphate in the structure. It can be seen through the analysis of infrared spectroscopy, the functional groups such as phosphate ester, carbonate ester, phenyl group, methyl group and ether group are included in the polycarboxylate superplasticizer with clay-inerting and slump retention, showing that the synthesized product has the expected molecular structure.

3.3. Adaptability test results of PCEs to different clays

Selection of polycarboxylate superplasticizer with clay-inerting and slump retention PCE-2, PCE-5, PCE-8, PCE-11 and polycarboxylate superplasticizer with general slump retention PCE-0, the fluidity of the slurry is tested by using calcium-montmorillonite, illite, kaolin (internal doping, instead of 3 g cement with the same quality) respectively, and under the same superplasticizer proportion and test conditions, the adaptability of polycarboxylate superplasticizer with clay-inerting and slump retention to different clays is investigated. The test results are shown in table 4.

Table 4. Adaptability test results of PCEs to different clays.

Sample	PCE	Clay	Dosage %	Fluidity/mm		
				0.5h	1h	2h
A	PCE-0	Ca-MTT	0.5	-	198	175
B	PCE-2			-	206	185
E	PCE-5			-	213	190
H	PCE-8			-	228	205
K	PCE-11			-	175	130
A	PCE-0	illite		-	208	185
B	PCE-2			100	213	190
E	PCE-5			105	223	205
H	PCE-8			110	236	219
K	PCE-11			-	190	170
A	PCE-0	kaolin		100	218	195
B	PCE-2			114	223	202
E	PCE-5			118	230	210
H	PCE-8			131	246	219
K	PCE-11			110	204	201

From the influencing rule of the three kinds of clay cement mixed with PCE to the fluidity of cement paste can be seen, the negative effect of calcium-montmorillonite on cement paste is the largest, which is not conducive to superplasticizer to play its role in water dispersion, especially the negative effect of the polycarboxylate superplasticizer with general slump retention PCE-0 in water reduction and dispersion is particularly serious. Illite to the fluidity of the cement slurry shows an overall downward trend, and the decreasing trend of the cement slurry fluidity of polycarboxylate superplasticizer with general slump retention is more obvious than that of polycarboxylate superplasticizer with clay-inerting and slump retention. The presence of kaolin has no negative effect on the cement slurry fluidity of the polycarboxylate superplasticizer with and slump retention, and does not affect the polycarboxylate superplasticizer to play its role in water dispersion.

As the clay crystal structure adsorbs a lot of metal cations, if the clay mineral crystal structure is unstable, clay-adsorbing cations such as Mg^{2+}, Ca^{2+} and other polyvalent metal cations react with a large number of anions such as -COO^- on the branch chain of the polycarboxylate superplasticizer to produce complexes that is insoluble in water, polycarboxylate superplasticizer is consumed, resulting in the polycarboxylate superplasticizer molecules used to disperse water decreased, which resulted in slurry fluidity losses [13]. From the date of table 4 can be obtained: The prohibition effect of polycarboxylate superplasticizer with clay-inerting and slump retention on the negative effect of clay is significant, especially the prohibition effect on calcium-montmorillonite is especially obvious. Polycarboxylate superplasticizer with clay-inerting and slump retention is a kind of polymer with strong adsorption property, its molecular structure contains a great deal of functional groups such as carbonic ester, hydroxyl group, phosphoric ester, initially it hydrolyzes a small amount of -COO^-, and only adsorbs Mg^{2+}, Ca^{2+} and other polyvalent metal cations, and not reacts with the metal cations to produce a complex that is insoluble in water and is consumed. With the cement hydration, the carbonic ester and phosphoric ester in the polycarboxylate superplasticizer with clay-inerting and slump retention are respectively turned into carboxylic acid and phosphonic acid by hydrolysis, the adsorption capacity of phosphonic acid group is stronger than-COO^- ,and preferential adsorbs to the soil surface, thereby further reduces the consumption of -COO^-. Hydrophilic groups dissolve in the solution to form an adsorption film, the hydrophobic groups reduce the surface tension and is more conducive to adsorbing the cement particle, so that they are closely connected together and like charges with the same electrical charge. According to the electrostatic force, the same kind of electric charges repel each other. Flocculent structure formed by cement-water is forcibly destroyed, and the free water inside the package is released, the free water content of the slurry increases, so greatly improves the flow of cement paste.

3.4. Electrical properties of cement particle in cement-PCE solution

In this study, using a water-cement ratio of 800: 1 system, 0.5g of reference cement mixes with 400ml various concentration of PCE, stirs for 20 min, and the mixture is well mixed. After standing for 10 min, take the upper suspension solution into the U-type electrophoresis tank, begin measuring zata potential. Select the appropriate cement particle for testing, and apply a positive and negative voltage each time, and take ten as a group, to determine the zeta potentials of cement particle with different concentrations of PCE.

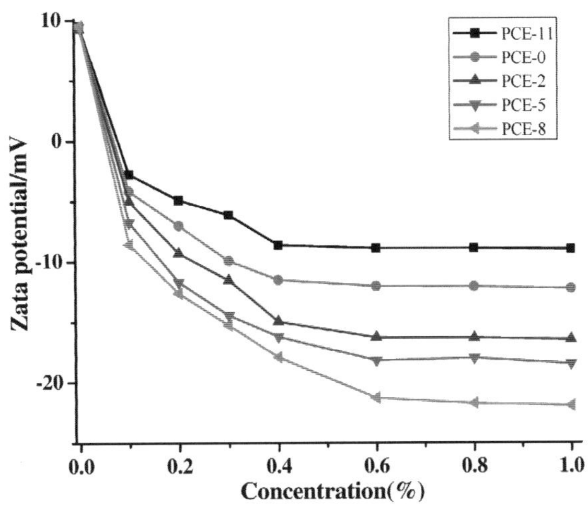

Figure 3. Influence of different concentration of PCE on Zata potential.

Figure 3. shows that the surface of pure cement particle is positively charged, while the surface of cement colloid particle with PCE is negatively charged. The absolute value of the zeta potential of cement particle at saturation content is: PCE-8 (-21.941mV) > PCE-5 (-18.507mV) > PCE-2 (-

16.463mV) > PCE-0 (-12.257mV) > PCE-11 (-8.997 mV). The larger the absolute value of zeta potential, the greater the anionic charge intensity on the cement particle surface, and the anion charge originates from the hydrolysis of the carbonic ester and phosphate ester in the molecular structure of the polycarboxylate superplasticizer. The hydrophobic group content in the molecular structure of PCE-11 exceeds a certain range, so that the association between molecules increases, the hydrophobic structure shrinks together, and the hydrophilic structure is distributed inside the aggregates, shields the part of the charge itself, so the zeta potential absolute is minimum. The molecular structure of PCE contains phosphate ester, which hydrolysis to produce phosphonate group with two negative charges, with a strong adsorption, and easy to adsorb in the cement particle surface, so the absolute value of zeta potential is greater than polycarboxylate superplasticizer with general slump retention PCE-0. The phenyl content in the branch-chain structure of PCE-2 and PCE-5 superplasticizer is large, that makes the stiffness of superplasticizer molecular increasing, and the steric hindrance effect is much greater than the PCE-8, so the adsorption probability of the adsorbed polar point of the polycarboxylate macromolecule and the surface of the cement particle is reduced.

3.5. Absorptive capacity of PCE in clay and cement paste

In order to find out whether there are differences in absorptive capacity of polycarboxylate superplasticizer with general slump retention and polycarboxylate superplasticizer with clay-inerting and slump retention for different types of clay minerals, and to determine absorptive capacity of polycarboxylate superplasticizer with clay-inerting and slump retention for cement particle and a variety of clay minerals, so as to reveal the root cause of different effects on concrete performance. In this paper, the TOC total organic carbon analyzer would be used to determine the absorptive capacity of five polycarboxylate superplasticizer with slump retention (PCE-0, PCE-2, PCE-5, PCE-8, PCE-11) for cement and cement respectively mixed with three kinds of monogenic clay mineral (calcium-montmorillonite, illite and kaolin) under the same conditions. The preparation process of TOC test sample is as follows, weighs 1 g powder (with the presence of a monogenic clay mineral, the doped quality is 1% of powder), adds to 50 ml concentration of 4 g / L of polycarboxylate superplasticizer solution prepared by pure water ,stirs 20 min. After standing for 10 min, the solution is filtered using a sand core active filtration device and a microfiltration membrane, and takes 5 ml filtrate liquid to dilute 10 times as a TOC test sample; And then using the TOC total organic carbon analyzer to measure the organic carbon content of the sample, from the total amount of mixed with polycarboxylate superplasticizer with slump retention minus the total organic carbon content in filtrate is the solid particle adsorbed by superplasticizer. The experimental results are shown in figure 4.

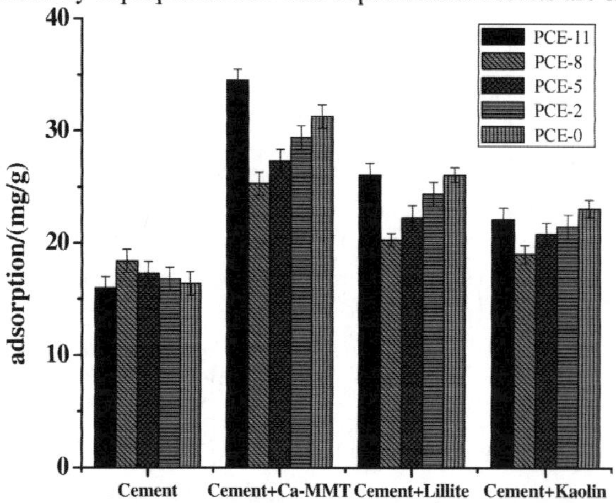

Figure 4. Adsorption of PCE on cement and clay.

As can be seen from figure 4, in the pure cement system, different polycarboxylate superplasticizers with slump retention have different adsorption capacities to cement, which is due to the polycarboxylate superplasticizer with clay-inerting and slump retention grafts the free phosphate ester structure to different positions of molecular structure of superplasticizers by chemically bonded by Arbuzov reaction, and the introduction of a 4-vinylbenzyl chloride and trimethyl phosphite (TMP) performs Arbuzov reaction will be produced two phosphate ester. The molar ratio of phosphate ester in PCE-2 is 33.2%, the phosphate ester content of PCE-5 is 28.6%, the phosphate ester content of PCE-8 is 25%, the phosphate content ester of PCE-11 is 22.2%, the hydrolysis of the part of carbonic ester into -COO- and of part of phosphate ester into phosphonic acid group in molecular structure adsorbs on the surface of cement or hydration products, and phosphonic acid group with two negative charges, its adsorption capacity is stronger than -COO-,so the adsorbing capacity is PCE-8>PCE-5>PCE-2>PCE-0>PCE-11. In the cement system with three kinds of 1% clay, the influence of different clay on the adsorption capacity of polycarboxylate superplasticizer is different, and the adsorbing capacity is PCE-11 > PCE-0 > PCE-2 > PCE-5 > PCE-8. On the one hand relative to the free phosphate material, the different molecular structure of polycarboxylate superplasticizer with clay-inerting and slump retention contains phosphate ester at different locations, hydrolysis and decomposition into phosphate group anions and adsorption with clay, and the polar end of the benzene ring group forms an organic layer on the clay surface,by preventing the contact between the clay and the water molecules to suppress the expansion of the clay so that the interlayer spacing is increased or refined into particle having a larger specific surface area, and to reduce the adsorption of clay to polycarboxylate superplasticizer and mixing water; On the other hand, the adsorption capacity of the phosphate group is stronger than that of the carboxylic acid group, which is preferentially adsorbed on the clay surface in the aggregate. The benzene ring has large steric hindrance and occupies the adsorption activity point on the clay, thus reduces the adsorption of the clay to the polycarboxylate superplasticizer.

3.6. Concrete application performance test

The properties of concrete with different polycarboxylate superplasticizer with clay-inerting and slump retention admixture are compared test. Concrete mix proportion and test requirements are implemented in accordance with GB / T8076-2008, concrete reference mix proportion is shown in table 5.

Table 5. Mix proportion of concrete (kg/m^3).

Cement	Sand	Coarse aggregate	PCE	Water
390	785	1040	3.9	165

According to concrete reference mix proportion, clay content (1% of the amount of cement) is instead of sand doped with the same quality, the effect of polycarboxylate superplasticizer on workability and strength of concrete with different clay content is studied. The results are shown in table 6.

As can be seen from table 6, the effect of PCE on the negative effect of clay on the workability of concrete mixed with polycarboxylate superplasticizer is remarkable. PCE introduces hydrophobic groups into its molecular structure, such as the introduction of part of the phenyl group to reduce its surface tension, it can be rapid dispersion and adsorption in the concrete; The introduction of part of the phosphate ester, phosphate ester hydrolysis into phosphonic acid with two negative charges, preferentially adsorbs and occupies the adsorption activity points of the clay surface, through the polar end of the benzene ring group to form a organic layer in the clay surface to prevent the clay and water molecules in contact, to inhibit the expansion of clay, and makes the interlayer spacing increased or refined into particle with larger specific surface area, and reduces the adsorption of clay to polycarboxylate superplasticizer and mixed water; Such as the introduction of part of the carbonate ester hydrolysis into carboxy, continue to play a role in the dispersion, and enhance the slump retention ability of superplasticizer. The introduction of the phosphate ester, carbonic ester and phenyl group in the molecular structure of polycarboxylate superplasticizer with clay-inerting and slump retention, so that it has remarkable soil adaptability.

Table 6. Polycarboxylate superplasticizer influent on properties of concrete mixed different clay.

Sample	PCE	Clay	Slump/Flow/mm/mm			Compressive strength/MPa	
			0.5h	1h	2h	7d	28d
A	PCE-0	-	110/-	210/500	210/480	33.7	42.6
B	PCE-2		150/-	215/510	210/490	34.2	43.5
E	PCE-5		165/-	215/520	210/505	35.3	45.3
H	PCE-8		180/-	220/540	215/520	36.4	46.3
K	PCE-11		90/-	210/480	205/465	33.4	42.9
A	PCE-0	Ca-MTT	-/-	200/450	195/430	32.5	41.3
B	PCE-2		110/-	210/475	205/450	34.2	42.4
E	PCE-5		125/-	210/505	210/485	34.3	44.3
H	PCE-8		150/-	210/515	210/500	35.4	45.4
K	PCE-11		-/-	200/400	180/320	32.6	42.9
A	PCE-0	illite	90/-	210/480	210/460	32.8	41.9
B	PCE-2		130/-	215/495	210/480	33.2	44.6
E	PCE-5		145/-	215/505	210/495	35.7	45.7
H	PCE-8		170/-	220/530	215/510	36.6	45.6
K	PCE-11		70/-	210/455	205/440	33.1	43.4
A	PCE-0	Kaolin	105/-	200/500	200/480	33.5	42.3
B	PCE-2		140/-	205/510	205/490	34.3	43.3
E	PCE-5		160/-	210/520	200/500	35.7	44.9
H	PCE-8		180/-	220/535	215/510	36.8	46.4
K	PCE-11		85/-	205/480	200/460	33.4	43.5

4. Conclusion

According to the molecular design principles, using oxidation - reduction free radical polymerization and Arbuzov reaction method, a kind of polycarboxylate high performance water reducing agent is synthesized, a kind of PCE is synthesized, the main conclusions are as follows:

(1) With hydroxyethyl acrylate (HEA), 4-vinylbenzyl chloride (VBC) and terminal alkenyl polyoxyethylene ether (TPEG) as reactants, based on different weight ratio, a series of polycarboxylate superplasticizer (PCE)with clay-inerting and slump retention by introduced phenyl group, phosphate ester and carbonic ester functional groups are synthesized by Arbuzov reaction.

(2) IR spectra prove the existence of the characteristic functional groups, and the synthesized products have the expected molecular structure. The results of GPC show that the molecular weight distribution of the synthesized product is narrow and the conversion rate of the reactants is high, and the resulting product has an ideal molecular structure.

(3) The experimental results of the surface tension, the zeta potential of slurry system and the adsorption capacity of the polycarboxylate superplasticizer to the powder show that, polycarboxylate superplasticizer with clay-inerting and slump retention can reduce the adsorption capacity of clay to polycarboxylate superplasticizer and inhibit the expansion of soil so as to reduce the adsorption of polycarboxylate superplasticizer and mixed water.

(4) The adsorption capacity of the polycarboxylate superplasticizer with slump retention in the presence of clay and cement is larger than that of the pure cement with the same quality. It is shown that the clay adsorbs some polycarboxylate superplasticizer. PCE is conducive to the dispersion and stability of cement particle to improve the flow effect, and to inhibit the expansion of clay to reduce the adsorption of polycarboxylate superplasticizer and mixed water, and has a very good adaptation with clay.

References

[1] Kazuo Yamada, Shoichi Ogawa and Shunsuke Hanehara 2001 Controlling of the adsorption and dispersing force of polycarboxylate-type superplasticizer by sulfate ion concentration in aqueous phase [J] *Cement and Concrete Research* **31(3)** 375–383

[2] Yamada K, Takahashi T and Hanehara S 2000 Effects of chemical structure on the properties of polycarboxylate-type superplasticizer *Cement and Concrete Research* **30(2)** 197–207

[3] J Plank, K P Ilmann, N Zouaoui, P R Andres and C Schaefer 2008 Synthesis and performance of methacrylic ether based polycarboxylate superplasticizers possessing hydroxyl terminated poly(ethylene glycol) side chains *Cement and Concrete Research* **38(10)** 1210 1216

[4] Norvell J K *et al.* 2007 Influence of Clays and Clay-Sized Particle on Concrete Performance *Journal of Materials in Civil Engineering* **19(12)** 1053–1059

[5] L Lei and J Plank 2012 A concept for a polycarboxylate superplasticizer possessing enhanced clay tolerance *Cement and Concrete Research* **42(10)** 129–1306

[6] S Ng and J Plank 2012 Interaction mechanisms between Na montmorillonite clay and MPEG-based polycarboxylate superplasticizers *Cement and Concrete Research*

[7] Kroyer H, Lindgreen H, JAakobsen H J and Skibsted J 2003 Hydration of Portland cement in the presence of clay minerals study by 29Si and 27Al MAS NMR *Advances in Cement Research* **15(3)** 103–112

[8] Quiroga P N, Ahn N and Fowler D W 2006 Concrete Mixtures with High Microfines *ACI Materials Journal* **103(4)** 258–167

[9] Muñoz J F *et al.* 2010 Effects of Coarse Aggregate Coatings on Concrete Performance *Journal of Materials in Civil Engineering* **22(1)** 97–103

[10] Fernades V A, Purnell P and Still G T C 2007 The effect of clay content in sands used for cementations materials in developing countries *Cement and Concrete Research* **37(4)** 751–758

[11] Norvell J K *et al.* 2007 Influence of Clays and Clay-Sized Particle on Concrete Performance *Journal of Materials in Civil Engineering* **19(12)** 1053–1059

[12] Norvell J K *et al.* 2007 Influence of Clays and Clay-Sized Particle on Concrete Performance *Journal of Materials in Civil Engineering* **19(12)** 1053–1059

[13] Kazuo Yamada, Shoichi Ogawa and Shunsuke Hanehara 2001 Controlling of the adsorption and dispersing force of polycarboxylate-type superplasticizer by sulfate ion concentration in aqueous phase *Cement and Concrete Research* **31** 375–383

[14] Stephen J Thompson, Colin White and Peter M Maitlis 1977 Pentamethyl cyclopentadienyl-rhodium and -iridium complexes XV. Cationic complexes with tertiary phosphines and phosphites and their rearrangements reactions *Journal of Organometallic Chemistry* **136(1)**

IOP Conf. Series: Materials Science and Engineering **182** (2017) 012033 doi:10.1088/1757-899X/182/1/012033

87

[15] T W Abraham, T I Kalman, E J McIntee and C R Wagner 1996 Synthesis and Biological Activity of Aromatic Amino Acid Phosphoramidates of 5-Fluoro-2'-deoxyuridine and 1-β-Arabinofuranosylcytosine: Evidence of Phosphoramidase Activity *J. Med. Chem Lett.* **39** 4569

[16] S Ikeda, J A Ashley, P Wirsching and K D Janda 1992 psi.[PO2-CH2N+], a new amide bond replacement: potent, slow-binding inhibition of the HIV protease *J. Am. Chem. Soc.* **114** 7604

[17] Jan Spengler and Klaus Burger 1998 An ecient synthesis of N-phosphinoylmethylamino acids and some of their derivatives. Journal of the Chemical Society *Perkin Transactions* **1** 2091–2095

[18] H Krawczyk and T J Bartczak 1993 New crystalline polymorphic form of glyphosate: synthesis, crystal and molecular structures of N-(phosphonomethyl)glycinePhosphorus Sulfur Relat *Elem.* **82** 117

[19] G H Hakimelahi, A A Moosavi-Movahedi, M M Sadeghi, S C Tsay and J R Hwu 1995 Design, Synthesis, and Structure-Activity Relationship of Novel Dinucleotide Analogs as Agents against Herpes and Human Immunodeficiency Viruses. *J. Med. Chem* **38** 4648

Mechanism and preparation of liquid alkali-free liquid setting accelerator for shotcrete

Ying Qiu, Bei Ding, Jiezhong Gan, Zhaolai Guo, Chunyang Zheng, Haidong Jiang

Jiangsu China Railway ARIT new materials Co. , Ltd. Nanjing 211505 China

E-mail: aritqy@163.com

Abstract. A new alkali-free liquid accelerator for shotcrete was prepared through normal temperature drop process by using the nano activated alumina and the modified alcohol amine as the main raw materials. The effect of alkali-free liquid accelerator on the cement setting time and the mechanical properties of mortar, the effect of the penetration strength on the shotcrete rebound were investigated. And the accelerating mechanism of the as-prepared alkali-free liquid accelerator was also analyzed via XRD and SEM characterization methods. The experimental results indicated that the hydration of C_3A was accelerated by the polyamine complexation of accelerator, resulting in forming a large number of acicular ettringite and reducing the amount of Ca(OH)2 crystal, which would not affect the later hydration of cement. When the content of alkali-free liquid accelerator was 6%, the initial setting time and final setting time were less than 3min and 8min respectively, and 1d and 28d compressive strength ratios reached 207.6% and 114.2% respectively; beside that, the shotcrete rebound was very low because of the high penetration strength within 30min.

Keywords: alkali-free liquid accelerator, penetration of strength, rebound, Mechanism

1. Introduction

Accelerator is a kind of indispensable admixture in the shotcrete construction. The main purpose of adding accelerator is promoting rapidly condense of the fresh concrete, increase the thickness of the primary spray layer, shorten the time interval between the two spraying, enhance the early strength of shotcrete, which can provide timely support resistance [1]. Thus it is possible that the shotcrete used on the wall and the top by introducing the accelerator. At present, most of powder and liquid accelerator are alkaline [2], which was mainly composed of sodium silicate, sodium aluminate and sodium carbonate and so on. Although they can meet the required standards of shotcrete, yet problems are not to be solved. For example, an accelerator with pH greater than 12 causes strong corrosiveness, seriously deducting the later intensity and the durability of concrete and cohesive property of shotcrete, consequently resulting in the increase of shotcrete rebound and serious dust pollution.

The property of accelerator has an impact on congeal time, resilience, adaptability, early and late strength development and durability of shotcrete. In order to bring out the potentials of accelerator in shotcrete, a large number of researches have been performed. The United States and European researchers [3-4] developed the alkali-free liquid accelerator by using the aluminum sulphate [5] and alcohol amine, which significantly improved the cohesion of concrete mixture and decreased its late compression strength loss, while the performance characteristics of the alkali-free accelerator and alkali accelerator were analyzed and compared [6]. Recently, most of accelerators are aluminum-sulfate, simultaneously introducing sulfur trioxide to exceed sulfur trioxide in cement. And owing to the existence of the sulfate and chloride ions, the durability of concrete will be reduced.

In the present work, the alkali-free liquid accelerator (AFL) was investigated. The alkali-free liquid accelerator was prepared by using the nanometer alumina and modified alkanolamine through complexometric titration process under room temperature. The effect of alkali-free liquid accelerator on the cement setting time and the compression strength of mortar, and the effect of the penetration strength on the shotcrete rebound of concrete have been investigated. Meanwhile the accelerating mechanism of the as-prepared alkali-free liquid accelerator and the morphology of hydration products of cement have also been analyzed via XRD and SEM characterization methods. It is believed that this work will provide theoretical instruction for the application of non-alkali liquid accelerator in shotcrete.

2. Experimental process

2.1. Materials

Nanometer activated-alumina (industrial grade, CHALCO Shandong Co., LTD), magnesium nitrate (analytical grade, Sinopharm Group Co., Ltd), modified polyhydroxy alcohol amine (self-made), phosphoric acid (analytical grade, Sinopharm Group Co., Ltd), PO 42.5 grade portland cement (Onoda Cement Co., Ltd. and its physical and mechanical properties, ISO standard sand are listed; whereas it is under the circumstance of the medium sand with the fineness modulus of 2.7 that the crushed stone with the continuous grading of 5-10mm were selected as raw materials.

Table 1. Physical and mechanical properties of P·O 42.5 cement.

Apparent density/(g/cm^3)	Water requirement of normal consistency/%	Flexural strength/MPa		Compressive strength/MPa		Setting time/min		Soundness
		3d	28d	3d	28d	Initial	Final	
3.12	26.8	5.6	8.8	26.8	51.3	220	330	Qualified

2.2. Preparation of Alkali - free Liquid Quick - setting Agents

Firstly, alcohol amines were modified by using of dodecyl quaternary ammonium salts. The domiphen was added into the absolute ethyl alcohol to form a solution with the concentration of 20%, and then the solution was heated to reach 60 °C while stirring, after that, N, N, N', N'-methylethylenediamine was dropped into the solution and this process continue for 2 hours, and then modified polyhydroxy alcohol amine solution was prepared.

After adding water and a certain mass of modified polyhydroxy alcohol amine solution in the reaction kettle, the phosphoric acid and magnesium nitrate were introduced in order, while stirring at normal temperature. And then the aqueous solution of nanometer active alumina with the concentration of 5% was dropped into the reaction kettle. After reaction for 3.5 hours, the AFL alkali-free liquid accelerator with the color of transparent light yellow was prepared, and the homogeneity index was shown in table 2.

Table 2. The homogeneity index of AFL liquid alkali-free accelerator.

Apparence	Alkali content/%	Chlorine ion content/%	Density/(g·cm^{-3})	pH	Solid content/%
Light yellow	0.010	0.001	1.41	3.0	40

2.3. Characterization

2.3.1. Test method of setting time and compressive strength. Tests of the paste setting time and the mortar strength were performed according to JC477-2005 "shotcrete accelerator". The size of mortar specimen was 40 mm×40 mm×160 mm.

2.3.2. Test method of penetration strength of concrete. It is confirmed that the sand rate is 45%, and concrete slump is 120 mm according to JGJ 55-2011 "ordinary concrete mix design rules" and the actual construction experience. The final mix proportion was shown in table 3. The P·O 42.5 conch cement, the M12 polycarboxylate high performance water reducing agent (PCE) (Jiangsu iron Aolaite Co,. Ltd.), the commercially available BASF167 alkali-free liquid accelerator and SL alkaline liquid accelerator were used as comparative samples. The concrete specimens with size of 150mm × 150mm × 150mm were prepared to test the penetration strength. The penetrating strength was measured at 10min, 20min, 30min, 60min and 180min using a concrete penetration resistance meter.

Table 3. Mix proportion of concrete.

Accelerator type	Accelerator content /%	Mix (kg/m^3)					
		Concrete	Sand	Stone	Water	PCE	Accelerator
KB	0	450	805	873	202	1.35	0
SL	4	450	849	849	202	1.35	18
AFL	6	450	849	849	202	1.35	27
BASF167	6	450	849	849	202	1.35	27

2.3.3. Microstructure analysis. The hardening paste with different ages (10min, 1d and 28d) was prepared. And the carbonized surface layers were removed, then the samples with the size of 3-5mm was obtained from the inner of hardening paste, after that the samples were washed with absolute ethanol and placed in acetone to prevent hydration. After immersed in acetone for 24 hours, the samples were dried in an oven at a temperature of 45 ° C. After cooling, 3-5 small pieces of each sample were selected for SEM analysis. The other samples were crushed and screened by 200 mesh sieve, and then the phase composition of each sample was characterized by XRD.

3. Experimental results and discussion

3.1. The setting time of AFL

The effect of additional amount of AFL accelerator on the c setting time of cement paste has been investigated by using Onoda cement, as shown in table 4.

Table 4. Setting time of cement paste mixed with accelerator.

Accelerator type	KB	SL	AFL	BASF167
Accelerator content /%	0	4	6	6
Initial setting time/min:s	235	4:40	2:25	3:30
Final setting time/min:s	405	9:35	8:00	9:20

3.2. Compressive strength of mortar prepared with AFL

Table 5. Compressive strength of concrete mixed with different accelerator.

Accelerator type	Accelerator content /%	Compressive strength					
		1d /MPa	Ratio/%	7d /MPa	Ratio/%	28d/MPa	Ratio/%
KB	0	10.5	100.0	32.5	100	40.8	100
SL	4	16.3	155.2	29.5	90.8	35.2	86.3
AFL	6	21.8	207.6	36.5	112.3	46.6	114.2
BASF167	6	18.9	180	34.6	106.5	44.1	108.1

3.3. Penetration strength of concrete

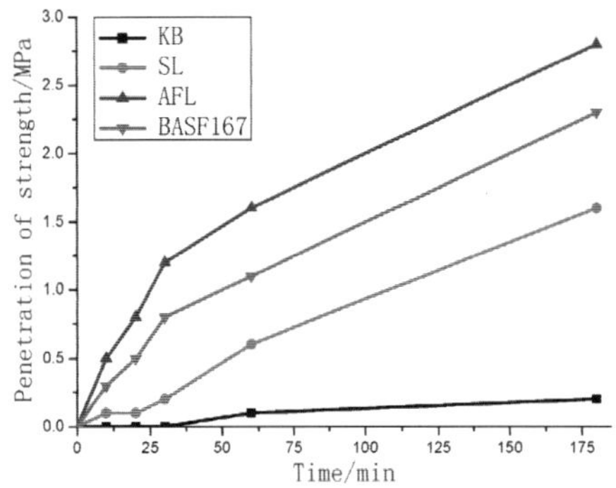

Figure 3. Penetration strength of concrete mixed with different accelerator.

It can be seen from figure 3 that in 30 minutes, the concrete using SL alkali liquid accelerator showed the lowest penetration strength, and the slow strength development. While the concrete using AFL alkali-free liquid accelerator showed the highest penetration strength. The complexation of activated alumina and polyhydroxy alcohol amine in AFL accelerator make the concrete has the property of promoting coagulation.

3.4. shotcrete rebound rate

According to the on-site shotcrete construction situation, the vault of upper step and the sidewall of middle step construction site were selected to calculate the corresponding rebound ratios by using the design amount, the super-cut amount and the actual amount. The effect of different accelerator on the rebound ratios of shotcrete were shown in table 6.

Table 6. Influence of quick setting admixture on rebound resilience of shotcrete.

Construction site	Accelerator type	design amount /m^3	super-cut amount/m^3	actual amount/m^3	rebound amount/m^3	Rebound ratio/%
the vault of upper step	SL	19.3	5.3	33.5	10.5	31.3
	AFL	18.5	5.1	26.2	3.5	13.4
	BASF167	18.9	5.1	28.6	4.8	16.8
Sidewall of middle step	SL	10.8	3.5	16.8	3.2	19.0
	AFL	9.5	2.6	12.5	0.8	6.4
	BASF167	9.3	2.8	13.7	1.5	10.9

It can be seen from table 6 that the vault and side wall with AFL alkali-free liquid accelerator showed a small rebound ratio, were 13.4% and 6.4% respectively. And the vault and side wall with BASF167 alkali-free liquid accelerator showed a lager rebound ratio, were 16.8% and 10.9% respectively. The rebound ratio of vault and side wall with SL alkali liquid accelerator showed a largest rebound ratio, were 31.3% and 19.0% respectively, which is related to the narrower additional amount of alkali liquid accelerator. Application of AFL alkali-free liquid accelerator showed the best results, which can increase the bond strength and early strength, decrease the coagulation time and the rebound amount.

3.5. XRD analysis

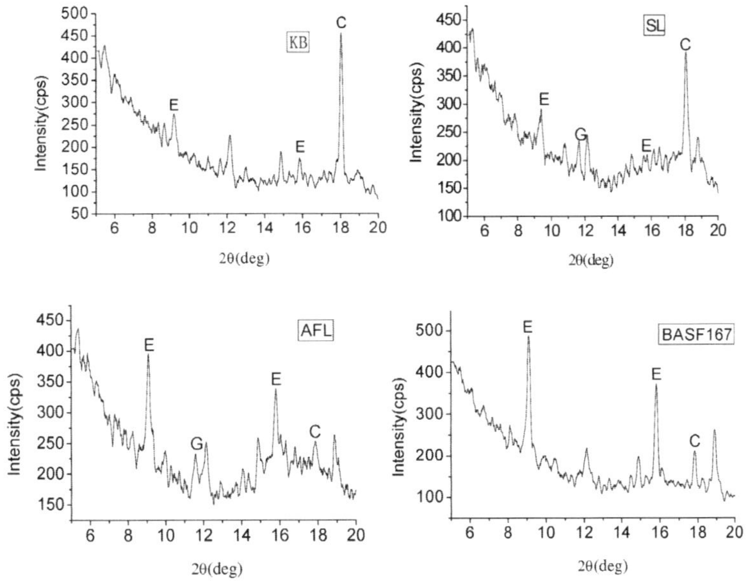

Figure 4. XRD patterns of the hardened paste mixed with different accelerators at the final setting stage. E: ettringite; G: gypsum; C: $Ca(OH)_2$.

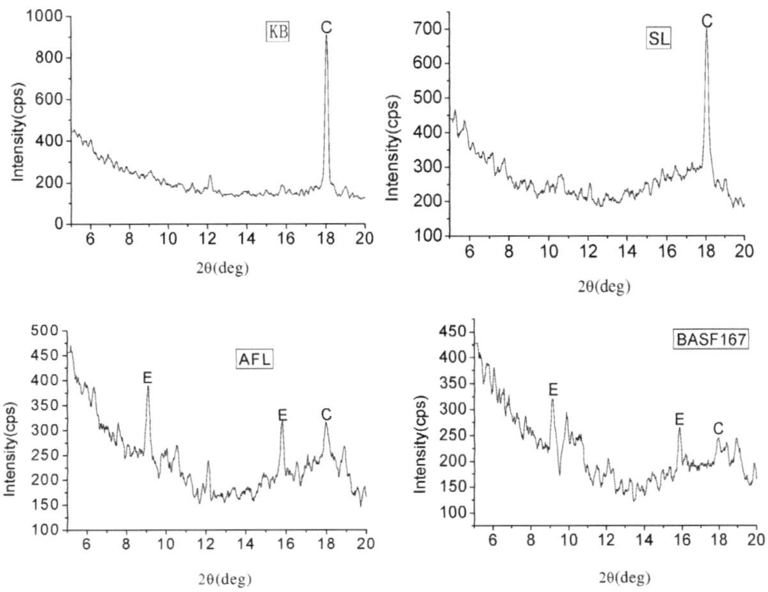

Figure 5. XRD patterns of the hardened paste mixed with different accelerators after 1 day. E: ettringite; G: gypsum; C: $Ca(OH)_2$

Figure 4 and figure 5 showed the XRD patterns of the hardened paste mixed with different accelerators at the final setting and after 1 day. At the initial stage of hydration, a large amount of $Ca(OH)_2$ and a small amount of ettringite formed in the hardened paste mixed with the SL alkaline liquid accelerators and the blank paste. Meanwhile, a certain amount of ettringite crystal and a small amount of $Ca(OH)_2$ formed in the hardened paste mixed with AFL and SA167.

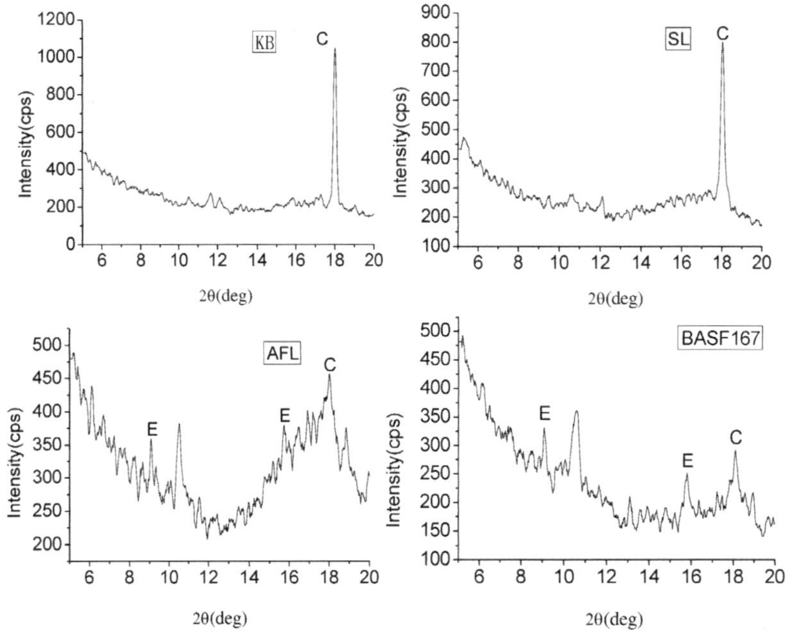

Figure 6. XRD pattern of the hardened paste mixed with different accelerators after 28 days. E: ettringite; G: gypsum; C: $Ca(OH)_2$

Figure 6 showed the XRD patterns of hardened paste mixed with different accelerators after 28 days. It can be seen from the figure that a large amount of $Ca(OH)_2$ formed in the hardened paste mixed with the alkaline liquid accelerators and the blank paste, while a certain amount of ettringite and a small amount of $Ca(OH)_2$ formed in the hardened paste mixed with AFL and SA167.

Table 7. XRD characteristics of the peak intensity table.

	d value	Blank			SL			AFL			SA167		
		final setting	1d	28d	final setting	1d	28d	final setting	1d	28d	final setting	1d	28d
Ettringite	9.68	65	16	22	52	30	35	119	123	49	242	88	47
$Ca(OH)_2$	4.91	272	546	635	155	376	480	35	55	28	83	35	40
C_3A	2.6	206	123	0	66	0	0	180	32	0	183	62	0
C_3S	1.75	296	104	17	330	155	115	315	145	18	328	150	28

The characteristic peak intensities of hardened cement paste mixed with different accelerator at the final setting, 1 day and 28 days were shown in table 7. It revealed the crystal products and the reaction between different mineral phases at the various stages of hydration. The early behavior of cement hydration is mainly determined by C_3A reaction and the development of condensation and early strength mainly depends on the hydration degree of silicate, especially the hydration degree of C_3S [7]. It can be seen from table 7 that a certain amount of $Ca(OH)_2$ crystals and a very small amount of ettringite formed in the cement paste mixed with SL alkaline liquid accelerator produced in final setting stage, while a certain amount of C_3S and a large number of C_3A was consumed. When the hydration time reached 1 day, the amount of $Ca(OH)_2$ crystals continued to increase, while C_3A was depleted and the amount of C_3S decreased, and the hydration product of C-S-H contributed to the 1 d strength. When the hydration time reached 28 days, there is still a considerable amount of C_3S in the paste mixed with SL alkaline liquid accelerator, which was significantly higher than that of other cement hardening paste, indicating the addition of alkaline accelerator delayed the late hydration of C_3S. This is one of reasons for low compressive strength of 28d.

A certain amount of ettringite and a very small amount of $Ca(OH)_2$ formed in the cement paste mixed with AFL and BASF167 accelerator produced in final setting stage. Compared with the paste mixed with alkaline liquid accelerator, fewer C_3A and comparable C_3S were consumed. When the hydration time reached 1 day, the amount of $Ca(OH)_2$ crystals was still very small, while the amount of C_3A and C_3S was reduced obviously, and the formation of hydrated gel of ettringite and C_3S enhanced the 1d strength of cement hardened paste. When the hydration time reached 28 days, there is still a little amount of C_3S in the paste, while C_3A was depleted and the residual amount of C_3S is very small, which is equivalent to the blank paste without any accelerator. According to the consumption of C_3S and C_3A, the incorporation of alkali-free liquid accelerator has little effect on the late hydration of cement.

The amount of $Ca(OH)_2$ and ettringite in the hardened paste mixed with AFL accelerator is less than that of paste mixed with BASF167D accelerator, but the consumption of C_3S is relatively high. This due to the $Mg(OH)_2$ precipitation was formed in paste mixed with AFL accelerator during the initial stage of hydration, which is more difficult to dissolve than $Ca(OH)_2$, resulting in promoting the hydration of C_3S and reduces the forming amount of $Ca(OH)_2$. The main constituent of SA167 is aluminum salt, and its mechanism of quick setting and early strength mainly promotes the formation of ettringite through aluminum salt. Therefore, AFL has better cement adaptability, which provides the ideal quick-setting early strength, effectively reduces the rebound of shotcrete.

3.6. SEM analysis

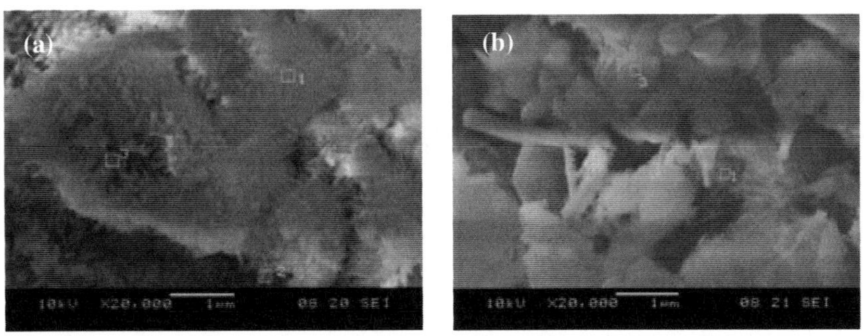

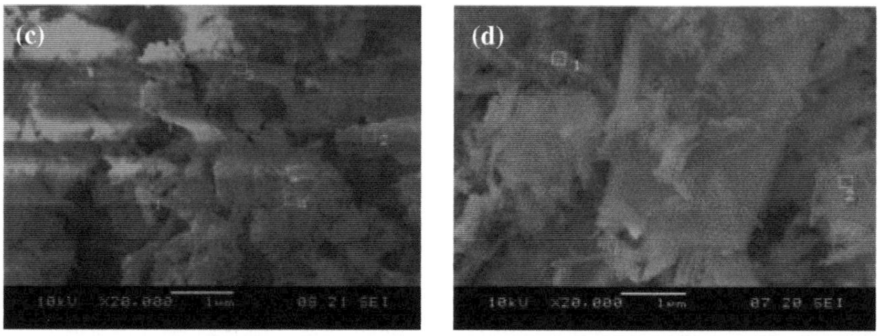

Figure 7. Hydration product morphology of hardened paste mixed with different accelerators at the final setting stage. (a) blank paste; (b) paste mixed with SL; (c) paste mixed with AFL and (d) paste mixed with BASF167.

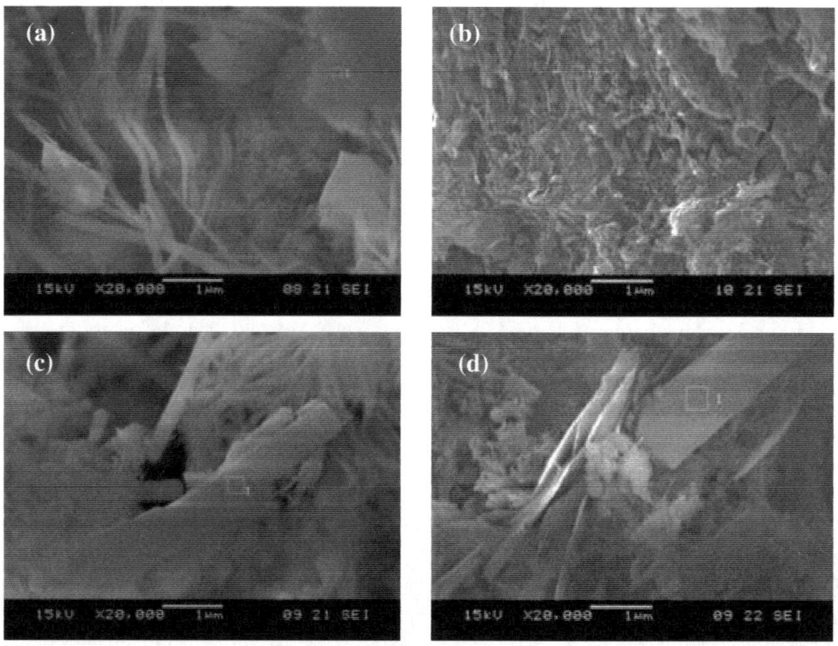

Figure 8. Hydration product morphology of hardened paste mixed with different accelerators after 1 days. (a) blank paste; (b) paste mixed with SL; (c) paste mixed with AFL and (d) paste mixed with BASF167.

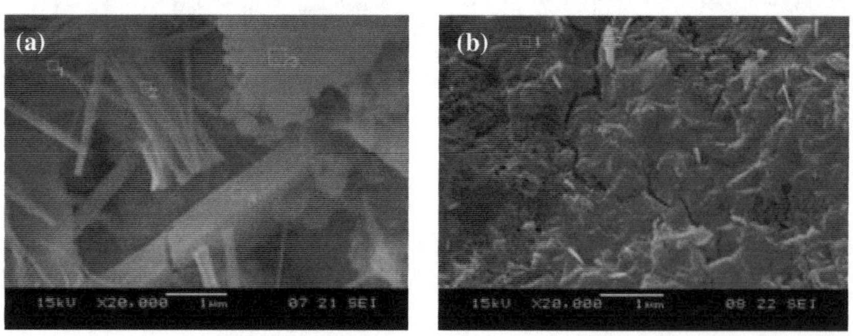

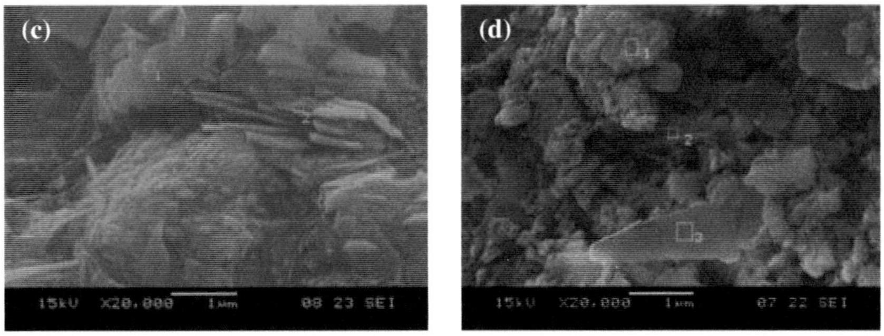

Figure 9. Hydration product morphology of hardened paste mixed with different accelerators after 28 days. (a) blank paste; (b) paste mixed with SL; (c) paste mixed with AFL and (d) paste mixed with BASF167.

It can be seen from figure7 that there was a significant difference in the microstructures between the pastes mixed with different accelerators at the initial stage of hydration (final setting). The blank paste showed a loose structure. And a large amount of hydration products of C_3A and another amorphous hydration product can be observed in the paste mixed with SL alkaline liquid accelerator, while a large number of acicular ettringite was formed in the paste mixed with AFL and BASF167 alkali-free liquid accelerator.

As indicated in figure 8, when the hydration time reached 1 day, the paste mixed with different accelerators showed a significant difference in the microstructures. A great deal of fiber-like CSH gels was formed in the blank samples. The paste mixed with KD alkaline liquid accelerator showed a loose structure, and some elongated ettringite and hydrated CSH gels were formed in it. And the morphology of the paste mixed with AFL and BASF167 alkali-free liquid accelerator was similar, a large amount of rod-like ettringite with a large size was distributed in them.

As shown in figure 9, when the hydration time reached 28 day, a great deal of CSH gels were formed in all samples those showed denser microstructures. However, the hydration products in the sample mixed with KD alkaline liquid accelerator appeared to be looser than other samples, resulting in a serious loss of 28 d strength. This phenomenon revealed that the addition of alkali-free liquid accelerator HAS is very beneficial to the development of 28 d compressive strength of cement.

4. Conclusions

(1) A new alkali-free liquid (AFL) accelerator without sulfate ion and chloride ions for shotcrete was prepared. When the content of alkali-free liquid accelerator was 6%, the initial setting time and final setting time were 2min 25s and 8min respectively, and 7d and 28d compressive strength ratios reached 112.3% and 114.2% respectively, which shows good procoagulant performance.

(2) The penetration strength of concrete prepared by using AFL accelerator is higher than that of concrete prepared by using BASF167 alkali-free liquid accelerator and SL alkali liquid accelerator, which shows a small rebound ratio of the vault and side wall, are 13.4% and 6.4% respectively. It also decreases the coagulation time, increases the bond strength and the early strength.

(3) XRD analysis show that a large amount of ettringite crystal and a small amount of $Ca(OH)_2$ form in the hardened paste prepared with AFL alkali-free liquid accelerator. According to the consumption of C_3S and C_3A, the incorporation of alkali-free liquid accelerator has little effect on the late hydration of cement.

(4) SEM analysis indicates that a large amount of CSH gel and needle-like ettringite form in the hardened paste prepared with AFL alkali-free liquid accelerator, showing a denser microstructure. The results reveal that the addition of alkali-free liquid accelerator HAS is very beneficial to the development of 28d compressive strength of cement.

References

[1] C Paglia, F Wombacher and H Bohni 2001 The influence of alkali-free and alkaline shotcrete accelerators within cement systems I.Characterization of the setting behavior *Cement and Concrete Research* **31(6)** 913–918
[2] Christian Paglia 2000 The Influence of Calciumsulfoaluminate as Accelerating Component within Cementitious Systems *Ph. D Thesis* (Swiss Federal Institute of Technology)
[3] Snyder R.C and Snyder P F 1977 Liquid concrete accelerator mixtures and methods for use thereof: US, 4046584[P]
[4] Al Manaseer A A and Nasser K W 1990 Laboratory and field tests with a new nondestructive apparatus *Can J Civil Engng* **176** 904–910
[5] C Paglia F Wombacher and H Bohni 2001 The Influence of Alkali-free and Alkaline Shotcrete Accelerators within Cement Systems-I. Characterization of the Setting Behavior *Cement and Concrete Research* **31** 913
[6] H Huber 1995 Practical Experiences with Shotcrete containing Alkali-free Acceleration *Shotcrete for Underground Support VII Telfs, Austria*
[7] Malmgren L Nordlund E and Rolund S 2005 Adhesion strength and shrinkage of shotcrete *Tunnelling and Underground Space Technology* **20(1)** 33–48

Determination of dynamic mechanical properties of silica aerogel by resonance and non-resonance method

Z P Qu and M P Sheng*

School of Marine Science and Technology, Northwestern Polytechanical University, Xi'an 710072, China

E-mail: *smp@nwpu.edu.cn

Abstract. Silica aerogel has great potential in aeronautics, building construction, and offshore oil transportation. The dynamic mechanical properties of silica aerogel were measured by the resonance and non-resonance method. The results of the two methods were compared. While the Young's modulus coincided well, the loss factor had big difference. To explain this phenomenon, the sensitivity of the two methods was analyzed. It shows that the loss factor determination of low-loss materials by the non-resonance method is quite sensitive to the experiment uncertainties. This probably results in the loss factor difference between the two methods.

Key words: silica aerogel, dynamic, Young's modulus, loss factor, resonance, non-resonance

1. Introduction

Silica aerogel, due to its excellent thermal insulating capability and low density, has great potential in aeronautics, building construction, and offshore oil transportation. The thermal properties of the material have been studied extensively [1,2]. However, the mechanical properties have only been studied by a few works, which mainly focused on the static modulus [3].

As noise-vibration engineers, we care most how the material properties vary within the frequency range, namely, the dynamic properties. The dynamic modulus and loss factor, representing the energy storage and dissipation capability respectively, are two most important parameters. However, few works about this could be dated.

In this work, we use the resonance and the non-resonance methods to measure the dynamic mechanical properties of silica aerogel. The two methods are usually used to measure the properties of polymers [4,5]. Lately they have also been applied to foam-like materials [6]. First the two methods were briefly described. Then they were used to measure the properties of silica aero, and the results were compared. Both the Young's modulus and the loss factor rise slightly as the frequency increases. The Young's modulus obtained by the two methods coincides well, but the lost factor has big difference. Finally the sensitivity analysis was made, which reveals the difference causes and helps to judge the data reliability.

2. Description of the resonance and non-resonance method

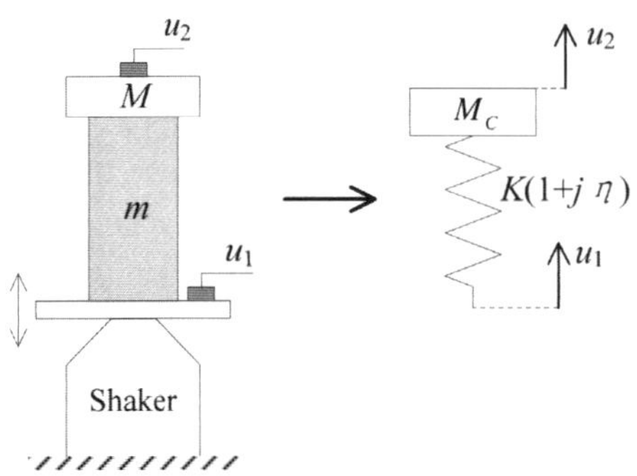

Figure 1. Set-up and equivalent model of the resonance method.

Figure 1 shows the set-up and equivalent model of the resonance method. One end of the specimen is excited by a shaker, while the other end is loaded by a mass. The displacements of the two ends are measured by two accelerometers. Just as the mass is heavy enough, the set-up can be modeled as a mass-spring system near the first resonance frequency. Thus the displacement transfer function from the excited end to the other is

$$H = |H| e^{j\varphi} = \frac{u_2}{u_1} = \frac{K(1+j\eta)}{K(1+j\eta) - M_C} \tag{1}$$

where the spring stiffness is related to the specimen's properties by $K = ES/l$, η is the loss factor of the material, and the equivalent mass is $M_C = M + m/3$. E, S and l represents the Young's modulus, the cross-section area and length of the specimen, respectively. At the resonance frequency, the phase of the transfer function satisfies $\varphi = -\pi/2$. Substituting it into equation (1) gives

$$E = \frac{\omega^2 l M_C}{S} \frac{|H|^2}{|H|^2 + 1} \tag{2}$$

$$\eta = \frac{1}{|H|} \tag{3}$$

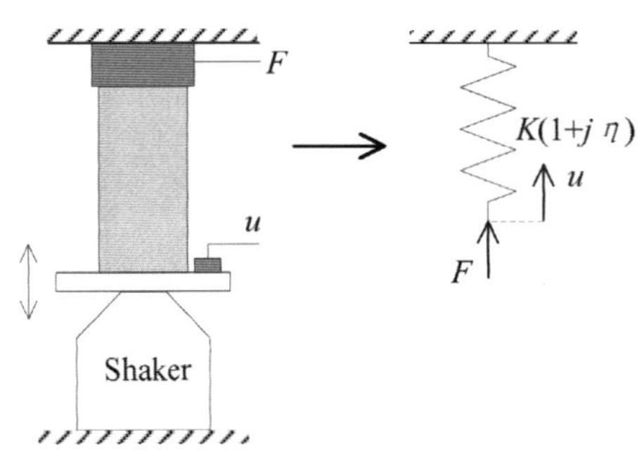

Figure 2. Set-up and equivalent model of the non-resonance method.

Figure 2 shows the set-up and equivalent model of the non-resonance method. One end of the specimen is excited by a shaker, while the other end is fixed to a rigid wall. The displacement of the moving end and the force of the fixed end are measured by an accelerometer and a fore transducer, respectively. The specimen can be modeled as a spring well below the first resonance frequency. Thus the transfer function from the displacement to force is

$$H = |H|e^{j\varphi} = \frac{F}{u} = K(1 + j\eta) \tag{4}$$

So the modulus and the loss factor are

$$E = \frac{|H| l \cos\varphi}{S} \tag{5}$$

$$\eta = \tan\varphi \tag{6}$$

3. Description of the resonance and non-resonance method

The Young's modulus and loss factor of the silica aerogel are investigated. Three cuboid specimens with length 20mm, 40mm and 60mm, as shown in figure 3, are measured by the resonance method. The cross-section of the specimens is a 10mm×10mm square. The 60mm specimen is also measured by the non-resonance method. The material density is 316 kg/m^3 and the loading mass is 11.8g. The ambient temperature is about 20℃.

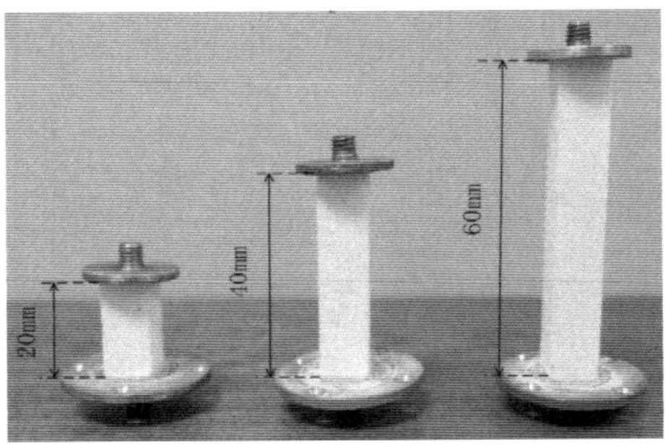

Figure 3. Three experimental specimens.

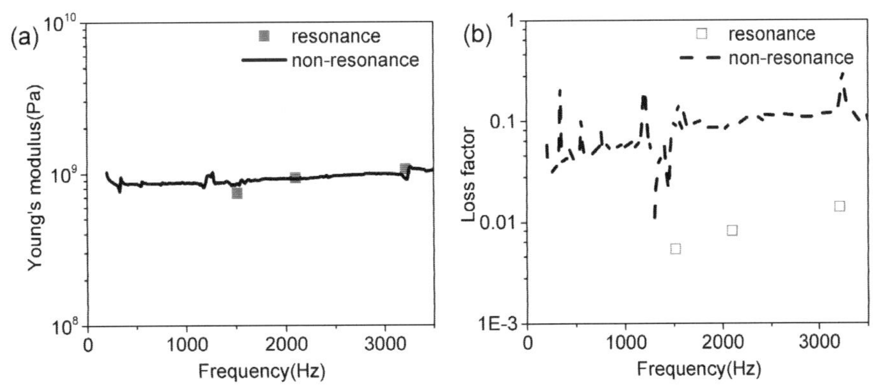

Figure 4. Experimental results, (a) Young's modulus and (b) loss factor of the silica aerogel.

Figure 4 shows the measured Young's modulus and loss factor of the silica aerogel. Three separate resonance frequencies are measured, each for one specimen. The continuous curves are the results of the 60 mm specimen. The Young's modulus coincides well. It increases slightly with the frequency increasing. However, the loss factor departs largely. The non-resonance result is about one order greater than the resonance result. In addition, the loss factor of the non-resonance method fluctuates more severely than the Young's modulus.

4. Discussion based on the sensitivity analysis

For the resonance method, the errors in E and η caused by the uncertainties in $|H|$, M_C, l, S and ω can be represented as

$$\begin{aligned} \Delta E/E &\approx D_{E|H|}\left(\Delta|H|/|H|\right)+D_{EM_C}\left(\Delta M_C/M_C\right)+D_{El}\left(\Delta l/l\right) \\ &\quad +D_{ES}\left(\Delta S/S\right)+D_{E\omega}\left(\Delta\omega/\omega\right) \\ \Delta\eta &\approx D_{\eta|H|}\left(\Delta|H|/|H|\right) \end{aligned} \tag{7}$$

where D_{yx} relates to the partial derivatives of y by x, and are called the error coefficients. They are computed as follows

$$\begin{cases} D_{E|H|}=\left(|H|/E\right)\left(\partial E/\partial|H|\right)=2\eta^2/\left(1+\eta^2\right), \quad D_{EM_C}=\left(M_C/E\right)\left(\partial E/\partial M_C\right)=1 \\ D_{El}=\left(l/E\right)\left(\partial E/\partial l\right)=1, \quad D_{ES}=\left(S/E\right)\left(\partial E/\partial S\right)=-1 \\ D_{E\omega}=\left(\omega/E\right)\left(\partial E/\partial\omega\right)=2, \ D_{\eta|H|}=\left(|H|/\eta\right)\left(\partial\eta/\partial|H|\right)=-1 \end{cases} \tag{8}$$

For the non-resonance method, the errors in E and η caused by the uncertainties in $|H|$, φ, l and S can be represented as

$$\begin{aligned} \Delta E/E &\approx D_{E|H|}\left(\Delta|H|/|H|\right)+D_{E\varphi}\Delta\varphi+D_{El}\left(\Delta l/l\right)+D_{ES}\left(\Delta S/S\right) \\ \Delta\eta &\approx D_{\eta\varphi}\Delta\varphi \end{aligned} \tag{9}$$

where D_{yx} are computed as follows

$$\begin{cases} D_{E|H|}=\left(|H|/E\right)\left(\partial E/\partial|H|\right)=1, \quad D_{E\varphi}=\left(\varphi/E\right)\left(\partial E/\partial\varphi\right)=-\eta \\ D_{El}=\left(l/E\right)\left(\partial E/\partial l\right)=1, \quad D_{ES}=\left(S/E\right)\left(\partial E/\partial S\right)=-1 \\ D_{\eta\varphi}=\left(1/\eta\right)\left(\partial\eta/\partial\varphi\right)=\eta+1/\eta \end{cases} \tag{10}$$

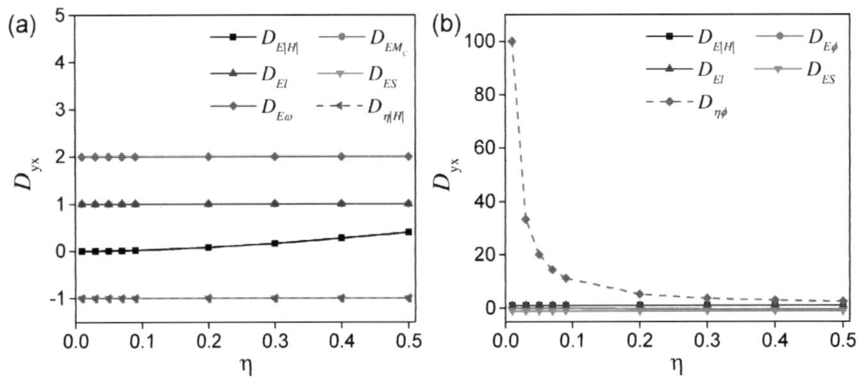

Figure 5. Error coefficients, (a) Resonance method and (b) non-resonance method.

As is evident from equations (8) and (10), the error coefficients D_{yx} are determined by the loss factor η individually. Figure 5 shows how D_{yx} vary with η for the two methods. From figure 5 (a) we see that the absolutes of D_{yx} are small for the resonance method. Apart from $D_{E|H|}$, all the coefficients are constant. $D_{E\omega}$ is the biggest with a value of 2, which means that the resonance frequency ω uncertainties will be magnified by 2 times. D_{EM_C}, D_{El}, D_{ES} and $D_{\eta|H|}$ are smaller with a value of 1 or -1. $D_{E|H|}$ rises slightly as η increases, but is always smaller than 1. So the experimental uncertainties are magnified by no more than 2 times. As the experimental uncertainties are generally little, we can say that it is quite reliable to investigate both the Young's modulus and the loss factor by the resonance method.

Figure 5(b) shows the error coefficients of the non-resonance method. We see that D_{yx} relating to E are quite small, with absolutes of no more than 1. So the Young's modulus can be determined quite accurately. However, $D_{\eta\varphi}$ rises sharply as η decreases. If $\eta > 0.1$, $D_{\eta\varphi}$ is only moderately big (less than 10). This will not result in too big errors, only if the phase φ uncertainties can be controlled at a low level. But if $\eta < 0.01$, $D_{\eta\varphi}$ can reach up to 100. This means tiny errors of φ will be magnified by 100 times or higher. And φ should be the main error origin in the experiment. So though it is suitable to investigate the loss factor of high-loss materials by the non-resonance method, the determination of low-loss materials may lead to big errors.

The above analysis helps to explain why the loss factor diverges for the two methods, and why the loss factor obtained by the non-resonance method fluctuates so much. The phase uncertainties, resulting probably from the rigid wall vibration, are very likely the cause of the errors. The fluctuations can be simply amended by removing the peaks and by curve fitting techniques. However, the large deviation cannot be corrected without full knowledge of the experimental set-up's energy dissipation mechanism. This may be a worthwhile subject for future work.

5. Discussion based on the sensitivity analysis

The Young's modulus and the loss factor of the silica aerogel are measured at room temperature by resonance and non-resonance method. The Young's modulus is about 10^9 Pa, and the loss factor is about 0.01, both rising slightly as the frequency increases. The Young's modulus obtained by the two methods coincides well, but the lost factor is quite different. The sensitivity analysis shows that the loss factor of the resonance method is reliable, but that of the non-resonance method is not trustworthy as the error coefficient is big for low-loss materials.

References

[1] Padmanabhan S K, Haq E U and Licciulli A 2016 *Ceram. Int.* **42** 7216
[2] Wei G, Wang L, Xu C, Du X and Yang Y 2016 *Energ. Build* **118** 226
[3] Wong J C H, Kaymak H, Tingaut P, Brunner S and Koebel M M 2015 *Micropor. Mesopor. Mat.* **217** 150
[4] Pritz T 1980 *J. Sound. Vib.* **81** 317
[5] Oyadiji S O and Tomlinson G R 1995 *J. Sound. Vib.* **186** 623
[6] Jaouen L, Renault A and Deverge M 2008 *Appl. Acoust.* **69** 1129

Synthesis, characterization and mechanism of polycarboxylate superplasticizer with slump retention capability

Mengting Li, Yi Wang, Haidong Jiang, Chunyang Zheng and Zhaolai Guo
Jiangsu China railway ARIT new materials Co., Ltd., Nanjing 211505 China

E-mail: aritlmt@163.com

Abstract. Two kinds of slump-retaining polycarboxylate superplasticizer with different molecular structures were synthesized via maleic anhydride(MAH), polyethylene glycol monoallyl ether(APEG) and hydroxyethyl acrylate (HEA)/ethyl acrylate (EA) according to molecular structure design principle. The comprehensive performances of each slump-retaining polycarboxylate superplasticizer were analyzed and compared with that of different molecular structure. The Gel Permeation Chromatography (GPC), surface tension, zeta potential, adsorption behavior, cement paste fluidity and application performance in concrete were measured. And the slump retention mechanism was analyzed. The results show that the polycarboxylate superplaticizer with hydroxyethyl group prepared in this study has good dispersing performance and high slump retention capability for concrete.

Keywords: polycarboxylate superplaticizer, hydroxyethyl acrylate/ethyl acrylate, dispersing performance, slump retention

1. Introduction

Polycarboxylate superplaticizers have been increasingly widely used in concrete engineering because of their low admixing volume, high water reducing rate, good slump retention capability, adjustable molecular structure, and environmental protection. However, the large slump loss of concrete often appears in hot weather or during long-time and long-distance transportation due to the existence of large differences in concrete raw materials, especially in the hot summer, the slump loss is faster [1-2]. During the construction process, it is often to increase the amount of water reducing agent to restore good work performance of the concrete, making the concrete have good liquidity. However, the increase of the amount of water reducing agent in the initial stage of mixing will lead to the segregation and bleeding of concrete, which will destroy the uniformity of concrete and make the strength of each part of concrete quite different, reducing the compressive strength of the whole concrete. In practical engineering, it is generally to compound the retarders such as sodium citrate, sucrose, sodium citrate and so on [3-4]. But the problem of slump loss still remains unsolved with the additon of these retarders, and the retarders are easily to deteriorate in high temperature environment. Thus affecting the operational performance of the water reducing agent. To solve this problem, domestic and foreign researchers have begun to study the slow-release polycarboxylate superplaticizers [5]. The polycarboxylate superplaticizers with slump retention capability can be prepared through the optimal combination of different functional units to solve the problem of slump loss, as the molecular structure of polycarboxylate superplaticizers can be designed easily [6].

In this paper, two kinds of polycarboxylate superplaticizers PC-1 and PC-2 were synthesized through free radical polymerization of maleic anhydride (MAH) and polyethylene glycol monoallyl ether(APEG), using hydroxyethyl acrylate (HEA) and ethyl acrylate (EA) as small molecule sustained-release monomers, respectively. The structure of hydroxyethyl ester/ethyl ester polycarboxylate superplasticizer was synthetically analyzed from the point of the synthesis of the polycarboxylate superplaticizers. The adsorption and dispersion mechanism of the two kinds of slump-retaining polycarboxylate superplaticizers PC-1 and PC-2 with the cement particle surface were studied by Gel Permeation Chromatography (GPC), surface tension test, Zeta potential, adsorption behavior, cement paste fluidity test and concrete test, which will provide a deep theoretical basis for the application of polycarboxylate superplaticizer in concrete.

2. Experiments

2.1. Main materials and instruments

Polyethylene glycol monoallyl ether(APEG), industrial grade, Zhejiang Huangma Chemical Group Co.Ltd .; small molecular ester monomers: hydroxyethyl acrylate (HEA) and ethyl acrylate (EA), analytical grade, Chinese Medicine Group Chemical Reagent Co.Ltd.; sodium methyl methacrylate (SMAS), industrial grade, Nanjing All-plus Chemical Co.Ltd.; maleic anhydride(MAH), ammonium persulfate (APS), sodium hydroxide, all of analytical grade, Nanjing Wanqing Chemical Classware Istrument Co.Ltd.; vitamin C, commercially available.

Cement: P•O42.5 benchmark cement; sand: medium sand in area II, fineness modulus of 2.6 to 2.9, mud content <1%.

2.2. Synthesis of the polycarboxylate superplaticizers

Ammonium persulfate (APS) and H_2O were weighed and added into a three-necked flask containing polyethylene glycol monoallyl ether(APEG) and stirred and heated in water bath. Component A and component B were weigned at the same time, component A: maleic anhydride (MAH)+small molecule ester monomers+H_2O; component B: Vc+sodium methyl methacrylate (SMAS). Component A and component B were dropwise added at the same time, finished within 3h and 3.5h respectively, and then the reaction was incubated for 1h. After cooling, the pH value of the reactant was adjusted to 6 ~7 with NaOH solution with concentration of 40% to obtain the desired the polycarboxylate superplaticizer with slump retention capability.

In this experiment, the slump-retaining polycarboxylate superplaticizer PC-1 and PC-2 were prepared by using hydroxyethyl acrylate (HEA) and ethyl acrylate (EA) as monomers, respectively.

2.3. Performance testing and characterization

2.3.1. Gel Permeation Chromatography (GPC). Gel permeation chromatography (GPC) separates the polymer molecules based on the principle that their hydrodynamic volume of the polymer molecular is different. The mobile phase was 0.10mol/L sodium nitrate solution with the flow rate of 0.5mL/min, and the stationary phase was gelatinous porous filler. 1.1mL of superplasticizer sample with solid content of 40% was dissolved into 6.0 mL of 0.10 mol /L sodium nitrate solution to make a gel chromatogram.

2.3.2. Surface tension test. The surface tension of the superplasticizer solutions, of which the concentrations were 0.2%, 0.5%, 1.0%, 1.5%, 2.0%, 2.5% and 3.0% respectively, was determined by Wilhelmy method. The superplasticizer solutions were prepared by superplasticizer and deionized water. The same group of solutions was measured continuously for three times and took the average.

2.3.3. Zeta potential. The Zeta potential of the cement paste was determined by the microscopic (cell) electrophoresis meter. Weighed 0.5g of cement, 0.1g of superplaticizer and 150mL of deionized water were weighed. After mixing well, the mixture was allowed to stand for about 5 minutes. The upper layer solution was taken at regular intervals to measure the Zeta potential. The voltage was 100V. 10 cement particles were taken as a group and took the average.

2.3.4. Adsorption behavior(TOC). 30g of superplasticizer solutions with different concentrations were prepared using the superplasticizer and deionized water, the concentrations of the solutions (%) were 0.2g/L, 0.4 g/L, 0.8 g/L, and 1.6 g/L. Then 5g of reference cement was added into the prepared solutions respectively, stirred evenly, and filtered centrifugally after standing. The adsorption capacity was tested at 10min and 1h respectively.

2.3.5. Cement paste fluidity. Cement paste fluidity was tested with reference to GB 8077-2012 "test methods of concrete admixture homogeneity", the admixture solid content of the polycarboxylate superplaticizer was 0.2% of cement quality, water-cement ratio of 0.29.

2.3.6. Concrete test. Concrete test was carried out with reference to GB8076-2008 "concrete admixtures".

3. Results and discussion

3.1. Gel Permeation Chromatography (GPC)

Molecular weight and molecular weight distribution is one of the most basic structural parameters of water reducing agent. If the molecular weight is too small, the molecular chain will be too short, which influences the water-retaining property and has the defect of easy bleeding; If the molecular weight is too large, bypass effect and flocculation will happen, influencing the disperse effect. Therefore, the molecular weight and molecular weight distribution have great influence on the performance of water reducing agent. The synthesized slump-retaining polycarboxylate superplasticizers PC-1 and PC-2 were analyzed by GPC, as shown in figure 1.

After the polycarboxylic acid molecules were injected into the column, the large molecules first came out through the eluent, and the small molecules occur later. It's known that the former larger peak is the molecule with lager molecular weight polycarboxylate superplasticizer, of which the peak area is larger, the peak height is also high, and the peak area ratio is about 95%. However, the shoulder peak is clearly observed in PC-2, while the smaller peak is probably the self-polymer of maleic anhydride monomer or the macro monomer not participating in reaction after analyzing. The peak area is smaller, the peak height is also lower, and the peak area ratio is 3.5%. Comparison of the two spectral lines shows that there is little difference between the GPC diagrams of the synthesized two kinds of polycarboxylate superplasticizers. GPC results show that the monomer conversion rate is higher in the polymerization of PC-1.

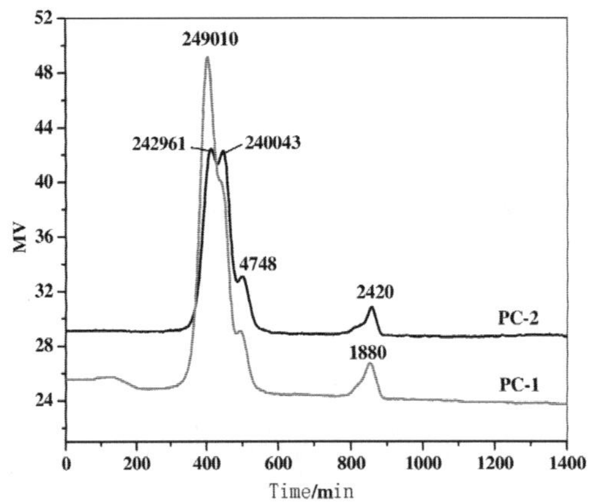

Figure 1. Gel Permeation Chromatography (GPC).

3.2. Surface tension test

The surface tension of the superplaticizer has a great influence on the dispersion of cement particles and the hydration process of cement. Hydroxyethyl acrylate-type / ethyl acrylate-type polycarboxylate superplaticizer can be arranged oriented on the solution surface as a surface active agent with carboxyl and hydroxyl hydrophilic groups, making the surface tension decrease significantly.

In this study, the synthesized two kinds of polycarboxylate superplasticizers were compared in the concentration gradient of 0.2% to 3%, and their influence variety on the surface tension of the solution is shown in Figure 2. From figure 2, it can be seen that the surface tension of the synthesized two kinds of polycarboxylate superplasticizers show the same trend, that is, the surface tension of the solution gradually decreased with the increase of the concentration, which significantly reduced compared with the surface tension of pure water. When the concentration of admixture was 1.0%, the difference of the surface tension reduction rate between PC-1 and PC-2 was little, about 12%. But when the concentration was from 1.0% to 3%, the surface tension reduction rate was up to 20%.The results showed that the hydroxyethyl acrylate-type polycarboxylate superplaticizer PC-1 was more effective than ethyl acrylate-type polycarboxylate superplaticizer PC-2 in obviously decreasing the surface tension of water, which indicated that PC-1 could increase the trendency of the superplaticizer molecule to wet, penetrate and adsorb the cement surface due to its strong adsorption ability of cement particles and the formation of strong electric field.The structure of the hydroxyethyl ester in the molecular chain can effectively increase the adsorption and alignment of the molecular chain to the polar surface, reduce the solid-liquid interface energy on the cement particles and increase the wetting effect between the water and the cement, so as to improve the thermodynamic stability of the dispersion system, and ultimately to achieve the purpose of improving the dispersion of cement particles.

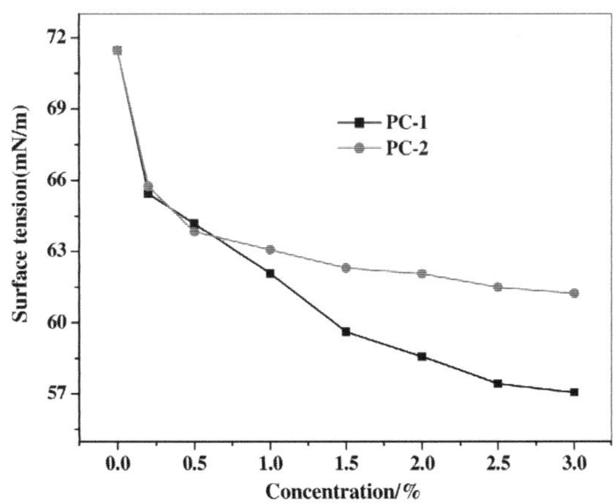

Figure 2. Effect of admixture concentration on the surface tension of the solution.

3.3. Zeta potential

In order to further understand the effect of the prepared polycarboxylate superplaticizer on the cement particles, two kinds of polycarboxylate superplasticizers PC-1 and PC-2 synthesized in this experiment were tested for the surface ζ-potential of the cement particle. The results are shown in figure 3.

It can be seen from figure 3, the surface ζ-potential of the cement particle without admixtures is positive. It gradually transits from positive to negative with the concentration of admixtures increasing, while the absolute value gradually increases to flatten. This phenomenon may be due to: pure cement particles have positive charge on the surface, so ζ-potential is positive, whereas the superplasticizer molecules are generally anionic surfactants, making the cement particles have negative charge after adsorption on the surface under the joint influence of electrostatic interaction and complex effect.Therefore, the electrostatic repulsion occurs between the particles, which makes the particles disperse from each other. As the amount of admixture increases, the amount of superplaticizer adsorbed on the surface of cement particles continuously increases, so that the negative charge on the surface of cement particles increases and the absolute value of ζ-potential increases. The traditional electrostatic repulsion theory holds that the higher the Zeta potential, the greater the electrostatic repulsion between the cement particles and the more dispersed the cement particles, which can explain the mechanism of high efficiency superplaticizer to a certain extent. From figure 3, it can be seen that the dispersion property of the polycarboxylate superplaticizer PC-1 is better than that of PC-2. This may be related to the molecular structure of PC-1. The structure of ethyl hydroxyl in the molecular chain of PC-1 can increase the wetting and adsorption trend of the superplaticizer molecules on the cement surface. The greater the adsorption capacity of cement particle surface, the stronger the electrostatic repulsion caused by the same charge between particles, which effectively increases the adsorption of the molecular chain to polar surface, thus improving the dispersion of cement particles.

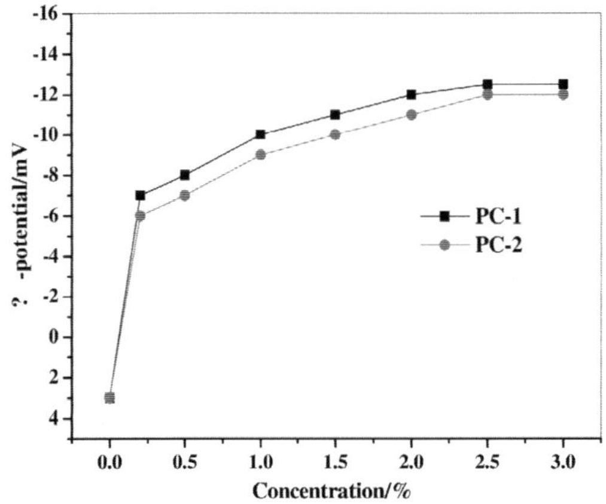

Figure 3. Effect of admixture concentration on the surface ζ -potential of cement particles.

3.4. Adsorption behavior

In order to further study the relationship between the molecular structure and the adsorption performance of the synthesized polycarboxylate superplaticizer, the adsorption capacity of the two kinds of slump-retaining polycarboxylate superplasticizers at different concentrations on the surface of cement particles was studied. The adsorption curve is shown in figure 4.

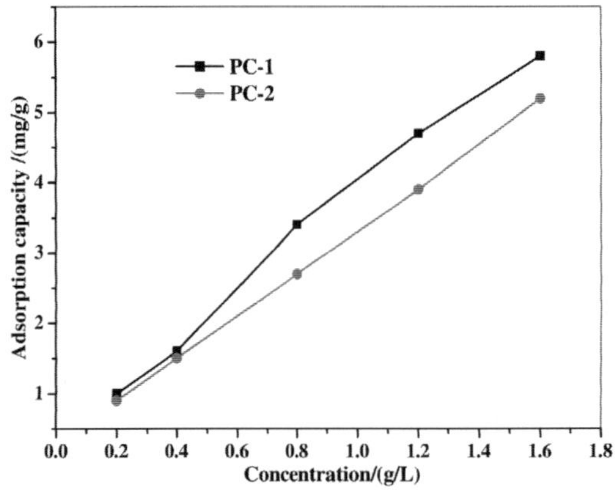

Figure 4. Adsorption curve of the admixtures on cement particles.

It can be seen from figure 4, with the increase of the concentration of the superplaticizer, the adsorption capacity of cement to superplasticizer is gradually increasing. The adsorption capacity of cement to the synthesized polycarboxylate superplaticizer PC-1 is higher than that of PC-2, Which also reflects the flowability and slump retention capability of the concrete mixture with polycarboxylate superplasticizer PC-1 is better than PC-2. This is consistent with the ζ-potential testing result that the synthesized polycarboxylate superplaticizer PC-1 has better dispersion retention capability than PC-2.

3.5. Cement paste fluidity

Cement paste fluidity test is the most direct way to characterize the ability of admixtures to disperse cement. Figure 5 is the cement paste fluidity and flow retention curve over time of the prepared two kinds of polycarboxylate superplasticizer, with the admixture solid content of 0.2% of cement quality, water-cement ratio of 0.29.

From figure 5, it can be seen that the initial cement paste fluidity of the prepared two kinds of polycarboxylate superplaticizers is very small, but gradually increases with time. The fluidity value reaches the maximum at 90min, and gradually decreases after 90min. And the synthesized polycarboxylate superplasticizer PC-1 has better dispersibility and retention than PC-2. This is because the hydroxy group of the hydrophilic hydroxyethyl ester of the superplaticizer PC-1 with a low surface tension is more likely to adsorb on the surface of the hydrophilic cement particles to enhance the steric hindrance effect, thereby making it possible to obtain a larger initial cement paste fluidity. At the same time, the hydroxy groups in the hydrophilic hydroxyethyl groups are easily associated with polar water molecules in the form of hydrogen bonds, forming a stable layer of solvated hydration membrane with a certain mechanical strength on the surface of the cement particles. The formation of the hydration membrane makes the cement particles wet and easy to slip, which prevents mutual coalescence of the cement particles, therby maintaining better cement paste fluidity.

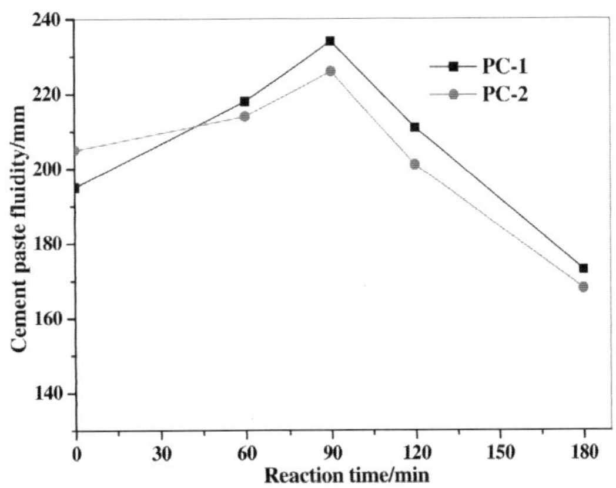

Figure 5. Cement paste fluidity and flow retention curve over time.

3.6. Concrete test

The properties of the concretes mixed with the synthesized two kinds of polycarboxylate superplasticizers PC-1 and PC-2 respectively were compared. Concrete mix ratio and test requirements are tested according to GB / T 8076-2008 "Concrete admixtures", the test results are shown in table 1.

It can be seen from Table 1 that under the same test conditions and the same admixture dosage, the synthesized polycarboxylate superplasticizer PC-1 has smaller slump loss than PC-2. The slump loss of the concrete is small at 2h, and the slump retention capability of PC-1 is better than that of PC-2.

The alkyl structure of the alkyl ester in PC-2, with smaller HLB value than that of the hydroxyl of hydroxyl ester in PC-1, has strong air-entraining performance, which is also confirmed from its concrete compressive strength values.The compressive strength ratio of PC-1 reached 150% at 28d. The higher gas content of PC-2 also affected the compactness of cement stone, which reduced the strength of concrete.

Table 1. Test results of concretes with different slump-retaining polycarboxylate superplasticizers.

Type of polycarboxylate superplasticizer	Slump/Divergence (mm)			Gas content /%	Compressive strength ratio/%		
	Initial	1h	2h		3d	7d	28d
PC-1	210/460	195/440	190/430	4.6	180	161	150
PC-2	210/430	180/400	170/370	7.0	174	154	144

3.7. The mechanism of superplaticizer with hydroxy ester structure

In the cement-water system, the polycarboxylate superplaticizer has the main functions of electrostatic repulsion, steric hindrance, complexation, hydration membrane lubrication and wetting [7-8]. Based on the existing theory and performance results, the mechanism of the effect of the polycarboxylate superplaticizer with hydroxy ester structure was discussed.

1) Electrostatic repulsion effect. According to the electrostatic repulsion theory of the polycarboxylate superplasticizer [9-10], when the polycarboxylate superplaticizer with hydroxy ester structure adsorbed to the surface of cement particles, the electrostatic repulsion is generated on the surface of the cement particles with the same charge due to the electrolysis of hydrophilic polar groups ,making the cement - water system in a relatively stable state of suspension, thereby improving the cement paste fluidity. The results showed that the hydrophilic hydroxylethyl ester of PC-1, the superplaticizer with low surface tension, more easily adsorbed on the hydrophilic surface of cement particles by generating electrostatic repulsion as well as reducing surface tension, and the steric hindrance was enhanced. This achieved a lower surface tension and a stable electrostatic repulsion of the superplaticizer, which is conducive to the realization of cement particles dispersion.

2) Steric hindrance effect. The main chain of the superplaticizer with hydroxy ester structure adsorbed on the surface of cement particles, and the hydrophilic side chain stretched to form thicker polymer molecular adsorbent layer in the liquid phase, thus forming a huge three-dimensional adsorption structure, resulting in steric hindrance effect [11], so that the cement particles repel each other and disperse. With the results of this study, the hydroxyl group of hydrophilic hydroxyethyl ester on the synthesized main chain with comb structure was easily associated with the polar water molecules in the form of hydrogen bonds,and the introduction of hydroxyl ester structure could also form strong hydrogen bonds with the polyether side chain, forming a stable layer of solvated hydration membrane with a certain mechanical strength on the surface of the cement particles,thus playing a role in strengthening the steric hindrance.

3) Complexation. The Ca^{2+} released from cement hydration process can form complexes with carboxyl groups hydrolyzed by hydroxyl ester [12]. The concentration of Ca^{2+} was reduced to delay the formation of $Ca(OH)_2$ and reduce the formation of C-S-H gel, thus delaying the cement hydration process. The incorporation of superplaticizer with hydroxy ester structure can delay the hydration process obviously.

4) lubrication. The hydrophilicity of the hydroxy ester is favorable to the adsorption of the superplaticizer with hydroxy ester structure on the cement particles, and the bonding with the surface charge of the cement, which plays the anchoring effect, is beneficial to the wetting and lubricating effect and the relative lubrication among the aggregate particles of the superplaticizer aqueous solution. According to the theory of Kim [13], the surface of the particles can be stabilized to form a solvated hydration membrane with a certain mechanical strength, generating the lubrication which can destroy the flocculation structure of cement particles and improve the slurry flow performance.

4. Conclusion

Two kinds of functional polycarboxylate superplaticizers PC-1 and PC-2 were synthesized through introducing hydroxyethyl acrylate / ethyl acrylate into the molecular structure respectively by free radical copolymerization. The molecular weight and distribution of the molecular structure were determined by GPC. The surface tension, Zeta potential and adsorption capacity testing results show that the polycarboxylate superplasticizer PC-1 can not only reduce the surface tension of the solution, but also increase the wetting, permeation and adsorption of the cement surface with good dispersion retention performance.

The cement paste fluidity and concrete performance of the synthesized two kinds of polycarboxylate superplasticizers PC-1 and PC-2 have been studied. The results show that under the same test conditions and the same dosage, the synthesized polycarboxylate superplasticizer PC-1 has better slump retention capability and compressive strength than PC-2.

Referrences

[1] Chunyan L, Ziwei W and Jianguo R 2012 Journal of Shanxi University: Natural Science edition 35 113–117
[2] Ziming W. 2009 Polycarboxylate superplasticizer: Synthesis, properties and application (Beijing: China Building Industry Press) p 45
[3] Zunming L and Lufeng F 2009 *Concrete* **12** 52–53
[4] Yonghui L, Shuxin J and Huisheng Y 2000 *Concrete* **6** 30–31
[5] Xiaofeng Y, Huqun W and Weifeng X 2013 *New Building Material* **3** 32–33
[6] Xinxiu C 2012 *Fujian Building Materials* **1** 19–22
[7] Uchikawa H, Sawaki D and Hanegara S 1995 *Cem Concr Res* **25** 353–364
[8] Yin S, Aishah F and Salmiah A 2001 *Cem Concr Res* **31** 1009–1015.
[9] Grabiec A M, Krzwo B and Ockallaurow R 1997 *10th International Congress on the Chemistry of Cement*(Goteborg, Sweden) p 6–14
[10] Sakai E, Kang J K and Daimon M 2000 *6th CANMET/ACI International Conference on Superplasticizers and Other Chemical Admixtures in Concrete*(Nice, France) p 75–86
[11] Yoshioka K, Sakai E and Daimon M 1997 *J Am Ceram Soc* **80** 2667–2671
[12] Yamada K and Hamejara S 2003 *Proceedings of the 11th International Congress on the Chemistry of Cement* (Durban, South Africa) p 538–549
[13] Kim B G, Jiang S P and Jolicoeur C 2000 *Cem Concr Res* **30** 887–893

Laboratory test method for dirt pickup resistance and stain removal

Shiwei Ren, Xueying Zheng, Yi Liu and Quan Jiang

National Building Material Testing Center, Beijing, 100024 China

E-mail: 2829294538@qq.com

Abstract. The pollution characteristics of current atmospheric particulates was summarized in the present investigation. The composition and proportion of the pollution sources used for dirt pickup resistance and stain removal test were adjusted, and the pollution sources used for new type dirt pickup resistance and stain removal test produced. In addition, a new dirt pickup method was adopted, and a set of new type laboratory dirt pickup resistance and stain removal tests developed by taking comprehensive consideration of the existing state and dirt pickup mode of actual atmospheric particulates. It verifies the rationality, feasibility and effectiveness of new test methods for dirt pickup resistance and stain removal based on the contrast test over the new and old test methods.

Keywords: atmospheric particulates, dirt pickup resistance, pollution source, test method

1. Introduction

Dirt pickup resistance and stain removal are one of the important indicators for evaluating the performance of coated metal plates. In the current test method for dirt pickup resistance and stain removal, the ash which is prepared mainly with graphite powder is adopted as the pollution source, and is brushed on the film test plate, and then the test plate is washed with water in uniform flow under the specified water pressure. In practical operation, the brushing method will cause uneven film layer due to human factors, thus the test result may be affected; in addition, the brushing method is inconsistent with the actual dust fall effect.

At present, the urban atmospheric pollution is becoming increasingly serious; and with the increase of motor vehicle population, motor vehicle exhaust pollution increases greatly. In many large and medium-sized cities, urban atmospheric environment pollution caused by motor vehicle exhaust emission has exceeded 60%, which shows that motor vehicle pollution has become the main source of atmospheric pollution. The organic pollution caused by motor vehicle exhaust emission is predominant and organic matter has become one of most primary components of atmospheric particulates. The pollutants used in the current test methods for dirt pickup resistance and stain removal do not involve any organic matter, thus this paper studies the preparation of new pollutant which better meets the composition of actual atmospheric particulates.

In dirt pickup resistance and stain removal test, the composition and proportion of pollution sources used are the key factors determining the rationality of test method for dirt pickup resistance and stain removal, and meanwhile, the dirt pickup mode determines the feasibility of such test method, a rational and feasible artificial accelerated test method for dirt pickup resistance and stain removal is the final object of this study. This paper carries out a series of studies on how to make the results of

IOP Conf. Series: Materials Science and Engineering **182** (2017) 012037 doi:10.1088/1757-899X/182/1/012037

laboratory artificial accelerated test be more approximate to those of outdoor exposure test and how to make the test results truly and rationally reflect the dirt pickup resistance and stain removal of film.

2. Method and theory

Atmospheric particulates belong to a kind of complex mixture and a group composed of the solid and liquid particulates of different particle sizes. The size and distribution mode of atmospheric particulates are closely related to their properties and their influence on environment. Therefore, the study on pollution characteristics of atmospheric particulates must take the size composition, particle size and distribution into account [1-2].

Based on the current findings of investigations on atmospheric particulates, PM10 concentration in atmospheric environment exceeds 80%. In addition, according to "Greenfield Window" theory, the particulates with particle size between 0.1μm and 10μm are hard to be effectively eliminated during the formation of rain from atmospheric water vapor, i.e., the particulates within this particle size range are the main components of atmospheric particulates [3-5].

According to the source apportionment of atmospheric particulates by environmentalists in China, fugitive dust sources are the important source of PM10 in most areas of China, therein the proportion of soil dust, road dust, cement and construction dust in PM10 may be as high as 80%. Secondly, the proportion of organic matter and carbon black caused by industrial emission and vehicle exhaust emission also reaches 20%~40% [6]. These components basically are the sources of main components of atmospheric particulates.

According to statistics, the motor vehicle population in Beijing in 2012 is about 5,200,000, which is a huge amount. According to the analytical prediction of experts, around 2020, the motor vehicle population in China will reach 200,000,000; the annual average growth rate of vehicle population in Shanghai in recent 10 years is also above 10%. The continuous growth of urban motor vehicle population brings great convenience for the life of residents, but also increases the environmental load. Based on the findings from studies on atmospheric particulates, the pollution source from motor vehicle exhaust emission is the main component of PM2.5 in Beijing. According to statistics, only in 2013, the proportion of pollution caused by motor vehicle exhaust emission in PM2.5 increased by approximately 10% [7], the contribution rate of motor vehicle exhaust emission to hydrocarbon compounds in atmospheric pollution reached 73.5%, this figure in the central area of Shanghai also reached 96%, and similar situation also happened in Tianjin, Chongqing, Guangzhou, etc. [8].

In the source apportionment of atmospheric pollution in Beijing, the State Environmental Protection Administration of China analyzes that, from the point of view of composition, the organic pollutants account for 26% of PM2.5 (proportion of PM2.5 in PM10 is about 70%); from the point of view of sources, 64%~72% of atmospheric pollution is locally caused in Beijing. While among the local sources, the proportion of motor vehicle emission (accounting for 31.1%) and coal burning (accounting for 22.4%) exceeds 50%; and the correlational studies show that the contribution rate of volatile organic compounds (VOCS) exhausted by motor vehicles to the atmospheric pollution in urban area of Shanghai exceeds 65% [9]. Cui Mingming et al [10]. have analyzed the atmospheric inhalable particulate matter (PM10) sample nearby the district of Guangzhou City and concluded that the mass concentration of organic matter in PM10 was 24%~32%.

The pollution source used for dirt pickup resistance and stain removal test of exterior wall coating shall follow certain selection principles:

1) As pollutants vary from area to area, the selected pollution source shall be with certain representativeness and universal applicability;

2) In order to ensure personal safety of experimenters, the prepared pollution source shall be nontoxic and nonhazardous;

3) This pollution source is used for commenting the dirt pickup resistance and stain removal performance of exterior wall coating, and it shall be with good reproducibility and stability in order to make the test results be more convincing.

As for dirt pickup resistance test and stain removal of external wall coating, China adopts GB/T 9780-2013 Test Method for Dirt Pickup Resistance and Stain Removal of Film of Architectural Coatings and Paint, and in this test, the standard ash sample prepared mainly with graphite powder is adopted as the pollution source [11]. The composition of current atmospheric particulates becomes complicated increasingly. As reported by the Ministry of Environmental Protection, the contribution rate of motor vehicle exhaust emission to atmospheric environment pollution reaches 50~60% in most large and medium-sized cities in China, and this figure may be much higher in some cities with rapid development. This shows that the motor vehicle emission pollution has already become the main source of part of atmospheric pollution, organic pollution is serious and organic matter has become one of the most primary components of atmospheric particulates.

Upon study, it is found that saturated hydrocarbon in motor vehicle exhaust mainly consists of n-alkanes, and n-alkanes containing less than 25 carbon atoms is mainly generated from such human activities as combustion of fossil fuel and coal, and the alkane containing more than 26 carbon atoms is from emission of biological source. In various cities, the distribution range of n-alkanes is approximately C10~C35, and the main carbon peak basically is C22~C27. Therefore, a kind of saturated hydrocarbon, i.e. n-Tetracosane (C24), is selected as the organic matter added in the pollution source used for dirt pickup resistance and stain removal test.

Upon comprehensive analysis as well as reference to the pollution sources used in current dirt pickup resistance and stain removal tests at home and abroad, the composition of the pollutant sources in laboratory dirt pickup resistance and stain removal test method is finally selected: loess dust (loess in Shaanxi) - the representative material of fugitive dust sources, n-Tetracosane - the representative material of organic matter, and carbon black.

There into, loess dust is the standard sample simulating the atmospheric dust, while n-Tetracosane and carbon black are provided by the product supply department. Fugitive dust sources are the important source of PM10 in most areas of China, and the proportion of dust in PM10 may be as high as 80%; secondly, the content of organic matters generated from industrial emission and vehicle exhaust emission also reaches 20%~30%, and the ratio of organic carbon to inorganic carbon is about 3 [12]. Reference to current dirt pickup resistance tests: the proportion of No.8 ash in pollutant of Japan Kansai Paint Co., Ltd. is 70%, that of carbon black is 17%; during simulation of atmospheric dust, Germany Sto Ges.m.b.H. determines the proportion of ash and carbon black in dust as 85% and 15% respectively. According to the aforesaid data, the proportional range of three components of the new pollution source are temporarily taken as 70%~80% for loess dust, 20%~30% for n-Tetracosane and 5% and 15% for carbon black.

As for the three components of pollution source powder, in addition to following the aforesaid proportion of mixture, appropriate mixing method also shall be selected. In this study, three kinds of powder are mixed in liquid state and solid state respectively, and comparison is carried out based on the color and state of mixture; thereinto, mixing in solid powder state improves the mixing uniformity and also meets existence of actual atmospheric particulates in nature; meanwhile, with consideration of simulating the actual pollution mode of atmospheric particulates - dust fall, the pollution source powder is hereby sprayed, and the proportion of these three pollution sources are finally determined, as shown in table 1.

Table 1. Components of New Pollution Source.

Component	Loess dust	n-Tetracosane	Carbon black
Proportion	75%	20%	5%

In this dirt pickup resistance and stain removal test, both loess dust and carbon black used are standard materials, of which the particle diameter is less than 100μm, and only n-Tetracosane has nonuniform particle size and is flaky. Dry powder mixing method is adopted and organic solvent cannot be added, thus n-Tetracosane (melting point: about 55℃) is heated to melt on heating plate at about 55℃, then is added with loess dust and carbon black and stirred sufficiently by manpower so as to mix the three components uniformly, and the mixture is sufficiently ground in mortar to obtain the pollution source powder.

Table 2. Test Results of Loss on Ignition.

Batch No.	No.	Weight of sample before ignition(g)	Weight of sample after ignition(g)	Loss on ignition(%)
	1	1.0633	0.7123	32.07
First batch	2	1.0192	0.6997	31.35
	3	1.0324	0.6973	32.45
	1	1.0536	0.7199	31.67
Second batch	2	1.0603	0.7258	31.55
	3	1.0507	0.7073	32.68
	1	1.0464	0.7192	31.27
Third batch	2	1.0250	0.6959	32.11
	3	1.0790	0.7328	32.09
	1	1.0761	0.7331	31.87
Fourth batch	2	1.0304	0.7003	32.04
	3	1.0756	0.7347	31.69
	1	1.0268	0.6921	32.59
Fifth batch	2	1.0341	0.7008	32.23
	3	1.0683	0.7266	31.99

Pollution source powder is stirred, ground and mixed manually, thus the generally applicable loss on ignition test requiring relatively simple operation is carried out to inspect the mixing uniformity of pollutant. Five groups of pollution sources prepared at different time are taken, and three samples are randomly taken from each group for the loss on ignition test. The test results are as shown in table 2-5.

The test method and data statistical method for uniformity are as follows:

As for the test results, variance analysis method is adopted to calculate the five groups of data, with group consisting of repeated results from three independent measurements, and the test results of uniformity of pollutant are as shown in table 3.

Table 3. Result of Uniformity.

Sample No.	Uniformity		
	Result 1	Result 2	Result 3
1	32.07	31.35	32.45
2	31.67	31.55	32.68
3	31.27	32.11	32.09
4	31.87	32.04	31.69
5	32.59	32.23	31.99
$\overline{\overline{X}}$		31.98	
SS_{within}		0.6943	
SS_{among}		0.3665	
MS_{within}		0.0916	
MS_{among}		0.0694	
F		0.76	
$F_{0.05(19,20)}$		2.14	

Conclusion: according to the test results, F value of loss on ignition is 0.76, which is less than the critical value $F_{0.05}(19, 20)=2.14$, and this shows that the uniformity of new pollutant meets the relevant requirements.

Upon particle size test of n-Tetracosane, loess dust and carbon black, three materials are mixed with the aforesaid method to prepare the pollution source used for new type dirt pickup resistance and stain removal test. It is found that the uniformity of this mixture meets the requirements upon uniformity test.

3. Material and experiment

The pollution source powder prepared with aforesaid method shall be sprayed with appropriate container. Simple spraying unit manufactured in laboratory: the top surface is screen mesh in hole diameter of 0.1mm, the whole container body is made of stainless steel, and pollution source powder is let "fall" onto the film test plate by this unit to simulate the actual dirt pickup mode of atmospheric particulates, i.e., dust fall.

Samples: new pollution source, three pieces of 70mm*150mm aluminium veneers with fluorocarbon film, with polyester film and with powder film respectively.

Instruments: electronic balance; self-prepared powder spraying unit; magnetic stirrer; dirt pickup-resistant washing device; electro-thermostatic blast oven; colorimeter.

Standard test conditions: (23±2)℃; relative humidity of (50±5)%.

The investigation shows that the pollution process of atmospheric particulates in Beijing presents obvious periodicity, with typical period of 4~10 days and average period of 5~7 days, and such a periodicity is essentially determined by the periodical typical cycle synoptic systems. In consideration of the influence brought by the periodicity of pollution as well as by reference to current dirt pickup resistance and stain removal test methods, this test adopts 5 dirt pickup-flushing cycles [13].

Test procedure: weigh 1g of pollutant powder into self-made powder spraying unit. Take the sample plate and test the initial reflection coefficient of this film test plate at three positions, i.e., upper, middle and lower parts, with colorimeter, and calculate their average value and record it as L_1. Spray pollutant powder with the self-made spraying unit to let the power uniformly "fall" down to the test plate, place the test plate covered with pollutant into a (60±2)℃ drying oven and keep it here for 2h, then take it out and keep it still for 2h under standard test conditions, then place the test plate on test plate holder of washing device and wash it by making reference to GB/T 9780-2013 *Test Method for Dirt Pickup Resistance and Stain Removal of Film of Architectural Coatings and Paint.* Carry out this spraying-flushing process for five cycles in total, wait till the moisture on surface of test plate is evaporated, then test reflection coefficients at three positions (at upper, middle and lower part) of test plate again, calculate their average value and record it as L_2.

Result is calculated according to the following formula:

$$X = \frac{|L_1 - L_2|}{L_1} \times 100\%$$

Where,

X - the decrease scale of the reflection coefficient of film;

L_1 - the initial average reflection coefficient of film;

L_2 - the average reflection coefficient of film after dirt pickup test.

As for the films of three kinds of materials, i.e., fluorocarbon, polyester and powder, the arithmetic mean value of three test plates is taken, and the result is retained with two significant figures.

3.1. Comparison with GB/T 9780-2013 Test Method for Dirt Pickup Resistance and Stain Removal of Film of Architectural Coatings and Paint

The test is divided into two groups. Group I is dirt pickup resistance and stain removal test in accordance with GB/T 9780-2013 *Test Method for Dirt Pickup Resistance and Stain Removal of Film of Architectural Coatings and Paint*, and Group II is dirt pickup resistance and stain removal test with the aforesaid new type test method.

The test results are as shown in table 4.

Table 4. Dirt Pick Resistance and Stain Removal of Film.

Film type	Group I Decrease scale of reflection coefficient after dirt pickup test *X*	Average	Group II Decrease scale of reflection coefficient after dirt pickup test *X*	Average
Fluorocarbon	0.42 0.52	0.47	5.99 6.25	6.12
Powder	1.33 1.19	1.26	13.08 18.95	16.05
Polyester	0.24	0.225	10.63	10.92

The two groups of test results show that the metal plate with fluorocarbon film has the best dirt pickup resistance and stain removal, followed by the metal plate with polyester film, with the metal plate with powder film ranked last, thus the law is consistent and conforms to actual situation. Upon comparison of two groups of test results, the average value of decrease scale of reflection coefficient in Group I tests is less than 2, and the contrast effect of film test plates is not obvious; the results of Group II tests show certain gradient interval and are convenient for distinguishing the dirt pickup resistance and stain removal of different film test plates.

3.2. Contact angle test

The contact angle test and dirt pickup resistance and stain removal test are carried out respectively with fluorocarbon, polyester and powder film test plates, and the contact angle is as shown in table 5.

Table 5. Test Result of Contact Angle.

Film type	Contact angle θ			Average value of contact angle	Decrease scale of reflection coefficient （%）
Fluorocarbon	78.50°	74.10°	73.25°	75.28°	6.87
Polyester	76.85°	74.96°	72.24°	74.68°	10.31
Powder	71.99°	71.33°	71.92°	71.75°	16.93

4. Results and discussion

Based on the research of atmospheric particulates, a new kind of pollution sources used in powder contamination resistance test was born, adjust the corresponding way at the same time, formed a new laboratory artificial accelerated fouling resistance test method.

Through the research of atmospheric particulates, found that the increase of motor vehicle ownership makes the motor vehicle emission pollution become the major source of atmospheric pollution, organic pollution is serious, and the organic matter becomes the important component of atmospheric pollution. And then through the research of vehicle exhaust emissions of organic ingredients found that the important part of the corresponding normal alkanes was developed, according to the survey, the corresponding normal alkanes was finally selected the 24 silane as a new pollution source is the representation of the organic material, at the same time reference fouling resistance test method at home and abroad, it finally determine the composition of pollution sources, namely the loess dust (shaanxi loess), soot, and organic compounds are twenty-four alkanes. Through the component proportion of debugging, and ultimately determine the composition proportion respectively: loess dust (75%), carbon black (5%), is 24 (20%) of three kinds of pollutants in accordance with the proportion mixing produced new contamination resistant test powder with sources. Due to the accuracy and simplicity, loss on ignition used to test the uniformity of the new dust, and the experimental results F value is 0.76, less than the critical value $F_{0.05}$ (19, 20) = 2.14, which shows that the new pollution sources is qualified from uniformity.

The atmospheric particulate contamination fallout on the roof and wool space, in order to simulate the atmospheric dust fall, a homemade fallout device was tried tomake the new pollution sources "fall" on the coating plate, which formed a new stain resistance test method.

In the first experiment, two groups of test results show that the metal plate with fluorocarbon film has the best dirt pickup resistance and stain removal, followed by the metal plate with polyester film, with the metal plate with powder film ranked last, thus the law is consistent and conforms to actual situation. Upon comparison of two groups of test results, the average value of decrease scale of reflection coefficient in Group I tests is less than 2, and the contrast effect of film test plates is not obvious; the results of Group II tests show certain gradient interval and are convenient for distinguishing the dirt pickup resistance and stain removal of different film test plates.

The second, which is contact angle test results show that, fluorocarbon contact angle > polyester contact angle > powder contact angle. Based on the aforesaid sticking-wetting theory, the larger the θ value is, the lower the surface energy of film is, and the better the dirt pickup resistance and stain removal are. This shows that the fluorocarbon film has the best dirt pickup resistance and stain removal, the decrease scale of reflection coefficient of fluorocarbon film is minimum among the results obtained with new type method for dirt pickup resistance and stain removal, i.e. the dirt pickup resistance and stain removal are the best, followed by polyester film, with powder film ranked last, and this is consistent with the test result of contact angle.

5. Conclusion and outlook

Dirt pickup resistance and stain removal are important indicators for evaluating the performance of coated metal plates. The composition and proportion of pollution sources are the key factors determining the rationality of test method for dirt pickup resistance and stain removal, and meanwhile, the dirt pickup mode determines the feasibility of such test method. At present, the increase of motor vehicle population changes the composition of atmospheric pollutants, and the organic pollution becomes serious, while in the current test method for dirt pickup resistance and stain removal, the ash which is prepared mainly with graphite powder is adopted as the pollution source and cannot sufficiently embody the composition features of current atmospheric pollutants; moreover, the dirt pickup mode with contaminating fluid adopted in current test method also has certain difference from the dust fall mode of atmospheric particulates. Therefore, a rational and feasible artificial accelerated laboratory test method for dirt pickup resistance and stain removal should make comprehensive consideration in such aspects as selection of pollution source and determination of dirt pickup mode, so as to be more in line with the actual dirt pickup condition.

As for the new type test method for dirt pickup resistance and stain removal, both the composition of pollution source and the dirt pickup mode are adjusted, but the preparation of pollution source and the dirt pick process are manually carried out, the standard and uniform preparation of pollution source powder requires further improvement, and the self-made dirt pickup unit also requires further standardization; in addition, the new type test method for dirt pickup resistance and stain removal covers verification test over three kinds of usual films only, does not include any comparison with the natural solarization test, and is even lack of universal applicability, thus this new type test method for dirt pickup resistance and stain removal will be further refined and improved in the next research work to form a standard system.

Acknowledgements

This work was financial supported by the National Science and Technology Ministry (2014BAB15B02).

References

[1] G S Zheng 2004 The Contamination Resistance of Exterior Latex Paint and The Test *Shanghai Coatings*

[2] S L Yang *et al.* 2007 Fractal Characteristics of Particle Size Distribution of Inhalable

Atmospheric Particulates in Shanghai *China Environmental Science* **27(5)** 594–598
[3] Z H Wang 2007 Size Distribution and Shape Characteristics of Subaerial Airborne Particles *Environmental Science* **28(9)** 1935–1940
[4] M Hu *et al.* 2011 Source and Feature Analysis of Atmospheric Particulates in China *Environment and Sustainable Development* **5** 15–19
[5] Q Y Lang *et al.* 1997 The Analysis and Evaluation of Atmospheric Organic Pollution III. Research Report - the Monitory and Analysis of the Atmospheric Organic Pollution at the Miami Area, USA *Rock & Mineral Analysis* **16(2)** 118–127
[6] Butler J D 1979 Air Pollution Chemistry (London: Academic Press)
[7] Z P Hao 2012 Study on Relationship between Particle Size Distribution and Atmospheric Visibility Variation of Atmospheric Particulates in Shijiazhuang (Shijiazhuang: Hebei Normal University)
[8] C M Zhang, S C Lv, Z H Song and Q Wang 2014 Study on Motor Vehicle Population under Urban PM2.5 Control *Highways & Automotive Applications* **9** 30–33
[9] Y Chen and K Peng 2011 Vehicle Exhaust Pollution in Urban China and Control Measures *Guangdong Chemical Industry* **7(38)** 95–96
[10] J Liu 2012 Motor Vehicle Emission Pollution in Shanghai and Control Measures *Quality and Standardization* **6** 20–22
[11] M M Cui, X S Wang, H Su and Y H Zhang 2008 Chemical Characters and Sources Identification of PM10 in Guangzhou Area *Acta Scientiarum Naturalium Universitatis Pekinensis* **5(44)** 459–466
[12] X Y Lin 2004 Discussion on Test Method for Dirt Pickup Resistance and Stain Removal of Film of Architectural Coatings and Paint *Building Coating and Painting* **6** 17–18
[13] R J Zhang, l Shi and Y Liu 2007 High-resolution Observation and Source Analysis of Organic Carbon and Elemental Carbon in PM10 in Winter in Beijing *China Powder Science and Technology* **6** 1–8

Synthesis of geopolymer composites from a mixture of ferronickel slag and fly ash

Yun Liu [1], Kang Zhang [1], Enjuan Feng[2,3], Hongyi Zhao[2,3] and Futian Liu [1,3,*]

[1] School of Material Science and Engineering, University of Jinan, Jinan 250022, China
[2] Shandong Hongyi Technology Co., Ltd., Linyi 276034, China
[3] Key Laboratory of cement additives of Shandong province（prepare）, Linyi 276034, China

E-mail: *mse_liuft@ujn.edu.cn

Abstract. The synthesis of geopolymers using ferronickel slag and fly ash under alkaline activation was studied. In order to study the effects of different fly ash content on the mechanical properties of the geopolymers produced, the compressive strength of samples was tested at 3, 7, 28 days. The results showed that when the fly ash content was 40%, the compressive strength reached the highest (110.32MPa) at 28 days. XRD analysis showed that the ferronickel slag geopolymers had amorphous aluminosilicate phase formation, indicating that the hydration reaction occurred. FTIR analysis showed the reaction of the geopolymers generated at Si-O-T (Si, Al) and Al-O-Si three-dimensional network. In SEM images, the structure of the geopolymers with 40% fly ash was more compact and cohesive.

Keywords: ferronickel slag, fly ash, alkali activation, geopolymers

1. Introduction

Davidovits initially reported geopolymer in 1978 [1]. Geopolymer is three dimensional inorganic polymer with ceramic-like properties and amorphous microstructure [2, 3]. Geopoly-meric reaction was applied to use solid aluminosilicate materials (naturally occurring minerals, industrial by-products or waste) [4-6] to produce a wide range of revolutionary materials. Geopolymerization involves a heterogeneous chemical reaction between aluminosilicate solid materials and alkali metal silicate solutions while the geopolymeric network consists of alternately linked SiO_4 and AlO_4 tetrahedra [7,8], under highly alkaline conditions. Geopolymer possesses excellent physical, chemical and mechanical properties such as good compressive strength, outstanding chemical resistance against acids, fire and heat, high thermal resistance, comparatively low emission of greenhouse gases, low shrinkage, quick setting and hardening time and more precise mouldability. In recent decades, geopolymer has attracted worldwide attentions for their low CO_2 emissions and excellent properties.

Ferronickel slag is the solid residues generated in production of ferronickel from laterites [9]. Compared with other metallurgical slag, large slag emissions and low utilization rate have made ferronickel slag becoming a serious problem of metallurgical slag processing. Fly ash is one of the solid residues composed of the fine particles that are driven out of the boiler with flue gases in coal-fired power plants [10]. The stockpiling of fly ash not only occupies a lot of land area, but also pollutes the environment [11,12]. Therefore, new technologies have to emerge aiming at the utilization of the ferronickel slag and fly ash as raw materials for the production of added-value products. The geopolymer technology provides a new good and green solution to the utilization of the slags, avoiding their negative impact on environment and ecology. Geopolymers synthesized by ferronickel slag and fly ash not only suppresses the environmental hazards, but also improves the ecological cycles [13].

2. Experimental

2.1. Materials

The ferronickel slag used in the present study comes from Shandong Xinhai Technology Co., Ltd with specific surface area of 461m²/kg and 45μm sieve of 7.03%. The fly ash used comes from Shandong Hongyi Technology Co., Ltd with specific surface area of 547m²/kg and 45μm sieve of 5.27%. Table 1 presents the chemical composition of ferronickel slag and fly ash in the form of oxides and trace elements. The mineralogical characterizations of the slags determined with the X-ray diffraction method are shown in figure 1. Ferronickel slag mainly contains dicalcium silicate, tricalcium silicate minerals and a large amount of vitreous, and the main crystalline component of fly ash is mullite and quartz.

The alkali activator is mixtures of industrial liquid sodium silicate with the chemical composition: 26.2% SiO_2, 8.2% Na_2O and 65.6% H_2O (silica molecular modulus, Ms = 3.3) and granular sodium hydroxide.

Table 1. Chemical analysis of raw materials/ wt %.

Composition	SiO_2	Al_2O_3	Fe_2O_3	CaO	MgO	SO_3	K_2O	Na_2O	Cr_2O_3	Loss
Ferronickel slag	48.85	15.94	9.71	17.45	5.98	0.71	0.23	0.15	0.12	0.76
Fly ash	51.12	28.31	4.38	6.92	1.21	0.32	1.11	0.44	0.89	5.29

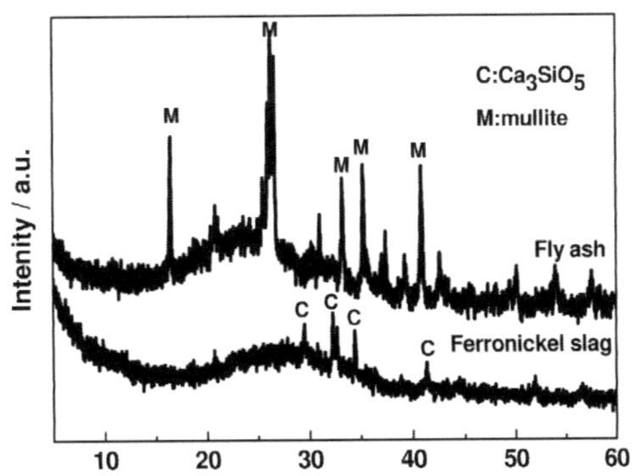

Figure 1. X-ray diffractogram of the raw materials.

2.2. Experimental procedure

2.2.1. Preparation of activator. The activator was prepared by premixing the liquid sodium silicate with sodium hydroxide 24h earlier before the preparation of geopolymer. The activator was mixed in certain proportions making the modulus fixed at 2.2.

2.2.2. *Preparation of geopolymer.* The alkaline activator content kept constant at 12% (calculated as solids content) and water/solid ratio kept constant at 0.30. Geopolymer samples were synthesized by following stages. Firstly, a homogeneous viscous paste was obtained by mixing the solid phase with the alkaline aqueous phase mechanically. The solid phase was consisted of ferronickel slag and fly ash. Then the fresh paste was poured in the cubic molds with size of 20 × 20 × 20 mm and vibrated for 1min on a vibrating table. Finally, the moulded paste was cured in a laboratory oven for 24h at temperature 20±2°C and relative humidity 95%. After curing, the geopolymer samples were de-molded and left more days at ambient conditions for further hardening.

3. Results and discussion

3.1. *Compressive strength*

The compressive strength of the geopolymer samples was tested at 3, 7 and 28 days. As shown in figure 2, the compressive strength of all the specimens grows fast as the curing period extending. Initially, with the increase of fly ash content, the compressive strength of the sample increases. When the fly ash content is 40%, the compressive strength of the geopolymer reaches the highest 110.32 MPa, which is 1.5 times of the strength of the ferronickel slag-based geopolymer without fly ash. But, when the fly ash content continues to increase, the compressive strength of the specimens begins to decrease. Fly ash with smaller particle size can fill the fine pores generated by the evaporation of water and make the structure more compact and cohesive without discontinuities. But, because of the low activity of fly ash, adding too much fly ash will reduce the compressive strength.

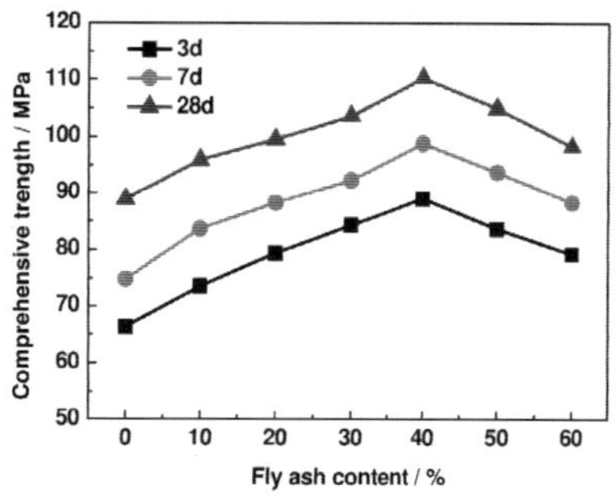

Figure 2. Compressive strength of the different fly ash content geopolymers.

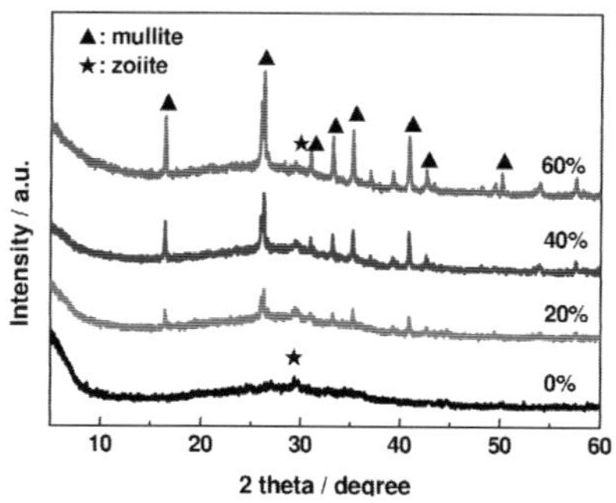

Figure 3. X-ray diffractograms of the different fly ash content geopolymers.

3.2. *XRD analysis*

In order to investigate the compositions of the reaction products, reacted pastes were analysed with X-ray diffraction (XRD) technique. Figure 3 shows the compared diffractograms for different fly ash content geopolymers. It is clearly observed that the only crystalline mineralogical phase of the geopolymer without fly ash is zoisite ($Ca_2Al_3SiO_4SiO_2 \cdot 7\ H_2O$) belonging to the zeolite substances. This peak is attributed to an aqueous aluminosilicate that was formed during the geopolymerization as the result of the polycondensation phenomena. The geopolymers added fly ash also contain mullite peaks, the crystalline mineralogical phase of the fly ash, and has not undergone any substantial change during the geopolymerization process. With the fly ash content increasing, the mullite peaks gradually become sharp.

3.3. *FTIR analysis*

The FTIR spectra of different fly ash content geopolymers is presented in figure 4. Significant broad bands of O-H stretching and O-H bending are observed in the range of 3410-3455 cm^{-1} and 1645 cm^{-1} in all FTIR spectra of geopolymers [6]. These bands indicate the presence of water molecules which

IOP Conf. Series: Materials Science and Engineering **182** (2017) 012038 doi:10.1088/1757-899X/182/1/012038

are surface absorbed or entrapped in the large cavities of the polymeric framework. The bands located between 1380 cm^{-1} and 1500 cm^{-1} are attributed to stretching vibrations of O–C–O bond [6]. These absorption bands indicate the presence of sodium bicarbonate which is suggested to due to atmospheric carbonation. The peaks located at wavenumbers lower than 1300 cm^{-1} and includes absorption bands ascribed to asymmetric stretching Si-O-Si (970-1090 cm^{-1}) [14-16], symmetric stretching Si-O-Si and Al-O-Si (500-750 cm^{-1}) vibrations [6]. All the above results proved the phase transformations that occurred during the geopolymerization.

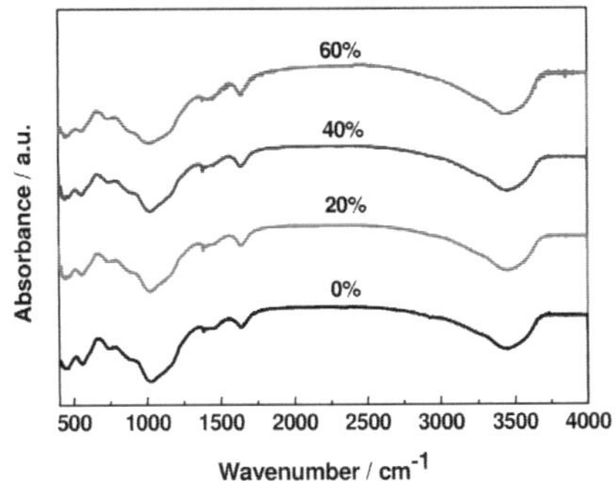

Figure 4. FTIR spectra of the different fly ash content geopolymers.

3.4. SEM observation

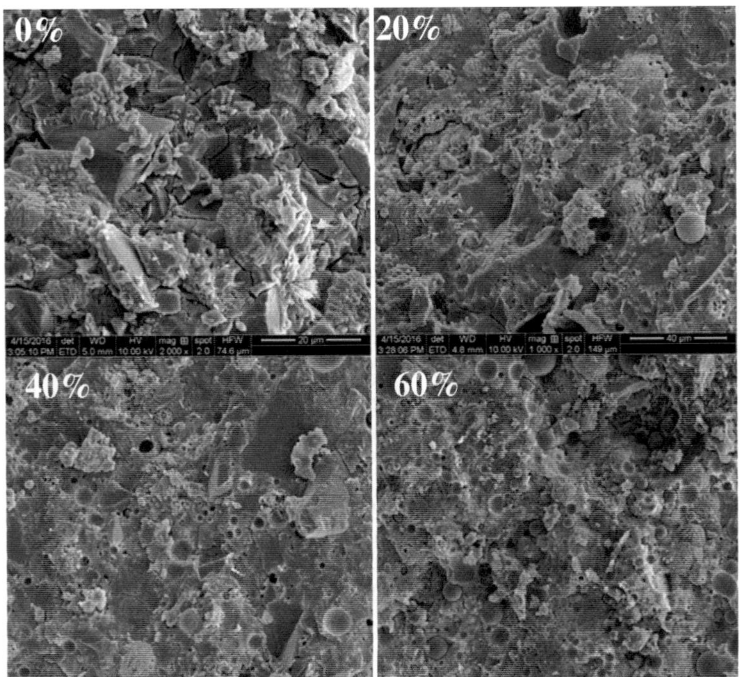

Figure 5. SEM images of the different fly ash content geopolymers.

The SEM images of the produced geopolymers are shown in figure 5. It appears that all samples are composed of a large amounts of gelatinous substance and trace particulate matter. With the fly ash content increasing from 0 to 40%, the structure of geopolymer gradually becomes more and more compact and cohesive. This is ascribed to the fact that fly ash with smaller particle size can fill the fine pores generated by evaporation of water and make the structure dense. When the fly ash content is 60%, from the figure it can be observed that the surface of samples decline in compactness and increase in unreacted particles. This is related to the fly ash with low activity makes the polymerization and hydration degree of geopolymer reduced.

4. Conclusions

The geopolymer has been prepared using ferronickel slag and fly ash as raw materials under activation of sodium silicate solution. The geopolymer technology provides a new good and green solution to the utilization of the slags. When the fly ash content was 40%, the compressive strength reaches the highest 110.32MPa. Amorphous aluminosilicate phase founded in the XRD analysis indicates that hydration also occurred in the geological polymerization reaction. The FTIR analysis shows that the Si-O-T (Si, Al) of three-dimensional network generates in the reaction of the geopolymers. In SEM microstructure images, the structure of the geopolymer with 40% fly ash is more compact and cohesive. Fly ash with small particle size can fill the fine pores generated by evaporation of water and make the structure dense.

Acknowledgement

This research has been supported by the National Natural Science Foundation of Shandong province (ZR2015EL005) and Independent Innovation and Achievement Transformation Project of Shandong Province (2014ZZCX05301). We would like to thank Mr. Futian Liu for the suggestion of this paper and Mr. Kui Li for his help with the modification of this paper.

References

[1] J Davidovits 1994 Geopolymers: man-made rocks geosynthesis and theresulting development of very early high strength cement *Mater.Educ.* **16** 91–139
[2] J Davidovits 1991 Geopolymers: Inorganic Polymeric New Materials *Journal of Thermal Analysis* **37** 1633–1656
[3] Xu H, Van Deventer and J S J 2000 The geopolymerization of alumino-silicate minerals *International Journal of Mineral Processing* **59** 247–266
[4] Cheng T W and Chiu J P 2003 Fire-resistant geopolymer produced by granulated blast furnace slag.[J] *Minerals Engineering* **16** 205–210
[5] Panias D, Giannopoulou I and Perraki T 2007 Effect of synthesis parameters on the mechanical properties of fly ash-based geopolymers *Colloids and Surfaces A: Physicochemical Engineering Aspects* **301** 246–254
[6] Palomo A, Grutzeck M W and Blanco M T 1999 Alkali-activated fly ashes: a cement for the future *Cement and Concrete Research* **29** 1323–1329
[7] Davidovits J 1998 Geopolymer chemistry and properties *Proceedings of the 1st International Conference Geopolymer '88* (Compiegne) **1** 25–48
[8] Duxson P, Fernández-Jiménez A and Provis J L 2007 Geopolymer technology: the current state of the art *Journal of Materials Science* **42** 2917–2933
[9] Kirillidi Y, and Frogoudakis E 2005 Electric arc furnace slag utilization *Proceedings of the 9th International Conference on Environmental Science and Technology* (Rhodes, Greece) 768–772
[10] Styzko-Grochowiak K, G a J and Jankowski H 2004 Characterization of the coal fly ash for the purpose of improvement industrial-line measurement of undurned carbon content *Journal of Fuel* **83** 1847–1853
[11] Mishra S B, Langwenya S P and Mamba B B 2010 Study on urface morphology and physicochemical properties of raw and activated South African col and col fly ash *Joournal of Physics and Chemistry of the Earth* **35** 811–814

[12] Temuujin J, William R P and Rieen A 2009 Effect of mechanical activation of fly ash on the properties of geopolymer cured at ambient temperature *Journal of Materials Procesing Technology* **209** 5276 –5280
[13] Z Yao, X Ji, P K Sarker, J Tang and L Ge 2015 A comprehensive review on the applications of coal fly ash *Earth-Sci. Rev* **141** 105–121
[14] Lee W K W and Van Deventer J S J 2002 Structural reorganization of class F fly ash in alkaline silicate solutions *Colloids and Surfaces A* **211** 49–66
[15] Fernandez-Jimenez A and Palomo A 2005 Composition and microstructure of alkali activated fly ash binder: effect of the activator *Cement and Concrete Research* **35** 1984–1992
[16] Fernandez-Jimenez A and Palomo A 2005 Mid-infrared spectroscopic studies of alkali activated fly ash structure *Microporous and Mesoporous Materials* **86** 207–214

Preparation and characterization of nano bismuth titanate powders with high reflectivity in near-infrared waveband

Y H Jia[1], Y B Wu[1], Y Huang[1, 2], Z H Zhou[1, 2, *] and S Shen[3]

[1] College of Materials, Xiamen University, Xiamen, 361005, China
[2] Fujian Key Laboratory of Advanced Materials, Xiamen, Fujian, 361005, China
[3] CSIRO Manufacturing, Clayton, VIC 3168, Australia

E-mail: *zzh@xmu.edu.cn

Abstract. Nano powders with high reflectivity in near-infrared waveband have broad applications as thermal insulation materials. In this study, nano bismuth titanate powders with the reflectance to near-infrared of as high as 89.5% in average were prepared via a sol-gel method by using tetrabutyl titanate, bismuth nitrate as raw materials and citric acid, acetic acid as reaction adjuvant reagents. Furthermore, to control the reflectivity in the visible light waveband, the as-prepared nano bismuth titanate powders were further coated with nano-Ag by using $NaBH_4$ as a reduction agent. The influence of different dispersants on reflectivity and on powder dispersibility has also been studied. SEM characterization demonstrates that PEG1000, worked as a dispersant, significantly enhances the dispersion of bismuth titanate powders comparing with non-dispersant system. UV-Vis-NIR spectra reveal that with addition amount of $AgNO_3$ of 1.5 ml and PEG1000 as the dispersant, the Ag-coated bismuth titanate nano powders can reach about 60% of reflectance to near-infrared, while the reflectance of visible light can be controlled as low as around 14%. It is very promising for such nano powders to be used in thermal insulation glass materials.

Keywords: nano bismuth titanate powders, nano-Ag, dispersant, PEG1000

1. Introduction

Inorganic heat–reflective nano powders are hot–spot topics in studies on energy conservation and emission reduction of both automotive industry and construction industry. Among them, TiO_2, a white pigment with a high solar reflectance of about 87%, is currently regarded as the best pigment for thermal insulation materials [1]. However, the very high reflectivity of TiO_2 in visible band (as high as 68%), which leads to light pollution [2], is the technological roadblock for thermal insulation applications in glass materials. Therefore, the exploration of preparation technology of inorganic heat-reflective nano powders and application with a high reflectance in near-infrared while a low reflectance in visible light is still a challenge and a need.

Bismuth titanate ($BiTi_2O_7$, light yellow powder) can be prepared through a co-precipitation method [3], a hydrothermal method [4], an inverse-emulsion method[5] and so on. However, the investigation of solar radiation properties of $BiTi_2O_7$ nano powder has rarely been reported. In the present work, nano bismuth titanate powders with reflectivity to near-infrared of as high as 90 % in average have been prepared via a sol-gel method. Furthermore, the as-prepared nano bismuth titanate powders have then been coated with nano Ag by using $NaBH_4$ as a reduction agent, which leads to a decrease of reflectivity to visible light from 78% to 14%. It is a new approach of controlling the reflectivity in visible band of heat-reflective inorganic nano powders, particularly for thermal insulation glass materials.

2. Experimental

2.1. Preparation of Bismuth Titanate Nano Powders

The chemical reagents used were tetra-n-butyl titanate ($Ti(C_4H_9O)_4$), bismuth nitrate($Bi(NO_3)_3 \cdot 5H_2O$), nickel nitrate($Ni(NO_3)_2 \cdot 6H_2O$), acetic acid (CH_3COOH), ethanol (C_2H_5OH), citric acid ($C_6H_8O_7$), silver nitrate ($AgNO_3$), polyethylene glycol (PEG1000), polyvinyl ketone (PVP), sodium boron hydride($NaBH_4$), and distilled water (H_2O). All reagents were analytical grade and used without further purification.

First, bismuth nitrate (0.01 mol) was dissolved in a mixed solvent (acetic acid : ethanol = 1 : 1 wt.%) by stirring and tetra-n-butyl titanate (0.01 mol) was added consequently and stirred for 30 min. Citric acid (0.01mol) was, then, added and stirred for 4h. The sol was dried at 80 °C after cured for 24h. Finally, $BiTi_2O_7$ nano particles were obtained by grinding the dried precursor and then calcination in air at 540 °C for 3h.

$BiTi_2O_7$ nano particles with Ag-coated were prepared as follow. The as-prepared $BiTi_2O_7$ was dissolved in 30g mixed solvent (dispersant: ethanol = 0.08 : 99.92 wt ratio , PEG1000 and PVP used as dispersants, respectively). The mixed solution was put in a 100 ml polytetrafluoroethylene ball milling tank, and milled for h at a rotating speed of 400 r/min. 1.5 ml $AgNO_3$ solution (2 wt.%) was added to the above solution and stirred for hours. Then, 10ml $NaBH_4$ solution (2 wt.%) was also added and stirred for 5h. Finally, after centrifugation and washing for several times, the powders were dried at 80 °C.

2.2. Characterization

Bruker AXS (D8 advance) X-ray diffraction was used to analyze the phase structure of the samples. The XRD patterns were collected under the following conditions: Irradiated with graphite monochromatic copper (Cu, Kα) = 0.154 nm, at 40 kV and 40 mA, over a range of 20 ~ 80° with a step of 0.0167° for 0.1s. Scanning electron microscope (SEM, Hitachi SU-70, 10 kV) was used for particle size and morphology characterization. The optical properties were detected using a UV-Vis-NIR spectrophotometer (Varian, Cary 5000) with an integrating sphere. The reflection spectra were scanned in the range of 380 ~ 2500 nm with a step of 600 nm/min.

3. Results and discussion

Figure 1 shows XRD patterns: the bottom two are the standard patterns of $Bi_2Ti_2O_7$ and Ag; and $Bi_2Ti_2O_7$ samples (a) without Ag coated; and (b) with Ag coated and EG1000 as the dispersant, (c) with Ag coated and PVP as the dispersant and (d) with Ag coated without any dispersant. It can be seen that the main diffraction peak positions of these particles agree well with the reflection of $Bi_2Ti_2O_7$ phase (JCPDFS No. 320188). Since the $Bi_2Ti_2O_7$ (662) diffraction peak (2θ=38.100°) is very close with Ag (111) primary peak (2θ=38.116°), the XRD peak of Ag coating may not be determined directly. To determine the existence of Ag, $NiTiO_3$ coated with Ag was prepared by the same method with PEG1000 as a dispersant. In figure 2, the XRD pattern of (e)$NiTiO_3$ coated with Ag shows clearly the diffraction peaks corresponding to the Ag (111) plane and (200) plane, therefore the existence of Ag in the samples of $Bi_2Ti_2O_7$ with the same Ag coating method is reasonable.

Figure 3 shows the SEM images of different $Bi_2Ti_2O_7$ samples: (a) without Ag coated; and (b) with Ag coated without any dispersant, (c) with Ag coated and PVP as the dispersant and (d) with Ag coated and PEG1000 as the dispersant. The grain size of $Bi_2Ti_2O_7$ powders is 50 ~ 200 nm (figure 3 (a)). Before coated with Ag, $Bi_2Ti_2O_7$ powders have good dispersion. After coated with Ag, the particles agglomerated seriously in the case of without any dispersant (figure 3 (b)), but the dispersion is slightly improved when PVP was used as the dispersant (figure 3 (c)), and the dispersion is a lot better when PEG1000 was used as the dispersant figure 3 (d)). Both the PEG and the PVP are non-ionic dispersants. In the solution, they could form a steric hindrance what reduce the agglomeration of particles. In the results, the dispersion is better when PEG was used as the dispersant, which may be due to PEG with larger molecular weight has better steric hindrance and better dispersion because of its higher degree of polymerization [6].

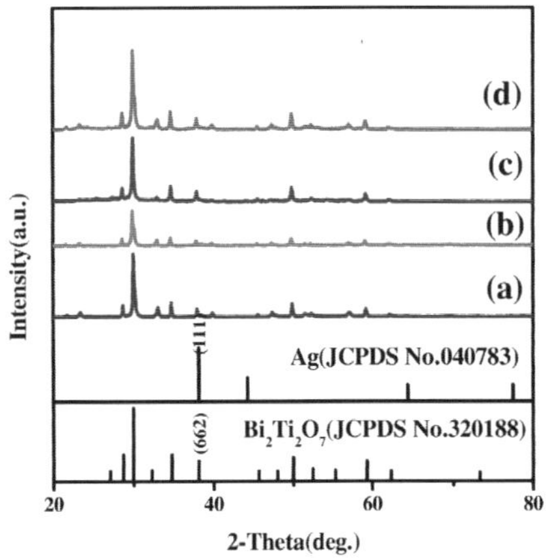

Figure 1. XRD patterns. From bottom, standard patterns of $Bi_2Ti_2O_7$ and Ag; $Bi_2Ti_2O_7$ samples (a) without Ag coated; and (b) with Ag coated and EG1000 as the dispersant, (c) with Ag coated and PVP as the dispersant and (d) with Ag coated without any dispersant.

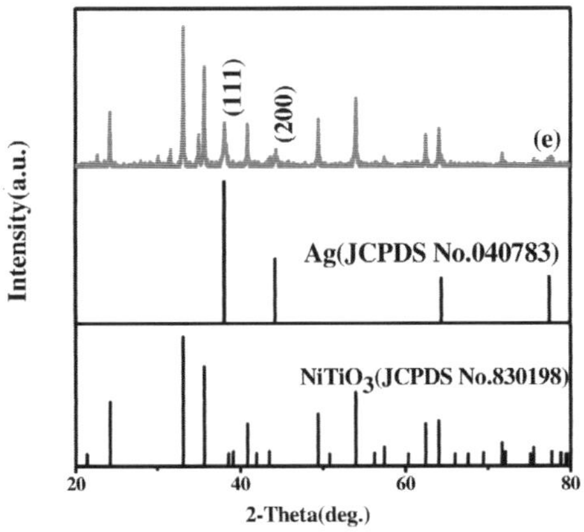

Figure 2. XRD patterns. From bottom, standard patterns of $Bi_2Ti_2O_7$ and Ag; (e)$NiTiO_3$ coated with Ag.

Figure 3. SEM of different $Bi_2Ti_2O_7$ samples: (a) without Ag coated; and (b) with Ag coated without any dispersant, (c) with Ag coated and PVP as the dispersant and (d) with Ag coated and PEG1000 as the dispersant.

Figure 4 shows Vis-NIR reflection spectra of the $Bi_2Ti_2O_7$ samples: (a) without Ag coated; and (b) with Ag coated and EG1000 as the dispersant, (c) with Ag coated and PVP as the dispersant and (d) with Ag coated without any dispersant. Table 1 presents the reflectance data of $Bi_2Ti_2O_7$ samples at range of visible light and near infrared range before and after coated with Ag. The NIR reflectance of $Bi_2Ti_2O_7$ without coated Ag reaches 89.50 %, however, its visible light reflectance (78.37 %) is also very high. After coated with Ag, the reflectance decreases at both visible light and near infrared band. Among $Bi_2Ti_2O_7$ with coated Ag, the sample without dispersant and the sample with PVP as the dispersant have similar reflectance in above two bands (figure 4 (c)and(d)); noteworthy, reflection curve of the sample with PEG1000 as the dispersant is higher in the range of 1000 ~ 2500 nm with NIR reflectance of 60.24 %, while its visible light reflectance is the lowest, only being 13.88 %.

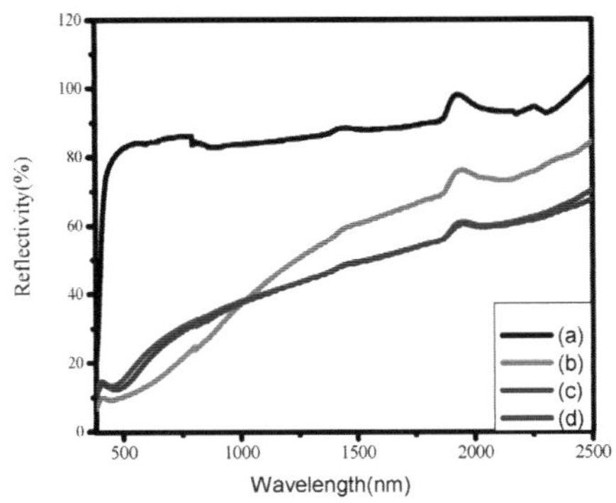

Figure 4. Vis-NIR reflection spectra of $Bi_2Ti_2O_7$ samples: (a) without Ag coated; and (b) with Ag coated and PEG1000 as the dispersant, (c) with Ag coated and PVP as the dispersant and (d) with Ag coated without any dispersant.

Table 1. Visible light and near infrared reflectance data of $Bi_2Ti_2O_7$ samples before and after coated Ag.

Reflectance(%)	Blank $Bi_2Ti_2O_7$	$Bi_2Ti_2O_7$ after Ag coated		
		Without using dispersant	Using PVP as dispersant	Using PEG1000 as dispersant
Vis	78.37	21.65	19.78	13.88
NIR	89.50	51.29	50.81	60.24

4. Summary

In this study, nano bismuth titanate powders with the reflectance of near-infrared as high as 90 % have been prepared via a sol-gel method. However, the reflectance in visible light is also very high (78 %). To decrease reflectance in visible light wave band, the as-prepared nano bismuth titanate powders have been coated with nano Ag, which leads to a significant decrease of reflectance in visible band. When PEG1000 is used as the dispersant, the Ag-coated bismuth titanate nano powders have good dispersion and their reflectance of near-infrared reaches up to 60%, while the reflectance of visible light is controlled to be merely 14%. Ag coating treatment is an effective method to keep the high infrared reflectivity of the thermal reflection nano powder materials as well as a reduced visible light reflectivity.

Acknowledgment

This work is supported by the Science and Technology Major Program (2014HZ0005) of Fujian Province, China.

References

[1] Thongkanluang T, Limsuwan P and Rakkwamsuk P 2011 Preparation and application of high

near- infrared reflective green pigment for ceramic tile roofs *Int. J. Appl. Ceram. Technol.* **8** 1451–58

[2] Lin Y C and Huang W X 2012 Light pollution in modern cities: a review *J. Sci Soc* **21** 21–23

[3] Kan Y M and Jin X H 2002 Preparation of bismuth titanate ultrafine powders by co precipitation method *J. Inorg. Mater.* **17** 51–55

[4] Lin X 2013 Controlled synthesis and photocatalytic activity of $Bi_4Ti_3O_{12}$ particles with different morphologies *J. Phys Chem.* **29** 411–17

[5] Hou J G 2009 Bismuth Titanate Materials: Synthesis, Characterization and Their Properties (Tianjin: Tianjin University)

[6] Li L 2014 Study on dispersion of ultrafine nickel powder prepared by ultrasonic and dispersant in the elephant reduction method (Kunming: Kunming University of Science and Technology)

Effects of glass fiber modified with calcium silicate hydrate (C-S-H(I)) reinforced cement

M Xin[1,2], L Zhang[1,2], S Ge[1,3] and X Cheng[1,2]

[1] School of Materials Science and Engineering, University of Jinan, Jinan 250022, China
[2] Shandong Provincial Key Laboratory of Preparation and Measurement of Building Materials, University of Jinan, Jinan 250022, China
[3] School of Chemistry and Chemical Engineering, University of Jinan, Jinan 250022, China

E-mail: ujn_chengxin@163.com

Abstract. In this paper, calcium silicate hydrate (C-S-H(I)) and glass fiber modified with C-S-H(I) (SiF) at ambient temperature were synthesized. SiF and untreated fiber (OF) were incorporated into cement paste. Phase composition of C-S-H(I), SiF and OF was characterized by XRD. The surface morphologies were characterized by SEM. Flexural performance of fiber reinforced cement (FRC) at different curing ages was investigated. Results indicated that both SiF and OF could reinforce cement paste. SiF had a more positive effect on improving the flexural performance of FRC than OF. The strength of SiF reinforced cement was 11.48MPa after 28 days curing when fiber volume was 1.0%, 12.55% higher than that of OF reinforced cement. The flexural strength increased with the addition of fiber volume. However, the large dosage of fiber might cause a decrease in flexural strength of FRC.

Keywords: glass fiber, cement, flexural strength, modification

1. Introduction

Cement structures are the most widely used construction materials in contemporary development of the society. It has been applied to industrial floors, elevated slabs, overlays, dam, bridge, road and many other fields for their distinct advantages of high compressive strength, good environmental adaptability and so on [1]. However, shortcomings including poor toughness and brittle behavior have restrained its further development, leading to a deterioration of cement structures and shortened service life [2-5]. The incorporation of fibers has great potential for improving the defects of cement structures with great success obtained up to date [6-8]. Polymer fiber [9], glass fiber [10], nature fiber and steel fiber [11] have been applied to various fields and performed well, playing a positive role on the brittle behavior and poor toughness of cement structures. However, the interfacial adhesion between fiber and matrix is weak which will affect the toughening effect of the fibers. Effective methods have been put forward to improve the interfacial adhesion, including sizing [12], coating [13] and plasma technology [14]. But out-controlled process or complex operation are the main shortcomings of above all methods.

Tobermorite, a kind of C-S-H(I) which crystallizes from the system of $CaO-SiO_2-H_2O$ [15], has been a focus for the similarity with C-S-H phase produced in the hydration of Portland cement [16-18]. The mineral could be synthesized with silicon dioxide (SiO_2) and calcium oxide (CaO) as the starting materials by hydrothermal reaction [19]. Lucie Galvankova [15] *et al.* synthesized C-S-H(I) at different reaction temperatures (170-190 °C) and the CaO/SiO_2 (Ca/Si) ratio was set as 0.83. H Youssef investigated the influence of reaction conditions including time, Ca/Si ratios and temperature on the formation of C-S-H (I) [20].

In this paper, C-S-H (I) modified glass fiber (SiF) is prepared by CaO and SiO_2 and characterized by XRD and SEM. SiF and untreated glass fiber (OF) are incorporated into cement. The effects of fiber volume and fiber types to the flexural performance of fiber reinforced cement (FRC) are studied.

2. Materials and methods

Glass fiber chopped is bought from Taishan Glass Fiber Co.,Ltd., China with the length of 4mm and one bundle has about 500 filament fibers with an average diameter of 11μm. P.O42.5 cement is produced by Landscape Cement Group Co., Ltd. CaO powder is calcined from $CaCO_3$ (A.G., Sinopharm Chemical Reagent Co. formula) and SiO_2 is bought from Alfa Aesar (BET S.A= 350~ 410 m^2/g).

2.1. *Synthesis of C-S-H (I) and Surface modification of SiF*

SiO_2, CaO and distilled water were added into a reactor in appropriate proportion and the mole ratio of Ca/Si was controlled at 0.5, 1.0 and 1.5. The reactor was shaken slightly to ensure the mixture fully mixed. Then, it was placed into an oven of 100 °C for 48h. After the reaction, the mixture was poured into a beaker and dried at 60 °C. The process of surface modified SiF was similar to that of C-S-H(I). The raw materials included glass fiber, SiO_2, CaO and distilled water. Reaction conditions were same as that of C-S-H(I). Ca/Si ratios were set as 0.5, 1.0, 1.3 and 1.5, respectively. The modified fibers were washed with distilled water to remove the unreacted calcium hydroxide (CH) and SiO_2 for 30 minutes. The washed fibers were then dried at 45 °C for 48h.

2.2. *Characterize of C-S-H(I) and SiF*

Synthesized C-S-H(I), OF and synthesized SiF were characterized by X-ray diffraction (XRD). Data was collected by X-ray diffractometer Empyrean (Pananalytical) on a D8 advance diffractometer system equipped with Cu Kα radiation (Bruker Co., Karlsruhe, Germany). The spectra of all synthesized phases were recorded in 2θ range of 10°-80° at the scanning speed of 0.3°/min. Surface morphologies of modified glass fiber were acquired by Quanta FEG 250 field emission scanning electron microscope (SEM) produced in FEI (American) whose resolution was 1.04 nm.

2.3. *Preparation and testing of FRC*

Table 1. Details of samples' information.

Curing ages (d)	Fiber volume (%)				
	0	0.50	0.75	1.00	1.25
3	A0-0	A0-1	A0-2	A0-3	A0-4
	A1-0	A1-1	A1-2	A1-3	A1-4
7	B0-0	B0-1	B0-2	B0-3	B0-4
	B1-0	B1-1	B1-2	B1-3	B1-4
28	C0-0	C0-1	C0-2	C0-3	C0-4
	C1-0	C1-1	C1-2	C1-3	C1-4

The water to cement ratio (w/c) of cement paste was 0.4 and fiber volume varied from 0.5% to 1.25%, respectively. Details of the samples' information (denoted as Xy-n) were listed in table 1. In table 1, X in Xy-n meant curing ages (A: 3d; B: 7d; C: 28d), y meant fiber type (0: OF; 1: SiF) and n signified fiber volume (1: 0.5%; 2: 0.75%; 3: 1.0%; 4: 1.25%). The size of FRC samples for flexural performance was 80 mm × 20 mm × 20 mm. Water, cement and fiber were weighted, mixed, molded and demoulded. Samples were cured for 3 days, 7 days, 28 days at the relative humidity of 95% and the curing temperature of 20±2 ℃. The flexural performance was tested referenced to GB/T 7897.2-1987 by MTS (370 Load Frame, MTS Systems Corporation 14000 Technology Drive Eden Prairie. MN. 55344). The loading rate was 0.2 mm/min.

3. Results and discussion

3.1. XRD analysis of C-S-H(I), OF and SiF

Figure 1 showed the XRD spectra of C-S-H(I) with different Ca/Si ratios. As was shown, there was only one intense peak in all samples at 2θ of 29.6°, indicating the formation of C-S-H(I) [20]. Also, the product of the hydrothermal reaction was C-S-H (I) in no matter with Ca/Si ratios. Figure 2 described the XRD spectra of OF and SiF. It has been observed that there only exists one broad peak in OF with a range of 10°~ 37°, amplifying its amorphous feature. As was known, glass was a kind of uncrystalline material and the results in figure 2 just reflected the amorphous properties of OF. A similar peak occurred in SiF, but a sharp peak was clearly visible at the position of 29.6° which is the same as C-S-H(I), indicating that the phase of C-S-H(I) could be synthesized on the surface of glass fiber referenced to figure 1.

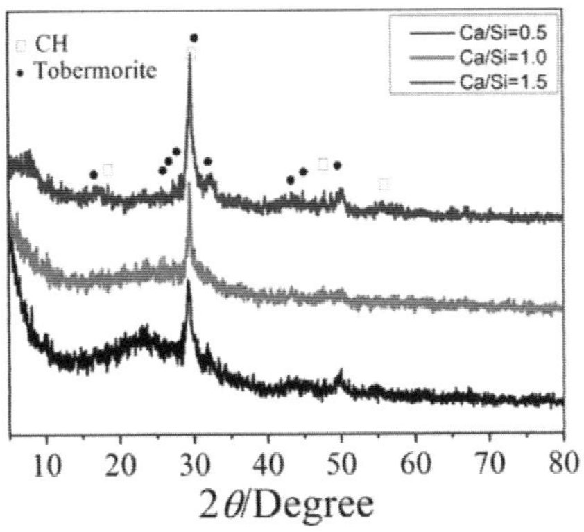

Figure 1. XRD spectra of C-S-H(I) synthesized with different Ca/Si ratios at 100 °C.

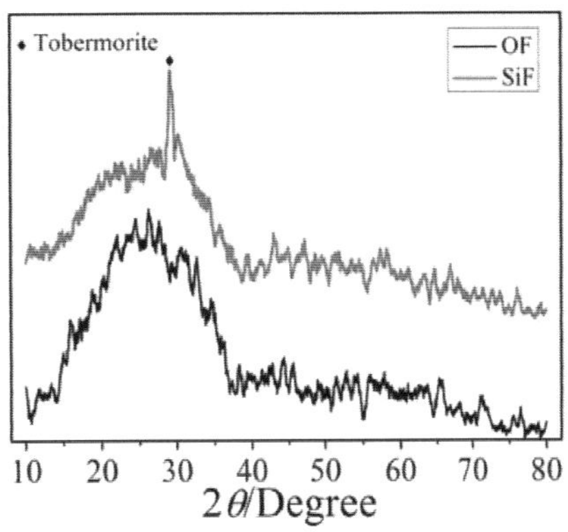

Figure 2. XRD spectra of OF and SiF.

3.2. EDS analysis of OF and SiF

Figure 3 showed the surface morphologies and EDS analysis of OF and SiF. The surface of OF was smooth (figure 3(a)) and SiF was rough (figure 3(b)). The element of OF were consistent with SiF. Most notably, the element of Si had the most percentage and other element with lower percentage including Na, O and Ca (as shown in figure 3(c) and figure 3(d)). The atomic percentages of Si in OF and SiF were 20.02% and 32.54% and for the element of Ca, they were 1.89% and 5.04%, respectively. The atomic percentage of Si and Ca from SiF was higher compared with OF, indicating the formation of C-S-H(I) product on the surface of SiF as well.

3.3. Optimum conditions of SiF

Figure 4 showed the surface morphologies of modified glass fiber with C-S-H(I) under varied Ca/Si. As we could see, when the ratio of Ca/Si was 0.5, no significant change was visible on the surface of glass fiber (as was shown in figure 4(a)). With the increase of the ratio of Ca/Si, more C-S-H(I) was synthesized. In figure 4(c), a visible but inhomogeneous nano material was synthesized on the surface of glass fiber and when the ratio of Ca/Si was 1.5. Coming to figure 4(d), a layer of homogeneous and compact C-S-H(I) with the size of 350~400 nm was synthesized, indicating the method of C-S-H(I) modified glass fiber was feasible and the optimum ratio of Ca/Si was selected to be 1.5 herein.

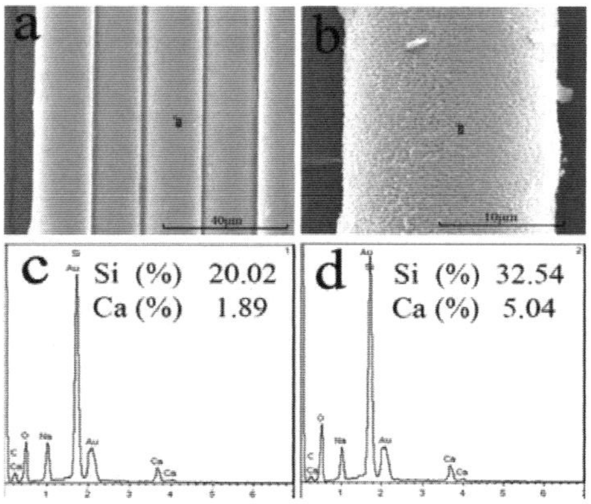

Figure 3. EDS analysis of OF and SiF (a, c: OF; b, d: SiF).

Figure 4. Surface morphology of glass fiber modified with C-S-H(I) (a: Ca/Si=0.5, b: Ca/Si=1.0, c: Ca/Si=1.3, d: Ca/Si=1.5).

3.4. Flexural performance of FRC

Figure 5 showed the flexural strength of FRC at different curing ages. Basically, the incorporation of fiber could improve the flexural strength of FRC. Flexural strength increased with the increase of fiber volume and SiF had more advantage than OF. It could also be seen that excessive fiber may decrease the flexural strength of FRC. As was shown in figure 5(b), the flexural strength of B1-4 was 9.81 MPa, 28.24 % higher than that of B1-0 (7.65 MPa) and the flexural performance of B0-4 improved 21.3 % than that of B0-0.This may be attributed to the fiber reinforcing effect. FRC cracked under the action of load and propagates. Fiber who distributes randomly in brittle matrix created a three dimensional network. The fiber bridged the cracks and it must overcome the interfacial friction and interfacial bond between fiber and brittle matrix in the process of propagation. However, a negative effect may generate for a larger dosage of fiber and lead to a decrease of flexural strength of FRC. As was shown figure 5(c) that flexural strength of FRC increased with the addition of fiber volume until 1 % and then, a sharply decrease of flexural strength was visible when the incorporated fiber was 1.25 %. The strength of C1-4 and C0-4 declined down to 10.8 % and 9.42 % than that of C1-3 and C0-3, respectively. Such may be related to the poor dispersion of glass fiber in cement paste. More incorporated fibers implied a reduction in relative water consumption and led to inhomogeneous distribution of fiber even local agglomeration. Both SiF and OF had a positive effect on reinforcing cement no matter with the curing age, however, SiF had more advantage than OF.As was shown in figure 5(a), the flexural strength of A1-3 was 7.9 MPa, 9.6 % higher than A0-3.

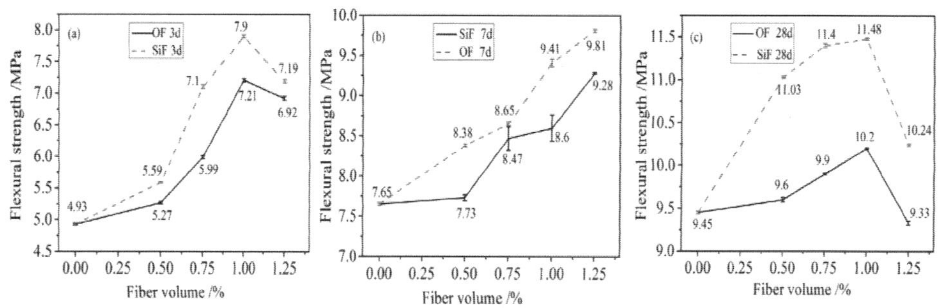

Figure 5. Flexure strength of FRC at different curing ages.

4. Conclusion

Synthesis of C-S-H(I) on the surface of glass fiber was successful and effective. The method was simple, and the complex experimental conditions and precision instruments were not required. After the modification, a homogeneous and compact layer of nano material was obtained on the surface of glass fiber when the Ca/Si ratio was 1.5. The results indicate that both SiF and OF can improve the flexural strength of FRC. To a certain extent, large dosage of fiber results in high flexural strength of FRC. However, it may also cause the decrease of flexural strength for a poor dispersion and local agglomeration. Compared to OF, SiF has a more excellent effect on the enhancement of the flexural strength of FRC, evidencing that the modification of glass fiber with C-S-H(I) is an effective manner to enhance mechanical properties. It is anticipated that such modified glass fibers can be a good candidate to promote the development of high-performance FRC in future.

Acknowledgments

This work was financially supported by the National High Technology Research and Development Program of China (2015AA034701), the National Natural Science Foundation of China (51502112), Natural Science Foundation of Shandong Province, China (ZR2014EMP010) and Key Research and Development Plan of Shandong Province (2016GGX102035).

References

[1] Kim J, Kim D J, Park S H and Zi G 2015 *J Coms. Struct* **122** 198–208
[2] Sayyar M, Soroushian P, Sadiq M M, Balachandra A and Lu J 2013 *J Constr Build Mater* **44** 458–463
[3] Beglarigale A and Yazıcı H 2015 *J Constr Build Mater* **75** 255–265
[4] He Q, Liu C and Yu X 2015 *J Constr Build Mater* **79** 311–317
[5] Khan MI and Abbass W 2016 *J Constr Build Mater* **125** 927–935
[6] Brandt A M 2008 *J Coms. Struct* **86** 3–9
[7] Sukontasukkul P and Jamsawang P 2012 *J Constr Build Mater* **29** 201–205
[8] Yoo D Y, Kang S T, Banthia N and Yoon Y S 2015 *J Int J Damage Mech* **12** 86–92
[9] Shafiq N, Ayub T and Khan S U 2016 *J Coms. Struct* **153** 30–41
[10] Enfedaque A, Gálvez JC and Suárez F 2015 *J Constr Build Mater* **754** 72–87
[11] Wang J Y, Chia K S, Liew J Y R and Zhang M H 2013 *J Cement Concrete Comp* **43** 39–47
[12] Scheffler C, Gao S L, Plonka R, Mäder E, Hempel S and Butler M 2009 *J Compos Sci Technol* **69** 905–912
[13] Gao S, Mader E and Plonka R 2008 *J Compos Sci Technol* **68** 2892–2901
[14] Bozaci E, Sever K, Demir A, Seki Y, Sarikanat M and Ozdogan E 2010 *J Fiber Polym* **10** 781–786
[15] Galvankova L，Masilko J,Solny T and St Cpankova E 2016 *J Procedia Eng* **151** 100–107
[16] Taylor H F W 1993 *J Cem. Based Mater* **1** 38–46
[17] Tunega D and Zaoui A 2012 *J Comput. Chem* **32** 306–314
[18] Grangeon S, Claret F, Lerouge C, Warmont F, Sato T, Anraku S, Numako Ch, Linard Y and Lanson B 2013 *J Cement Concrete Res* **52** 31–37
[19] Pei L Z, Yang L J, Yang Y, Fan C G , Yin W Y, Chen J and Zhang Q F 2010 *J Mater Charact* **61** 1281–1285
[20] Youssef H, Ibrahim D, Komarneni S and Mackenzie K J D 2010 *J Ceram Int* **36** 203–209

Ice melting properties of steel slag asphalt concrete with microwave heating

Bin Li[1], Yihan Sun[1], Quantao Liu[1,*], Hao Fang [1], Shaopeng Wu[1], Jin Tang[1] and Qunshan Ye[2]

1 State Key Laboratory of Silicate Materials for Architectures, Wuhan University of Technology, Wuhan 430070, China
2 Key Laboratory of Road Structure and Material of Ministry of Transport (Changsha), Changsha University of Science & Technology, Changsha, China.

E-mail: *liuqt@whut.edu.cn

Abstract. The ice on the surface of asphalt pavement in winter significantly influences the road transportation safety. This paper aims at the improvement of the ice melting efficiency on the surface of asphalt pavement. The steel slag asphalt concrete was prepared and the high ice melting efficiency was achieved with the microwave heating. A series of experiments were conducted to evaluate the ice melting performance of steel slag asphalt concrete, including the heating test, ice melting test, thermal conductivity test and so on. The results indicated that the microwave heating of steel slag concrete can improve the efficiency of deicing, mainly because the heating rates of steel slag asphalt mixture are much better than traditional limestone asphalt mixture. According to different thickness lever of ice, the final temperatures of each sample were very close to each other at the end of melting test. It is believed the thickness of the ice has a limited impact on the ice melting efficiency. According to the heating tests results, the bonding of ice and asphalt concrete is defined failure at the moment when the surface temperature of the ice reached 3 °C.

Keywords: ice melting, steel slag, asphalt concrete, microwave heating

1. Introduction

As is known, the ice on the surface of asphalt pavement in winter significantly influenced the road transportation safety. Pervious investigations indicated that 25%-30%of the traffic accidents were caused by the ice on the pavement surface in the cold winters, especially on some sections like bridges and railroad [1]. Several kinds of deicing methods were investigated and recommended, including chemical method, mechanical method, Joule heating method, hydronic method and so on[2]. Nowadays, the chemical method and mechanical method were widely used, such as chemical salt and gritty materials. In China, the annual amount of deicing salt was 100 thousand tons on average and seemed to be increasing in recent years [3].However, chemical salt usually leads to the corrosion of steel and pollution of water [4], while some mechanical methods would surely lead to the surface damages and high maintenance cost for mechanical devices. Another two methods were also popular for removing ice. One is the Joule heating method by electrically conductive asphalt concrete [5]. The other is the hydronic heating with pipe systems interred in the pavement to induce heating [6]. Both the experiences of those two systems confirmed that it was feasible and controllable to melt the ice with heating. But the applications of those two methods were limited due to the practical conditions.

The successful heating of Joule heating method depends on the conductive materials inside the asphalt pavement, and the hydronic heating required the pervious building of pipe systems.

On the other hand, self-healing technology, as a kind of advanced maintenance method of asphalt pavement, has been world-wide researched and applied in recent year [7].With the temperature increase of asphalt pavement, the asphalt binder will flow autonomously, and the crack would be cured [8-9]. Compared with self-healing method, the deicing technique shared the same temperature control mechanism and performance. However, few researches had been conducted on this field.

The purpose of this paper is to investigate the ice melting performance of self-healing asphalt concrete with microwave heating. Both the ice melting efficiency and properties of the deicing were characterized. A series of steel slag asphalt concrete were prepared, and microwave heating technology was applied to study their ice melting efficiency.

2. Materials

Both limestone and steel slag were used as aggregates in this research. The coarse aggregate of limestone was provided by Keyuan Stone Material Factory from Inner Mongolia Province. The coarse steel slag aggregate was provided by Wuhan Iron and Steel (Group) Company from Hubei Province. The properties of the aggregates were shown in table1. AH-70# base bitumen was also used in this research, which provided by SK Energy Co. Ltd. of South Korea. The properties of the bitumen were shown in table 2.

Table 1. Properties of limestone aggregate and steel slag aggregate.

Tests	Limestone aggregate	steel slag aggregate	Specification	Test method
Water absorption%	0.6	1.8	≤3.0	T0304-2005
Crushed value%	13.2	14.5	≤28	T0316-2005
Los Angeles abrasion value%	15.6	12.9	≤30	T0317-2005
Apparent specific gravity %	2.901	3.413	≥2.50	T0308-2005
Acicular and flat particle%	14.4	11.3	≤18	T0312-2005

Table 2. Properties of AH-70 asphalt.

Tests	Results	Specification	Test method
Penetration (25°C , 100g, 5s) [0.1 mm]	68	60-80	T 0604-2011
Softening point (ring & ball) [°C]	49	44-54	T0605-2011
Ductility (5cm/min, 15°C) [cm]	>150	≥100	T0606-2011

3. Experiments

3.1. Temperature Raising Characteristic of Mixture

Marshall Specimens of limestone and steel slag asphalt mixtures were prepared in this stage. Temperature raising performances of asphalt mixtures were measured during the heating process. Each Marshall specimen was placed in the microwave oven for heating 20s per time and rest for 10s. Both the flank surface and internal temperatures were measured by an infrared gun and an optical fiber temperature sensor respectively between each two heating periods. The temperature of flank surface were taken in three equally space directions and the average value was defined as the final result. The internal temperatures of optical fiber temperature sensor were taken until the data was stably. This process was repeated six times, and the total heating period reached 120s for each time.

3.2. Thermal Conductivity Tests

In this study, the thermal conductivity of three different kinds of mixtures was tested, including limestone asphalt mixtures, steel slag asphalt mixtures and steel slag asphalt mixtures with moisture. A thermal constant analyzer was applied in this test. The thermal conductivity, thermal diffusivity and specific heat of samples were recorded.

3.3. Ice Melting Experiments

Different thicknesses of ice with Marshall Specimens were prepared (seen in figure 1). The thicknesses of ice were 1.5mm, 3.5mm and 5.5mm respectively. A microwave oven with output power of 700W was applied as the microwave heating equipment. The thickness and mass of the ice was measured before the experiment. Each sample was heating for 30s and rest for 30s alternately. The temperatures of ice surface and Marshall Samples were measured with an infrared gun during each rest period (seen in figure 2). The temperatures were measured in three equally spaced directions. Then the sample was inverted to test if the ice could peel off from the Marshall Sample under its own gravity (seen in figure 3).When the ice peeled off the Marshall Sample, the heating time was recorded as the final melting time. In this study, ice melting experiments were conducted at room temperature of 9 ℃.

Figure 1. Ice on Marshall Specimens.

Figure 2. Heating in microwave oven.

Figure 3. The sample was inverted.

4. Results and Discussion

4.1. Heating Rate of Mixtures

The temperature performance of limestone and steel slag asphalt mixtures was shown in figure 4.

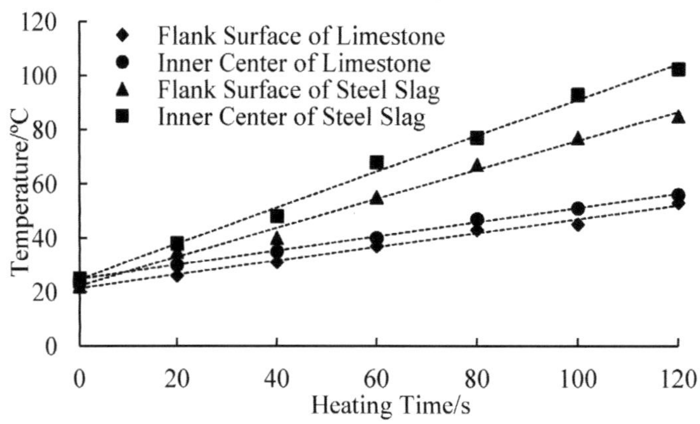

Figure 4. Heating rate of asphalt mixtures.

Microwave is a kind of wave with frequency in the range from 0.3GHz to300 GHz, corresponding to the wavelength of 1 mm to 1 m of electromagnetic waves[10].When the microwave effect on most waves absorbed materials, like metal oxides, semiconductors and polar substances, it can be absorbed, and thus converted into heat to increases the temperature of objects. For non-magnetic materials, the absorbing capacity related to its dissipation factor tan δ_E of complex permittivity tan δ [11]. As for the magnetic substances, its absorbing capacity is also related to its loss factor tan $\delta_M$of complex permeability. Because of the loss factor tan δ_E and tan δ_M (non-magnetic material can be considered as 0) of each substance is not the same, so the wave absorbed capacity of each kind of material is changed. In general, the greater the loss factor tan δ, the stronger absorbing ability of a substance achieved, and more heat was generated as a consequence. From some previous researches [12-13], it is indicated that the microwave would be a good technique to heat asphalt mixtures. The result of this test also agreed with this conclusion.

From figure 4, it is clear that both the steel slag and limestone could be heated by microwave, and the steel slag asphalt mixture is more effective than traditional limestone asphalt mixture. After six times of discontinuous heating, the inner center temperature of steel slag asphalt mixture could reach 102.5 °C, which is approximately two times faster than the heating rate of limestone surface. The flank surface temperature of those two kinds of asphalt mixtures has the same trend. One thing should be pointed out that, the inner center temperature was always higher than the flank surface temperature of each sample. This phenomenon indicated that more heat was absorbed in the middle of the sample, which means that during microwave heating asphalt pavement process. So in practical applications, the inner center temperature should be taken into consideration to avoid the overheating in the center of asphalt pavement while melting process.

4.2. Thermal Conductivity

The result of thermal test of asphalt mixtures were shown in table 3. As is known, the steel slag contains some parts of metallic elements. Some small holes also existed in the steel slag aggregate. Both the metallic elements and small holes may have some influence on the performance of asphalt mixtures thermal conductivity in this test.

Table 3. Results of thermal conductivity test.

Types	Limestone Asphalt Mixtures	Steel Slag Asphalt Mixtures	Steel Slag Asphalt Mixtures After Water Damage
Thermal Conductivity/[W/mK]	1.583	1.421	1.514
Thermal Diffusivity/[mm^2/s]	1.758	1.213	1.149
Specific Heat/[MJ/m^3K]	0.900	1.171	1.318

From table 3, it is clear that the thermal conductivity was slightly decreased from 1.583W/mK to1.421W/mK. The thermal diffusivity was also decreased by about 0.5mm^2/s. As a consequence, the specific heat was increased from 0.900MJ/m^3K to 1.171 MJ/m^3K. The reasonable explain might be small holes play a more important role in thermal conductivity than metallic elements. As the amount of decrease was limited, and the speed of microwave heating doesn't depend on the thermal conductivity of materials, it is believed that this difference between steel slag asphalt mixture and limestone asphalt mixture will not influence the heating rate of asphalt mixtures seriously.

It was found that thermal conductivity of steel slag concrete increased from 1.421W/mK to 1.514W/mK as the moisture was involved in the thermal properties test. The interpreted might be water contribute to thermal conductivity by reason of water filled in holes. With the water replaced holes, thermal diffusivity decreased from 1.213mm^2/s to 1.149mm^2/s and specific heat raised by 0.147 MJ/m^3K. It is indicated that the increased moisture will contribute to the heating of the asphalt pavement and improve the efficiency of the ice melting.

4.3. Analysis of Ice Melting Efficiency

The temperature raising characteristic of mixture with ice was also measured in this paper. The surface temperatures of ice with limestone mixture and steel slag mixture were shown in figure 5.

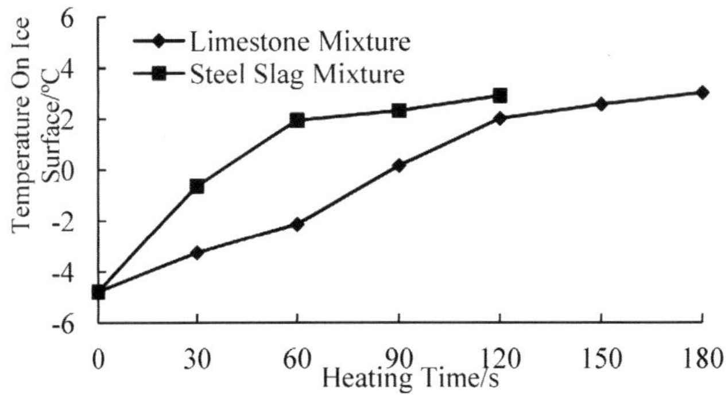

Figure 5. Ice melting of Limestone and Steel Slag with same thickness.

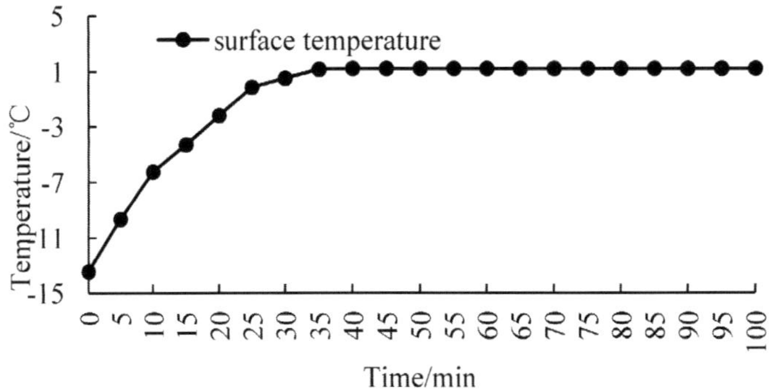

Figure 6. Temperature on Ice surface at natural condition.

In figure 5, the samples of limestone mixture and steel slag mixture have the same thickness of 3.5mm. From the figure 5, it is obvious that the ice-melting heating rate of steel slag asphalt mixture was much higher than the limestone. After four times of discontinuous heating, the increase rate of temperature of ice on steel slag asphalt mixture could reach 3.8°C/min, and the heating rate of limestone surface was only2.6℃/min. Besides, the limestone asphalt mixtures cost six times of discontinuous heating to melting. The ice melting with steel slag asphalt mixture was about one minute quicker than the limestone asphalt mixture. This result shown that, steel slag asphalt concrete had better heating efficiency and ice melting with microwave irradiation compared with traditional concrete.

Figure 6 reflects the ice melting performance of natural melting conditions. It took about 100 minutes for ice to peel off the Marshall sample with its own gravity. However, as shown in figure 5, the microwave heating only needs around 2mins to separate the ice from the Marshall sample. It can be concluded that use the microwave heating can significantly improve the ice melting efficiency compared with natural melting conditions.

As can be seen from figure 7, the ice melting performance of steel slag asphalt mixtures with different thickness was evaluated. The thickness of ice was 1.5mm, 3.5mm and 5.5mm respectively. The final temperatures of each sample were very close (near 3 °C) to each other at the end of melting process. It is concluded that the thickness of the ice have a limited impact on the ice melting efficiency. The temperature of 3 °C could be used to determine the finish timing of melting.

It could be explained that the heat transfer between the asphalt mixture and the ice would become a state of dynamic equilibrium when the temperature of the interface of those two materials reached at 3℃. The amount of heat absorbed by the ice is equal to the heat absorbed by the concrete with microwave. This dynamic process ended until interface-bound ice completely melted into water and the ice layer peeled away from the concrete surface.

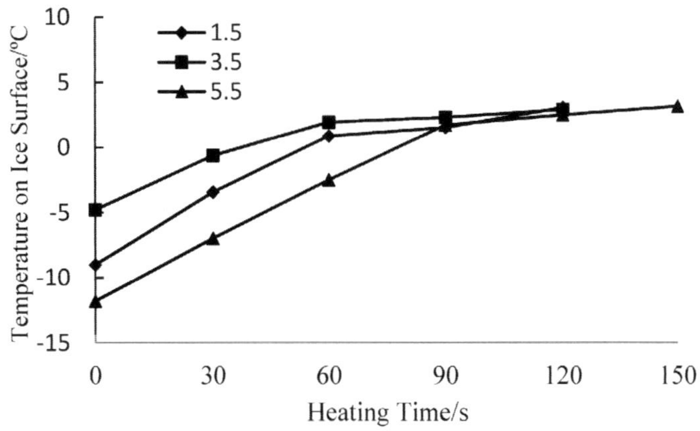

Figure 7. Temperature on Ice Surface.

5. Conclusions

- The application of microwave irradiation heating could significantly improve the ice melting efficiency compared with nature conditions.
- Compared with traditional asphalt concrete, steel slag asphalt concrete has better heating efficiency. The heating rates of asphalt mixture are much better than traditional limestone asphalt mixture. Thermal conductivity of steel slag concrete is slightly decreased as there existed some small holes with cold air in steel slag, but the influence is limited.
- The thickness of the ice has a limited impact on the ice melting efficiency. The temperature of 3 °C could be used to determine the finish timing of heating.
- In practical applications, the inner center temperature should be taken into consideration to avoid the overheating in the center of asphalt pavement while melting process.

Acknowledgements
The project was supported by the Natural Science Foundation of China (No. 51508433), the International Science & Technology Cooperation Program of China (No. 2013DFE83100), the Open Fund of Key Laboratory of Road Structure and Material of Ministry of Transport (Changsha University of Science & Technology), the Doctoral Fund of Ministry of Education of China (No. 20130143120008), and the Scientific Research Foundation for Returned Overseas Chinese Scholars, State Education Ministry.

References

[1] Kim M S, Jang D, Hong J and Kim T 2015 Thermal modeling of railroad with installed snow melting system *Cold Reg Sci Technol* **109** 18–27
[2] Blackburn R R 2004 Snow and ice control :guidelines for materials and methods (Washington, D.C.: Transportation Research Board)
[3] Wang H, Zhao J and Chen Z 2008 Experimental investigation of ice and snow melting process on pavement utilizing geothermal tail water *Energ Convers Manage* **49** 1538–1546
[4] Ramakrishna D M and Viraraghavan T 2005 Environmental impact of chemical deicers - A review *Water Air Soil Poll* **166** 49–63
[5] Pan P, Wu S, Xiao Y and Liu G 2015 A review on hydronic asphalt pavement for energy harvesting and snow melting *Renew Sust Energ Rev* **48** 624–634
[6] Chen M, Wu S, Wang H and Zhang J 2011 Study of ice and snow melting process on

conductive asphalt solar collector *Sol Energ Mat Sol C* **95** 3241–3250
[7] Ayar P, Moreno-Navarro F and Rubio-Gámez M C 2015 The healing capability of asphalt pavements: a state of the art review *J Clean Prod*
[8] Liu Q, Schlangen E and van de Ven M 2013 Characterization of the material from the induction healing porous asphalt concrete trial section *Mater Struct* **46** 831–839
[9] Garcia A, Schlangen E, van de Ven M and Liu Q 2009 Electrical conductivity of asphalt mortar containing conductive fibers and fillers *Constr Build Mater* **23** 3175–181
[10] Mutyala S, Fairbridge C, Pare J R J, Belanger J M R, Ng S and Hawkins R 2010 Microwave applications to oil sands and petroleum: A review *Fuel Process Technol* **91** 127–135
[11] Mishra R R and Sharma A K 2016 Microwave - material interaction phenomena: Heating mechanisms, challenges and opportunities in material processing *Composites Part A: Applied Science and Manufacturing* **81** 78–97
[12] Sun Y H, Liu Q T, Wu S P and Shang F 2014 Microwave Heating of Steel Slag Asphalt Mixture *Key Engineering Materials* **599** 193–197
[13] Gallego J, Del Val M A, Contreras V and Paez A 2013 Heating asphalt mixtures with microwaves to promote self-healing *Constr Build Mater* **42** 1–4

Ice melting properties of steel fiber modified asphalt mixtures with induction heating

Hao Fang[1], Yihan Sun[1], Quantao Liu[1,*], Bin Li[1], Shaopeng Wu[1] and Jin Tang[1]

1 State Key Laboratory of Silicate Materials for Architectures, Wuhan University of Technology, Wuhan 430070, China

E-mail: *liuqt@whut.edu.cn

Abstract. In this paper, the ice melting performance of asphalt concrete with steel fibers was studied. Steel fiber modified asphalt mixtures were prepared, five different fiber amount of steel fiber modified asphalt mixtures were mixed to study their induction heating rate. The samples covered with different thickness of ice were heated with induction heating to study their ice melting efficency. It was proved that the induction heating of steel fiber modified asphalt mixtures could significantly improve their ice melting efficency compared with the natural condition. And it was found that the thickness of the ice had little influence on the induction heating rate of the asphalt concrete.

Keywords: asphalt mixtures, steel fiber, ice melting, induction heating

1. Introduction

In most areas of the earth, the chemical salt was used to solve the pavement icing problem in winters [1]. The snowfall, sleet and freezing rain would usually lead to the formation of ice on the surface of road, which could result in serious transportation safety problems [2]. Eliminating ice effectively is important for traffic safety. From the late 1940s, deicing salts have been applied extensively in the North American countries [3]. By far, over 30 million tons of chemical salt were produced and applied all over the global world [4]. Unfortunately, this traditional method to melt ice will result in concrete corrosion and environmental pollution, and this problem has been paid much attention by International Energy Agency (IEA) and World Health Organization (WHO) [5]. And ice will not be melted by the most popularly used salt (sodium chloride) if the temperature falls below -3.9 ℃ [6]. On the other hand, hydronic and heating with electrical cables has been used for a number of decades as a means of controlling snow and ice formation on a variety of pavement surfaces [7,8]. However, the low energy efficiency of the old technologies limited their application and promotion. In recent years, the study of induction heating was more and more popular. Induction heating is a heating method based on electromagnetic induction and Joule heating [9,10]. It has been applied in many fields such as self-healing techniques. For instance, induction heating was applied to increase the temperature of the asphalt concrete in order to enhance its self-healing capacity [11-13]. As far as I know, few researches had been studied on the feasibility of using induction heating for ice melting.

The objective of this research was to study the ice melting properties of asphalt concrete with steel fibers through induction heating. The influence of steel fibers amount to the rate of induction heating was investigated. Marshall samples of asphalt concrete with different steel fibers content were prepared with this purpose. And the ice melting properties of the asphalt concrete with steel fiber were studied.

2. Materials

AC-13 asphalt mixtures based on Marshall design procedure was used in this research. The raw materials involved andesite aggregates, limestone mineral filler, AH-70# base bitumen and steel fibers. Five different percentages of steel fibers were added to the asphalt mixture by volume of asphalt: 2%, 4%, 6%, 8% and 10% respectively. The properties of AH-70# base bitumen and steel fibers were shown in table 1-2.

Table 1. Properties of AH-70# base asphalt.

Tests	Results	Specification
Penetration (25 °C, 100g, 5s) [0.1mm]	68	60-80
Softening point (ring and ball) [°C]	49	44-54
Ductility [cm]	>150	≥100

Table 2. Properties of steel fibers.

Properties	Results
Length [mm]	4.2
Equivalent diameter [μm]	70-130
Oil content [%]	<0.2
Melting point [°C]	1530

3. Experiments

3.1. Heating test of asphalt specimens with different steel fibers content

Marshall Specimens with different steel fibers content were prepared in this study, including 0%, 2%, 4%, 6%, 8% and 10% respectively. The influence of steel fibers content on the heating rate of asphalt mixtures with induction heating was measured. The asphalt-aggregate ratio of all specimens was kept at 5.3% in this test.

3.2. Ice melting test of asphalt specimens

In this test, only the specimens with 6% steel fibers content were used to investigate the ice melting property of mixture. As shown in figure 1, different thickness of ice was attatched to the top surface of asphalt specimens by freezing. The height and weight of the specimens were measured before the melting test.

Figure 1. Marshall Specimen covered with ice.

Figure 2. Ice melting test with induction heating machine.

The ice melting proformance of asphalt specimen in the natural conditions was also conducted for comparison purpose. And all these tests were conducted at the indoor temperature of 9 °C.

An induction heating machine with a maximum output of 2.7 kW and at a frequency of 123 kHz (seen in figure 2) was applied in this stage. The asphalt specimens were placed under the center of coil of the induction heating machine. The gap between the coil and the surface of asphalt specimen for ice melting was kept at a constant distance of 15 mm. Then, each specimen was heated for several times, each time included 30s of induction heating and another 30s of rest. During this stage, the surface temperature of the samples was recorded. When the ice could separate from the asphalt specimen with its own weight, the melting test would come to its end.

3.3. Ice melting test with different thickness of the ice

Different thickness of the ice were perpared in this stage, including 2.0mm, 3.0mm, 4.5mm and 5.5mm respectively. The ice melting experiment was also performed by using the induction heating machine mentioned before. The samples were placed as the same the position in 3.2. Each studied sample was heated for 30s and then rest for another 30s. While the induction heating machine was turned off, the temperature of both the surface and flank side of the test sample were measured. Each temperature was tested for three times, and the average data were recorded. Then the sample was turned upside down to test the bonding state between the ice and the asphalt specimen. The heating test was continued until the ice could separate from asphalt specimen and fall down due to its own gravity.

4. Results and discussion

4.1. Induction heating properties of mixtures with fibers

As shown in figure 3, the surface temperature of specimens with different steel fibers mix ratio was tested. It can be seen that the temperature of specimen without fibers did not increase in this condition. It was also clear that the surface temperature of samples increased as the heating time or the amount of steel fibers in the mixtures was increased.

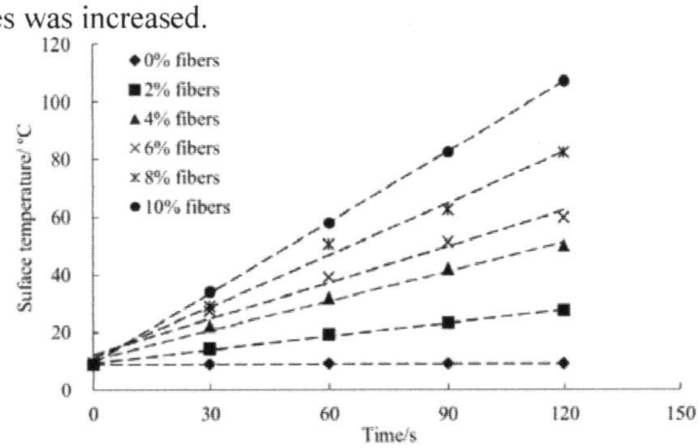

Figure 3. Effect of heating time on the surface temperature of test samples due to induction heating.

In figure 3, it can be observed that the induction heating rate of each content of steel fibers is a kind of linear relationship. With the increase of the mix ratio of steel fiber, the heating rate of asphalt mixture was also increased. From Figure 3, it is calculated that the heating rates of asphalt mixtures were 0 °C/min, 9.2 °C/min, 20.5 °C/min, 25.1 °C/min, 35.9 °C/min, and 49.0 °C/min of the 0%, 2%, 4%, 6%, 8% and 10% mix ratio of steel fibers respectively. It is obvious that the mixtures with higher percentage of fibers showed higher heating performance of heating rate. While the andesite asphalt mixtures without steel fiber couldn't be heated by the induction heating. The reason may be that the principles of induction heating are electromagnetic induction and Joule heating. This kind of heating needs a closed circuit in a changing magnetic field. In this condition, the more steel fiber mixed, the more closed circuit formed. The increase mix ratio of steel fiber will result in the increase of heating rate as a consequence.

4.2. The ice melting performance in natural condition

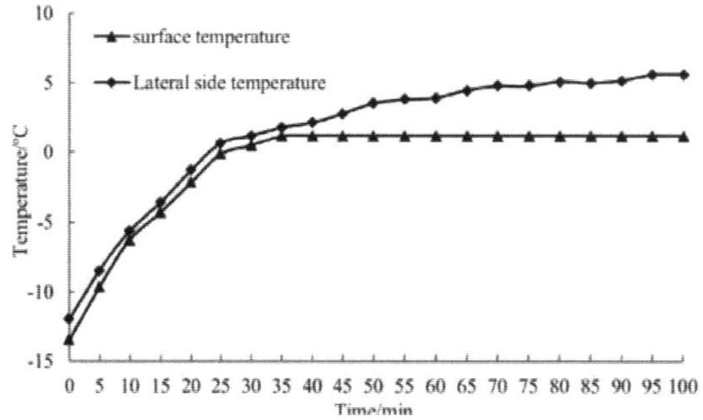

Figure 4. Evolution of specimen temperature in natural condition.

Figure 4 showed the temperature variation of the specimen in natural condition of 9 °C. It took about 100 min until the failure of the bonding state between the ice and the Marshall Specimen. It can be seen that the final temperature of the ice surface was kept steadily at 1.2 °C. Compared with the result of induction heating shown in figure 5, it is clear that the induction heating for deicing was much more efficiency.

4.3. The ice melting performance with induction heating

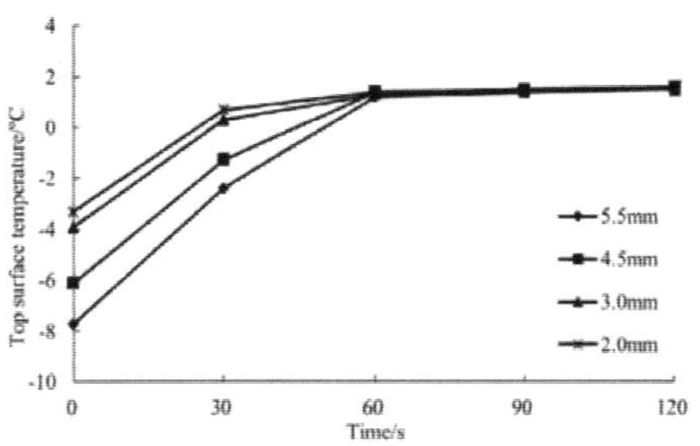

Figure 5. Ice melting performance with induction heating.

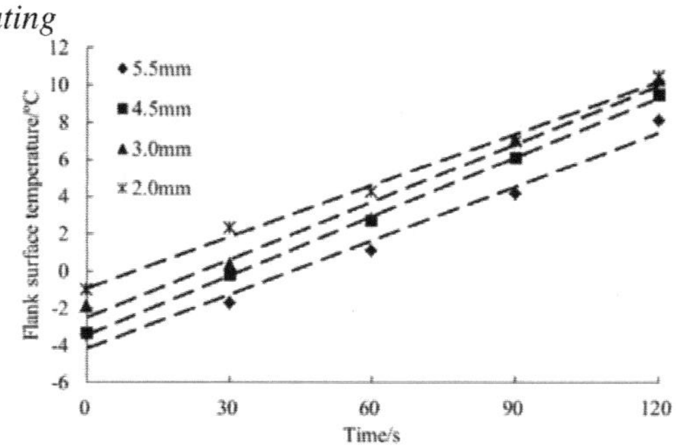

Figure 6. The flank surface temperature of the test specimen.

As mentioned in 3.3, different thickness of the ice were prepared on the surface of the Marshall Specimen. The evolutions of temperature in surface of ice during induction heating were shown in figure 5. From figure 5, it can be observed that the induction heating time with four different thickness of ice until bonding state between the ice and the Marshall Specimen failure was the same. And after four times of discontinuous heating (120s of heating in total), the surface temperature of ice was increased from subzero temperature to above freezing, this is mainly because heat transfer through the interface area to the ice.

In the figure 5 it was also clear that the final temperature of the four different samples were almost the same (1.6 °C), and during the ice melting process, the heating rate of ice surface was stayed at 0 °C/min. In this condition, the low heating rate and 1.6 °C can be regarded as the bonding failure moment between the ice and the asphalt mixtures. This may be because that when ice layer was about to failure, a large amount of ice in contact with the sample melted into water and retains in the interface area, so a lot of heat transfered to the water in this moment. As a result, the ice surface temperature tends to be the same.

Besides, the flank surface temperature of each sample during the induction heating was also recorded. As shown in figure 6, it can be seen that the heating rate of four different samples was almost the same. And the heating rate of each specimen was around 6 °C/min. The reason for this is that the steel fiber content of each sample was the same.

From figure 5 and figure 6, it was also obvious that the thickness of ice has no or little influence on the induction heating rate of the asphalt concrete. In other words, the thickness of ice has no or little influence on the ice melting efficiency when suffered with induction heating.

5. Conclusions

Based on the experimental results, the following conclusions can be drawn:

- The andesite asphalt mixtures without steel fiber couldn't be heated by the induction heating. But with the amount of steel fibers in the asphalt mixture increases, the heating rate of asphalt mixtures increases under the induction heating, the heating rate of asphalt mixtures with 10% steel fibers can reach as high as 49.0 °C/min. And the induction heating rate of each content of steel fibers is a kind of linear relationship.
- Compared with the ice melting in natural condition, the use of induction heating for deicing is much more efficiency. The heating rate of samples with different thickness of the ice is almost the same, the thickness of ice has no or little influence on the ice melting efficiency when suffered with induction heating.
- The temperature of 1.6 °C was regarded as the bonding failure temperature between ice and asphalt mixtures. It is feasible to use this temperature to determine the induction heating time needed in the actual road deicing.

Acknowledgements

The project was supported by the Natural Science Foundation of China (No. 51508433), the International Science & Technology Cooperation Program of China (No. 2013DFE83100), the Open Fund of Key Laboratory of Road Structure and Material of Ministry of Transport (Changsha University of Science & Technology), the Doctoral Fund of Ministry of Education of China (No. 20130143120008), and the Scientific Research Foundation for Returned Overseas Chinese Scholars, State Education Ministry.

References

[1] Wang X, Zhu Y, Zhu M, Zhu Y, Fan H and Wang Y 2017 Thermal analysis and optimization of an ice and snow melting system using geothermy by super-long flexible heat pipes *Appl Therm Eng* **112** 1353–1363

[2] Pan P, Wu S, Xiao Y and Liu G 2015 A review on hydronic asphalt pavement for energy harvesting and snow melting *Renewable and Sustainable Energy Reviews* **48** 624–634

[3] Wang H and Chen Z 2009 Study of critical free-area ratio during the snow-melting process on pavement using low-temperature heating fluids *Energ Convers Manage* **50** 157–165

[4] Wang H, Zhao J and Chen Z 2008 Experimental investigation of ice and snow melting process on pavement utilizing geothermal tail water *Energ Convers Manage* **49** 1538–1546

[5] Sanzo D and Hecnar S J 2006 Effects of road de-icing salt (NaCl) on larval wood frogs (Rana sylvatica) *Environ Pollut* **140** 247–256

[6] Chen M, Wu S, Wang H and Zhang J 2011 Study of ice and snow melting process on conductive asphalt solar collector *Sol Energ Mat Sol C* **95** 3241–3250

[7] Liu X, Rees S J and Spitler J D 2007 Modeling snow melting on heated pavement surfaces. Part I: Model development *Appl Therm Eng* **27** 1115–1124

[8] Lai J, Qiu J, Chen J, Fan H and Wang K 2015 New Technology and Experimental Study on Snow-Melting Heated Pavement System in Tunnel Portal *Adv Mater Sci Eng*

[9] Liu Q, Schlangen E, García Á and van de Ven M 2010 Induction heating of electrically

conductive porous asphalt concrete *Constr Build Mater* **24** 1207–1213

[10] García Á, Schlangen E, van de Ven M and Liu Q 2012 A simple model to define induction heating in asphalt mastic *Constr Build Mater* **31** 38–46

[11] Liu Q, Wu S and Schlangen E 2013 Induction heating of asphalt mastic for crack control *Constr Build Mater* **41** 345–351

[12] Garcia A, Bueno M, Norambuena-Contreras J and Partl M N 2013 Induction healing of dense asphalt concrete *Constr Build Mater* **49** 1–7

[13] Garcia A, Norambuena-Contreras J, Bueno M and Partl M N 2014 Influence of Steel Wool Fibers on the Mechanical, Termal, and Healing Properties of Dense Asphalt Concrete *J Test Eval* **42** 1107–1118

Influence of silica-based hybrid material on the gas permeability of hardened cement paste

R Li[1,2], P Hou[1,2], N Xie[1,2], Z Zhou[1,2] and X Cheng[1,2]

[1] School of Materials Science & Engineering, University of Jinan, Jinan 250022, China
[2] Shandong Provincial Key Laboratory of Preparation and Measurement of Building Material, Jinan 250022, China

E-mail: pkhou@163.com

Abstract. Surface treatment is one of the most effective ways to elongate the service life of concrete. The surface treatment agents, including organic and inorganic types, have been intensively studied. In this paper, the silica-based hybrid nanocomposite, which take advantages of both organic and inorganic treatment agents, was synthesized and used for surface treatment of hardened cement-based material. The effectiveness of organic and inorganic hybrid nanocomposite was evaluated through investigations on the gas permeability of cement-based materials. The results showed that SiO_2/PMHS hybrid nanocomposite can greatly decrease the gas transport properties of hardened cement-based materials and has a great potential for surface treatment of cementitious materials.

Keywords: hybrid silica-based materials, cement-based materials, surface treatment, gas permeability

1. Introduction

Durability of concrete is of great concern in modern society [1, 2]. As the surface of the concrete is the most vulnerable part of the structure, it is the most effective way for achieving a longer service life of concrete structures by improving the surface quality [3]. There are several ways to improve the surface quality of concrete structures, and for the existing structures, surface coating is a commonly used technique [4]. In general, there are two types of coating materials that are widely used. One is the organic coating, including silane, siloxane [5], which makes the surface of the concrete structures water-repellent. The other one is the inorganic coating, like water glass, nano-SiO_2 [3, 6], which could refine the pore structure of the cement-based materials due to its in-situ pozzolanic reactivity with the hydration products of cement.

In our previous studies, nano-SiO_2 showed a good pozzolanic reactivity [7], but its reduction of water absorption rate is still far less than that of silane/siloxane-treated sample. Currently, silica-based hybrid materials bridged with organic groups has drawn particular research interests, which takes the advantages of both organic and inorganic treatment agents [8]. In this paper, a silica-based hybrid nanocomposite was synthesized and used for surface treatment of cement-based materials. Its effectiveness on the transport properties has been evaluated through gas permeability and SEM measurements. The results indicated that the hybrid materials have a good potential for surface treatment of cement-based materials.

2. Materials and methods

2.1. Materials

Commercially available Ordinary Portland cement complying with the Chinese standard GB 175-2007 was used. And its physiochemical properties are listed in table 1.

Polymethylhydrosiloxane (PMHS) was supplied by Aladdin Industrial Corporation. Tetraethoxysilane (TEOS) and other chemicals, i.e. Tetrahydrofuran (here after THF), ethanol, anhydrous ethylenediamine with the analytical reagent (AR) grade were all purchased from Tianjin Kermel Chemical Reagent Factory.

Table 1. Chemical and physical properties of Ordinary Portland cement.

Composition	**Cement**
SiO_2	21.1
Al_2O_3	4.7
Fe_2O_3	3.5
SO_3	3.3
CaO	62.9
MgO	2.8
Density, g/cm^3	3.1
Fineness, m^2/kg	322

2.2. Sample preparation and surface treatment

2.2.1. Synthesis of hybrid material. The hybrid nanoparticles were synthesized by sol-gel methods. A certain amount of PMHS and TEOS was dropped into tetrahydrofuran solutions in the presence of anhydrous ethylenediamine and stirred for over 12 h to get the final product.

2.2.2. Sample cutting and surface treatment. In this work, cement pastes samples were prepared at a w/c ratio of 0.4 in centrifuge tube with the diameter of 35mm and cured for 1 year in standard curing chamber (21°C, 95%RH). Before surface treatment the samples were cut into piece with the thickness of 3mm and were then oven-dried at 60 °C for 24h before been soaked with treatment agents for a certain time. After that, samples were cured in standard curing condition for 3d. Before gas permeability test, the samples were oven-dried at 60 °C for 72h to remove the moisture.

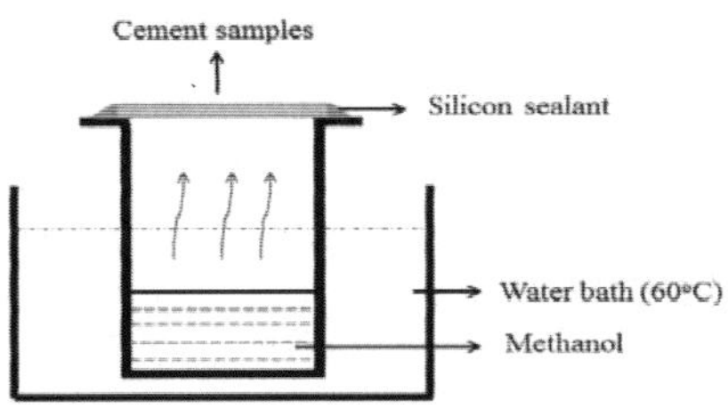

Figure 1. Schematic illustration of gas permeability test setup.

2.2.3. Gas permeability test. Gas permeability tests of cement pastes were used to evaluate the effectiveness of various surface-treatment agents in accordance with the procedures proposed by Alshamsi and Imran [9], and the set-up is shown in figure 1, a predetermined amount of methanol was kept in the cell, then cement paste samples was used to seal the cell with silicon sealant. When the silicon sealant gets dried, the mass of cell system was measured and then placed under 60 °C. The mass loss of cell was recorded at different time. The average value of three replicates was taken as the representative value.

2.2.4. SEM. Chemical grade calcium hydroxide (CH) was mixed with the hybrid material at a w/b ratio of 2.0 to simulate a cement-hybrid material system. And the morphology was observed by a field emission scanning electron microscope (SEM, QUANTA 250 FEG, FEI, America).

3. Results and discussion

3.1 Gas permeability

The variations in the mass loss of methanol with time are shown in figure 2. All the cement paste samples within or without surface treatment have comparable trends, i.e., the loss of methanol increase with time. In comparison to that of control samples, the samples treated by hybrid material have less mass loss. It indicates that the cement paste samples could be densified by treated with hybrid material, and this could be attributed to the pozzolanic reactivity of the hybrid material that hydrolysized from TEOS. As proposed by Taylor that reaction between hydrous silica and CH in aqueous suspension at ordinary temperature gives a product called C-S-H (I). In this case, the silica-based hybrid materials have a reaction with CH which was products of cement hydration, and form additional C-S-H gel to refine the pore structure of cementitious materials, leading to a reduced methanol loss at 60 °C.

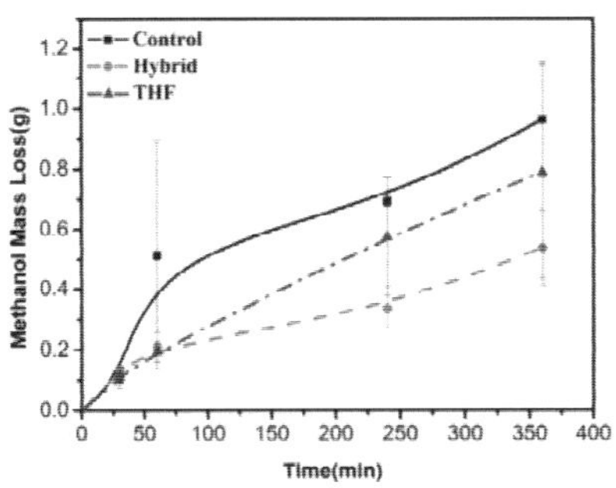

Figure 2. Mass loss vs. time of samples with and without surface treatment.

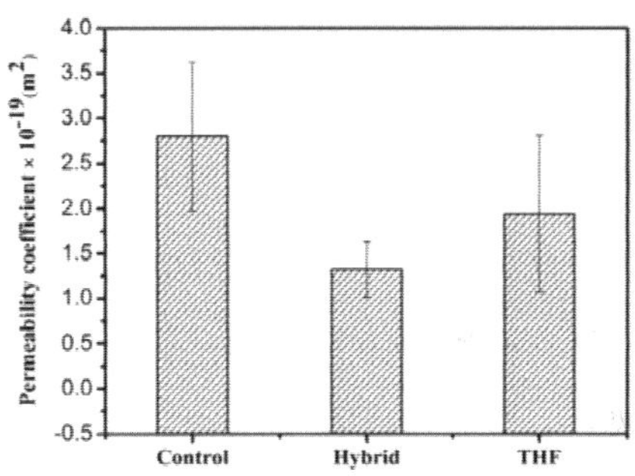

Figure 3. Gas permeability coefficient of samples.

The gas permeability coefficient k (m^2) was calculated using the equations mentioned by Han [10]. As shown in figure 3, the hybrid material provides a good protection from gas penetration, the reduction of gas permeability of cement paste treated by hybrid material is about 36.5% compared to that of control samples, which indicates a highly impermeable, dense microstructure of the hybrid materials treated surface. It suggests that the dense of the cement paste could be attributing to the pozzolanic reactivity of hybrid material, which will be discussed next.

3.2 SEM

The imagines of dried-calcium hydroxide powder before and after mixed with hybrid materials (cured for 3 d with the w/b=2) are demonstrated in figure 4(a-b) and the influence of hybrid material on the morphology of CH can be observed. It can be seen in the control sample that there are a lot of sharp hexagonal plates, which were calcium hydroxide. Meanwhile, the hexagonal plates can be hardly found in hybrid material-treated sample. It clearly shows that the hybrid material can react with calcium hydroxide and densify the surface structure of hardened cement-based materials.

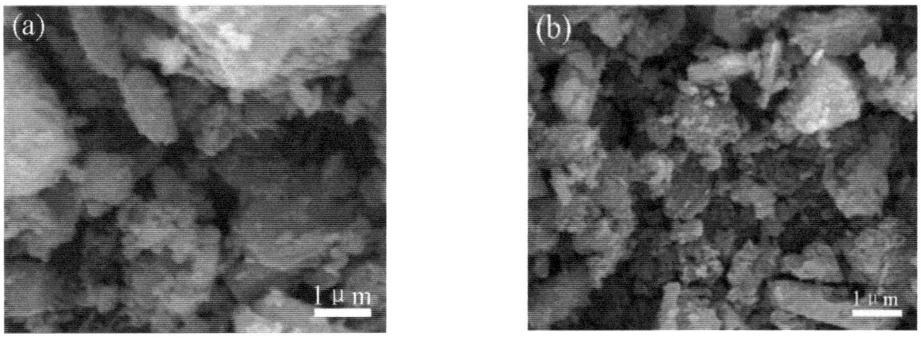

Figure 4. (a) Images of CH (w/c=0.4) before and after hybrid material treated.

4. Conclusions

- The silica-based hybrid materials are effective in decreasing the gas permeability of cement-based materials.
- The silica-based hybrid material could densify the cementitious materials.
- The silica-based hybrid material has the potential to be used for surface treatment of cement-based materials.

Acknowledgements

Supports from National High Technology Research and Development Program ("863 Program", 2015AA034701) and Natural Science Foundations of China (Grant Nos. 51302105 and 51272092) are greatly acknowledged.

References

[1] Alexander M G and Magee B J 1999 *Cement and Concrete Research* **29** 917–922

[2] Yagüe A, Valls S, Vázquez E and Albareda F 2005 *Cement and Concrete Research* **35** 1064–1073

[3] Cai Y, Hou P, Duan C, Zhang R, Zhou Z, Cheng X and Shah S 2016 *Construction and Building Materials* **117** 144–151

[4] Hou P, Cheng X, Qian J, Zhang R, Cao W and Shah S 2015 *Cement and Concrete Composites* **55** 26–33

[5] Kong X, Liu H, Lu Z and Wang D 2015 *Cement and Concrete Research* **67** 168–178

[6] Wang D, Yang P, Hou P, Zhang L, Zhou Z and Cheng X 2016 *Cement and Concrete Research* **87** 22–30

[7] Hou P, Qian J, Cheng X and Shah S 2015 *Cement and Concrete Composites* **55** 250–258

[8] Zhai S, Song Y, Zhai B, An Q and Ha C 2012 *Microporous and Mesoporous Materials* **163** 178–185

[9] Imran A 2002 *Cement and Concrete Research* **32.**

[10] Han B, Yang Z, Shi X and Yu X 2012 *Journal of Materials Engineering and Performance* **22** 184–189

Freezing resistance of high iron phoasphoaluminate cement

S X Zhang[1,2], L C Lu [1,2], S D Wang[1,2], P Q Zhao[1,2] and C C Gong[1,2]

[1] Shandong Provincial Key Lab. of Preparation and Measurement of Building Materials, University of Jinan, Jinan 250022, China
[2] School of Materials Science and Engineering, University of Jinan, Jinan 250022, China.

E-mail: mse_lulc@ujn.edu.cn

Abstract. The influence of freeze-thaw cycle on the mechanical properties of high iron phoasphoaluminate cement was investigated in the present study. The visual examination was conducted to evaluate the surface damage. The deterioration considering the weight loss, modulus loss of relative dynamic elastic and strength loss of mortar were also investigated. The morphology of hydration products were analysed by SEM. Compared with ordinary Portland cement and sulphoaluminate cement, the frost resistance of high iron phosphoraluminate cement is better. Hydration products of high iron phoasphoaluminate cement contain sheet crystals, and a lot of gel form a dense three- dimensional network structure, which results in a lower porosity. Different from ordinary Portland cement, the hydration product of high iron phoasphoaluminate cement does not contain $Ca(OH)_2$, and low alkalinity reduces its osmotic pressure. The lower porosity and osmotic pressure are the two main reasons which causes in the higher frost resistance of high iron phoasphoaluminate cement.

Keywords: high iron phosphatealuminate cement, freeze-thaw cycles, frost resistance, mechanical properties

1. Introduction

There exists a lot freezing marine area in the north of China. In such regions, a large number of concrete structures are severely deteriorated and their service life is largely shortened mainly because of the freeze–thaw damage and erosive ions in the sea [1,2]. Thus, studying the freezing resistance of cement is necessary for the development of marine engineering. Many researchers have studied the freezing resistance of sulphoaluminate cement (SAC) and ordinary Portland cement (OPC). Due to the low durability of silicate mineral of C_2S, these two types of cement had some limit when they were used to prepare marine concrete. After the appearance of sulphoaluminate cement, phoasphoaluminate cement (PAC) is another new type of special cement, which has excellent performances [3,4]. Shiqun L. and Jiashan H. carefully researched the alumina-rich area in $CaO\text{-}Al_2O_3\text{-}P_2O_5\text{-}SiO_2$ and found that the a new ternary phase formed by the addition of P_2O_5. Electron probe microanalyser results showed that the chemical composition of new phase L was $CaO\cdot(1\text{-}X\text{-}Y)Al_2O_3\cdot XSiO_2\cdot YP_2O_5$, X=0.146~0.206, Y=0.048~0.081. It's a solid solution transformed from the monocalcium aluminate ($CaO\cdot Al_2O_3$), and they named it LHss phase. The paste of that phase showed hydration performances of good early strength and increasing long-term strength. The concrete can produce a useful strength in 6h and a strength at 14 days equal to that given by Portland cement at 28 days [5,6]. Besides high early and long-term mechanical strength [7] and high temperature resistance [8], PAC had excellent corrosive resistance [9,10]. In this paper, basic experimental research on the performance of PAC mortars subjected to freeze–thaw cycles was conducted based on the macroscopic and microscopic tests.

2. Experiment procedure

2.1. Specimen preparation

Two types of reference cement, 42.5 grade OPC and 42.5 grade SAC, were used in experiment, which came from Shanshui Corporation and Qufuzhonglian Corporation. In contrast, PAC used here was synthesized by the regents of $Ca(PO_4)_3$, Al_2O_3, $Ca(OH)_2$, Fe_2O_3 and SiO_2 in laboratory, whose physical properties was list in table 1.

2.2. Test methods

For freeze-thaw cycles test, three types of mortar specimens with size of 40×40×160 cm^3 were cast and the water–binder ratio (W/B) was 0.5. Then under the condition of temperature in 20±1℃ and the relative humidity in 90±4 %, they were cured for different time (28 d for OPC, 14 d for SAC and PAC). After that, according to the GB [11,12], the specimens experienced different times of freeze-thaw cycles with variation temperature from -18℃ to 18℃. The experiment was terminated when 300 freeze–thaw cycles were completed, the result showed that 60% loss rate of relative dynamic elasticity modulus was reached or 10% weight loss ratio was obtained.

Then the influences of freeze-thaw cycles on the elasticity modulus, mass loss, flexural strength and compressive strength were test according to the GB subsequently. All the final value was the average of text result for three specimens. The weight loss was calculated by the Eq(1):

$$\Delta Wn = \left[\frac{(Wo - Wn)}{Wo}\right] \times 100 \quad (1)$$

In the equation ΔWn is the weight loss of specimens at every 25 freeze–thaw cycles (%), W0 is the average weight of mortars specimens before freeze–thaw cycles (kg) and Wn is the average weight of concrete specimens at every 25 freeze–thaw cycles (kg). The dynamic modulus of elasticity loss, which was determined by Eq (2), was measured with a high-accuracy nonmetal ultrasonic analyzer as the initial value.

$$\Delta E = \left(\frac{Edn}{Ed0}\right) \times 100 \quad (2)$$

In the equation ΔE is the dynamic modulus loss of specimens at every 25 freeze–thaw cycles (%).Ed0 is the dynamic modulus of elasticity of concrete specimens before freeze–thaw cycles (GPa) and Edn is the dynamic modulus of elasticity of concrete specimens at every 25 freeze–thaw cycles (GPa).

The following Eq (3) was used to calculate the compressive strength, where P is the compressive strength (MPa), F is the maximum load (N) and A is the area of the cube loading face (mm^2).

$$P = F/A \quad (3)$$

The test of SEM was used to observe the morphology of hydration products. The hydration of specimens was discontinued by alcohol and vacuum dried first and SEM images were acquired by Quanta FEG 250 field emission scanning electron microscope produced in FEI (American) whose resolution is smaller than 1 nm.

Table 1. Physical properties of PAC.

Cement	density g/cm^3	soundness 3h	Fineness residue /%	Setting time /min Initial setting	Final setting
PAC	3.06	steadily	3.1(74 μm)	230	250

3. Results and discussion

Figure 1 showed the surface deterioration characteristic of mortars subjected to 70 freeze–thaw cycles in water. The degree of surface deteriorations for three types of mortars is obviously different. The mortar of PAC exhibits slight attack with only a surface layer of mortar scaled. On the contrary, mortar of OPC and SAC suffers badly from the corners and edges, which shows that their damage is much worse than PAC. Visual inspection revealed that the specimens of OPC and SAC were more severely damaged than that of PAC.

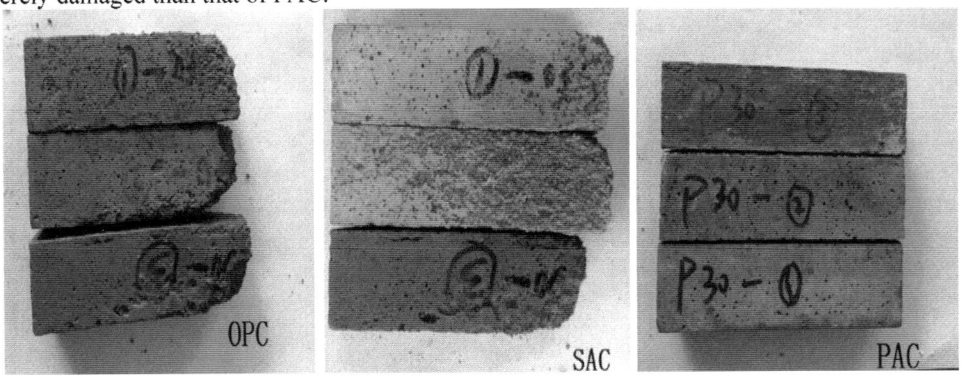

Figure 1. Surface damage characteristic of mortar subjected to70 freeze–thaw cycles.

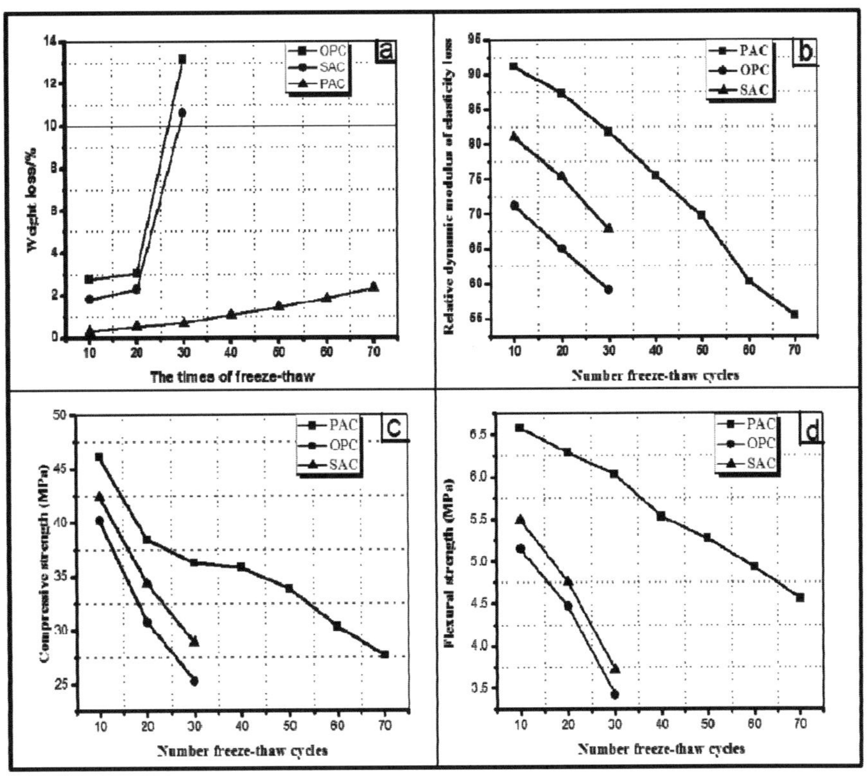

Figure 2. Results of freeze-thaw cycles.

The weight loss of OPC, SAC and PAC mortars subjected to the freeze-thaw are presented in figure 2(a). The weight loss of OPC and SAC mortars subjected to the freeze-thaw exhibited two distinct stages. In stage I, from the initial immersion to 20 freeze-thaw cycles, the weight loss increased gradually. In the stage II, from 20 to 30 freeze-thaw cycles, the weight loss of OPC mortars increased from 2.76% to 13.16% and SAC increased from 1.81% to 10.61%. Since the weight loss exceed 10%, the freezing resistance test was stopped after 30 freeze-thaw cycles for OPC and SAC mortars. As for the specimens of PAC, from 10 to 70 freeze-thaw cycles, the weight loss increased gradually from 0.31% to 2.31%.

The modulus loss of relative dynamic elastic of mortars subjected to the freeze-thaw cycles are presented in figure 2(b). As freeze–thaw cycle increased, the relative dynamic elasticity modulus increased. The relative dynamic modulus of elasticity loss in three types of mortars presented different characteristic. For the mortars of OPC, from 10 to 20 cycles, the modulus loss of relative dynamic elastic increased gradually from 71.26% to 59.06% and it is from 81% to 67.85% for SAC. As for the specimens of PAC, from 10 to 70 cycles, the modulus loss of relative dynamic elastic increased from 91.23% to 55.43%. The results agree with the weight loss presented in figure.1.

Figure 2(c) and figure (d) show the results of compressive strength and flexural strength tests for the mortar samples subjected to freeze-thaw cycles. It can be seen from figure2(c) that after 30 freeze–thaw cycles, the compressive strength loss ratios of mortars for PAC, SAC and OPC are 0.7%, 10.61% and 13.16%, respectively. As for PAC mortar, after 70 freeze-thaw cycles, the compressive strength loss is 2.31%. It can be seen from the figure 2(d) that the flexural strength loss of OPC and SAC developed rapidly, while for PAC mortar the flexural strength loss developed gradually. The results agree with the compressive strength loss presented in figure 2(c).

Figure 3, figure 4 and figure 5 present the microstructure of mortars subjected to freeze–thaw cycles. As shown from figure 4, needle-like crystals can be observed in the pores for SAC mortar, which are ettringite crystals. The main production of PAC hydration is sheet crystals, which connect each other and many gels filled in the gap. The sheet crystals and gels could form a more compact mortar structure.

Figure 3. SEM image of OPC cement mortar.

Figure 4. SEM image of SAC cement mortar.

Figure 5. SEM image of PAC cement mortar.

Table 2. Physical properties of PAC.

CEMENT	alkali content	PH
OPC	0.77	12.9
SAC	0.59	12.1
PAC	<0.4	10.9

Table 2 shows the alkali content and PH of three types of mortar. The PH of OPC, SAC and PAC is 12.9, 12.1 and 10.9, respectively. Since $Ca(OH)_2$ is not the production of PAC hydration, the alkali content of PAC cement is much lower than OPC and SAC. The lower alkali content may have positive effects on the mortar subjected to freeze–thaw cycles. The positive effect could be that the lower ion concentration result in small permeation pressure during the freeze-thaw cycles. The small permeation

IOP Conf. Series: Materials Science and Engineering **182** (2017) 012044 doi:10.1088/1757-899X/182/1/012044

pressure cannot break the three- dimensional structure network which can maintain the compressive strength of PAC. The effect of lowing permeation results in cement mortar damaged lightly when exposed to freeze–thaw cycles.

4. Conclusions

In this paper, the freezing resistance of high iron phoasphoaluminate cement was analyzed by comparing the OPC and SAC. The following conclusions can be drawn.

The experimental result of weight loss, compressive strength loss and flexural strength loss show that the freezing resistance of PAC is better than SAC and OPC. The test of SEM indicates that the production of sheet crystals and gels of PAC form a more compact cement mortar structure and may result in a lower porosity, which could maintain the compressive strength. The alkali content and PH can result in small permeation pressure during the freeze-thaw cycles. The small permeation pressure cannot break the three-dimensional structure network, which can maintain the compressive strength of PAC. The lower porosity and osmotic pressure are the two main reasons which causes in the higher frost resistance.

Acknowledgements

This work was supported by the Natural Science Foundations of China (Grant 51472109, 51672108) and the Shandong Province Bold Talent Program (ZR2015EM002). We also thank the Program for Scientific Research Innovation Team in the Colleges and Universities of Shandong Province.

References

[1] Smith A *et al.* 2002 Correlation between hydration mechanism and ultrasonic measurements in an aluminous cement: effect of setting time and temperature on the early hydration. *J. J eur Ceram Soc.* **22(12)** 1947–1958

[2] Al-Amoudi and O S B 1995 Performance of 15 reinforced concrete mixtures in magnesiumsodium sulphate environments *J .Constr.Build.Mater.***9(3)** 149–158

[3] Weiping Ma and P W Brown 1992 Mechanical beheavior and microstrctural development in phosphate modifided high alumina cement *J. Cement Concrete Re.* **22(6)** 1192–1200

[4] Li Shiqun *et al.* 1999 Fundamental study on aluminophosphate cement *J. Cement Concrete Re* .**29(10)** 1549–1554

[5] Jia L, Shiqun L and Jiashan H 2001 Study on the aluminophosphate glass-rich cement *J. Cement Concrete Re.***31(6)** 949–952

[6] Shuai Yang, Shoude Wang and Lingchao Lu 2015 Constituent phases and mechanical properties of iron oxide-additioned phosphoaluminate cement *J. Mater Construcc.* **65(318)**

[7] Beshir M Mohamed and John H Sharp 1997 Kinetics and mechanism of formation of monocalcium aluminate CA *J. J Mater Chem.* **7(8)** 1595–1599

[8] Shahwana Rashid and Xavier Turrillas 1997 Hydration kinetics of $CaAl_2O_4$ using synchrotron energy-dispersive diffraction *J. Thermochim Acta.* **302(1)** 25–34

[9] V S Rarnachandran and R F Feldman 1973 Hydration characteristic of monocalcium aluminate at low water/solid ratio *J. Cement Concrete Re.* **3(6)** 729–750

[10] Jia L, Shiqun L, Jiashan H and Biao L 2001 Study on the aluminophosphate glass-rich cement *J .Cement Concrete Re.* **31(6)** 949–952

[11] GB/T 50081 2002 Standard for test method of mechanical properties on ordinary concrete *Chinese standard Institution.*

[12] GB/T 50082 2009 Standard for test method of long-term performance and durability of ordinary concrete *Chinese standard Institution.*

A novel kind of concrete superplasticizer based on aryl isocyanate polycondensates

Bei Ding[1], Shanshan Qian[1], Ying Qiu[1], Yi Wang[1], Chunyang Zheng[1] and Gaoming Wang[1]

[1] Jiangsu China Railway Arit New materials Co., Ltd. Nanjing 211505 China

E-mail: 31954694@qq.com

Abstract. A novel superplasticizer was synthesized by polycondensation of Alkyl phenol phosphate and toluene diisocyanate grafting with methoxy polyoxyethylene ether (MPEG). The chemical structure and molecular weight of polycondensates molecules were determined by ^{1}H nuclear magnetic resonance (NMR) and gel permeation chromatography respectively. The experimental results indicated that the polycondensates of Alkyl phenol phosphate and toluene diisocyanate grafting with MPEG not only exhibited good water-reducing properties but also demonstrated effective anti-clay abilities alone. Furthermore, the polycondensates showed good fluidity maintaining abilities within 1–3 h and good workability of concrete. The results of T500 Time experiments show that lower plastic viscosity of the polycondensates leads to fresh concrete much "looser" than conventional PCEs in self-compacting concrete.

Keywords: superplasticizer, polycondensation, isocyanate, phosphoric acid group, plastic viscosity

1. Introduction

Polycarboxylate superplasticizers (PCEs) are widely used in different industrial fields to improve the performance of concrete. With the development of the construction industry and requirements of environmental protection, modern concrete technology aims towards minimizing the needs for Portland cement without sacrificing the important engineering properties of hardened concrete [1-3]. PCEs reduce the dosage of water to increase the strength of hardened concrete. However, it can be observed that some PCEs lead to fresh concrete with a very "sticky" and cohesive consistency [4]. In china, there has been a growing dependency on manufactured and dramatic decline of availability of naturally local sands as fine aggregates for concrete and mortar because of environmental and resource pressures. The sands used in concrete and mortar are getting less and less pure. The presence of clay as impurities increasing the water demand to provide a concrete or mortar of given workability, compromising the rheological property of concrete or mortar after absorbing water and swelling, weakening the bond between the cement and aggregates, and dramatic increasing the dosage of PCEs especially [5-8].

In this work, we synthesized and characterized a novel superplasticizer by polycondensation of Alkyl phenol phosphate and toluene diisocyanate grafting with methoxy polyoxyethylene ether (MPEG) and compare it with commercially available conventional PCEs. The backbone of this superplasticizer consists of benzene ring structure which is the same as poly (β-naphthalene sulfonates) (BNS).

2. Experimental investigation

2.1. Materials
The chemicals were used without further purification: Methoxy polyoxyethylene ether (MPEG, Mw=1000, Clariant corporation), Toluene-2,4-diisocyanate (TDI, Kumho Mitsui Chemical Inc.), Polyphosphoric acids (85%, Shanghai ziyi-reagent corporation), Phenoxyethanol (Shanghai Ziqibio corporation), Formaldehyde (37%, Ningbo Xingye corporation), Montmorillonite (MMT, Aladdin Reagent corporation), the effective content ≥95%, average particle size=1.52μm, specific surface area=10.86 m^2/g.

2.2. Measurements
NMR spectra were performed on Bruker Avance DPX 400 MHz for ^1H NMR. The residual solvent signals were uses as reference; chemical shifts are given on the TMS scale (D_2O: δ=4.79 ppm). Molecular weight was determined by gel permeation chromatography (GPC) measurements. The samples were analyzed using a 0.1mol/L $NaNO_3$ aqueous solution as eluent at a flow of rate of 0.5 mL/min.

2.3. Cement paste tests
An ordinary Portland cement (52.5, Jiangnan Onoda Co., China), which meets the requirements of GB8076 standard was used throughout all the experiments. The cement fluidity tests were conducted with the following procedure. The fresh cement pastes were prepared at 20℃～25℃, the water to cement ratio (w/c) is 0.29. The amount of 87g water containing the defined amount of polymer was added to 300g cement and mixed for 2 min at low speed and a further 2 min at high speed. The cement fluidity took the average of two measurements.

2.4. T_{500} Time of self-compacting concrete [9]
This test is used to assess the flow rate of self-compacting concrete (SCC). The slump cone with above described dimensions was filled with fresh concrete. Upon lifting the cone, the time for the concrete to flow to diameter of 500 mm is recorded. The concrete mixture proportions are shown in table 1.

Table 1. Concrete mixture proportions (kg/m^3).

Cement	Fly ash	Sand	Gravel	Water
360	195	840	835	178

2.5. Polycondensates synthesis

2.5.1. Preparation of Phenoxyethanol Phosphate. Preparation of Phenoxyethanol Phosphate was according to the preparation method of literature [10]. Phenoxyethanol was esterified azeotropically in the presence of polyphosphoric acids in toluene on a water separator at 120℃～125℃. The reaction product which precipitated on cooling was filtered off, washed with petroleum ether and dried to give a white powder. The structure of Phenoxyethanol Phosphate was shown as figure 1.

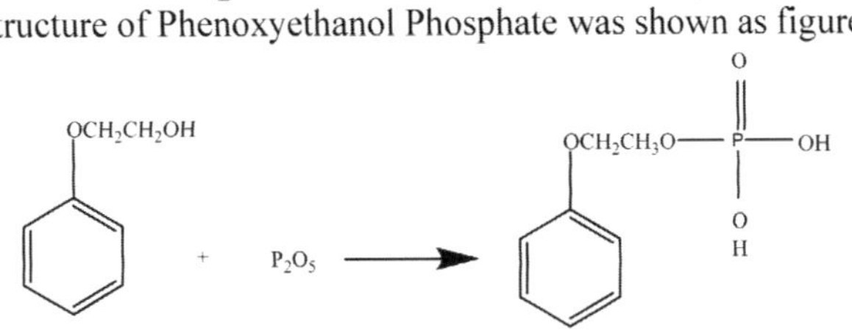

Figure 1. The reaction of preparation of phenoxyethanol phosphate.

IOP Conf. Series: Materials Science and Engineering **182** (2017) 012045 doi:10.1088/1757-899X/182/1/012045

2.5.2. Grafting of toluene diisocyanate with methoxy polyoxyethylene ether (MPEG-TDI). Toluene-2,4-diisocyanate was reacted with MPEG (Mw=1000) at 80℃～85℃ for several hours. The reaction product was diluted by water. The structure of MPEG-TDI was shown as figure 2.

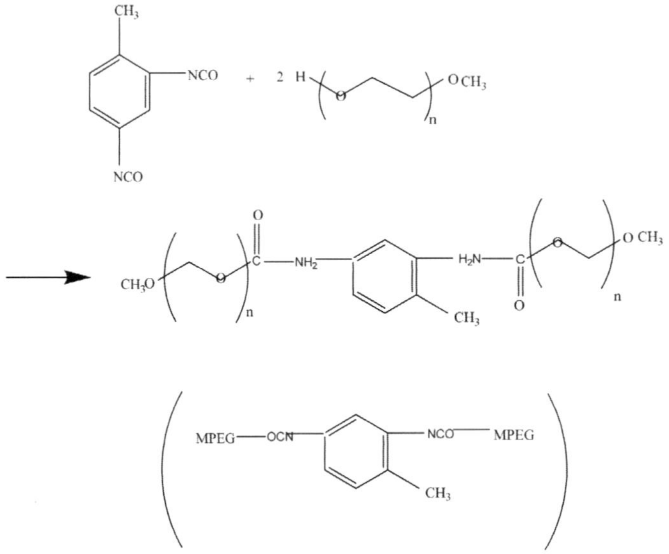

Figure 2. Reaction of preparation of MPEG-TDI.

2.5.3. Polycondensation of Phenoxyethanol Phosphate and MPEG-TDI. Phenoxyethanol Phosphate, MPEG-TDI, water and H_2SO_4 were uniform mixing, formaldehyde in the form of a 37% aqueous solution were added dropwise to the solution formed in this way. The polycondensation reaction is complete at 105℃. After the end of the reaction, the reaction mixture is brought to a pH of 7.5 by means of 32% aqueous NaOH solution. The structure of Polycondensates was shown as figure 3.

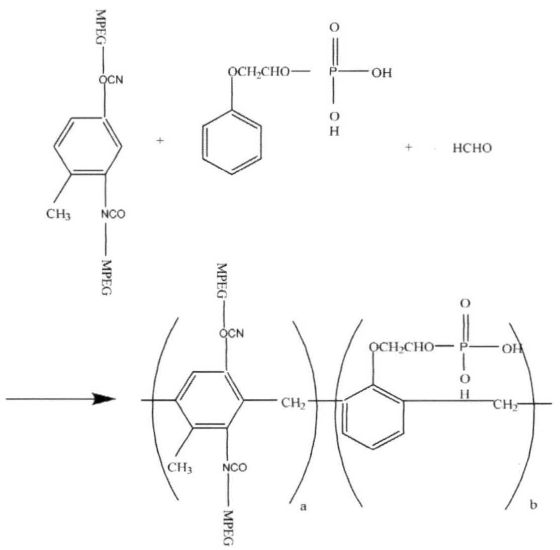

Figure 3. Reaction of preparation of MPEG-TDI.

3. Results and discussion

Polycondensates of phenoxyenthanol phosphate and MPEG-TDI were synthesized and characterized by ^{1}H-NMR. The performance of polycondensates was investigated by cement paste and T500 time of fresh concrete.

3.1. Structures of the polycondensates

Figure 4 shows the 1H NMR spectrum of the typical polycondensates of phenoxyenthanol phosphate and MPEG-TDI. (a)-(c) peaks are ascribed to methylene protons in –O-CH_2- group, - $(CH_2)2$- protons, (d) are ascribed to the chemical shift value of CH_3O- proton characteristic peaks in the MPEG chain segment moiety.

The characteristic properties of polycondensates obtained from GPC are shown in figure 5. The results indicate that the polycondensates have molecular weight 18336, with the PDIs are 1.22.

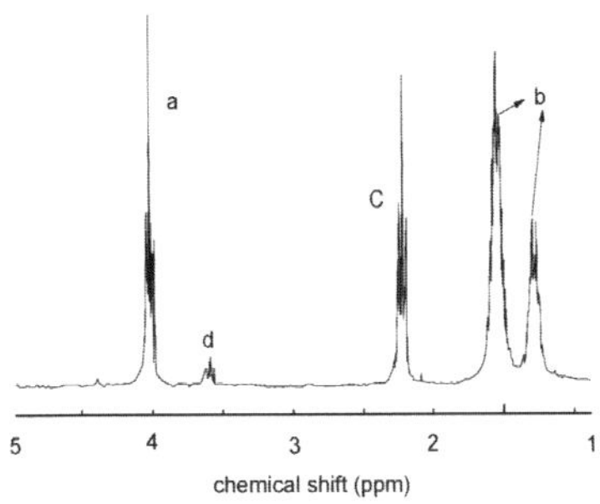

Figure 4. ^{1}H NMR spectrum of the polycondensates.

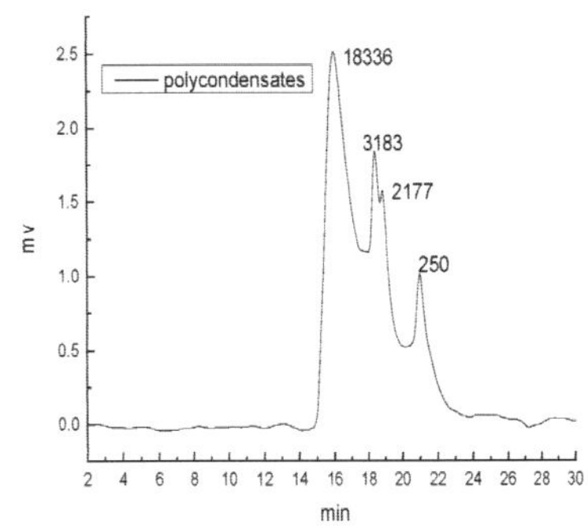

Figure 5. Gel permeation chromatography.

3.2. Cement dispersion test

Cement dispersion test evaluates the dispersing performance of synthesized compounds. In figure 6, the results indicated that polycondensates demonstrated good water-reducing effects compared with conventional PCEs, despite the dosage of polycondensates is higher about 30% than conventional PCEs. The fluidity of polycondensates reached 200 mm at a 0.2 wt.% and 240 mm at a 0.25wt.% dosage. Impressively, the spread flow decreased only from 200 to 165 mm for polycondensates after 1 h, even maintained at 130 mm after 3 h. The loss of polycondensates was obviously less than conventional PCE used here (from 205 to 100 mm after 3 h). The polycondensates showed good fluidity maintaining abilities within 1～3 h here.

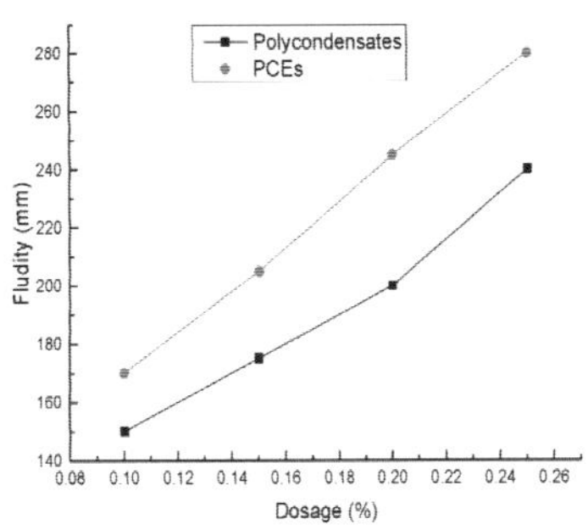

Figure 6. Dispersion properties.

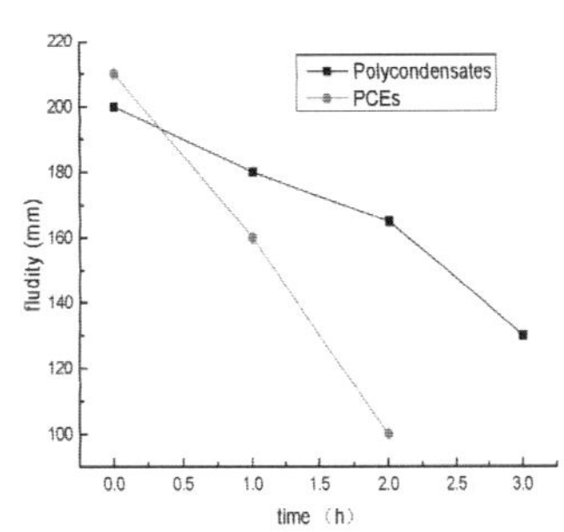

Figure 7. Fluidity loss of cement paste.

3.3. Effect of clays on cement fluidity
Figure 8 indicates the impact of MMT on the conventional PCE and polycondensates. As shown in figure 8, in the cement/MMT system, the conventional PCE exhibited a strong decrease in performance caused by the addition of MMT. With approximately 1 wt.% MMT, the spread flow decreased by 36 % and became less than 155 mm at 0.2 wt. % dosage, compared with figure 6. The newly synthesized polycondensates demonstrated good resistance property of MMT under the same condition. With approximately 1 wt.% MMT, the fluidity decrease from 200 mm to 180 mm at 0.2 wt.% dosage, the spread flow reductions were 10 %, respectively.

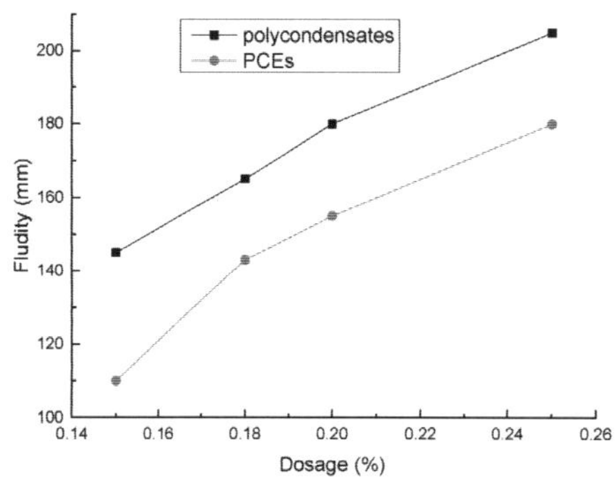

Figure 8. Effect of clays (MMT) on cement fluidity.

3.4. T_{500} time of concrete
Concrete viscosity can be assessed by the T_{500} time during the slump-flow test. The time value of concrete fluidity with 500 mm is recorded to describe the rate of flow. Concrete with a low viscosity will have a very quick initial flow and then stop. Concrete with a high viscosity may continue to creep forward over an extended time. T_{500} time test results were described in table 2. There were the same slump and spread of fresh concrete with PCEs and polycondensates by adjusting the dosage of admixture. T_{500} time of polycondensates was only 7.8s. Compared with 11.4s of PCEs at the same spread flow, the spread flow velocity of polycondensates increased by 36 %. It means fresh concrete with polycondensates has lower plastic viscosity than concrete with PCEs at the same plastic stress.

Table 2. T_{500} time test results.

admixture	Slump /mm	Spread flow /mm	T_{500} /s	Admixture dosage (%)
PCEs	270	625	11.4	0.45
Polycondensates	268	630	7.8	0.80

4. Conclusions
A novel kind of concrete superplasticizer has been synthesized based on aryl isocyanate polycondensates. The chemical structure and molecular weight of desired products have been determined by ^1H NMR and GPC, respectively. The results of cement fluidity tests indicate that the polycondensates of Alkyl phenol phosphate and toluene diisocyanate grafting with MPEG not only exhibite good water-reducing properties but also demonstrate effective anti-clay abilities alone. Furthermore, the polycondensates show good fluidity maintaining abilities within 1–3 h and good

workability of the concrete. The results of T500 Time experiments show the lower plastic viscosity of the polycondensates than conventional PCEs at the same plastic stress.

References

[1] Schober I and Flatt R J 2006 Optimizing polycarboxylate polymers. *8th CANMET/ACI International Conference Superplasticizers and Other chemical Admixtures in Concrete Corento* ACI 16
[2] Yamada K, Takahashi T, Hanehara S and Matsuhisa M 2000 Effects of the chemical structure on the properties of polycarboxylate-type superplasticizer *Cement and Concrete Research* **30(2)** 197–207
[3] Papayianni I, Tsohos G, Oikonomou N and Mavria P 2005 Influence of superplasticizer type and mix design parameters on the performance of them in concrete mixture *Cement and Concrete Composites* **27(2)** 217–222
[4] Banfill P F G 2011 Additivity effects in the rheology of fresh concrete containing water-reducing admixtures *Construction and Building Material* **25(6)** 2955–2960
[5] Yool A I G, Lees T P and Fried A 1998 Improvements to the methylene blue dye test for harmful clay in aggregates for concrete and mortar *Cement and Concrete Research* **28(10)** 1417–1428
[6] Topcu I B and Ugurlu A 2003 Effect of the use of mineral filler on the properties of concrete *Cement and Concrete Research* **33(7)** 1071–1075
[7] Norvell J K, Stewart J G, Juenger M C G and Fowler D W 2007 influence of clays and clay-sized on concrete performance *Journal of Materials in civil engineering* **19(12)** 1053–1059
[8] Leslie A Jardine *et al.* 2002 admixtures and method for optimizing addition of EO/PO superplasticizer to concrete containing smectite clay-containing aggregate US 6352952
[9] JGJ/T 283-2012 2012 Technical specification for application of self-compacting concrete
[10] Philipp Wieland *et al.* 2011 Polycondensation product based on aromatic or heteroaromatic compounds, method for the production thereof, and use thereof US 7919640

IOP Conf. Series: Materials Science and Engineering **182**(2017) 012046 doi:10.1088/1757-899X/182/1/01204(

Effects of TEA·HCl hardening accelerator on the workability of cement-based materials

Wenhao Pan[1,*], Zhaoyang Ding[2] and Yanwen Chen[1]

[1] The School of Materials Science and Engineering, Shenyang Jianzhu University, Shenyang 110168, China
[2] Test and Analysis Center, Shenyang Jianzhu University, Shenyang 110168, China

E-mail: *pwh@sjzu.edu.cn

Abstract. The aim of the test is to research the influence rules of TEA·HCl on the workability of cement paste and concrete. Based on the features of the new hardening accelerator, an experimental analysis system were established through different dosages of hardening accelerator, and the feasibility of such accelerator to satisfy the need of practical engineering was verified. The results show that adding of the hardening accelerator can accelerate the cement hydration, and what's more, when the dosage was 0.04%, the setting time was the shortest while the initial setting time and final setting time were 130 min and 180 min, respectively. The initial fluidity of cement paste of adding accelerator was roughly equivalent compared with that of blank. After 30 min, fluidity loss would decrease with the dosage increasing, but fluidity may increase. The application of the hardening accelerator can make the early workability of concrete enhance, especially the slump loss of 30 min can improve more significantly. The bleeding rate of concrete significantly decreases after adding TEA·HCl. The conclusion is that the new hardening accelerator can meet the need of the workability of cement-based materials in the optimum dosage range.

Keywords: hardening accelerator, setting time, hydration, cement, bleeding rate

1. Introduction

In concrete construction, hardening accelerator can accelerate the hardening of concrete, can improve the early strength of the concrete [1,3], and accelerate concrete the construction schedule, so it is widely used in the modern concrete construction. In order to speed up the tension reinforcement [1,2], accelerate the turnover of template and pedestal utilization in production of concrete member, the hardening accelerators were more widely used.

Hardening accelerator is the admixture that can adjust the setting time and earlier strength of concrete, mortar and cement paste. There are many types of hardening accelerator, mainly include, inorganic salts hardening accelerator, such as salt-based chlorine series, sulfate series etc. organics hardening accelerator [1,5,12], such as lower organic acid salt(calcium formate, sodium acetate, calcium oxalate, etc.), triethanolamine, triisopropanolamine, and urea [3,9], compound hardening accelerator, such as compound of inorganic salts and organic classes, compound of many kinds of inorganic salts, and compound many kinds of organic matters [13,14].

There are three ways to improve the early strength of concrete; the first way is the use of special cement, this cement can reach the predetermined strength in a short time [4,10]. However, due to its low production and high prices are restricted not to be widely used. The second way is improving the construction and maintenance of concrete, such as hot-mix concrete, steam curing treatment methods [6]. However, this method is also the inconvenience in actual industrial production, therefore not been widely adopted. The third way is to use the early strength admixture; practice has proved that this method is the easiest and least costly method, so it has been widely adopted in the construction project [8].

On the other hand, the use of hardening accelerator has significant influence on the workability of the cement-based materials [7,11]. In order that the hardening accelerator is better applied in practical engineering, we should to have a good grasp the influence rules of hardening accelerator on the workability of the cement-based materials. TEA·HCl, as a new type hardening accelerator, prompt us to research the influence rules of it on the workability of cement paste and concrete.

2. Experimental

2.1. Raw Materials

2.1.1. Cement. P·O 42.5 ordinary portland cement was used, produced by Jidong Cement Factory, the performance of which is reported in table 1.

Table 1. Performance of the cement.

Fineness（80μm sieve residue）%	Setting Time/min		Soundness（boiling method）	Bending Strength /MPa		Compressive Strength /MPa	
	Initial	Finnal		3d	28d	3d	28d
1.7	210 >45	252 <390	Qualified	4.1 4.0	8.4 7.3	21.4 17.2	49.3 48.8

2.1.2. Superplasticizer. Polycarboxylate superplasticizer, self made polycarboxylic superplasticize was used in this test, the solid content of which is 40%. Choose the recommended dosage, the adding amount is usually 0.2% to 0.5%, tentative dosage is 0.2% (equivalent to solid dosage calculation).

2.1.3. Hardening Accelerator. TEA·HCl hardening accelerator: the hardening accelerator used in this test was self made products, 98% purity, white, crystalline powder.

2.1.4. Fine Aggregate. River sand, Region II medium sand, its fineness modulus is 2.65, mud content is less than 2%.

2.1.5. Coarse Aggregate. Coarse aggregate: gravel, 5 mm~25 mm, continuous grading.

2.1.6. Water. Tap water was used in this test, comply with the technical requirements of standard concrete mixing water.

2.2. Experimental Methods

The water requirement of normal consistency of the cement was pre-determined according to the national standard. The initial setting time and final setting time were tested for blank samples and which was added TEA·HCl hardening accelerator with the water requirement of normal consistency. According to preliminary tests that the optimal dosage range of the hardening accelerator, the fluidity of cement paste and the loses through time were tested when the dosage of hardening accelerator is 0.03%, 0.04%, 0.05%, 0.06% and 0.07%. Thus we analyzed the influence rules of TEA·HCl on the workability of cement paste.

According *JGJ55-2011 "Specification for mix proportion design of ordinary concrete"* designed C30, C40 and C50 three strength grades of concrete mix proportion, which included blank and adding

TEA·HCl concrete. Concrete slump and the loss through time, bleeding of concrete were tested, we analyzed the influence rules of TEA·HCl on the workability of concrete.

3. Results and Discussion

3.1. Effects of Different Dosage on Setting Time

Figure 1 shows the curves for the initial setting time and final setting time of cement samples with different dosage of TEA·HCl. From figure 1, we can see that the incorporation of the hardening accelerator can accelerate the hydration of the cement, and the setting time of cement samples is not gradually shortened with the increase of the hardening accelerator in the dosage selected range. The setting time is the shortest when the dosage of 0.04%. As can be seen from table 2, the initial setting time and final setting time are 130 min and 180 min respectively, which meet the requirements of GB175-2007 "Common Portland Cement". However, they are shortened by 38.1% and 27.0% compared with the blank sample.

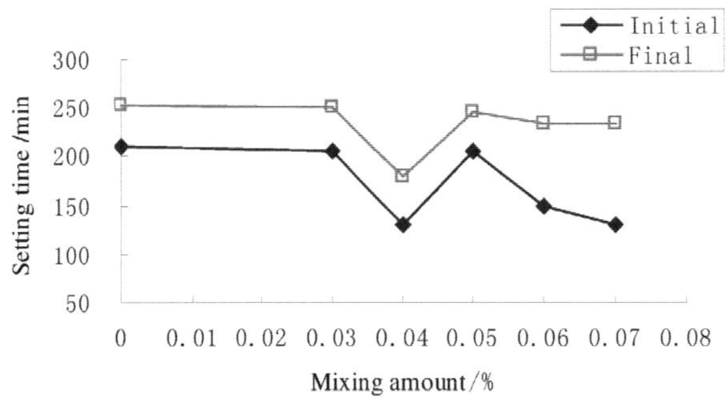

Figure 1. Setting time of cement samples with different mixing amount of TEA·HCl.

Table 2. Setting time of cement samples with different mixing amount of TEA·HCl.

Mixing Amount/ %	Initial Setting Time /min	Final Setting Time /min
0	210	252
0.03	205	250
0.04	130	180
0.05	185	245
0.06	150	235
0.07	133	215

3.2. Effects on Fluidity of Cement Paste

Figure 2 and figure 3 show the curves for the fluidity of cement paste samples with different dosage of TEA·HCl. We can see that the effects of incorporation TEA·HCl on the initial paste fluidity is not significant, but has significant improvement on the fluidity of after 30 minutes, which even larger than the initial fluidity when the dosage is more than 0.06%, the fluidity after 60 minutes would be significantly decreased with the dosage of accelerator increasing, this showed that the accelerator had plasticity-maintaining effect on the cement hydration in initial period (30 min). With time passing by (after 60 min), there would be occurred the hydration reaction accelerated, and the fluidity significantly decreased phenomenon.

3.3. The concrete slump and slump loss

The test had designed three strength grades of concrete, which included blank and adding 0.04% TEA·HCl concrete. As can be seen from table 3, with the adding of accelerator, all the workability of concrete were improved. The initial slump had some improvement, especially for the slump loses after

30 min was improved obviously. There is not much improvement for the slump loses after 60min, which is roughly equivalent compared with that of blank samples.

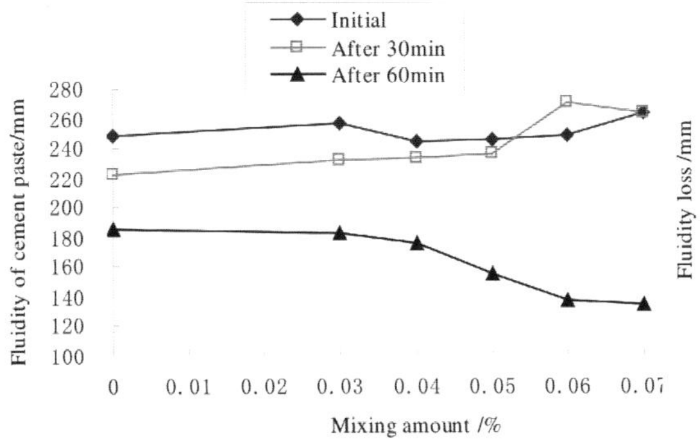

Figure 2. Fluidity of cement samples with different .

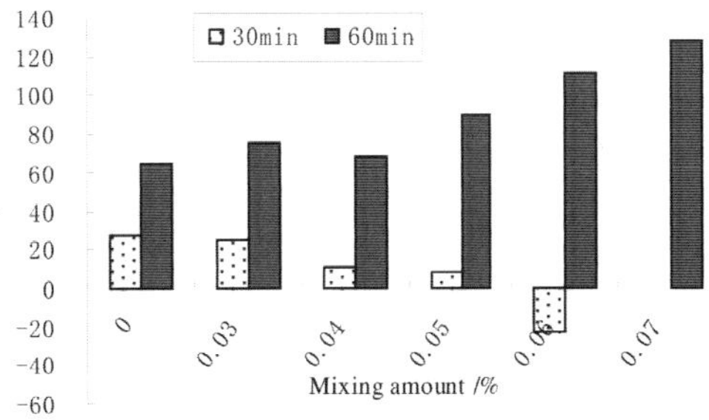

Figure 3. Fluidity loss after 30 min and 60 min mixing amount of TEA·HCl.

Table 3. Concrete slump and slump loss.

No.	Mix Proportion C:S:G:F:W	TEA·HCl /C × %	Slump/mm Initial	30 min	60 min
T3	1:2.01:3.58:0.25:0.58	0	93	68	46
T3'		0.04	95	87	53
T4	1:1.54:2.74:0.25:0.47	0	87	60	38
T4'		0.04	92	88	42
T5	1:1.22:2.17:0.25:0.39	0	99	73	39
T5'		0.04	106	95	35

This conclusion was basically consistent compared with that obtained from the cement paste fluidity test; it was that the accelerator had plasticity-maintaining effect on the cement hydration in initial period (30 min).

3.4. Bleeding Test

Bleeding test results of six groups of concrete (T3, T3', T4, T4', T5 and T5') as shown in figure 4~6, we could seen that the bleeding rate of concrete was significantly decreased after adding TEA·HCl hardening accelerator, particularly more obviously after 3 hours. Taking T3 and T3' as an example, the bleeding rate of T3' concrete is only 42.5% of T3 concrete. This may be because that TEA·HCl accelerated the hydration of cement after mixed 1 hours.

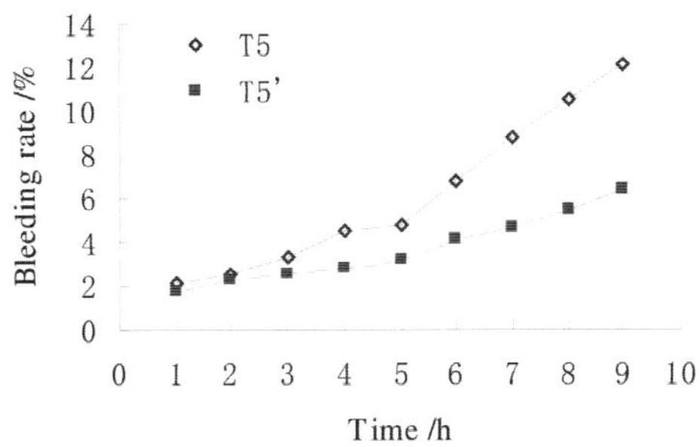

Figure 4. Bleeding rate of T3 and T3'.

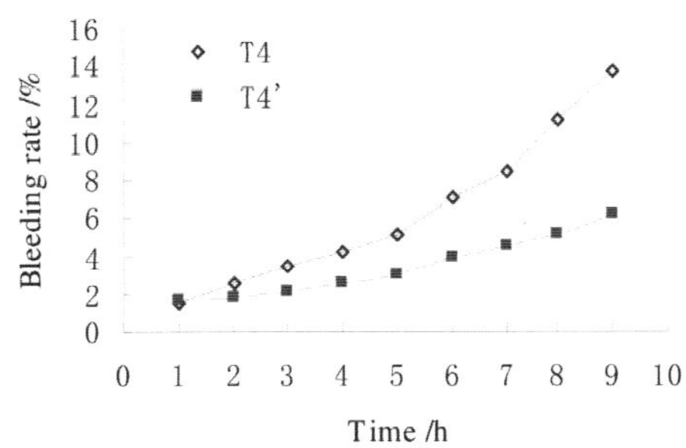

Figure 5. Bleeding rate of T4 and T4'.

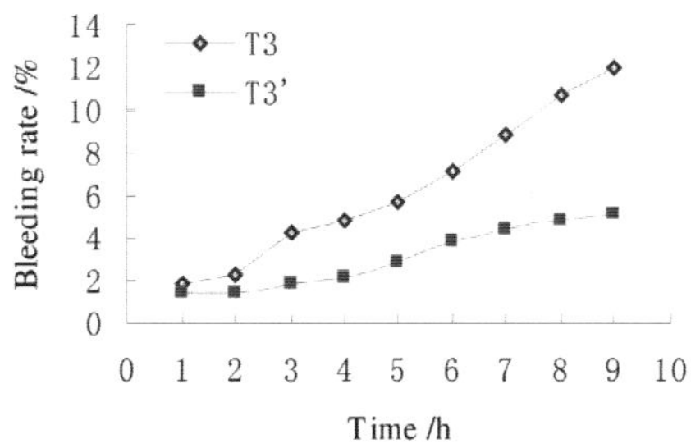

Figure 6. Bleeding rate of T5 and T5'.

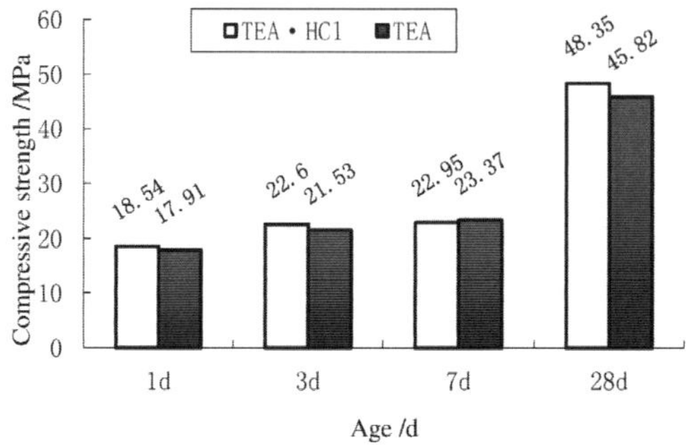

Figure 7. Comparison of compressive strength of cement mortar specimens.

3.5. Comparison of compressive strength with TEA

The strength comparative experiments were performed between the cement mortar mixed with TEA·HCl and the one with TEA. The experimental result is shown, as in figure 7.

As is shown in figure 7, we can see that the specimen strength of mixed TEA·HCl was slightly stronger than that of mixed TEA. However, the cost of TEA·HCL only accounts for about half of TEA. As for 28-day strength, the specimen strength of mixed with TEA·HCl was obviously higher than that of mixed with TEA.

4. Conclusions

Adding of the TEA·HCl hardening accelerator can accelerate the cement hydration, and what is more, when the dosage was 0.04%, the setting time was the shortest.

Adding of the TEA·HCl hardening accelerator has little effect on the initial fluidity paste, but has significant improvement on the fluidity of after 30 minutes, which is advantageous to the construction. The fluidity loss after 60 minutes increases.

After added TEA·HCl hardening accelerator, the initial slump has some improvement, especially for the slump loses after 30 min improves obviously. There is not much improvement for the slump loses after 60 min, which is roughly equivalent compared with that of blank samples.

The bleeding rate of concrete significantly decreases after adding TEA·HCl hardening accelerator.

References

[1] Aggoun S *et al.* 2008 Effect of some admixtures on the setting time and strength evolution of cement pastes at early ages *Construction and Building Materials* **22(2)** 106–110

[2] Angelskaar T and Iwata R 2007 *U.S. Patent No. 7198*669 (Washington DC: U.S. Patent and Trademark Office)

[3] Dransfield J 2003 Admixtures for concrete, mortar and grout. Advanced Concrete Technology - 1 *Constituent Materials* **1** 69

[4] Hardjito D *et al.* 2005. Fly ash-based geopolymer concrete *Australian Journal of Structural Engineering* **6(1)** 77

[5] Heikal M 2004 Effect of calcium formate as an accelerator on the physicochemical and mechanical properties of pozzolanic cement pastes *Cement and Concrete Research* **34(6)** 1 051–1056

[6] Hofmann T 2004 *U.S. Patent No. 6723163* (Washington DC: U.S. Patent and Trademark Office)

[7] J Pourehez, P Grosseau, R Guyormet and B Ruot HEC 2006 Influence on cement hydration measured by conductometry *Cement and Concrete Research* **36** 1777–1780

[8] Nmai CharleS K 1998 Cold weather concreting admixtures *Cement & Concrete Composites* **20**

121–128
[9] Peschard A, Govin A and Grosseau P 2004 Effect of polysaccharides on the hydration of cement paste at early ages *Cement and Concrete Research* **34(2)** 153–158
[10] Price W H 1951 Factors influencing concrete strength *In ACI Journal Proceedings* **47(2)**
[11] Reinhardt H W and Grosse C U 2004 Continuous monitoring of setting and hardening of mortar and concrete *Construction and building materials* **18(3)** 145–154
[12] Xu Sailong *et al.* 2009 Facile preparation of pure CaAl-layered double hydroxides and their application as a hardening accelerator in concrete *Chemical Engineering Journal* **155(3)** 881–885
[13] Singh N K *et al.* 2003 Effects of hydroxyethyl cellulose and oxalic acid on the properties of cement *Cement and Concrete Research* **33(9)** 1319–1329
[14] Sommer M, Wombacher F, and Bürge T A 2003 *U.S. Patent No. 6537367* (Washington DC: U.S. Patent and Trademark Office)

Luminescent properties of tetracyanoethylene combined with triphenylphosphine oxide

Lv Yuguang [1, a], Song Shanshan[1], Guo Qiang [1, 2], Yu Liansheng[1], Yang Libin [1, b], Song Di[1], Qin Yushan[1] and Wang Bo[1]

[1] College of Pharmacy, Jiamusi University, Jiamusi, 154007, China
[2] College of Pharmacy, Mudanjiang Medical University, 154007, China

E-mail: [a]yuguanglv@163.com, [b]ylb76@163.com

Abstract. The spectral properties of tetracyanoethylene triphenylphosphine oxide complex were investigated. The reaction mechanism was studied deeply, and the luminescent properties were explored through fluorescence spectroscopy, UV spectroscopy and Raman spectroscopy. The experimental results showed that under certain conditions, the triphenylphosphine oxide with tetracyanoethylene can form stable charge transfer complex.

Keywords: triphenylphosphine oxide; tetracyanoethylene; charge transfer complex

1. Introduction

The alias of tetracyanoethylene (TCNE) is four cyano supersede ethylene. Middleton found that TCNE can almost react with all the electron-rich substance in 1958 [1]. With the study of functional organic materials, it is found that as a kind of strong stable electron acceptor TCNE expands the range of applications [2]. In the field of nonlinear optics, magnetism, electricity, etc, TCNE has broad application prospects. The stable radical anion formed by the tetracyanoethylene is a strong π acid [3], TCNE can interact with multiple types of electron donors, then formed a stable charge transfer complex. Triphenylphosphine oxide is mainly used for the synthesis of organic compounds, phosphate salts and other phosphorus compounds. The rigidity of the triphenylphosphine oxide skeleton and the basicity of the oxygen atoms can induce crystallisation of other compounds which are difficult to crystallize [4].

The charge teansfer [5], also known as charge exchange. It is a charge transfer process occcuring when a positive ion collides with a neutral atom, and in recent years, through the discovery of molecular structure, it made the appropriate adjustments and realized the luminous efficiency of the intramolecular charge transfer compounds is enhance, it has been widely applied in the field of organic electroluminescence [6].

The charge transfer complex, also known as electron donors-receptor complex [7]. In recent decades, due to the fact that many scientific researchers pay more attention to the concepts and theories of the charge transfer complex, the study of the relevant charge transfer complex is more in-depth. In general, according to the differences of receptor molecules the charge transfer complex can be divided into three types, σ-acceptor reagents [8] charge transfer complex, n-acceptor reagents [9,11] charge transfer complex and π-acceptor reagents [12-15] transfer complex. Compared with the original material, the molecular solubility, viscosity, absorption, stability and molar volume [16] of charge transfer complex will change in a certain extent. The application of charge transfer complex in medicinal chemistry, analytical chemistry, semiconductor, biochemistry and other fields are more widely [17]. In this paper, the structural characterization of the triphenylphosphine oxide-tetracyanoethylene charge transfer complex was tested by using fluorescence spectrum [18,19], Raman spectrum and ultraviolet spectrummeans [20].

2. Experimental

2.1. Instruments and reagents

2.2.1 The UV spectra. The mixture of triphenylphosphine oxide solution, tetracyanoetliylene solution(0.3mL) and anhydrous ethanol were added to a stopper colorimetric tube. The sample was detected using UV-visible spectrophotometer, and then detected the buffer solution as a bland control.

2.2.2 The fluorescence spectra. The mixture of triphenylphosphine oxide solution, tetracyanoetliylene solution(0.3mL) and anhydrous ethanol were added to a stopper colorimetric tube. The sample was placed in constant temperature water bath for 40 minutes at 30℃, and cooled to room temperature. The sample was measured using fluorescence spectrophotometer.

2.2.3 The Raman scattering. The liquid was injected into the OTA-LK-1 liquid sample cell using a 2 mL glass vial. The spectral range was from 300 cm^{-1} to 3200 cm^{-1}, using carbon tetrachloride as the solvent. Based on the silver glass plate, the excitation source is an air-cooled argon ion laser (Spectra-Physics Model 163-C4260). The sample was detected using Roman spectrophotometer at 532nm, and Integration time is 30 s for 3 times, then take a parallel testing for 3 times.

3. Results and Discussion

3.1. UV spectra

The sample solution was prepared and tested at room temperature. The UV-visible absorption spectra were shown in figure 1.

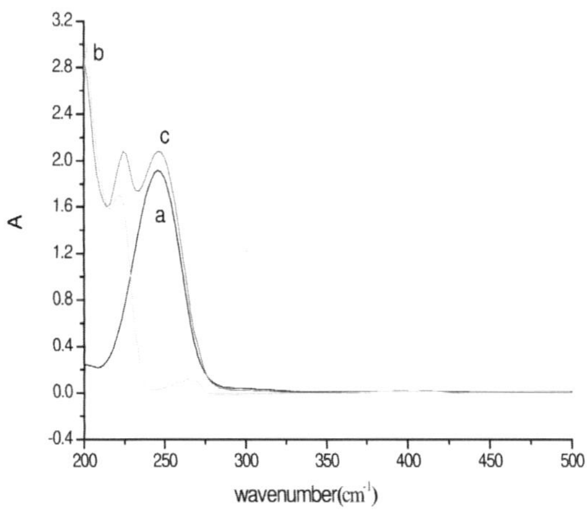

Figure 1. UV absorbance spectra of TCNE(a), triphenylphosphine oxide(b) and TCNE-triphenylphosphine oxide(c).

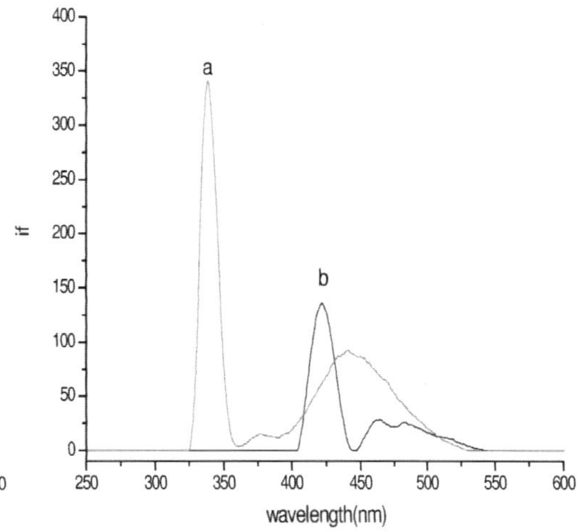

Figure 2. Fluorescence emission spectra of TCNE-triphenylphosphine oxide(a) and TCNE(b).

The results (figure 1) showed that (c) had the double absorption peaks at 224nm and 246nm, they were the characteristic absorption peaks of (a) and (b). And the maximum absorption peak of (c) at

246nm (the absorption value was 2.1). The absorption value of (c) compared with (a) and (b) had a siginificant increase, caused by π -π *transition, and it belongs to B absorption band.

3.2. fluorescence emission spectra

The fluorescence intensity of (a) and (b) with the excitation wavelength of 315nm was monitored at room temperature. The fluorescence curves were shown in figure2. The maximum fluorescence emission peak of curve (a) is at 327nm. Compared with the maximum emission peak position of curve (b), the fluorescence emission peak of curve (a) is redshift. The phosphorusatom of triphenylphosphine oxide has a lone pair electrons, it can form a transition of the electron energy level with TCNE. And the fluorescence intensity of (a) enhanced significantly. It indicated that triphenylphosphine oxide and TCNE can form a charge-transfer complex which has good fluorescence properties.

3.3. Raman scattering

The resonance Raman scattering spectrum of TCNE-Ph_3P=O complex in CCl_4 solvent was measured experimentally at the incident light of 532nm, and for comparison, the static Raman scattering spectrum of TCNE in CCl_4 solvent was also detected (figure 3 and figure 4).

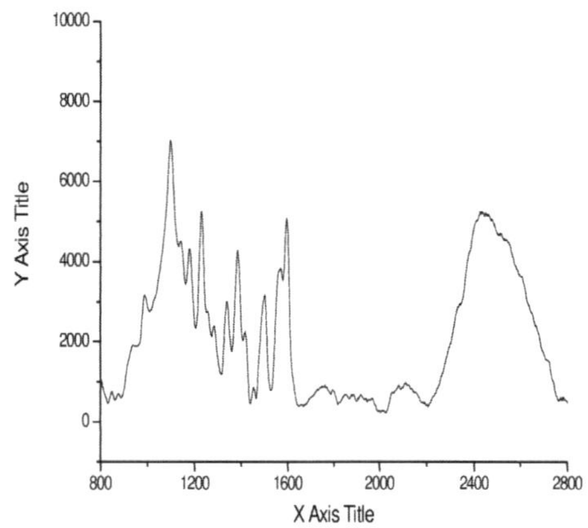

Figure 3. Raman scattering spectrum of TCNE.

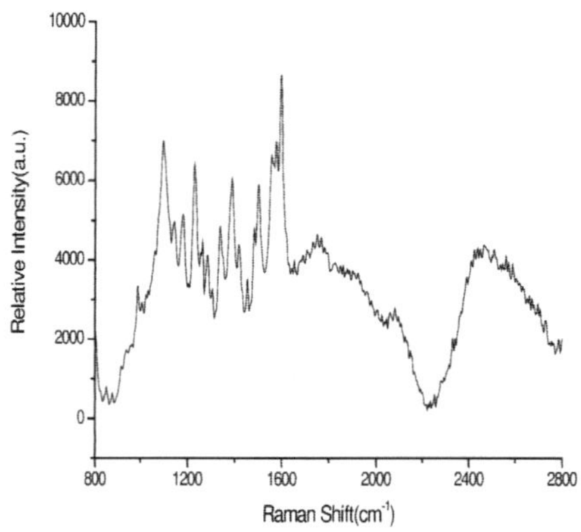

Figure 4. Raman scattering spectrum TCNE triphenylphosphine oxide.

It is found that the Raman peak at 2512cm^{-1} was the strongest in the static Raman scattering spectrum, while the Raman peaks around 1200cm^{-1}and 1600cm^{-1} were much stronger than these at 2512cm^{-1} in the resonance Raman scattering spectrum. The vibrational modes were strongly coupled at 1200cm^{-1} and 1600cm^{-1} because of the intermolecular charge transfer, and that the intensity of the Raman peak enhanced remarkably.

3.4. Preliminary Reaction Mechanism

Tetracyanoethylene is an important strong electron acceptor, and can form charge transfer complexes with a variety of types of electron donor. The phosphorusatom of triphenylphosphine oxid has a lone pair of electrons, it reacted with tetracyanoethylenebe as an electron donor, and produced a charge transfer complex of type n-π for the ratio of 1 to 1. The optimized structure was shown in figure 5.

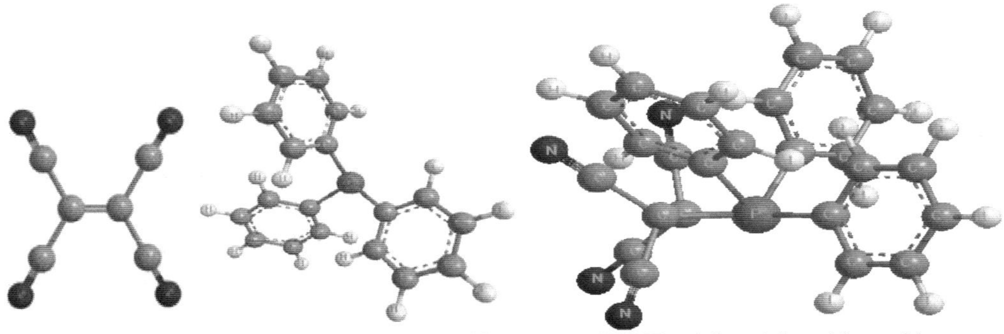

Figure 5. Molecular structure of TCNE, Triphenylphosphine oxide and TCNE triphenylphosphine oxide.

4. Conclusions

In this paper, the influence of the intermolecular charge transfer (electronic transition) has been studied through the Fluorescence spectra, UV spectra and resonance Raman scattering. The results show that in carbontetrachloride and ethanol, tetracyanoethylene with triphenylphosphine oxide forms stable charge-transfer complex.

Using intermolecular charge transfer, a series of tetrahydro-ethylene complexes has been, prepared, a multi-functional complex fluorescent probes built, and a new applications of nucleic acid probes explored and developed. The drug probe is the main tool to improve the accuracy of diagnosis and analysis with composite probe.

Acknowledgments

This study is financially supported by National Science Foundation of China(No.21473078 and 21346006), Department of scientific research project in Heilongjiang province(No.B201015), the national project training project of Jiamusi University (No. JMSUJCGP 2016-003), Scientific research project of Jiamusi University(JMSUJCMS2016-013), graduate scientific and technological innovation project of Jiamusi University (YM2016_092 and YM2016_093), Support was also given by the Key Laboratory of Photochemistry, Institute of Chemistry, Chinese Academy of Sciences, and the Key Laboratory of Luminescence Optical Information and Beijing Jiaotong University.

References

[1] Liya Peng *et al.* 2014 *Journal of Hangzhou Normal University(Natural Science)* The Synthesis, Characterization and Properties of O-anilino substituted 1,1,4,4-cyano-butadiene **13(3)** 291–196

[2] Vicker E B, Giles I D and Miller J S. 2005 *Chemistry of materials* M[TCNQ]y-based magnets(M=Mn, Fe, Co, Ni, TCNQ=7,7,8,8-tetracyano-p-quinodimethane) **17(7)** 1667–1672

[3] Webster O W 2002 *Journal of Polymer Science Part A: Chemistry* Cyanocarbons: a classic example of discovery-driven research **40(2)** 210–212

[4] Yingdong Lv *et al.* 2015 *NewJ.Chem* Substitution reaction of triphenylphosphine oxide with rare-earth metal phosphido methyl complexes **39** 7582–7588

[5] Jianzhang Zhao, Shaomin Ji, Yinghui Chen, Huimin Guo and Pei Yang. 2012 *Journal of the Owner Societies* Excited state intramolecular proton transfer (ESIPT): from principal photophysics to the development of new chromophores and applications in fluorescent molecular probes and luminescent materials **14** 8803–8817

[6] Dandan Liu *et al.* 2012 *Theoretical studies on the intramolecular charge-transfer luminescent materials and related molecular design* (Changchun: Jilin University) pp 1–3

[7] Wen J D *et al.* 2010 Chinese *Chemical Letters* Fluorescence quenching method for the determination of catechol with gold nanoparticles and tyrosinase hybrid system **21(3)** 346–348

[8] Rabie U M, Abou-El-Wafa M H and Mohamed R A 2007 *J Moled Struc* Interaction of 2-aminopyrimidine with σ-and π-acceptors involving chemical reactions via initial charge transfer complexation **871(1)** 6–13

[9] ZhaoY H, Gary H 2006 *Spectrochimica Acta Part B: Atomic Spectro-scopy* A spectral study of charge transfer and Penning processes for Cu, Zn, Ag, and Cd in a glow discharge **61(6)** 660–673

[10] Cheng S M and Chan W H 2006 *Tetrahedron* Hg^{2+} sensing in aqueous solutions: an intra-molecular charge transfer emission quenching fluorescent chemosensors **62** 8379–8383

[11] Jason Smith Henry Guly 2004 *Journal Of China Prescription Drug* Pentazocine and sustained-release morphine **11(3)** 87–91

[12] Ai H Z, Sheng H L and Yong W L 2003 *Chinese Chemical Letters* Two New Abietane Quinones from Isodon lophan-thoides var. Micranthus **14(6)** 591–593

[13] EL-Mossalamy E H 2006 *J.Moled Liq* Molecular spectroscopic studies of charge transfer complexes of thiourea derivatives with benzoquinones **123(16)** 118–123

[14] Liming Du, Mi Fu and Yaqin Du 2005 *Chinese Journal of Analytical Chemistry* The Norfloxacin of charge transfer complex fluorescence properties and applications **33(3)** 381–384

[15] Liming Du, Junpu Wang and Min Li 2006 *Chinese Journal of Analytical Chemistry* The measurement of Cephalosporinsr by Charge Transfe spectrophotometric method **34(1)** 142–145

[16] Gorindan K, Mark A T and Michael Z 1996 *J solution Chem* Effects of a Polarizable Medium on the Charge-Transfer States of the Photosynthetic Reaction Center from Rhodopseudomonas virid **1(25)** 75–78

[17] Na An *et al.* 2013 *Spectrophotometry study and application of Charge Transfer Reation of Pharmaceutical Analsis and interaction between Drugs and Bovine Serum Albumin* (Hohhot: Inner Mongolia University) pp 3–5

[18] Michael Borsch 2011 *J.Biol. Chem* Single-molecule fluorescence resonance energy transfer techniques on rotary ATP synthases **392** 135–142

[19] Chia-Pin Pan, Pedro L. Muino, Mary D Barkley and Patrik R Callis 2011 *J. Phys. Chem* Correlation of Tryptophan Fluorescence Spectral Shifts and Lifetimes Arising Directly from Heterogeneous Environment **115** 3245–3253

[20] Fushen Dai, Huizhen Lin, Junyong Jiang and Wuhua Yin 1995 *Chinese Journal of Pharmaceuticals* Preparation and clinical application of ofloxacin ear drops **30(4)** 213–216

IOP Conf. Series: Materials Science and Engineering 182 (2017) 012048 doi:10.1088/1757-899X/182/1/012048

Effect of Mn on microstructure and corrosion properties of extruded Mg-1%Zn alloy

Z M Zhang[1,2], Y Ma[1], Z Z Xi[2], C J Xu[1] and Z L Lv[1]

[1] School of Materials Science and Engineering, Key Laboratory of Electrical Materials and Infiltration Technology of Shaanxi Province, Xi'an University of Technology, Xi'an 710048, China
[2] School of Materials and Chemical Engineering, Shaanxi Key Laboratory of Optoelectronic Functional Materials and Devices, Xi'an Technological University, Xi'an, 710032, China

Email: rainmy905@163.com

Abstract. The microstructure of the extruded Mg-1Zn alloy doped with different content of manganese was analyzed by optical microscopy, scanning electron microscopy with energy dispersive X-ray spectroscopy and X-ray diffractometry. The mass-loss immersion method and electrochemical test were used to evaluate the corrosion properties. The results show that the microstructure of the extruded Mg-1%Zn-x%Mn (mass fraction, x=0.4, 0.8, 1.2) alloys consists of α-Mg and α-Mn, the grain size of α-Mg decreases with increasing Mn content. Electrochemical corrosion behavior of the alloys is similar. Mn has considerable effect on the corrosion rate, the corrosion process is exacerbated by the galvanic corrosion occurred at interface between α-Mg and α-Mn. The corrosion rate increases as the Mn content increases. Mg-1%Zn-0.4% Mn alloy exhibits the best corrosion resistance between the Mg-1%Zn-x%Mn alloys

1. Introduction

Magnesium alloys show potential application as degradable hard tissue implant materials due to their good biocompatibility and mechanical properties. Meanwhile, Mg alloys have the advantage that they do not show the undesirable effect of stress shielding, which occurs normally in conventional metallic implants made of stainless steel or titanium alloy which are widely used for bone implants [1,2]. It is expected that Mg alloys can be applied as lightweight, degradable load bearing orthopedic implant materials [3]. Thus, more researches have been focused on the study of biomedical Mg alloys [3,4]. However, the mechanical property of pure Mg is too low. Furthermore, pure magnesium corrodes too quickly in physiological system, and may lose mechanical integrity before tissues have sufficiently healed. Alloying is an important way to promote mechanical properties and corrosion resistance of Mg alloys [5]. The common alloying elements include Al, Zn, Mn, Ca and rare-earth. While toxicity researches about Zn, Mn have shown no harm to the human health [6].

The equilibrium solid solubility of Zn in Mg at eutectic temperature is around 6.2%, and it decreases with temperature [7]. In smaller amounts, Zn contributes to strength due to solid solution strengthening, it also acts, to some extent, as a grain refining agent and additionally contributes to grain boundary strengthening [6]. A study conducted by Song et al showed that the addition of Zn in pure Mg can significantly reduce corrosion rate [5]. But larger amounts (>2.0%) of Zn leads to a decrease of the corrosion resistance [6]. Zn is a essential element for bone metabolism [8]. As an active center for hundreds of enzymes, Zn element can promote cell renewing and healing of surgical wound [9]. However, it is believed that the released Mg ions may show a toxic effect on cells at high concentration. As a result, Zn content in biodegradable Mg alloys should not excess 2.0% [10].

Mn is mainly used to enhance ductility and strength [6,11], and can also improve heat resistance of Mg alloys and refine the grain size [7]. More important, Mn addition transforms Fe and other impurities into relatively harmless intermetallic compounds and can therefore be used to control the corrosion of magnesium alloys [6-7]. Mn can replace Zr additives in Mg-Zn-based alloys to reduce cost[7]. Mn is also one of the essential trace elements in bodies.

Mg-Zn alloys and Mg-Mn alloys have been investigated as biodegradable materials and have shown good biocompatibility [2,12]. Mg-Zn based alloys possess high yield strength and load capacity, and are especially suitable for high strength applications. Mn addition can improve casting property and plastic forming ability of Mg-Zn alloys [13]. Considerable attention has been focused on Mg-Zn-Mn wrought magnesium alloys, since high strength can be achieved after thermal processing [14]. Furthermore, Mg-Zn-Mn wrought alloys have advantage of low density, relatively high specific strength and stiffness, and can meet the requirement of light-weight construction. On the other hand, Mg-Zn-Mn alloys contain no rare earth elements and noble metal elements such as Zr, so lower cost can simply make them more widespread [14].

Mg-Zn-Mn alloys exhibite good biocompatibility [2]. Xu et al.[1] fabricated a Mg-Mn-Zn alloy with good mechanical and corrosion resistance properties. The study on corrosion behavior of hot extruded Mg-1Mn-1Zn alloy in 0.9% sodium chloride water solution showed that extrusion deformation can improve corrosion resistance [15]. Zhou et al. investigated the microstructure, mechanical property, corrosion behavior and biocompatibility of extruded Mg-Mn-Zn-Nd alloys [16]. The extruded alloys show good biocompatibility and much higher corrosion resistance than that of cast pure Mg. The extruded Mg-1Mn-2Zn-1.0Nd alloy shows a great potential for biomedical applications due to the combination of enhanced mechanical properties, high corrosion resistance and good biocompatibility [16]. The addition of cerium in Mg-2Zn-0.5Mn-1Ca alloy can enhance its in-vitro biocorrosion resistance and reduce its degradation rate. The alloy containing 1.35 % Ce has the potential to be used as future biomaterials [6]. Wang et al. fabricated a novel interpenetrating β-TCP/Mg-Zn-Mn composite by infiltrating Mg-Zn-Mn alloy into porous β-TCP [2]. The ultimate compressive strength of the composite is near with the natural bone. The electrochemical and immersion tests indicated that the corrosion resistance of the composite is better than that of the Mg-Zn-Mn matrix alloy [2].

Up to now, most researches have being concentrated on the effect of Mn on mechanical properties of Mg-Zn based alloys. A few of studies reported the role of Mn in Mg alloys, whereas the investigation of its effect on corrosion resistance is less, especially in physiological environment. On the basis of the previous studies, the corrosion behavior of extruded Mg-1%Zn-x%Mn alloys (mass fraction, x=0.4, 0.8, 1.2) in SBF solution (one of the simulated body fluids) was investigated by mass-loss immersion method and electrochemical test in the present paper, so as to provide reference for the application of the Mg alloys as degradable hard tissue implant materials.

2. Experimental

2.1. Alloy preparation

Commercial high-purity Mg ingot (>99.98%), zinc chip (>99.99%) and Mg-10%Mn master alloy were used to prepare the Mg-1%Zn-x%Mn alloys. All alloys were melted in a pit type electronic resistance furnace in a titanium crucible. When the melts were prepared, they were cast into a water-cooled copper mold with an ingot diameter of 51 mm. The melting, pouring, freezing and cooling process were all conducted under argon atmosphere. The ingots were then processed by a homogenization treatment, i.e. homogenized at 330 ℃ for 10 h and then furnace cooled. After that, the ingots were machined into rods of 49.5 mm diameter and 95 mm length, and then hot extruded into round bars of diameter of 16.5 mm at 320 ℃ with a ram speed of 20 mm/min and extrusion ratio of 9:1.

2.2. Corrosion property testing

Corrosion property of the experimental alloys was investigated by static immersion method and electrochemical test using SBF solution as corrosion medium.

The dimension of the immersion samples is 3 mm in height and 15 mm in diameter. The samples were abraded with SiC paper down to grade 1200, ultrasonically rinsed with acetone and washed completely with distilled water, and then dried at room temperature. Then the weights were measured by a electric balance with precision of 0.1 mg. All samples were immersed in SBF solution at 37 ℃ for 72 h. Then the samples were weighed again after the corrosion products were removed completely. Corrosion rate R was calculated according to equation (1):

$$R = \frac{8.76 \times 10^7 \times (m_0 - m_1)}{St\rho} \tag{1}$$

where R is corrosion rate, mm/a; m_0 and m_1 are weights of the samples before and after immersion, respectively, g; S is surface area of the samples exposed to the corrosive media, cm^2; t is the exposure time, t=72 h; ρ is the density of the samples, ρ=1.74 g/cm^3.

Cylindrical samples (12 mm diameter and 10 mm height) were used to perform the potentiodynamic polarization test. The end surfaces of the samples were vertical to extrusion direction. Except for one end surface, the samples were molded into mixed resin (epoxy resin/polyamide resin=1:1). A three-electrode cell was used for potentiodynamic polarization test by Electrochemical Workstation CS-310. The working electrode was the sample, the counter electrode was made of platinum and the reference electrode was saturated calomel electrode (SCE). The test was conducted at 37 ℃ in open air in a three necked round bottom flask containing 250 ml SBF solution. Prior to potentiodynamic polarization test, the samples were soaked in the solution for 10 min. All experiments were carried out at a constant scan rate of 0.5 mV/s from -100 mV SCE to 100 mV SCE.

2.3. Microstructure and corrosion morphology observation

Cross-sectional microstructures of the extruded bars were examined by optical microscopy (Olympus GX71) after etching with 4% nital solution. The distribution and surface morphology of secondary phase were observed by the scanning electron microscope JSM-6390A with an energy dispersion microanalysis system (EDS). A Shimadzu XRD-7000 X-ray diffractometer (XRD) was used to identify chemical constitution.

The observation of corrosion product morphology was carried out by SEM after the samples were removed from the SBF solution, gently rinsed with distilled water and air dried completely. For the characterization of the surface appearance of corroded specimens, it was conducted after the corrosion products were completely removed.

3. Results and discussion

3.1. Microstructure and phase constitution

Figure 1 shows the OM microstructure of the hot extruded Mg-1%Zn-*x*%Mn alloys with different Mn content. It can be seen obviously that the microstructure consists of nearly equiaxed grains, and the grain size decreases with increasing Mn content. In addition, the streamlines resulted from the extrusion deformation are also discernible (figure 1a). This phenomenon indicates that the incomplete dynamic recrystallization occurred during extrusion and thus resulted in equiaxed grains.

Figure 2 shows that the predominant phase of the Mg-1%Zn-*x*%Mn alloys is α-Mg phase. According to the binary Mg-Zn diagram[17], it can be observed that up to 2.0%Zn can completely dissolve in α-Mg. Thus, no any compound was found in Mg-Zn alloy. Meanwhile, no Mn peaks appeared in figure 2, this is because the amount of Mn is too little to be detected. Comparing the 2θ angles of α-Mg with that of pure Mg[1], one will see that the 2θ angles of α-Mg are smaller than that of pure Mg. Evidently, the difference of 2θ angles arises from the lattice deformation of Mg that was resulted from the solid solution of Zn which enlarged the crystal plane spacing.

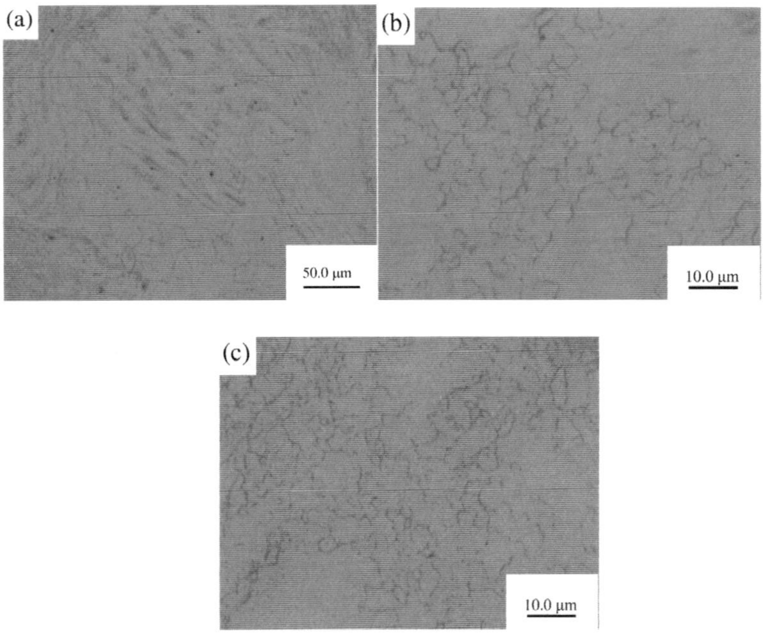

Figure 1. OM microstructures of the hot extruded Mg-1%Zn-x%Mn alloys with different content of Mn. (a) x=0.4Mn, (b) x=0.8Mn, (c) x=1.2Mn.

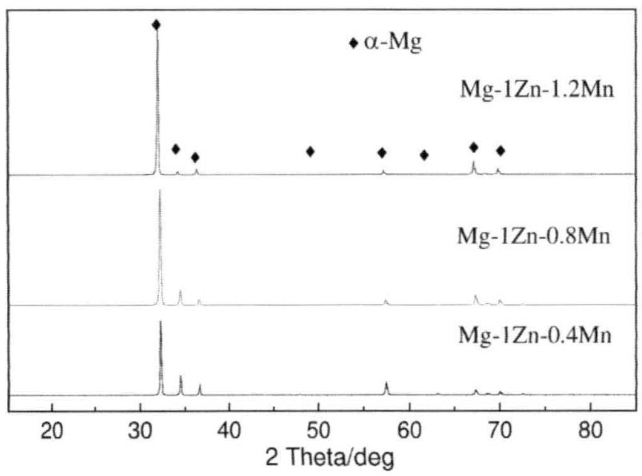

Figure 2. XRD pattern of Mg-1%Zn-x%Mn alloys with different content of Mn.

Figure 3 shows the SEM microstructures of the alloys under high magnification and EDS results. Bright secondary phases with small size are clearly visible. EDS results shown that except for Mg, higher content of Mn exists in the secondary phase. Detecting trace compositions in small sized particles by microstructural observation is often associated with high difficulties[17]. Considering small size of the bright secondary phases (less than 0.5 μm), the microzone of EDS analysis may cover part of α-Mg, this is the reason why high content of Mg was detected in the secondary phases. According to Mg-Mn binary phase diagram, very limited solubility of Mn in Mg and negligible solubility of Mg in Mn are observed, and no intermetallic compound is reported in Mg-Mn system[17], therefore it is certain that the secondary phase in the alloys is α-Mn.

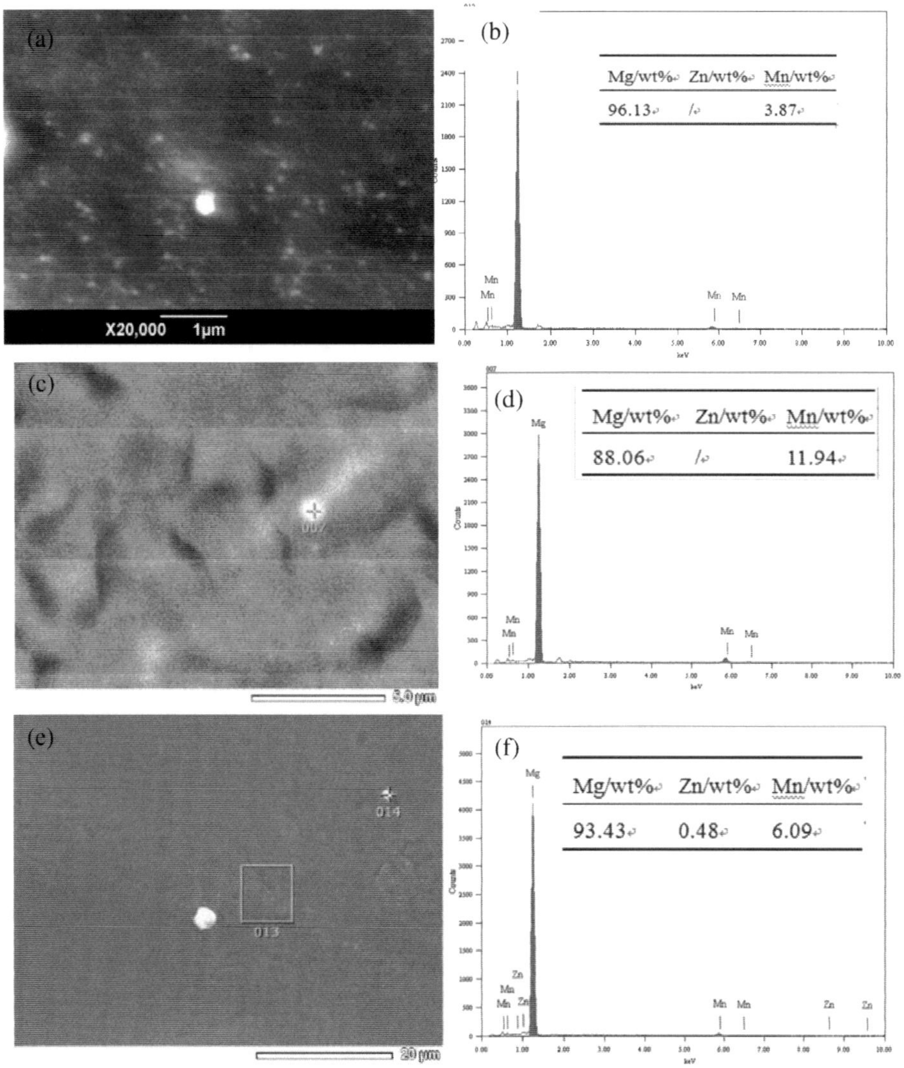

Figure 3. Secondary phases (shown by arrows) and their EDS results in the Mg-1%Zn-x%Mn alloys with different Mn content. (a,b) x=0.4Mn; (c,d) x=0.8Mn; (e,f) x=1.2Mn.

According to Mg-Zn binary phase diagram [17], the maximum solubility of Zn in Mg is about 6.2%. As temperature decreases, the solubility decreases prominently and changes to around 2.0% at ambient temperature. Consequently, the microstructure of Mg-1Zn alloy is composed of α-Mg. No any other secondary phase form by further processing. It can be known from Mg-Mn binary phase diagram [17] that the maximum solubility of Mn in Mg is about 2.2%, and decreases to below 0.1% at room temperature. As a result, α-Mn may precipitate from α-Mg during cooling process.

Very few experimental works on the Mg-Mn-Zn alloys are available in the literature, no ternary compound could be found in the literature [17]. According to Mg-Mn-Zn ternary phase diagram [17], the primary phase of the Mg-1%Zn-*x*%Mn alloys is α-Mg. The solubilities of Zn and Mn in Mg all decrease with temperature decreasing. For example, the solubility of Zn and Mn in Mg is 6.2% and 0.52% at 347 ℃, and decrease to 3.9% and 0.07% respectively at 250 ℃. Therefore, the equilibrium microstructure of the Mg-1%Zn-*x*%Mn alloys at ambient temperature consists of α-Mg and α-Mn. In the present study, Mg-1%Zn-*x*%Mn alloys solidified and cooled in a water-cooled copper mold, the corresponding freezing rate and cooling velocity were relatively high, and Mn and Zn were all dissolved in α-Mg after solidification. Therefore, α-Mn phases in the alloys (figure 3) would be precipitated during cooling period. Another possibility is that α-Mn phase may also precipitate during homogenization treatment, hot extrusion and following cooling process. Whereas, because of small amount of α-Mn and its small size, no diffraction peak of α-Mn was detected in XRD analysis.

It is proved that the dispersed tiny α-Mn particles can impede dynamic recrystallization and hinder growth of recrystallized grains [8,11]. As a result, the grain size of Mg-1%Zn-*x*%Mn alloys decreased as Mn content increasing.

3.2. Immersion corrosion behavior

Figure 4 reveals the average corrosion rate of the extruded Mg-1%Zn-*x*%Mn alloys measured by mass-loss method after immersion in SBF solution at 37 ℃ for 72 h. It can be seen that the corrosion rate of Mg-1Zn-0.4Mn alloy is the lowest (4.23 mm/a). As Mn content increases, the corrosion rate increases and thus the corrosion resistance declines. The highest corrosion rate is observed for Mg-1Zn-1.2Mn alloy (4.92 mm/a).

Figure 5 shows the potentiodynamic polarization curves of the alloys in SBF solution. In general, cathodic polarization curve is assumed to indicate hydrogen evolution via water reduction, however the anodic polarization curve exhibites the dissolution of Mg [18]. It appears that the shape of the polarization curves is similar, this indicates that the dynamic behavior of electrochemical corrosion reaction in the alloys is similar too. It can be found from the shape of anodic polarization curves that no passive region appears. This means that the Mg alloys were electrochemically active during the early stage of corrosion in SBF solution. No protective film formed on the surface of the sample.

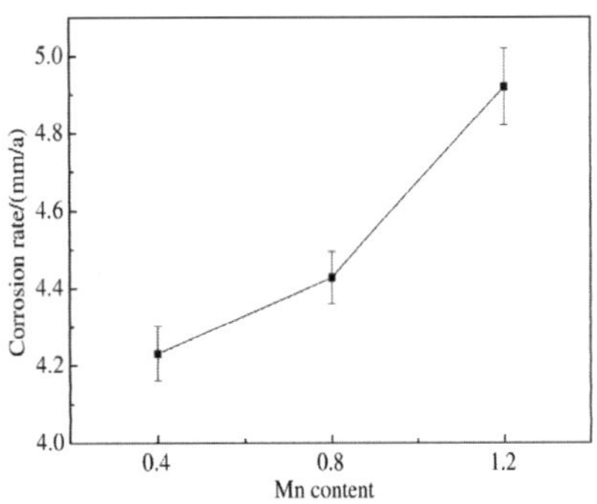

Figure 4. Average corrosion rate of the extruded Mg-1%Zn-*x*%Mn alloys with different Mn content immersed in SBF solution for 72h.

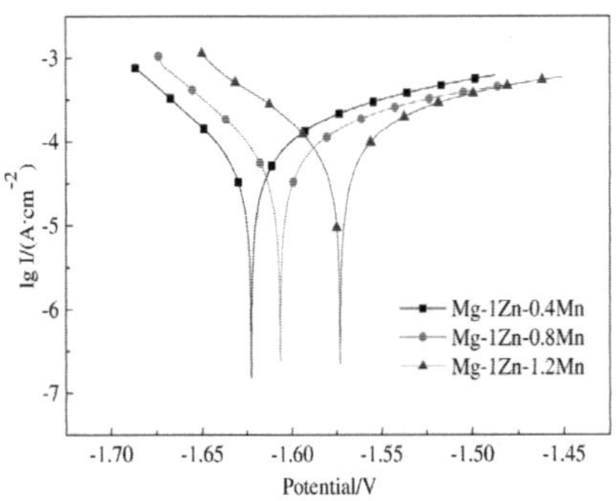

Figure 5. The potentiodynamic polarization curves of the extruded Mg-1%Zn-*x*%Mn alloys with different Mn content.

Table 1 lists the electrochemical parameters calculated from the polarization curves (figure 5) by traditional Tafel fitting. It shows that the self corrosion potential of the alloys shift towards nobler direction with increasing Mn content. Mg-1Zn-0.4Mn alloy exhibits the lowest corrosion potential of about -1.619 V. By addition of 1.2% Mn, the corrosion potential shifts towards nobler direction (-1.578 V SCE), which is about 41 mV nobler than that of Mg-1Zn-0.4Mn alloy. The corrosion current density of Mg-1Zn-0.4Mn alloy is 0.111 mA/cm^2, which is nearly the same with that of Mg-1Zn-0.8Mn alloy (0.122 mA/cm^2). It should be noted that all of them are much lower than that of Mg-1Zn-1.2Mn alloy (0.213 mA/cm^2). This indicates that Mn has considerable effect on the corrosion rate of the alloys. The corrosion rate of Mg-1Zn-1.2Mn is 4.72 mm/a which is nearly twice than that of Mg-1Zn-0.4Mn alloy (2.44 mm/a). The electrochemical test illustrates that the corrosion resistance of Mg-1Zn-0.4Mn alloy is the highest, this result is consistent with the mass-loss immersion test. It was reported [19] that the corrosion current density and corrosion rate of rolled Mg-1Zn alloy are 0.041 mA/cm^2 and 0.92 mm/a, respectively. By comparison with Mg-1Zn-0.4Mn alloy, Mg-1Zn has less current density and lower corrosion rate. This phenomenon confirms the significant effect of Mn on the corrosion behavior of the alloys. As Mn content increases, the corrosion rate increases.

Table 1. Electrochemical parameters of Mg-1%Zn-x%Mn alloys in SBF solution.

Alloy	Corrosion potential (V)	Corrosion current density I (mA.cm^{-2})	Corrosion rate (mm.a^{-1})
Mg-1Zn-0.4Mn	-1.619	0.111	2.455
Mg-1Zn-0.8Mn	-1.602	0.122	2.699
Mg-1Zn-1.2Mn	-1.578	0.213	4.720

3.3. Corrosion morphology observation

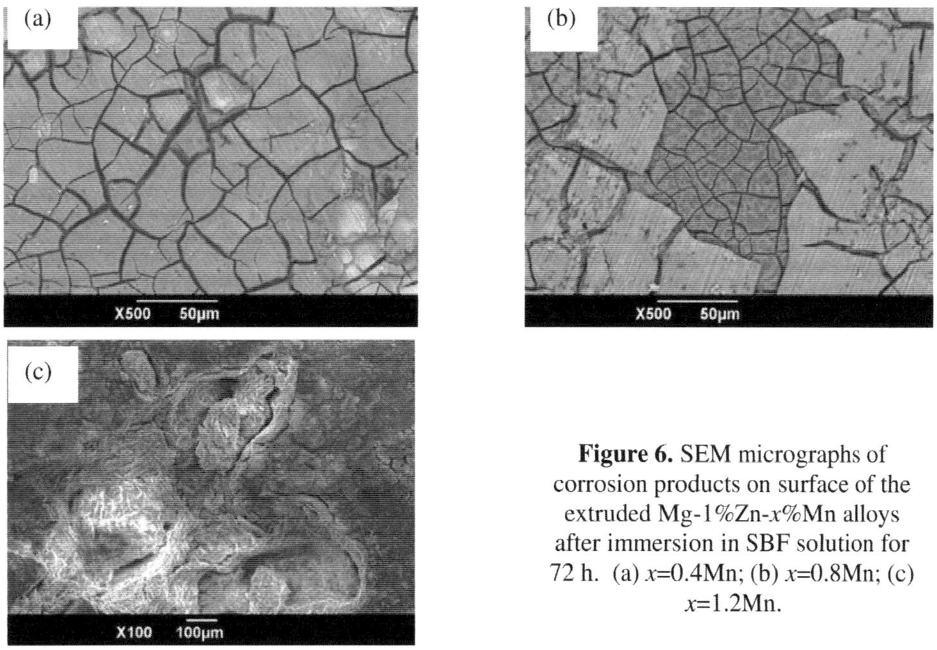

Figure 6. SEM micrographs of corrosion products on surface of the extruded Mg-1%Zn-*x*%Mn alloys after immersion in SBF solution for 72 h. (a) *x*=0.4Mn; (b) *x*=0.8Mn; (c) *x*=1.2Mn.

Figure 6 shows the morphologies of corrosion product of the Mg-1%Zn-*x*%Mn alloys. During the early stage, the alloys were electrochemically active, no protective layer formed. However, as corrosion proceeded, more and more corrosion product produced and accumulated which led to the formation of a protective layer on the surface of the samples. It can be seen from figure 6 that the surface of Mg-1Zn-1.2Mn was covered with a layer of corrosion product accompanied by deep and large pits, which indicates that serious corrosion occurred. Whereas, no obvious corrosion pits were found on the corrosion product layer of Mg-1Zn-0.4Mn alloy, and slight peeling of corrosion products were observed for Mg-1Zn-0.8Mn alloy. The network-like cracks can probably be attributed to the dehydration of corrosion product layer after removal of the samples from the test solution and drying in warm air [1].

The surface morphologies of the soaked samples after removal corrosion product are depicted in figure 7. Obviously, corrosion pits and filiform corrosion took place on the alloys. The filiform corrosion on Mg-1Zn-0.4Mn alloy is clearly visible and distributes along the deform streamlines. This phenomenon demonstrates that corrosion proceeded in a certain direction. For Mg-1Zn-0.8Mn alloy, there are a few of shallow corrosion pits and superficial filiform corrosion. However, besides the filiform corrosion, a few deep corrosion pits appear on Mg-1Zn-1.2Mn sample surface. Therefore, it can be concluded that the corrosion severity is in order of Mg-1Zn-0.4Mn, Mg-1Zn-0.8Mn and Mg-1Zn-1.2Mn after 72 h of immersion in SBF solution. Mg-1Zn-1.2Mn alloy corroded heavily as indicated by deep corrosion pits, this led to the highest corrosion rate.

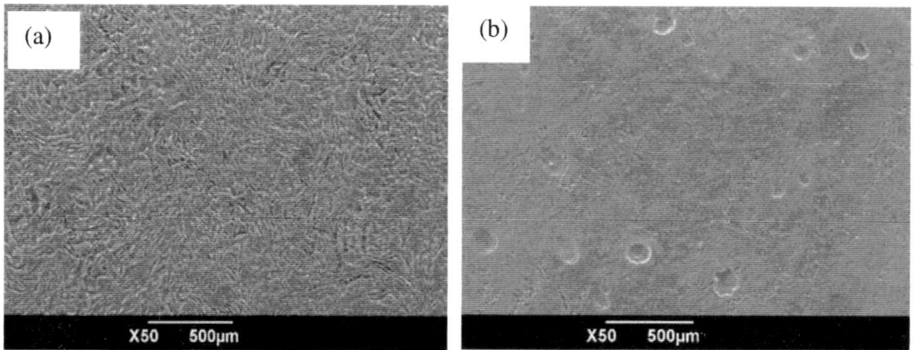

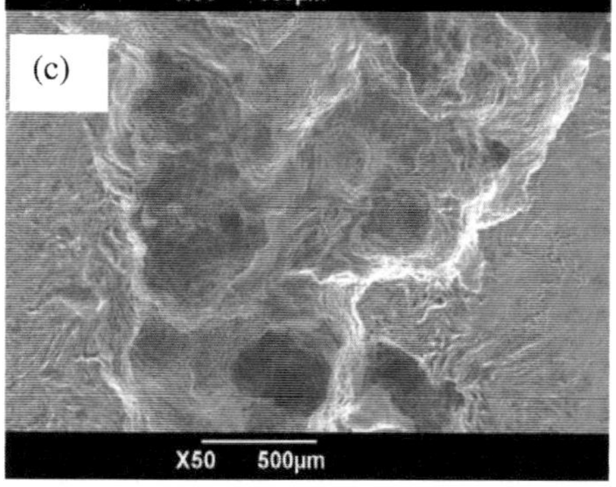

Figure 7. SEM surface micrographs of the extruded Mg-1%Zn-x%Mn alloy after immersion in SBF solution for 72 h. (a) x=0.4Mn, (b) x=0.8Mn, (c) x=1.2Mn.

The standard corrosion potentials of Mg, Mn, Zn are -2.37 V, -1.182 V and -0.762 V. The corrosion potential of α-Mg containing small amount of Zn or Mn is slightly higher than that of pure Mg [15,20], thus the addition of Mn reduced the corrosion rate of the Mg alloys. Microgalvanic corrosion plays a dominate role in corrosion. The corrosion potential of α-Mn is much higher than that of α-Mg. Therefore, α-Mn acted as cathode in microgalvanic corrosion. Precipitation of α-Mn promoted cathode process and resulted in the movement of corrosion potential of α-Mn to a nobler direction. As a result, microgalvanic corrosion was exacerbated.

From the result of the electrochemical experiment (figure 5), one can see that no protective layer formed at early corrosion stage. Whereas, as the corrosion proceeded, more and more corrosion products accumulated, and eventually the sample was completely coated. As a result, the corrosion product layer could protect the sample from being corroded to some extent, and decreased the corrosion rate to some extent. The quantity and area of minor corrosion pits under corrosion product layer increased with immersion period, which led to the formation of filiform corrosion. The corrosion product layer was porous and loose, and could not prohibit the corrosion reaction effectively. Consequently, corrosion continued, and thus large and deep corrosion pits were formed.

It was reported in reference [15] that the corrosion rate of Mg alloy decreases with immersion time, i.e., corrosion rate is time-dependent. This result explains why the corrosion rates from electrochemical test and immersion test are different, the reason is that corrosion rate in electrochemical test is a instantaneous value, while corrosion rate in immersion test is an average value for 72h.

4. Conclusions

(1) The Zn and Mn elements in Mg-1%Zn-*x*%Mn alloys (mass fraction, *x*=0.4, 0.8, 1.2) can dissolve completely in α-Mg during solidification, whereas Mn precipitates as α-Mn during cooling period. The microstructure of Mg-1%Zn-*x*%Mn alloys is composed of α-Mg and α-Mn. The grain size decreases with increasing Mn content.

(2) Mn has considerable effect on the corrosion rate of the alloys. The corrosion process is exacerbated by the galvanic corrosion occurred at interfaces between α-Mg and α-Mn. As Mn content increases, the corrosion rate increases. Mg-1Zn-0.4Mn exhibits best corrosion resistance.

(3) Electrochemical corrosion behavior of the Mg-1%Zn-*x*%Mn alloys is similar, no passivation occurs during corrosion process. The alloys are highly electrochemically active, and the effectively protective layer cannot form during corrosion.

Acknowledgments

This work was financially supported by the Shaanxi provincial project of special foundation of key disciplines, the research project of Shaanxi Key Laboratory of Optoelectronic Functional Materials and Devices (No.ZSKJ201403) and the scientific research project of Education Department of Shaanxi Province (No.2013JK0906)

References

[1] Wang X, Dong L H, Ma X L and Zheng Y F 2013 *Mater. Sci. Eng.* C **33**(2) 618–25
[2] Wang X, Zhang P, Dong L H, Ma X L, Li J T and Zheng Y F 2014 *Mater. Design* **54** 995–1001
[3] Mark P S, Alexis M P, Jerawal H and George D 2006 *Biomaterials* **27**(9) 1728–34
[4] Kannan M B and Raman R K 2008 *Biomaterials* **29**(15) 2306–14
[5] Song G L 2006 *Corrosion and protection of magnesium alloys* (Beijing: Chemical industry press) p238
[6] Zhang F, Ma A B, Song D, Jiang J H, Lu F M, Zhang L Y, Yang D H and Chen J Q 2015 *J. Rare Earths* **33**(1) 93–101
[7] Yuan J W, Zhang K, Zhang X H, Li X G, Li T, Li Y J, Ma M L and Shi G L 2013 *J. Alloys & Comp.* **578**(6) 32–36
[8] Zhuang Y, Yu Y D and Pei T 2013 *J. Harbin Univ. Sci. Tech.* **18**(1) 55–60
[9] Zhang S X, Li J N, Song Y, Zhao C L, Zhang X N, Xie C Y, Zhang Y, Tao H R, He Y H, Jiang Y and Bian Y J 2009 *Mater. Sci. Eng.* C **29**(6) 1907–19
[10] Yu G, Liu Y L, Li Y, Ye L Y, Guo X H and Zhao L 2002 *The Chinese J. Nonferrous Met.* **12**(6) 1087–1098
[11] Zhang D F, Qi F G, Shi G L and Dai Q W 2010 *Rare Met. Mater. Eng.* **39**(12) 2205–10
[12] Zhang S X, Zhang X N, Zhao C L, Li J N, Song Y, Xie C Y, Tao H R, Zhang Y, He Y H, Jiang Y and Bian Y J 2010 *Acta Biomaterialia* **6**(2) 626–40
[13] Zhang D F, Shi G L, Zhao X B and Qi F G 2011 *Mater. Review* **25**(9) 1–10
[14] Zhang D F, Shi G L, Zhao X B and Qi F G 2011 *Trans. Nonferrous Met. Soc. China* **21**(1) 15–25
[15] Zhang Z M, Zhang J, Ma Y, Wang T and Xu C J 2012 *Foundry Technology* **33**(3) 257–60
[16] Zhou Y L, Li Y C, Luo D M, Ding Y F and Hodgson P 2015 *Mater. Sci. Eng.* C **49**(1) 93–100
[17] Ghosh P and Medrag M 2013 *CALPHAD: Computer Coupling of Phase Diagrams and Thermochemistr* **41**(6) 89–107
[18] Chang J W, Guo X W, Fu P H, Peng L M and Ding W J 2007 *Electrochim Acta* **52**(9) 3160–67
[19] Gu X N, Zheng Y F, Cheng Y, Zhong S P and Xi T F 2009 *Biomaterials* **30**(4) 484–498
[20] Chen X Y, Wang J C, Yang G C and Guo Y N 2008 *Foundry Technology* **29**(5) 593–596

Mechanical and corrosion properties of biodegradable Mg-1.5Mn-1Ca-xSr alloys

X Sun[1,*], S Y Sun[2], Y H Ning[1] and Y T Ning[3]

[1] Datang northeast electric power test and research institute, Changchun, Jilin, China, No. 3195 Weishan Street, Changchun, 130000, China.
[2] Brand School of Engineering and Technology, Changchun Vocational Institute Of Technology, Changchun 130033, China
[3] State Grid Jilin Electric Power Company, Changchun, Jilin, China, No. 4969 Renmin Street, Changchun, 130000, China.

E-mail: 380541075@qq.com

Abstract. The mechanical properties and corrosion mechanism of both as-cast and solution-naturally age (T4) treated Mg-1.5Mn-1Ca-xSr alloys were investigated. The results showed that Sr is helpful to decrease grain size and increase the strength. The corrosion process of alloys was mainly determined by the quantity and distribution of second phases. $Mg_{17}Sr_2$, α-Mn and Ca-Sr phases acted as cathodes accelerated the corrosion of Mg_2Ca anodic phase and α-Mg matrix. However, continuous distributed $Mg_{17}Sr_2$ was beneficial to resist the happening of localized corrosion because of its barrier effect. T4 treatment could significantly improve the mechanical properties and corrosion resistance of Mg alloys because of the dissolution of Mg_2Ca phase and the dispersive distribution of $Mg_{17}Sr_2$ and α-Mn phases.

Keywords: Biodegradable; T4 treatment; Microstructure; Mechanical properties; Corrosion.

1. Introduction

The Mg and its alloys have attracted considerable attention in recent years as potential biodegradable implant materials due to their good biocompatibility [1-3]. The most obvious advantage of magnesium compared with the other biomedical metal material is the relatively negative standard electrode potential (-2.37 V at 25 °C) and its biological safety, which make magnesium able to be biodegraded in human body environment [4,5]. Hence, the second removal surgical operation for implants could be avoided. The suitable elastic modulus of magnesium (40~45GPa) is similar to natural bone, which will effectively prevent the happening of "stress-shield" phenomenon [6]. However, there are also some concerns in using Mg and Mg alloys for load-bearing implant materials. One is the relative poor mechanical performance compared with the other commonly used metallic implants. The second is the uncertain capability of Mg alloys due to its low corrosion resistance [4,7]. Thirdly, the biocompatibility of current Mg alloys is still inadequate for biomedical applications.

T4 treatment and alloying are two promising ways to overcome the disadvantages mentioned above [8]. As a common means to refine the microstructure of castings, T4 treatment can help to dissolve and disperse second phases, which is helpful to eliminate the negative effect of coarse phases on the mechanical and corrosion properties. The addition of alloying elements into Mg alloys would increase the electrode potential of Mg matrix and support solution strengthening and second phase strengthening effects. Base on the toxicological date issued by World Health Organization (WHO) [9, 10], Ca, Zn, Mn (essential trace elements for human body) and Si, Sr, Al, RE (probably essential for human body) can be chosen as potential healthy alloying elements for the application of biological metal. Mg alloys with Al or RE had been proved to have good mechanical properties and corrosion resistance. However, due to the uncertain biological safety of Al and RE, there is always a lot of dispute revolved about their addition into Mg alloys [11]. Mg-Ca system alloys are the most promising biomedical magnesium alloys due to the good biocompatibility of Ca [12,13], however, the mechanical properties and corrosion resistance of Mg-Ca system alloys are not good enough. Mg-Zn system alloys with the addition of Ca or/and Mn as alloying elements had also been investigated [14-17]. It is believed that the corrosion rate of Mg alloys increased with increasing content of Zn and the second phases have great influence on corrosion resistance [15-18], which indicating that a newly developed alloy without controversial alloying elements and has less amount of second phase may be a promising degradable biomedical material with a suitable mechanical and corrosion properties.

Ca is one of the most promising biocompatible alloying elements for Mg alloys. It is regarded as the most abundant mineral in the body and plays a crucial role in the formation of bone. The addition of Ca can not only improve the metallurgical quality of magnesium alloys, but also reduce oxidation during smelting and heat treatment processing [19]. Mn has little effect on the tensile strength of Mg alloys. However, Mn can significantly improve the corrosion resistance of Mg alloys by removing iron and other heavy-metal elements. Mn has no toxic and plays a primary role in the activation of multiple enzyme system, i.e., hydrolases, kinases, transferases, decarboxylases and mitochondrial respiration [20,21]. Sr is always known as a grain refiner for magnesium and is beneficial to improve the mechanical properties and corrosion resistance for certain Mg alloys [22,23]. Moreover, Sr can also improve bone strength and bone mineral density, and is used in the treatment for osteoporosis. Sr has been shown to decrease the osteoclast activity and increase the replication of pre-osteoblast cells, thereby decreasing the bone resorption while stimulating bone formation [24-26]. In consideration of human safety, Mn, Ca and Sr were finally chosen as alloying elements in this work to improve both the corrosion resistance and mechanical properties of Mg. The probability of Mg-Mn-Ca and Mg-Mn-Ca-Sr alloys used as biodegradable implant materials with low corrosion rate and high mechanical property was invested.

2. Materials and methods

2.1. Preparation of magnesium alloys

Two alloys with targeted compositions of Mg-1.5Mn-1Ca and Mg-1.5Mn-1Ca-1Sr alloys (wt. %, hereafter) were prepared in a self-assembly non-ferrous alloy melting equipment. The alloying components were melted around 1053 K (780 °C). After proper treatment, magnesium melt poured into a copper mold at 1023 K (750 °C) and cooled to room temperature under the protection of high purity argon atmosphere (99.999 %). During the melting process, the melt was mechanical stirred. Cast bars that cut from the ingot were then solution-treated at 480 °C for 16 h, and quenched rapidly in water. The analyzed chemical compositions of prepared alloys were determined by inductively coupled plasma-atomic emission spectrometer (ICP-AES), the results are given in table 1.

Table 1. Chemical compositions of Mg-1.5Mn-1Ca-xSr alloys (wt. %).

Alloys	Mn	Ca	Sr	Mg
Mg-1.5Mn-1Ca	1.585	1.049	-	Bal
Mg-1.5Mn-1Ca-1Sr	1.614	1.048	0.9867	Bal

2.2. Microstructure and Mechanical test

Samples for microstructure observation were cut from the ingot and the solution-treated bar. For metallographic observation, the specimens were mechanically polished and etched with a dilute solution of 4% nitric acid in methanol. Microstructural analyses were examined by using Optical microscopy and scanning electronic microscope (SEM) equipped with an energy dispersive spectrometer (EDS). Phase identification was performed on the bulk samples by using EDS and X-ray diffraction (XRD). Mechanical testing samples were prepared according to ASTM B557-06. Tensile tests were processed by MTS810 universal testing machine with a strain rate of 1×10^{-3} s^{-1} at room temperature. At least 3 samples were tested for each condition.

2.3. *Preparation of corrosion media*

In order to evaluate the degradation properties, electrochemical measurements and immersion tests were performed in simulated body fluid (SBF), containing NaCl 8.035 g/l, KCl 0.225 g/l, CaCl2 0.292 g/l, $NaHCO_3$ 0.355 g/l, $MgCl_2\cdot6H_2O$ 0.311 g/l, Na_2SO_4 0.072 g/l, $KH_2PO_4\cdot3H_2O$ 0.231 g/l [16]. The pH value of SBF was adjusted to a neutral level (pH = 7 ± 0.1). The solution temperature was controlled around 37 ± 0.5 °C, which is aim to simulate human body environment. Rectangular specimens for corrosion tests were molded into epoxy resin with only one side of 1 cm^2 exposed for the test

2.4. *Electrochemical test*

The potentiodynamic polarisation test was carried out at 37 ± 1 °C in a beaker containing 300 ml of SBF solution by using a standard three-electrode configuration: the saturated calomel as a reference, a graphite electrode as the counter and the sample as the working electrode. The samples were encapsulated into epoxy resin with a working surface of 1 cm^2. Prior to the experiment, the rectangular samples were ground up to 2000 grit by SiC paper, followed by washing with distilled water and acetone. The polarization curve was measured at a scan rate of 0.5 mV/s.

2.5. *Immersion test*

After grinded with SiC paper up to 2000 grit, the rectangular samples were rinsed with distilled water and dried. Every sample was placed in a beaker containing 250 mL of SBF solution. Hydrogen bubbles resulting from corrosion of each specimen were collected in a burette, so the volume of the generated hydrogen could be measured through reading the solution level in the burette. Detailed procedures for the hydrogen measurement have been reported elsewhere [27]. After short-term and long-term immersion test, samples were taken out of the soaking medium and cleaned in boiling 15% CrO_3 + 1% $AgCrO_4$ solution for 10 min to remove the corrosion products. Then the surface morphology without corrosion products could be observed by using laser scanning confocal microscope (LSCM) and SEM.

3. Results and discussion

3.1. *Microstructure and phase constitution of Mg-1.5Mn-1Ca-xSr alloys*

The microstructures of as-cast and T4 treated Mg-1.5Mn-1Ca-*x*Sr alloys are shown in figure 1. As can be seen in figure 1, Mg-1.5Mn-1Ca-*x*Sr alloys are mainly consisted of dendrites. Grain size decreases with the addition of Sr. The second phases are mainly arranged along the grain boundaries, some lies within the grains.T4 treatment partly dissolved the second phases and refined grain boundaries (figure 1 c and d). Figure 2 shows the XRD patterns of as-cast and T4 treated Mg-1.5Mn-1Ca-*x*Sr alloys, which showed the existence of Mg_2Ca phase, $Mg_{17}Sr_2$ phase and α-Mn phase.

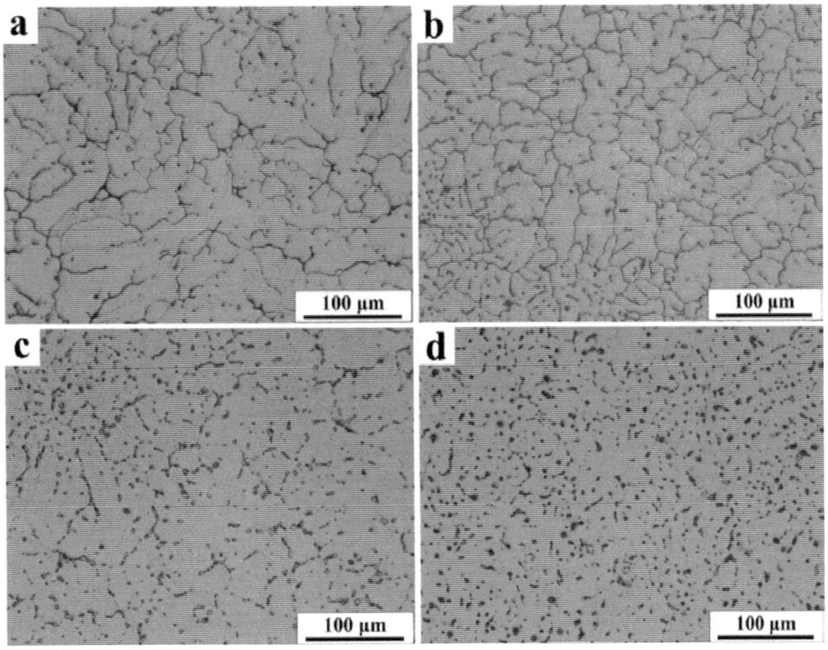

Figure 1. Optical micrograph of (a) as-cast Mg-1.5Mn-1Ca alloy, (b) as-cast Mg-1.5Mn-1Ca-1Sr alloy, (c) T4 treated Mg-1.5Mn-1Ca alloy, (d) T4 treated Mg-1.5Mn-1Ca-1Sr alloy.

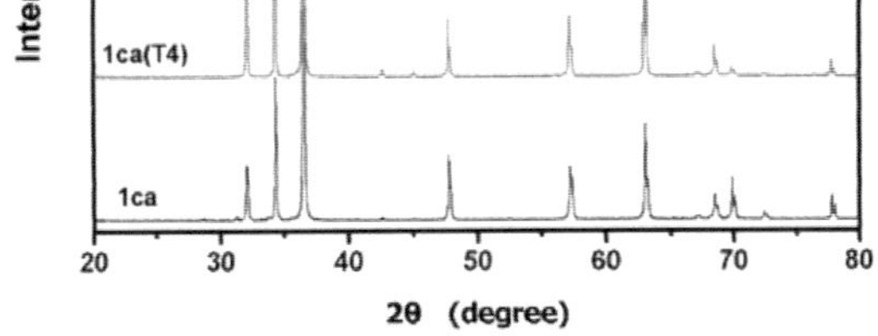

Figure 2. XRD analysis of as-cast and T4 treated Mg-1.5Mn-1Ca-*x*Sr alloys.

As for the reason that the grain refinement in figure 1, it was widely accepted that alloying elements process a strong segregation ability in melt alloy and form intensive constitutional undercooling in a diffusion layer ahead of the advancing solid/liquid interface which restricts the grain growth and promotes the nucleation of the primary Mg, thus refines the grain size [28]. The refinement efficiency of an element can be determined by the calculation of a growth restriction factor (GRF) which can be calculated by using binary phase diagrams. In GRF formula: $\sum m_i c_{o,i}(k_i - 1)$, where m_i is the slope of the liquidus line (assumed to be a straight line), k_i is the distribution coefficient, and $C_{o,i}$ is the initial concentration of element i. According to Mg/Sr binary phase diagram, Sr shows a high GRF, which indicated that Sr could effectively refine grains.

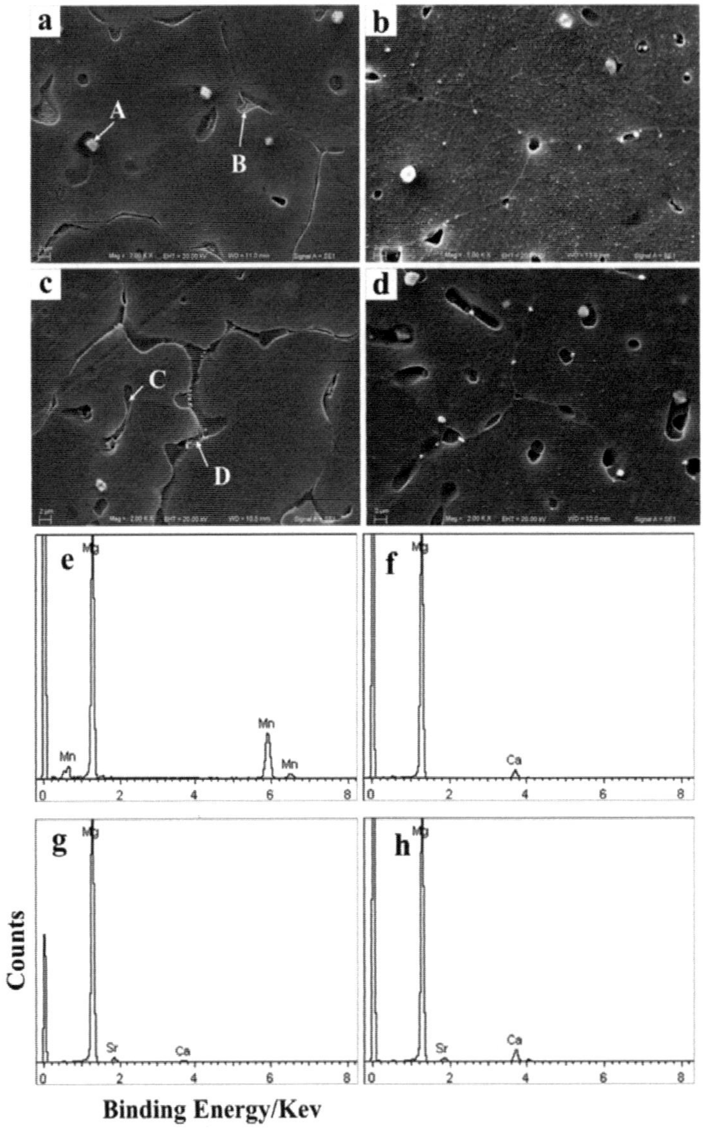

Figure 3. SEM microstructure of (a) as-cast Mg-1.5Mn-1Ca alloy, (b) T4 treated Mg-1.5Mn-1Ca alloy, (c) as-cast Mg-1.5Mn-1Ca-1Sr alloy, (d) T4 treated Mg-1.5Mn-1Ca-1Sr alloy, (e - f) EDS result of the sites from A to D in Fig. 3 (a) and (c).

Figure 3 (a-d) show the distribution and morphology of α-Mg + α-Mn peritectic structure (Site A), α-Mg + Mg_2Ca lamellar eutectic (Site B) and α-Mg + Mg_2Ca + $Mg_{17}Sr_2$ divorced eutectic structure, which may also be the Ca/Sr-rich phases [29], (Sites C and D). Figure 3 (e-h) shows the EDS results of sites from A to D. From the morphology of different eutectic structures, it could be distinguished that the phase which contains more $Mg_{17}Sr_2$ is more compact. Compared with as-cast alloy, T4 treated alloy has finer grain boundaries. By T4 treatment, Mg_2Ca partly dissolved in α-Mg matrix and α-Mn phase had the phenomenon of dissolution-re-precipitation which was caused by the effect of naturally age. The supersaturated α-Mn precipitated dispersedly from α-Mg in rod-like shape, which is consistent with the results observed in earlier study [22] and contributes to the microstructure homogenization. T4 treatment had little effect on the dissolution of $Mg_{17}Sr_2$ because of the low solubility of Sr in Mg.

3.2. Mechanical properties of Mg-1.5Mn-1Ca-xSr alloys

Figure 4 shows the stress-strain curves of as-cast and T4 treated Mg-1.5Mn-1Ca-*x*Sr alloys. The detail data of tensile test are listed in table 2. The addition of Sr is helpful to improve the yield strength (YS) and ultimate tensile strength (UTS), but decreases the elongation. After T4 treatment, both the UTS and the elongation of Mg-1.5Mn-1Ca-*x*Sr alloys increase significantly, except the YS decrease slightly. The mechanical properties of Mg-1.5Mn-1Ca-*x*Sr alloys are mainly determined by their microstructure. In Figure 1 the grain size decreased with the addition of Sr, which contributes to the increase of yield strength according to Hall-Petch law [30]. The second phases that distributed along the boundaries and in grains provided second phase strengthening effect which increased the strength but decreased the elongation of alloys. However, too much volume fraction of second phase is also harmful. It had been reported that the lamellar Mg_2Ca phase decreased the ductility of Mg-0.9Ca alloy. It was thought that Mg_2Ca phase dispersed at the grain boundary could be the crack source for Mg-Ca alloy [31]. By treatment, Mg_2Ca phase dissolved into matrix and the distribution of $Mg_{17}Sr_2$ phase and α-Mn phase are more diffuse, which provided solution strengthening effect and reduced the cracking tendency. The combined effect of solution strengthening and second phase strengthening contributes to the significant increase of the tensile properties of T4 treated Mg-1.5Mn-1Ca-*x*Sr alloys.

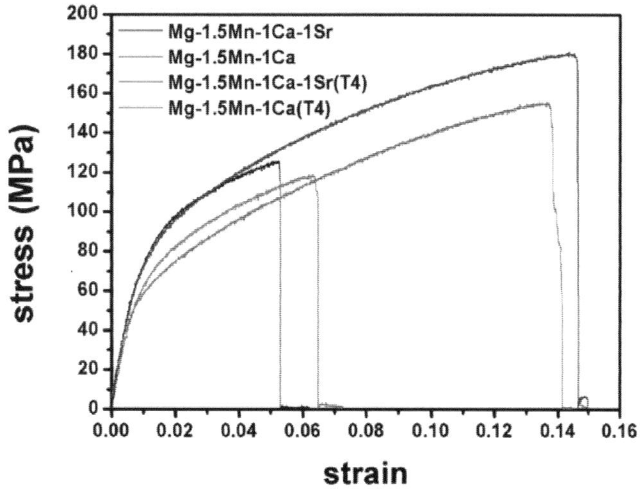

Figure 4. Typical stress-strain curves of as-cast and T4 treated Mg-1.5Mn-1Ca-*x*Sr alloys.

Table 2. Mechanical properties of as-cast and T4 treated Mg-1.5Mn-1Ca-*x*Sr alloys.

Alloy	YS (MPa)	UTS (MPa)	Elongation (%)
Mg-1.5Mn-1Ca	58±4	119±6	6.3±0.62
Mg-1.5Mn-1Ca(T4)	55±3	155±5	13.6±0.70
Mg-1.5Mn-1Ca-1Sr	78±2	125±4	5.3±0.42
Mg-1.5Mn-1Ca-1Sr(T4)	76±2	180±3	14.5±0.66

3.3. Electrochemical properties of Mg-1.5Mn-1Ca-xSr alloys

The electrochemical polarization curves of as-cast and T4 treated Mg-1.5Mn-1Ca-*x*Sr alloys in SBF solution are shown in figure 5. Generally, the cathodic polarization curves are assumed to represent the cathodic hydrogen evolution through water reduction, while the anodic polarization curves represent the dissolution of magnesium. In figure. 5, no current plateau is observed in the anodic region of Mg-

1.5Mn-1Ca-xSr alloys. The addition of Sr moves the corrosion potential (E_{corr}) to a more negative direction. By T4 treatment, the corrosion potential of both alloys increased slightly and followed by the decrease of corrosion current density, which indicated the enhancement of corrosion resistance. The detail electrochemical parameters of the electrochemical polarization curves in figure. 5 are listed in table 3. The E_{corr} and I_{corr} were derived from the plots by Tafel extrapolation. The corrosion rate (P_i) can be calculated using the conversion equation according to ASTM-G102-89: $P_i = 22.85 \times I_{corr}$, where the P_i is given in mm/yr and I_{corr} in mA/cm^2.

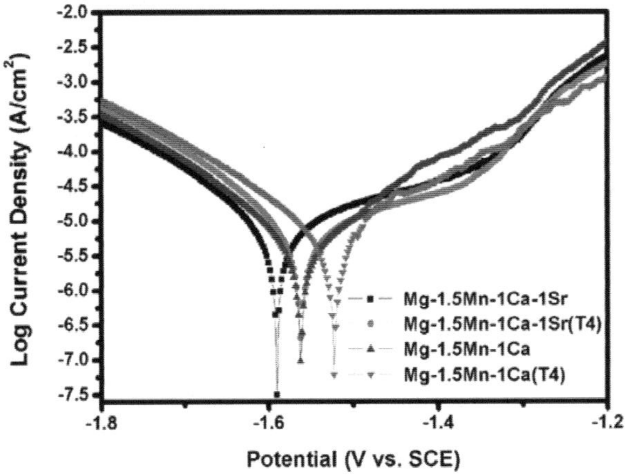

Figure 5. Tafel curves of as-cast and T4 treated Mg-1.5Mn-1Ca-xSr alloys immersed in SBF.

Table 3. Electrochemical parameters of as-cast and T4 treated Mg-1.5Mn-1Ca-xSr alloys in SBF solution.

Alloys	E_{corr} (mV)	i_{corr} (μA/cm^2)	P_i (mm/yr)
Mg-1.5Mn-1Ca	-1562	9.29	0.21
Mg-1.5Mn-1Ca(T4)	-1524	7.36	0.17
Mg-1.5Mn-1Ca-1Sr	-1596	17.8	0.4
Mg-1.5Mn-1Ca-1Sr (T4)	-1565	6.61	0.15

3.4. *Immersion tests*

Because of the "negative difference effect", hydrogen evolution is more effective than means of Tafel extrapolation on polarization curves to measure the corrosion rate of magnesium alloys [32]. Figure 6 presents the hydrogen evolution results of immersion test in SBF, which shows that the average hydrogen evolution rates of T4 treated Mg-1.5Mn-1Ca-xSr alloys are much less than that of as-cast alloys. The addition of Sr decreases the corrosion rate slightly. All specimens exhibit a decrease in hydrogen evolution rate with increasing immersion time. It is shown that all specimens had a relative high hydrogen evolution rate and undergone a quickly corrosion period in the first 7h, and then hydrogen evolution rate gradually go to a relative steady value, which indicated that the corrosion process enter the stable period. This result could be explained by the electrochemical corrosion mechanisms of magnesium in neutral corrosive mediums proposed according to G. Song [27]. At the anode regions:

$$Mg + H_2O \rightarrow Mg^{2+} + OH^- + 1/2H_2 + e^-$$

The phenomenon of anodic hydrogen evolution was only special for magnesium alloys. At the cathode regions:

$$H_2O + e^- \rightarrow + OH^- + 1/2H_2 \uparrow$$

The total reaction:

$$Mg + 2H_2O \rightarrow Mg(OH)_2 + H_2$$

So, in the beginning of corrosion process, the anode quickly dissolved and H_2 released at the cathode. The corrosion products, which are mainly composed of $Mg(OH)_2$, is loose and could support little protective effect. So in this period, Mg^{2+} could still diffuse toward the exterior surface of corrosion product layer and electrolytes diffuse toward the interior surface of corrosion product layer. Mg^{2+} and electrolytes meet inside the film layer and continue to form corrosion products. With the increase of immersion time, the corrosion product layer becomes more and more compact which restrict the diffusion of ions inside the layer. When the dissolution and formation of corrosion product layer keeps dynamic balance, the corrosion process advanced into the stable period in which the corrosion product layer stop growing and could support a certain protective effect. So the corrosion rates of the specimens decrease noticeably and keep on a relative low standard after the initial 7h.

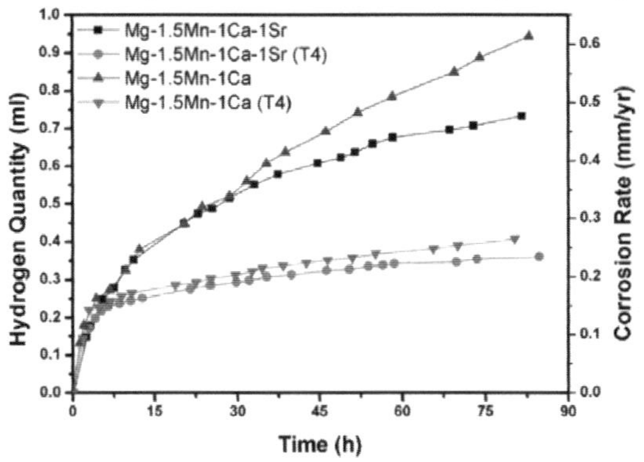

Figure 6. Hydrogen evolution for as-cast and T4 treated Mg-1.5Mn-1Ca-xSr alloys immersed in SBF.

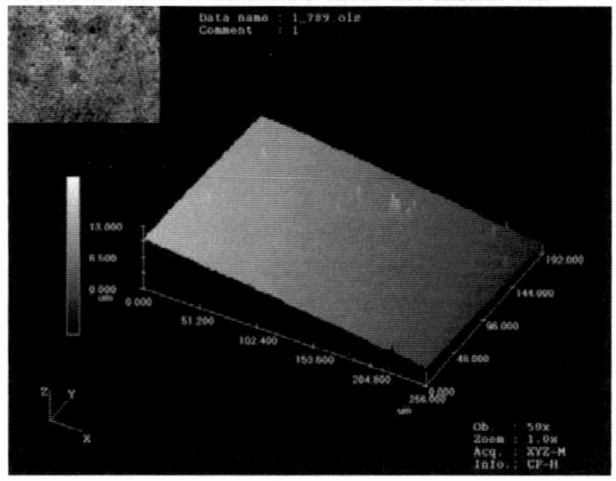

Figure 7. Laser scanning confocal microscope (LSCM) image of as-cast Mg-1.5Mn-1Ca-1Sr alloys immersed in SBF for 3 min.

3.5. Corrosion morphology analysis

In order to investigate the corrosion process and corrosion mechanism of Mg-1.5Mn-1Ca-xSr alloys in detail, the corroded surfaces of short-term (3 min) and long-term (52 h) exposure samples were analyzed under LSCM and SEM. Selected LSCM and SEM micrographs are shown in figure 7 and figure 8. Figure 7 shows the short-term corrosion morphology of as-cast Mg-1.5Mn-1Ca-1Sr alloy, from which it could be seen that the α-Mg matrix act as anodic was priority corroded. The raised parts are the second phases and the distinct peaks are α-Mn phase. Figure 8 shows the long-term corrosion morphology of as-cast and T4 treated Mg-1.5Mn-1Ca-xSr alloys after removed the corrosion products. From figure 8, it could be observed that the corrosion sites mainly located around the second phases, especially Mg_2Ca act as anodic phase was first corroded. Pitting occurs both in boundaries and grains. After T4 treatment, the corrosion degree of Mg-1.5Mn-1Ca-xSr alloys is weakened and the corrosion morphologies become smooth (figure 8c, d, f and g). But the residual Mg_2Ca and dispersedly distributed α-Mn phases still have a serious effect on the corrosion process. In figure 8, it could be observed that the T4 treated Mg-1.5Mn-1Ca-1Sr alloy has the most evenly morphology and there is no apparent corrosion around $Mg_{17}Sr_2$ and α-Mn phase.

The corrosion mechanism of Mg-1.5Mn-1Ca-xSr alloys is significantly controlled by the changes in quantity, distribution, configuration and size of the second phases. The effective cathodes phases $Mg_{17}Sr_2$ and α-Mn would cause micro-galvanic corrosion with anode phase Mg_2Ca and α-Mg matrix [12, 23]. Mg_2Ca would be corroded preferentially at the beginning of the corrosion process. The seriously localized corrosion surround α-Mn particles would also accelerate the spalling of α-Mn and the formation of pitting holes (see in figure 8f). After the addition of Sr, the microstructure of alloy was refined and the quantity of second phase is larger. Grain boundaries and second phases express less negative electrode potential compared with α-Mg matrix because of the formation of Ca/Sr-rich phases [29], and the larger quantity of second phases did not cause more sites of micro-galvanic reaction. In the mean time, the compact grain boundaries and continuous distribution of $Mg_{17}Sr_2$ along boundaries also block the corrosion of α-Mg matrix because of the barriers effect. That means during the process of corrosion, hydrogen ions need to go around the grain boundaries to reach the corrosion interface and react with the active α-Mg matrix, which increases the resistance of corrosion process. The similar phenomenon had also been reported in Mg-Al alloys [17, 33]. So, to Mg-1.5Mn-1Ca-1Sr alloy, the addition of Sr induced uniform anode consumption and decreased the corrosion rate.

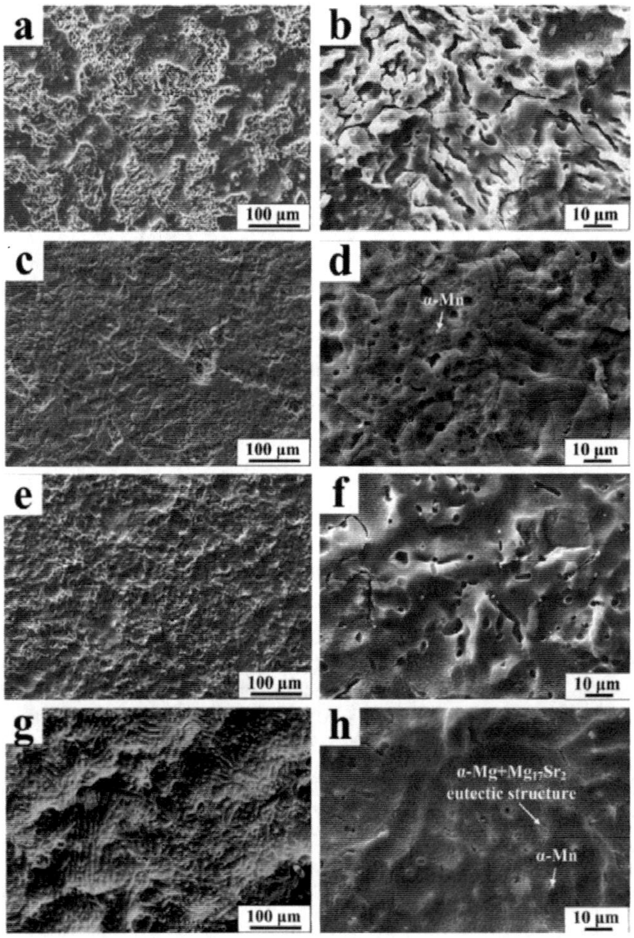

Figure 8. Surface morphology of Mg-1.5Mn-1Ca-*x*Sr alloys immersed in SBF for 52 h after removed the corrosion products: (a & b) as-cast Mg-1.5Mn-1Ca alloy, (c & d) as-cast Mg-1.5Mn-1Ca-1Sr alloy, (e & f) T4 treated Mg-1.5Mn-1Ca alloy, (g & h) T4 treated Mg-1.5Mn-1Ca-1Sr alloy.

4. Conclusions

Sr, Ca and Mn are naturally present in human body and the addition of these three alloying elements helped to reduce the grain size and formed Mg_2Ca, $Mg_{17}Sr_2$, Ca/Sr-rich and α-Mn phases in alloys, which reduced grain size and contributed to the increase of strength and corrosion resistance.

T4 treatment is a good way to improve both the mechanical properties and corrosion resistance of Mg-1.5Mn-1Ca-*x*Sr alloys. By T4 treatment, solution strengthening effect and second phase strengthening effect simultaneously contribute to the improvement of tensile properties. The dissolution of Mg_2Ca in α-Mg matrix not only decreases the fracture tendency but also reduces the sites of micro-galvanic reaction. The distribution of α-Mn phase is more uniform and dispersive after T4 treatment, which help to resist severe localized attack.

By investigating on the corrosion behavior of Mg-1.5Mn-1Ca-*x*Sr alloys, it could be summarized that corrosion first occurs in Mg_2Ca anode phase and α-Mg matrix that surround the cathode phase. The corrosion behavior of Mg-1.5Mn-1Ca-*x*Sr alloys is mainly controlled by the changes in quantity, distribution, configuration and size of the Mg_2Ca, $Mg_{17}Sr_2$ and α-Mn phases. The $Mg_{17}Sr_2$ and α-Mn phases act as effective cathode induces micro-galvanic corrosion with α-Mg matrix. The addition of Sr induced rather uniform anode consumption, which is mainly because the corrosion barrier effect caused by the compact boundaries and the uniformly distributed second phases. Finally, the suitable performances make T4 treated Mg-1.5Mn-1Ca-1Sr alloys possible to be potential biodegradable implant materials.

Acknowledgements
This work is supported by the "985 Project" of Jilin University, the Science and Technology Program of Jilin Province (201105007), the Open Subject of State Key Laboratory of Rare Earth Resource Utili zation (RERU2011001), the Science and Technology Support Project of Jilin Province (20130305008 GX) and the Science and Technology Support Project of Jilin Province (20140325003GX).

References

[1] Z Marciniak and K Kuczyński 1967 *International Journal of Mechanical Sciences* **9(9)** 609–620
[2] R S Busk 1987 *Magnesium products design* M Dekker
[3] B B Clow 1996 *Advanced materials & processes* Magnesium industry overview **150(4)** 33–34
[4] M P Staiger, A M Pietak, J Huadmai and G Dias 2006 *Biomaterials* **27(9)** 1728–1734
[5] N.-E Saris and E Carafoli 2005 *Biochemistry (Moscow)* **70(2)** 187–194
[6] D Raftopoulos, E Katsamanis, F Saul, W Liu, and S Saddemi 1993 *Journal of biomedical engineering* **15(1)** 60–66
[7] M Znamenskii 1945 *Khirurgiia* **12(1)** 60–63
[8] G Song and A Atrens 2007 *Advanced Engineering Materials* **9(3)** 177–183
[9] W H Organization 1996 Trace elements in human nutrition and health
[10] W H Organization 1994 Assessing human health risks of chemicals: derivation of guidance values for health-based exposure limits/published under the joint sponsorship of the United Nations Environment Programme, the International Labour Organisation, and the World Health Organization
[11] C K Yuen and W Y Ip 2010 *Acta biomaterialia* **6(5)** 1808–1812
[12] N T Kirkland, N Birbilis, J Walker, T Woodfield, G J Dias and M P Staiger 2010 *Journal of biomedical materials research. Part B, Applied biomaterials* **95(1)** 91–100
[13] H R B Rad, M H Idris, M R A Kadir and S Farahany 2012 *Materials & Design* **33** 88–97
[14] F Rosalbino, S De Negri, A Saccone, E Angelini and S Delfino 2010 *Journal of materials science. Materials in medicine* **21(4)** 1091–1098
[15] B Zhang, Y Hou, X Wang, Y Wang and L Geng 2011 *Materials Science and Engineering: C* **31(8)** 1667–1673
[16] E Zhang and L Yang 2008 *Materials Science and Engineering: A* **497(1-2)** 111–118
[17] L Wang, T Shinohara and B-P Zhang 2010 *Journal of Solid State Electrochemistry* **14(10)**

1897–1907
[18] Y Jang, Z Tan, C Jurey, Z Xu, Z Dong, B Collins, Y Yun and J Sankar 2015 *Materials science & engineering. C, Materials for biological applications* **48** 28–40
[19] M Salahshoor and Y Guo 2012 *Materials* **5(12)** 135–155
[20] K Kannan, J Vetrano and C Hamilton 1996 *Metallurgical and Materials Transactions* **27(10)** 2947–2957
[21] J A Helsen and H Jürgen Breme 1998 *Metals as biomaterials* （Wiley-VCH）pp
[22] M Celikin, A A Kaya and M Pekguleryuz 2012 *Materials Science and Engineering: A* **550** 39–50
[23] M Bornapour, N Muja, D Shum-Tim, M Cerruti and M Pekguleryuz 2013 *Acta biomaterialia* **9(2)** 5319–5330
[24] S Dahl, P Allain, P Marie, Y Mauras, G Boivin, P Ammann, Y Tsouderos, P Delmas and C Christiansen 2001 *Bone* **28(4)** 446–453
[25] P Marie 2005 *Osteoporosis international* **16(1)** S7–S10
[26] A Taylor 1985 *Clinics in endocrinology and metabolism* **14(3)** 703–724
[27] G Song and A Atrens 2003 *Advanced engineering materials* **5(12)** 837–858
[28] Y Lee, A Dahle and D StJohn 2000 *Metallurgical and Materials Transactions A* **31(11)** 2895–2906
[29] M Bornapour, M Celikin, M Cerruti and M Pekguleryuz 2014 *Materials science & engineering. C, Materials for biological applications* **35** 267–282
[30] B Brindley and P Worthington 1970 *International Materials Reviews* **15(1)** 101–114
[31] Y Chino, M Kobata, H Iwasaki and M Mabuchi 2002 *Materials Transactions* **43(10)** 2643–2646
[32] G Yu, Y-L Liu, Y Li, L-Y Ye, X-H Guo and L Zhao 2002 *Chinese Journal of Nonferrous Metals* **12(6)** 1087–1098
[33] Y J Ko, D Y Chang, J D Lim and K S Shin 2003 *Materials Science Forum* 851–856

IOP Conf. Series: Materials Science and Engineering **182** (2017) 012050 doi:10.1088/1757-899X/182/1/012050

Characteristics of MAO coating obtained on ZK60 Mg alloy under two and three steps voltage-increasing modes in dual electrolyte

Jun Yang[1], Ze-Xin Wang[1]*, Sheng Lu[1], Wei-gang Lv[2], Xi-zhi Jiang[1] and Lei Sun[1]

[1]School of Materials Science and Engineering, Jiangsu University of Science and Technology, Zhenjiang 212003, China

[2] Zhenjiang Naisi New materials Co, Ltd. Zhenjiang, 212000, Jiangsu, China

E-mail: 790209828@qq.com

Abstract. The micro-arc oxidation process was conducted on ZK60 Mg alloy under two and three steps voltage-increasing modes by DC pulse electrical source. The effect of each mode on current-time responses during MAO process and the coating characteristic were analysed and discussed systematically. The microstructure, thickness and corrosion resistance of MAO coatings were evaluated by scanning electron microscopy (SEM), energy disperse spectroscopy (EDS), microscope with super-depth of field and electrochemical impedance spectroscopy (EIS). The results indicate that two and three steps voltage-increasing modes can improve weak spark discharges with insufficient breakdown strength in later period during the MAO process. Due to higher value of voltage and voltage increment, the coating with maximum thickness of about 20.20μm formed under two steps voltage-increasing mode shows the best corrosion resistance. In addition, the coating fabricated under three steps voltage-increasing mode shows a smoother coating with better corrosion resistance due to the lower amplitude of voltage-increasing.

Keywords: Magnesium alloy; Micro-arc oxidation; Characteristic; step voltage-increasing

1. Introduction

Mg alloys are the lightest materials among those metals used for structural or mechanical applications [1]. Besides, Mg alloys have many excellent properties such as high specific tensile strength, high dimensional stability, good stiffness and vibration absorption [2, 3]. Thus Mg alloys are widely used in many industries such as aerospace, automobile and communication fields [4].Though ZK60 is a typical wrought alloy characterized with superior properties [5-7], the poor corrosion resistance, like most Mg alloys, has limited its widespread use [8].

Micro-arc oxidation (MAO), also referred as plasma electrolytic oxidation, is a new electrochemical surface treatment which is considered an attractive method for generating protective ceramic coatings, particularly on light alloys such as Ti, Al and Mg alloys [9-11]. The corrosion and wear resistance can be remarkably improved after the MAO process [12, 13].

Most MAO processes are carried out under constant current mode which is more efficient to obtain superior MAO coatings[14]. However, MAO process under constant voltage mode is hardly used in applications because of its insufficient breakdown ability acting on coating in later period [15]. In order to solve this problem, two steps voltage-increasing mode was developed to improve the coating quality in previous work [16]. However, the coating characteristics are different under different step voltage-increasing modes. In the present work, MAO coating was obtained under two and three steps voltage-increasing modes, the growth characteristic, thickness, growing rate, microstructure and corrosion resistance were investigated respectively.

2. Experimental

Rectangular coupons (20 mm×20 mm×5 mm) of ZK60 magnesium alloy (Zn 4.8-6.2, Zr >0.45 impurities ≦0.30, Mg balance) were prepared by means of electric arc CNC wire-cut machine and polished with alumina waterproof abrasive paper up to 1200 grits. The coupons were then cleaned in an ultrasonic bath with acetone and ethyl alcohol respectively at room temperature for 10 min before MAO process.

WHD-20 MAO system with a bipolar pulsed power was employed in MAO experiments. Dual electrolytes system was prepared from distilled water containing 14 g/L $NaAlO_2$, 6g/L Na_3PO_4, 4 g/L NaOH, 3g/L NaB_4O_7, and 4g/L $C_6H_5Na_3O_7$ [14]. Electric parameters including negative voltage, duty cycle and frequency kept constant with the value of 20V, 20%, 600 Hz respectively while the positive voltage was the main parameter to be investigated. All MAO experiments were conducted for 15 min at electrolyte temperature of less than 35℃ and the current-time curves were recorded automatically during MAO process. Spark discharge phenomenon were captured by digital camera.

In the experiment, positive voltage could be regulated on line during MAO process to achieve the increasing of voltage by two or three steps. Under two steps voltage-increasing mode, the first step was required for 6 min with the initial voltage value of 280V, while the second step was applied with increase value of 80V for the rest 9 min. When MAO process was carried out under three voltage-increasing mode, the first step was also required for 6 min with the initial voltage value of 280V, then the voltage rose to 320V for 360-630s, the voltage was changed to 360V for 630-900s at the third step, the initial voltage and the final voltage of each mode is identical(280V-360V).

The microstructural characteristics of coating and coating thickness phase were investigated by JSM−6480 SEM. The macro morphology and surface roughness of the coating were investigated by Digital Camera and Keyence VHX-900 Digital Microscope. Tests of electrochemical impedance spectroscopy were conducted at a 3.5%NaCl solution to assess the corrosion resistance of the coatings.

3. Results and discussion

3.1. Current-time curves and experimental phenomenon of MAO process

The current-time responses under two and three steps voltage-increasing mode are shown in figure 1. Under two steps voltage-increasing mode, according to the analysis of current changes during MAO process, it can be seen that MAO coating experiences the growth of passivation layer, spark discharge and rapid film-forming before voltage-increasing (0-360s). In this stage, the current rises rapidly at the beginning and then appears a sharp decline with a peak value 9.0A at 32s. In the previous work, basic width was put forward, which plays an important role in the growth of MAO coatings, the greater the width is the stronger the growth momentum will be. The basic width is 250s in this stage and later the current has a little fluctuation between 0.5A and 0.7A from 250s to 360s. According to the experimental phenomenon, the discharge sparks are brighter at first and then it becomes finer and weaker, which lead to a shortage of driving force of breakdown, as shown in figure 2(a). After increasing the voltage (360s-600s), MAO process comes into second stage in which the coating grows fast. The current quickly increase linearly and then get down with a lower speed compared with the first stage. At 367s, it get the peak value 5.2A which is significantly lower than the first stage, for the former coating has a large insulation resistance [17] which lead to the reduction of current peak when it is broken down again. In this stage, the baseline peak width is 240s, the growth momentum became stronger, while the spark quickly became brighter, larger and more intensive which replace the former weak sparks as shown in figure 2(b).

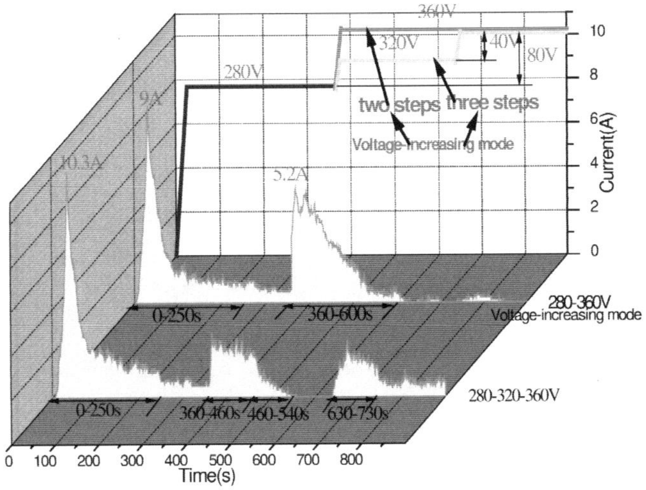

Figure 1. Current-time curves under two and three steps voltage-increasing mode.

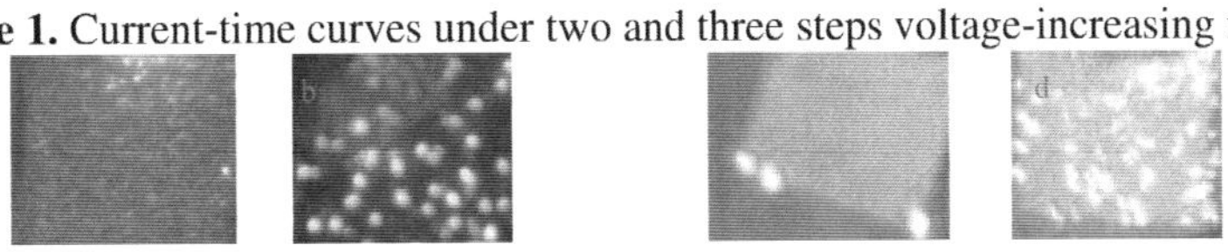

Figure 2. Spark discharge characteristics under two and three steps voltage-increasing mode two steps voltage-increasing mode: a, b; three steps voltage-increasing mode: c, d.

For three steps voltage-increasing mode, the samples experiences the same process as two steps voltage-increasing mode, however, it can be obviously seen that there is an additional current peak which results from an additional step of voltage-increasing, so the MAO coatings experience twice breakdown and growth with a lower spark discharging due to the reduction of amplitude of voltage-increasing. Before increasing the voltage, the current peak value is 10.3A and the baseline peak width is 250s, the result is roughly consistent with two steps voltage-increasing mode. At the first step of voltage-increasing, voltage increases to 320V, the current peak value is 2.8A and the baseline peak width is 100s which is apparently reduced, about half of what was in the former mode, compared with the two steps voltage-increasing mode. At this stage, the corners of the samples arc first as shown in figure 2(c), which is due to the lower amplitude of only 40V and the voltage now is only 320V, then the driving force of arc will be slightly insufficient.

However, the MAO process always occur first in the weakest point, therefore the corners of the samples which are weaker area are always the first place to be broken down, then the spark discharging in the surface of the samples is gradually homogenized as well as the coating growth. The second step of voltage-increasing which the voltage is adjusted to 360V at 630s. In this period, the current peak value is 2.5 A and the baseline peak width is 100s which is similar with the first period. Due to the first step of voltage-increasing, the coating growth rate is lower of only 0.53 μm/min as shown in Figure 3, the increase of coating thickness is less as well as the increase amplitude of coating insulation resistance, while the second step of voltage-increasing with the voltage of 360V, in the case of little variation of coating thickness, it will provide more adequate driving force of arc compared with the first step, then the coating growth rate reaches 1.10 μm/min, therefore the reaction is rather intense and the surface of the samples resume a new round of breakdown and spark discharging, as show in figure 2(d).

It can be found that, under two steps voltage-increasing mode, the voltage-increasing amplitude is higher (80V) as well as the variation amplitude of the corresponding current, at which the corresponding reaction is also more intensive; under three step voltage-increasing, the voltage-increasing amplitude is lower(40 V) in different period, then the MAO process is gentler compared with the two steps voltage-increasing and the effect of diuretic recovering of MAO coating is better based on the observation of the coating morphologies.

3.2. The MAO coating growth rate and thickness

Figure 3 shows the MAO coating growth rate of different period under different voltage-increasing modes. It can be seen that the growth rate of two kinds of mode is 1.38 μm/min. We choose the average value of 1.38μm/min under three different situations and the MAO coatings experience a main growing period. Under two steps voltage-increasing mode (280-360 V), after increasing the voltage (360-900s), the growth rate of MAO coatings accelerates obviously with the increase of voltage and the MAO coatings go through a rapid growth period which last for about 4min with a rate of 2.49 μm/min, therefore the electric energy provided by the increase of voltage enhance the growth momentμm of MAO coatings. However, when the current decreases slightly, the MAO process becomes smooth and steady with a tiny spark discharging and a low growth rate of only 0.37 μm/min.

Under three steps voltage-increasing mode, it makes the arc energy provided by the voltage-increasing amplitude of 80 V divided into two parts with 40 V per time, the coating growth rate of 1.10μm/min in the second time of voltage-increasing increases one time than the first time with the rate of 0.53 μm/min, the result is corresponding to the discharging phenomenon in MAO process mentioned above which the reaction in the second time is more intense than the first one as well as the coating growth rate. The three steps voltage-increasing mode add a voltage-increasing period with the total voltage-increasing amplitude remain unchanged, it can be known that the coating growth rate tends to be gentler with a value of only 0.53 μm/min as apparently shown in the first step of voltage-increasing which is due to the weakening arc driving force, and also it will have an impact on the final coating thickness under this mode.

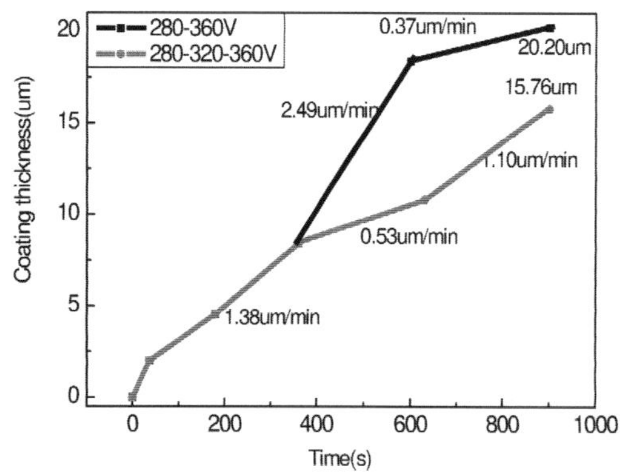

Figure 3. Growing rate of MAO coating between two and three steps voltage-increasing mode.

Comparing the thickness variation under two and three steps voltage-increasing modes, it can be seen that, the coating thickness of three steps voltage-increasing mode is thinner than that of two steps voltage-increasing mode in the condition of same final voltage and total amplitude of voltage-increasing. The coating thickness formed under two steps voltage-increasing mode is 20.20 μm and the other is 15.76 μm which is obviously thicker than one of 11.22 μm formed under constant voltage mode.

3.3. The morphologies of MAO coating

The morphologies of MAO coating under two and three steps voltage-increasing mode are shown in figure 4. It can be observed that MAO coating is characterized with a typical porous structure as shown in figure 4(a) and figure (c) [18-21]. Under steps voltage-increasing mode, due to the higher final voltage as well as the electric energy of MAO process, the surfaces of the samples are covered by the melt sprayed from the discharging channels during MAO process which is characterized with a uneven and rough surface with the roughness of 7.5μm and 5.1μm under two and three steps voltage-increasing modes respectively as shown in Figure 5. According to the cross-section morphologies of MAO coating, the coatings formed under the two different modes are relatively uniform and compact as shown in figure 4(b) and (d), the juncture between the coating and the substrate is characterized with a zigzag shape and P. Bala [22] Srinivasan et al. think that way of metallurgical bonding can enhance the bonding force between the coating and the substrate. The distance of two dashed lines is used to indicate the maximum waviness of joint surface of the coating and the substrate which is defined as the bonding degree of the coating and the substrate.

Therefore it can be known that the bonding degrees of two and three steps voltage-increasing mode are 4.53 μm and 6.53 μm respectively. Comparing the surface roughness of outer layer and the bonding degree of internal layer and the substrate, it can be seen that the surface roughness of outer layer is large and the bonding degree of internal layer and the substrate is low under two steps voltage-increasing mode while the result of three steps voltage-increasing is opposite, the results above show that the growth mechanism of MAO coating is different under different voltage-increasing modes. Under the two steps voltage-increasing mode, the melt accumulation on the surface of outer layer is uneven which can lead to a large surface roughness and the outward growth of the coating is more. While under the three steps voltage-increasing mode, the melt accumulation on the surface of outer layer is uniform which contribute to a low surface roughness with a flat surface and the bonding degree of the layer and the substrate is in crescent which can account for the increased inward growth.

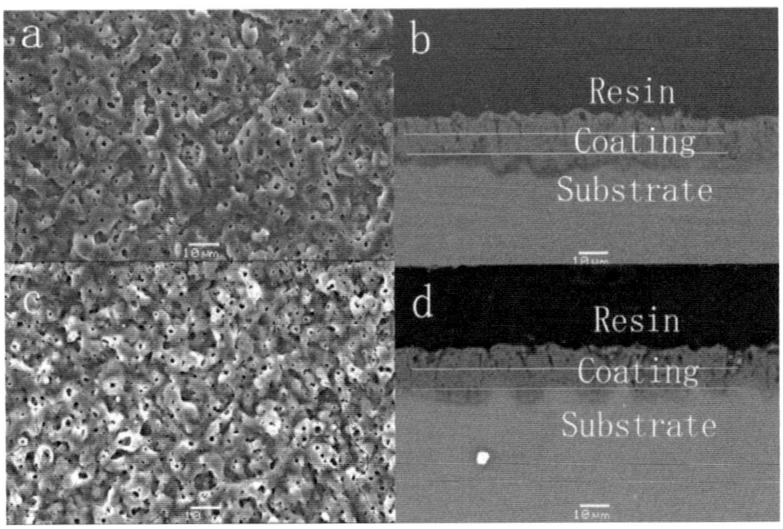

Figure 4. Surface and cross-section morphologies of coatings: (a), (b) 280V-360V; (c), (d) 280V-320V-360V.

According to the comparison of the surface morphologies of the two voltage-increasing modes, It could be found, under two steps voltage-increasing mode, the surface molten coverings of MAO coating connect each other and have a large area and also accumulate obviously with a larger aperture of 5.4 μm. While under three steps voltage-increasing mode, the surface molten coverings of MAO coating are little dispersive and the area is decreased, while the number of micro-pores is increased with the aperture of 3.5 μm, the result analysis of surface morphologies is consistent with that of roughness. Based on the analysis of the current-time curves of MAO process and the discharging phenomenon of experiment, it can be known that, under three steps voltage-increasing mode, although the total amplitude of voltage-increasing is the same as two steps voltage-increasing mode, due to the additional voltage-increasing step, the voltage-increasing process become more stable, the driving force of once again breakdown and melt-solidification process of MAO coating is decreased, the melts inside the discharging pores of the surface spray little and the accumulation beside the micro-pores is less, therefore the coating formed on the surface is relatively flat.

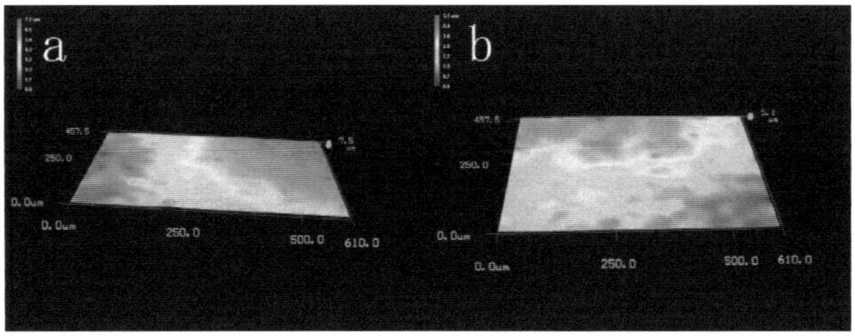

Figure 5. Roughness of coatings (a) 280V-360V, (b) 280V-320V-360V.

3.4. The corrosion resistance of MAO coating

The results of electrochemical impedance spectroscopy (EIS) of ZK60 magnesium MAO coatings formed under different voltage-increasing modes are shown in figure 6. As the Nyquist shown in figure 6 (a), all the impedance plots presented in Nquist plots are characterized with a small capacitive loop at high frequency and a large capacitive loop at low frequency. Compared with constant forward voltage mode, the MAO coating formed under steps voltage-increasing mode has a larger radius of capacitive loop which indicate a good corrosion resistance and the MAO coating formed under two steps voltage-increasing mode exhibits the largest radius of capacitive loop. Corrosion resistance assessed by low frequency impedance shown in figure 6 (b) indicates that MAO coating formed under two steps voltage-increasing mode would show best resistance with highest value of low frequency impedance of 1.414×105 ohm.cm^2, also the coating thickness is thickest and densest which can be seen from the morphologies of MAO coating. The value of low frequency impedance under three steps voltage-increasing mode is 6.143×104 ohm.cm^2 which decreases by 1 order of magnitude compared to 1.414×105 ohm.cm^2 under two steps voltage-increasing mode. Although the coating is still denser, the coating thickness is decreased compared to the former which shows a poorer corrosion resistance, however the value of impedance is increased much compared with the constant forward voltage mode.

Based on the general research [23], the coating thickness and compactness is the main factor related to the corrosion resistance of MAO coating which mainly consider the barrier effect of MAO coating against the outside corrosive medium. Also the other consider [24] that the coating corrosion resistance is certainly related to the different ceramic phases. The two steps voltage-increasing modes researched in this article which think the amplitude of voltage-increasing has a great effect on coating thickness, comparing the two and three steps voltage-increasing modes, it can be seen that: under two steps voltage-increasing mode, the amplitude of voltage-increasing is higher, the process of once again breakdown and melt-solidification is more intense and the driving force of coating growth provided is sufficient so that the coating is thicker and denser which also shows an improved corrosion resistance. While under three steps voltage-increasing mode, the amplitude of voltage-increasing is lower, the process of voltage-increasing is stable which contributes to a flat coating, however, the coating thickness is thinner with a poorer corrosion resistance.

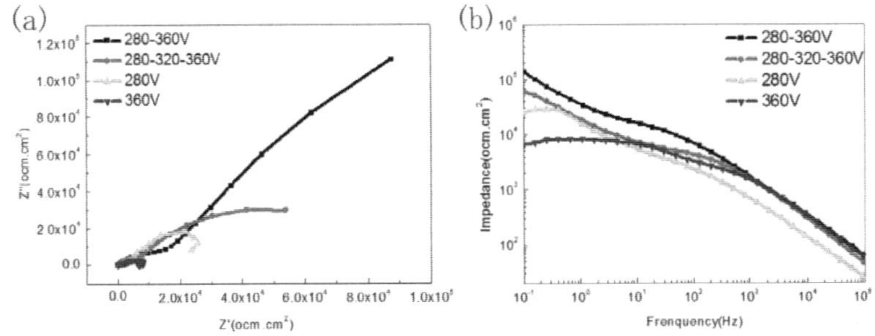

Figure 6. Electrochemical impedance spectroscopy of MAO coated ZK60 alloy specimens in 3.5% NaCl solution (a) Nyquist (b)Bode.

4. Conclusions

1) Under two steps voltage-increasing mode, the MAO process experience a new round of breakdown and regrowth period, the higher amplitude of voltage-increasing can provide a larger driving force of coating growth with a growth rate of 2.5μm/min which realize a rapid growth of MAO coating with a rather thick coating, while under three steps voltage-increasing mode, there is an additional regrowth stage in the MAO process, the amplitude of voltage-increasing is reduced to half of the former, the process is relatively stable, the coating is uniform and dense but the thickness is thinner than the former.

2) Due to the lower amplitude of voltage-increasing under three steps voltage-increasing mode, the MAO process is relative stable, the coating formed under this mode is uniform and dense compared with one of two steps voltage-increasing mode, the aperture is smaller of 3.5μm and the coating roughness is lower of 5.1μm.

3) The corrosion resistance of MAO coating can be improved much by steps voltage-increasing mode. Corrosion resistance assessed by low frequency impedance indicates that MAO coating formed under two steps voltage-increasing mode would show best resistance with highest value of low frequency impedance of 1.414×105 ohm.cm^2 which increases by 1 order of magnitude compared to the single voltage mode.

Acknowledgments

The present work was financially supported by a project funded by the priority academic program development of Jiangsu Higher Education Institutions and the key laboratory of advanced welding technology of Jiangsu Province, China (JSAWT-11).

References

[1] Fan X, Wang Y and Zou B 2014 Preparation and Bond Properties of Thermal Barrier Coatings on Mg Alloy with Sprayed Al or Diffused Mg-Al Intermetallic Interlayer *Thermal Spray Technology* **23** 304

[2] Jiao Miao, Bing Ye and Qudong Wang 2013 Mechanical properties and corrosion resistance of Mg–10Gd–2Y–0.5Zr alloy by hot extrusion solid-state recycling *Journal of Alloys and Compounds* **561** 184–192

[3] Shuyan Wang *et al.* 2014 Preparation and performance of MAO coatings obtained on AZ91D Mg alloy under unipolar and bipolar modes in a novel dual electrolyte *Ceramics International* **40(1)** 93–99

[4] Shang J *et al.* 2012 Effect of joining temperature on microstructure and properties of diffusion bonded Mg/Al joints *Transactions of Nonferrous Metals Society of China* **22(22)** 1961–1966

[5] Huang Z H 2013 Microstructures and Mechanical Properties of ZK60-XHo Wrought Magnesium Alloys *Materials Science Forum* **36(3)** 437–445

[6] Yu W *et al.* 2008 Microstructure and mechanical properties of ZK60–Yb magnesium alloys

Materials Science & Engineering A **478(1)** 101–107

[7] Yan-Long M A, Pan F S and Zuo R L 2004 Review on the Research of High-strength Wrought Magnesium Alloy ZK60 *Journal of Chongqing University* **27(9)** 80–85

[8] Chen X B, Birbilis N and Abbott T B. 2011 Review of Corrosion-Resistant Conversion Coatings for Magnesium and Its Alloys *Surface&CoatingsTechnology* **67(3)** 35005–35001

[9] Terleeva O P *et al.* 2010 Effect of microplasma modes and electrolyte composition on micro-arc oxidation coatings on titanium for medical applications *Surface & Coatings Technology* **205(6)** 1723–1729

[10] S Xin, L Song, R Zhao and X Hu 2006 Influence of cathodic current on composition, structure and properties of Al_2O_3 coatings on aluminium alloy prepared by micro-arc oxidation process *Thin Solid Films* **515** 326–332

[11] Z Song, Z Xie and G Yu 2015 A novel palladium-free surface activation process for electroless nickel deposition on micro-arc oxidation film of AZ91D Mg alloy *Journal of Alloys and Compounds* **623** 274–281

[12] Li C *et al.* 2014 Effect of the Micro-arc Oxidation and Cerium Salt Sealing on the Corrosion and Wear Resistance of New 7A85 *Mechanical Science & Technology for Aerospace Engineering* **33(1)** 127–132

[13] Wei-wei Chen, Ze-xin Wang, Lei Sun and Sheng Lu 2015 Research of growth mechanism of ceramic coatings fabricated by micro-arc oxidation on magnesium alloys at high current mode *Journal of Magnesium and Alloys* **3** 253–257

[14] S Lu, Z X Wang, J Chen and X S Zhou 2011 *The Transactions of Nonferrous Metals Society of China.* **21** 929–935

[15] Ping Wang, Jianping Li 2010 Growth process and corrosion resistance of ceramic coatings of micro-arc oxidation on Mg-Gd-Y magnesium alloys *Journal of Rare Earths* **28** 798–802

[16] S Lu, L Tang, X Z Jiang, Z X Wang and J Chen 2011 *Advanced Materials Research* **337** 101–105

[17] Jiang Bai-ling and Liu Dongjie 2011 Scientific aspects of restricting development and application of micro-arc oxidation technology *The Chinese Journal of Nonferrous Metals* **21(10)** 2402–2407

[18] Y M Wang, L X Guo, J H Ouyang, Y Zhou and D C Jia 2009 Interface adhesion properties of functional coatings on titanium alloy formed by Micro-arc oxidation method *Applied Surface Science* **255** 6875–6880

[19] Jun Liang, Litian Hu and Jingcheng Hao 2007 Improvement of corrosion properties of Micro-arc Oxidation coating on magnesium alloy by optimizing current density parameters *Applied Surface Science* **253** 6939–6945

[20] Chen Xian-ming, Luo Cheng-ping, Liu Jiang-wen and Li Wen-fang 2006 Structure of ceramic coating produced by micro-arc oxidation on Mg alloy *Journal of Central South University* **37 (6)** 1065–1069

[21] Ge Yangfeng, Jiang Bailing, Li Yulei and Yagn Zhiyuan 2011 Microstructure and Corrosion Resistance of Composite Micro-arc Oxidation and SiO2 Coatings on Magnesium Alloy *Chinese Journal of Materials Research* **25 (1)** 79–83

[22] P Bala Srinivasan, J Liang, C Blawert, M Störmer and W Dietzel 2009 Effect of current density on the microstructure and corrosion behaviour of plasma electrolytic oxidation treated AM50 magnesium alloy *Applied Surface Science* **255(7)** 4212–4218

[23] Hu Hui-li, Gao Ning-ning, Yu Yuan-chun and Li Ning 2009 Research progress on formation process of micro-arc oxidation ceramic coatings on magnesium alloy *Electroplating & Finishing* **28(12)** 35–42

[24] Qu L *et al.* 2013 Microstructure and corrosion resistance of ultrasonic micro-arc oxidation biocoatings on magnesium alloy *Journal of Advanced Ceramics* **2(3)** 227–234

IOP Conf. Series: Materials Science and Engineering **182** (2017) 012051 doi:10.1088/1757-899X/182/1/012051

Microstructure evolution and mechanical properties of biomedical Mg-Zn-Gd alloy wires

Gan Chunlei[1,2], Li Xiaohui[1,2], Nong Deng[1,2], Zhang Xiang[1,3], Zheng Kaihong[1,2] and Huang Zhenghua[1,2]

[1] Guangdong institute of materials and processing, Guangzhou 510650, China
[2] Guangdong provincial key laboratory for technology and application of metal toughening, Guangzhou 510650, China
[3] College of Materials Science and Engineering, Hunan University, Changsha 410082, China

E-mail: ganchunlei@163.com

Abstract. In order to manufacture the Mg-Zn-Gd alloy fine wires for the development of new biomedical Mg alloy implant devices, a hot extrusion and cold drawing process which is used to develop the Mg-Zn-Gd alloy fine wires were investigated. The results demonstrate that the Mg-Zn-Gd alloy has good formability. The microstructure and properties of the Mg-Zn-Gd alloy wires were studied by the observations of optical microscopy and scanning electron microscopy. The results show that the process is successfully developed to manufacture the high-quality wires with 3.00 mm diameter. The achievement of the high-quality Mg-Zn-Gd alloy wires is ascribed to the refined microstructure due to dynamic recrystallization during hot extrusion. Additionally, the grain morphology can play an important role in affecting the subsequent cold drawing performance.

1. Introduction

Mg alloys have attracted more and more attentions as promising implant materials due to their good biocompatibility, biodegradability and appropriate mechanical properties, etc.. Mg alloy fine wires are considered to be one of the huge application potential of Mg alloys in medical devices [1-3]. However, since Mg with the hexagonal close packed crystal structures displays poor ductility and cold formability, it is difficult to produce Mg alloy wires. In order to enhance the workability of Mg alloys, rare earth (RE) elements such as Gd and Y are usually added to significantly improve the comprehensive mechanical properties of Mg alloys. Furthermore, these rare earths are considered to be suitable for the preparation of medical Mg alloy wires as major alloying elements due to relatively high bio-safety [4–6]. And Gd is usually used for improving the properties of Mg alloys as one of important rare earth elements. It has been demonstrated that Gd as the addition of biomedical Mg alloys is feasible [7].

In the present study, Gd was added to the Mg-Zn alloy to fabricate the high-quality wires through a combination of hot extrusion and cold drawing. In addition, the microstructures and mechanical properties of the experimental Mg-Zn-Gd alloy wires were examined at different states. It is anticipated that the present experimental results will be helpful for future efficient production of Mg alloy fine wires by cold drawing processing.

IOP Conf. Series: Materials Science and Engineering **182** (2017) 012051 doi:10.1088/1757-899X/182/1/012051

2. Experimental procedures

2.1. Fabrication of wires

The chemical compositions of the experimental Mg-Zn-Gd alloy by chemical analysis can be listed in table 1.

Table 1. Chemical compositions of the experimental alloy.

Alloy	Zn	Gd	Mg
Mg-Zn-Gd	1.18	1.02	Bal.

Based on the previous works [8-11], we carefully selected the processing parameters for Mg-Zn-Gd alloy wires as follows: i) The Mg alloy was melted at 750 °C for 20 min, and then the molten Mg alloy was cooled to 720 °C and poured into an iron mold; ii) The obtained billets were machined into a size of 100.0 mm in diameter and 250.0 mm in length, and extruded into wires with 4.0 mm in outer diameter at 400 °C with the extrusion rate of 20:1; iii) The extruded wires were then drawn at ambient temperature with an area reduction between 9.5% and 12.5% every pass. After every two passes of cold drawing, annealing is performed at 400 °C for 60 min to remove the work hardening. In the drawing processes, vegetable oil was used as the lubricant. Eventually, the Mg alloy wires, whose outer diameter was reduced to 3.00 mm, were fabricated. These working procedures can be summarized as follows: melting→hot extrusion(4.0 mm in diameter)→cold drawing(3.8 mm in diameter)→cold drawing(3.6 mm in diameter, as-drawn 1#)→annealing(3.6 mm in diameter)→cold drawing(3.2mm in diameter)→cold drawing(3.0 mm in diameter, as-drawn 2#).

2.2. Microstructure analysis

The wires for microstructure observation were cut from the longitudinal sections, and ground with different grits of silicon carbide papers starting from 200 grit to 1500 grit and polished with 1 μm and 3.5 μm diamond pastes, and then chemically etched by a solution of 1.5 g picric acid, 25 ml ethanol, 5 ml acetic acid and 10 ml distilled water. The grain structures and morphologies of the samples were examined using the optical microscope (LEICA-DMI3000M).

2.3. Mechanical property test

The tensile mechanical property test was carried out on a universal material test machine (DNS200) at a strain rate of $1\times10\ s^{-1}$ at room temperature. For each group of tests, three wires were chosen as a function of Chinese standard (GB/T 228.1-2010) with an over length of 150 mm and a gage length of 100 mm. The fracture surfaces of the wires were examined using the scanning electron microscopy (JSM-5610LV).

3. Results and discussion

3.1. Processing of wires

Figure 1 shows the processing product from the extruded billets of the Mg-Zn-Gd alloy to drawn wires. The extruded wires and drawn wires exhibit no deformation. It can be further observed that the wires have smooth surfaces and high straightness, which are suitable for subsequent processing.

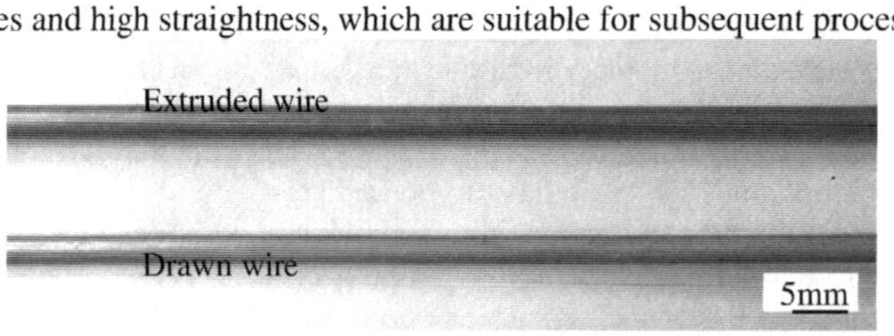

Figure 1. Experimental Mg-Zn-Gd alloy wires: extruded wire and drawn wire.

3.2. Microstructure observation

Figure 2 shows the microstructures from longitudinal sections of the wires for the Mg-Zn-Gd alloy under different conditions. For figure 2a, equiaxed, completely recrystallized grains of the as-extruded alloy are observed with the average grain size of approximately 6.9 μm. It demonstrated that the dynamic recrystallization occurred in the hot extrusion working processing of Mg-Zn-Gd alloy. When cold drawing deformation of 19% for the as-extruded alloy was employed, the microstructure of as-drawn wire was remarkably different from that of the as-extruded alloy, as shown in figure 2b, where the tendency of the fibrous structures along drawing direction was developed and grains became smaller in comparison with original equiaxed grains. Meanwhile, the work-hardening rate of the Mg alloy with the hexagonal close packed crystal structures was quickly increased during cold drawing deformation. Therefore, static recrystallization through annealing treatment is necessary to achieve subsequent cold drawing. Thus, annealing was carried out at 400 °C for 60 min in the present study. The results showed that equiaxed, static recrystallized grains formed during annealing treatment, as displayed in figure 2c. Measurement of grain size revealed that there was a fully recrystallized microstructure of the annealed Mg-Zn-Gd alloy wire with the mean grain size of approximately 12.7 μm, which was about twice as big as that of the as-extruded wires. The annealing treatment leads to the improvement in regularity and uniformity of recrystallization grains while there is a rise in average grain size. It hints that the microstructure of the wires grows enough during the present annealing treatment. Apparently, it is relatively easy for the Mg-Zn-Gd alloy wires with static recrystallized grains to further draw into finer wires with 3.0 mm in diameter, whose microstructure can be seen in figure 2d.

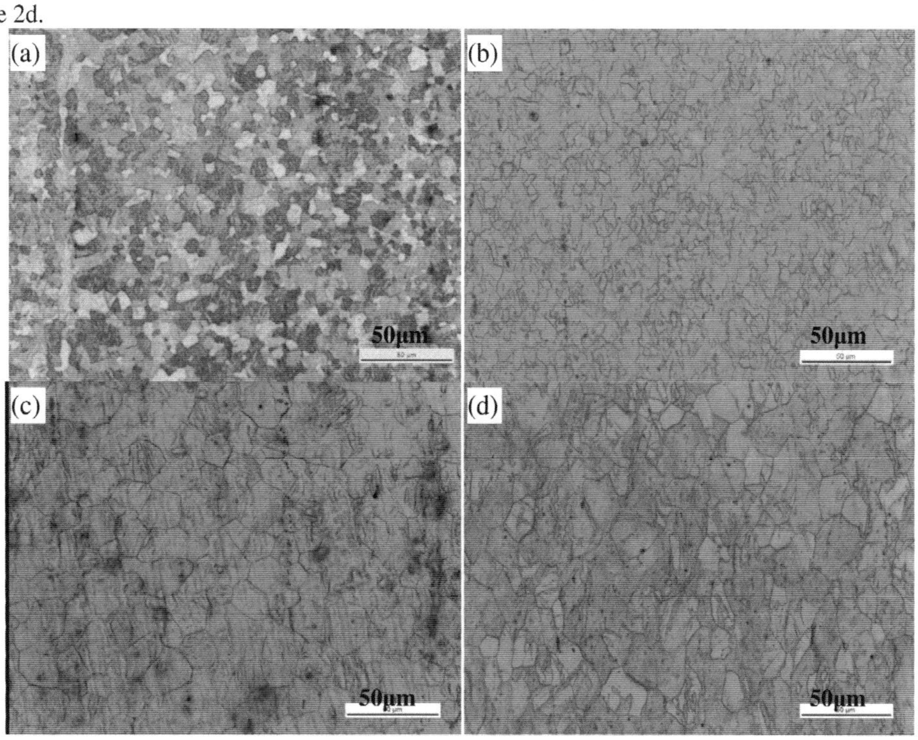

Figure 2. Optical microstructure of Mg-Zn-Gd alloy wires under different conditions. (a) as hot extruded(4.0 mm in diameter); (b) as-drawn 1# (3.6 mm in diameter); (c) as-annealed(3.6 mm in diameter); (d) as-drawn 2# (3.0 mm in diameter).

3.3. Tensile mechanical property of wires

Figure 3 illustrates the tensile mechanical property of Mg-Zn-Gd alloy wires under different conditions. The corresponding mechanical properties including ultimate tensile strength (UTS), 0.2% yield strength (YS) and breaking elongation (ε) are depicted, respectively. It is revealed that the yield-tensile ratio of as-drawn 1# and as-drawn 2# wires retains 0.58 and 0.79, respectively. A higher yield-tensile ratio of as-drawn 2# wires can be beneficial to application of implant materials. After annealed at 400 °C, the Mg alloy wires exhibit a significant decrease in the UTS and YS, but a significant increase in the breaking elongation. The elongation of as-drawn 1#, as-drawn 2# and as-annealed wires reaches 8%, 5% and 23%, respectively.

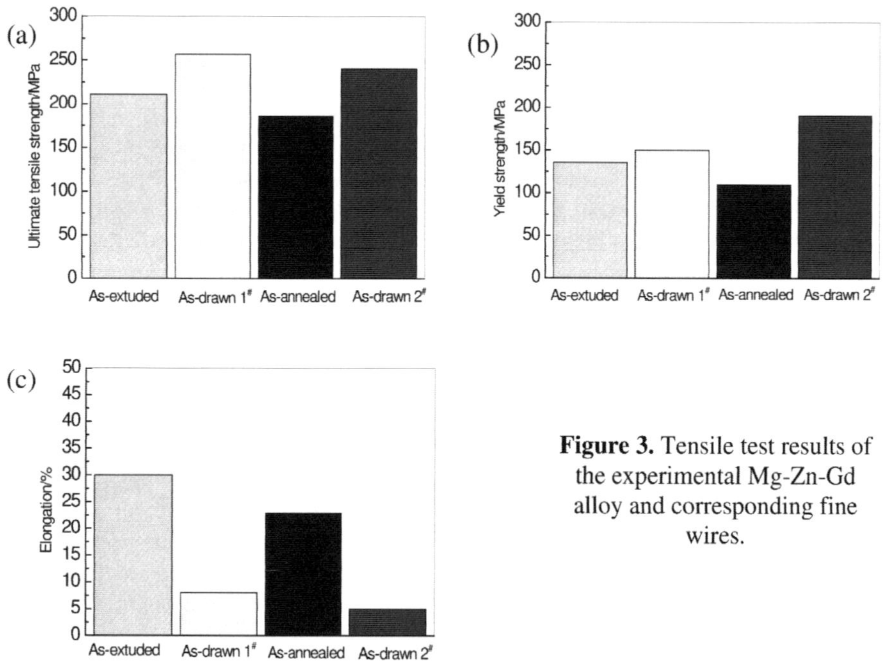

Figure 3. Tensile test results of the experimental Mg-Zn-Gd alloy and corresponding fine wires.

Further analysis shows that the room temperature tensile strength and elongation of as-extrude and 400 °C annealed wires assume the highly significant difference. The extruded wires have higher ultimate, yield strength, and even much higher elongation in comparison than as-annealed alloys, suggesting that the refinement of grains may be still a primary controlling factor on the mechanical properties of Mg-Zn-Gd wires during metal working.

3.4. Fractography

Figure 4 reveals the SEM images of the tensile fracture surfaces of the Mg-Zn-Gd alloy wires under different conditions. For all the Mg-Zn-Gd wires, the fracture surfaces appear to vary from one to another. The SEM image of as-extruded wires in figure 4a shows a complicated fracture surface, where there are shallow dimples with or without particles. After the as-extruded wires were drawn at room temperature, dimples become shallow, and brittle rupture surfaces are also observed (figure 4b). This is an explanation that the elongation of the as-drawn 1# wire was reduced. When the wires were annealed at 400 °C for 1 h, the dominant deep dimples came into being in the failure surfaces (figure 4c), which maybe account for the high ductility of the as-annealed wires. Figure 4d depicts the rupture surfaces of as-drawn wires. It can be seen that, the as-drawn wires have the combination of unevenly distributed brittle and ductile fracture characteristics. Therefore, it can be concluded that the Mg-Zn-Gd alloy wires have the combination of brittle and ductile fracture characteristics irrespective of the heat treatment and deformation.

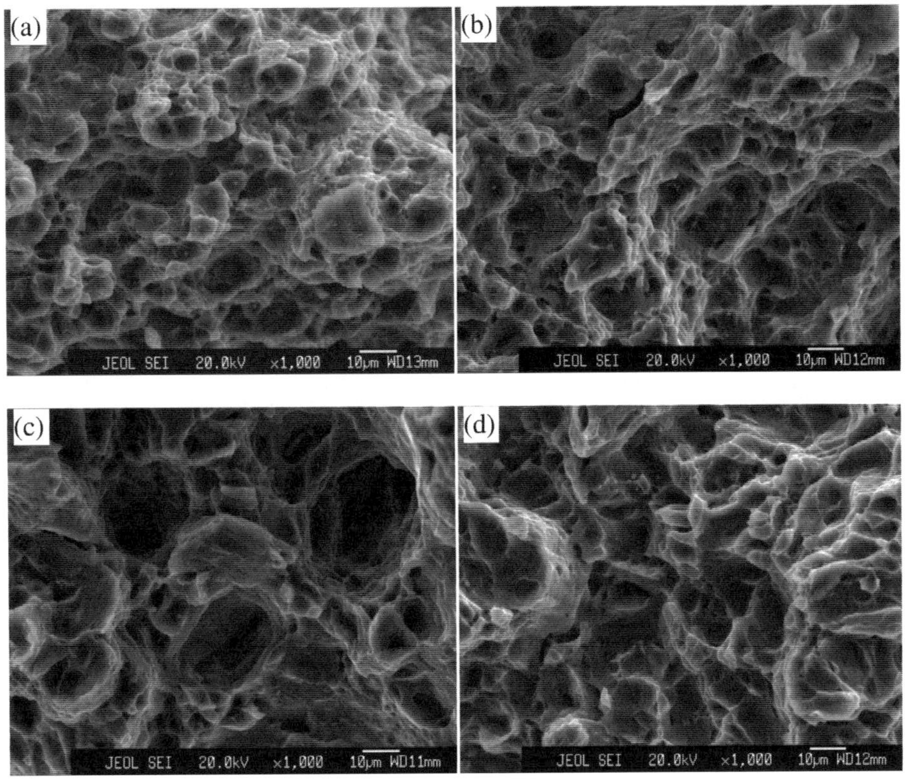

Figure 4. SEM images of the tensile fracture surfaces of Mg-Zn-Gd alloy wires under different conditions. (a) as-extruded(4.0 mm in diameter); (b) as-drawn 1# (3.6 mm in diameter); (c) as-annealed(3.6 mm in diameter); (d) as-drawn 2# (3.0 mm in diameter).

4. Conclusions

1) In comparison with the as-extruded wires, the as-annealed wires exhibit a greater average size of 12.7 μm.
2) The extruded wires exhibit nearly the breaking elongation of 30%. The as-drawn 1#, as-drawn 2# and as-annealed Mg-Zn-Gd alloy wires respectively show the YS of 150 MPa, 190 MPa and 110 MPa, and the breaking elongation of 8%, 5% and 23%. The yield-tensile ratio of as-drawn 1# and as-drawn 2# wires retains 0.58 and 0.79, respectively.
3) The tensile fracture surfaces of Mg-Zn-Gd alloy wires reveal that fracture mechanisms under different conditions are brittle and transgranular ductile fracture.

Acknowledgments

The present research has been fully supported by Guangdong province science and technology plan projects (Nos. 2013B010102025, 2014B010105009, 2014B070706027, 2014B090908006, 2014B090907008, and 2013B091602002), Guangdong provincial key laboratory for technology and application of metal toughening (No. 2014B030301012) and Guangzhou science and technology plan project (No. 201509010003).

References

[1] Wu Y H, Li N, Y Cheng, Zheng Y F and Han Y 2013 *J. Mater. Sci. Technol.* **29** 545
[2] Seitz J M, Utermohlen D, Wulf E, Klose C and Bach F W 2011 *Adv. Eng. Mater.* **13** 1087
[3] Chu C L, Han X, Xue F, Bai J and Chu P K 2013 *Appl. Surf. Sci.* **271** 271

[4] Feyerabend F, Fischer J, Holtz J and Witte F 2010 *Acta Biomater* **6** 1834
[5] Hort N, Huang Y, Fechner D and Stormer M 2010 *Acta Biomater* **6** 1714
[6] Heinrich M C, Kuhlmann M K, Kohlbacher S and Scheer M 2006 *Radiology* **242** 425
[7] Srinivasan A, Huang Y and Mendis C L 2014 *Mater. Sci. Eng. A* **595** 224
[8] Feng X W, Qi W J, Li X H and Li Z C 2016 *J. Chin. Soc. Corros. Prot* **36** 267
[9] Feng X W, Qi W J, Li X H and Li Z C 2016 *Chin. J. Rare. Met.* **40** 14
[10] Gong X Y, Qi W J, Huang Z H and Deng Y L 2013 *Foundry Techn.* **34** 1435
[11] Li X H, Qi W J and Zhou N 2013 *J. Magnes. Alloys* **1** 54

Effects of crystallinity and surface modification of calcium phosphate nanoparticles on the loading and release of tetracycline hydro-chloride

Huaizhi Zhang[1,2], Dong Yan[1,2], Sriyani Menike Korale Gedara[3], Sajith Sudeepa Fernando Dingiri Marakkalage[3], Jothirathna Gamage Kasun Methlal[3], YingChao Han[1,2] and HongLian Dai[1,2]

[1] State Key Laboratory of Advanced Technology for Materials Synthesis and Processing, Wuhan University of Technology, Wuhan 430070, China
[2] Biomedical Materials and Engineering Research Center of Hubei Province, Wuhan University of Technology, Wuhan 430070, China
[3] Faculty of Applied Sciences, Sabaragamuwa University of Sri Lanka, Belihuloya, 70140, Sri Lanka

E-mail: hanyingchao@whut.edu.cn

Abstract. The influences of crystallinity and surface modification of calcium phosphate nanoparticles (nCaP) on their drug loading capacity and drug release profile were studied in the present investigation. The CaP nanoparticles with different crystallinity were prepared by precipitation method under different temperatures. CaP nanoparticles with lower crystallinity exhibited higher drug loading capacity. The samples were characterized by XRD, FT-IR, SEM, TEM and BET surface area analyzer respectively. The drug loading capacity of nCaP was evaluated to tetracycline hydro-chloride (TCH). The internalization of TCH loaded nCaP in cancer cell was observed by florescence microscope. nCaP could be stabilized and dispersed in aqueous solution by poly(acrylic acid) surface modification agent, leading to enhanced drug loading capacity. The drug release was conducted in different pH environment and the experimental data proved that nCaP were pH sensitive drug carrier, suggesting that nCaP could achieve the controlled drug release in intracellular acidic environment. Furthermore, nCaP with higher crystallinity showed lower drug release rate than that of lower crystallinity, indicating that the drug release profile could be adjusted by crystallinity of nCaP. nCaP with adjustable drug loading and release properties are promising candidate as drug carrier for disease treatment.

Keywords: calcium phosphate,drug carrier,pH sensitive, nanoparticle, drug controlled release

1. Introduction

Calcium phosphate (CaP) bio-material has been used for several decades in clinic due to its good biological activity and compatibility [1-3]. However, the implantation of calcium phosphate normally requires anti-infection treatments [4,5]. Local drug delivery is an alternative way to treat infection by loading drug on calcium phosphate material [6-8]. The primary advantage of local antibiotic administration is the ability to obtain high local drug concentrations on the surface and in intimate vicinity of the device without increasing systemic toxicity[9]. Calcium phosphate is the major inorganic component of bones and teeth and has abundant adsorption sites (such as Ca^{2+} and PO_4^{3-}) on the surface for drug adsorption [10,11]. In addition, CaP can achieve controlled drug release based on its pH responsiveness. The surface property and specific surface area are critical for drug loading, which are varied with crystallinity. Therefore, it is important to study the influences of crystallinity and surface modification of CaP on the drug loading capacity and drug release profile.

Herein the drug loading capacity and drug release profile of CaP nanoparticles with different crystallinity were discussed in detail, which were controlled by adjusting reaction temperature. The surface modification was carried out by poly(acrylic acid) (PAA). The drug loading capacity and drug release profile of CaP nanoparticles to tetracycline hydro-chloride (TCH) was evaluated.

2. Materials and methods

2.1. Preparation and characterization of CaP

First, the aqueous solutions of $Ca(NO_3)_3 4H_2O$ with 0.02mol/L and $(NH_4)_2HPO_4$ with 0.0334mol/l concentration were prepared. Next, 20ml $(NH_4)_2HPO_4$ aqueous solution was introduced to 20 ml $Ca(NO_3)_2 4H_2O$ aqueous solution under strong stirring. The pH value was adjusted to 9-10 range using NH_3H_2O. The reaction solution was maintained for 1h at different temperatures(25°C,37°C,60°C,80°C) while stirring. Then precipitation was separated and washed 3 times by centrifugation. For PAA stabilized nCaP, the washed precipitation was re-dispersed in water with addition of PAA (0.3mg/mL) and the stable nCaP suspension was obtained after ultrasound treatment.

Several instruments were required for the characterization of the prepared materials. In this experiment, powder X-ray diffractometer (XRD, D8 Advance,Germany) was used to determine the phase composition. Fourier transform infrared spectroscopy (FT−IR, Thermo Nicolet 6700, U.S.A.) to obtain information about the function groups and bonding forms. The scanning electron microscope (SEM, Zeiss Ultra Plus, Germany) and high resolution transmit electronic microscopy (HRTEM, JEM2100F, U.S.A.) were used to illustrate the morphologic information and to determine element content. BET surface area analyzer (BET, ASAP 2020M , U.S.A.) was employed to determine the surface area of the prepared material. The internalized TCH loaded nCaP in HepG2 cancer cells was observed using fluorescence microscope (970CRT, China).

2.2. Drug loading

The washed precipitate was re-dispersed in water and drug was added with resulting concentration of 0.05-1 mg/mL which was kept under stirring for 1h. The drug loaded nCaP was separated after removing the unloaded drug by centrifugation. The drug concentration of the supernant was determined using UV-spectrophotometer to calculate the amount of drug loaded on to nCaP. The loaded drug concentration (DC) was calculated by deducting the unloaded drug from the feeding drug. The drug loading efficiency (LE) and drug loading capacity (DLC) were calculated respectively by the following equations: $DLC = 100\% \times M_1/(M_2+M_3)$, $LE = 100\% \times M_2/M_1$, where M_1, M_2 and M_3 are the weight amounts of feeding drug, loaded drug and nCaP. The independent experiment was repeated three times and result was shown as mean ±SD (n=3).

2.3. Drug release

The drug release profile was measured in PBS (pH 7.4) and NaAc-HAc buffer solution (pH 5.0) at 37°C under dark conditions. The drug loaded nCaP was re-dispersed in 5mL buffer solution and put into dialysis tube (3500) with external 80mL buffer solution. At different time points 3mL external buffer solution was taken out for the determination of drug concentration and then re-poured into external buffer solution for the following experiment. The independent experiment was repeated three times and the result was shown as mean ±SD (n=3).

3. Results and discussion

3.1. Characterization of nCaP

Figure 1 illustrates XRD patterns and FT-IR spectra of nCaP synthesized at different temperatures. As shown in figure 1 (A), at 25°C sample was amorphous and no crystal was formed. At 37°C the weak diffraction peaks were observed which were assigned to hydroxyapatite (HAP) crystal. With the temperature increased to 60°C and 80°C the diffraction peaks were becoming sharper, indicating the crystallinity of HAP was enhanced. In addition, the crystal size was increased shown in figure 1 (B). The FT-IR spectra also provide some information about HAP as shown in figure 1 (C). The peaks at 1048cm-1, 958cm-1, 890cm-1 and 605cm-1 were attributed to characteristic vibrations of PO_4^{3-} group. The peak at 3570cm-1 is due to stretching vibration of -OH group in the HAP crystal. The vibration of -OH group at 3570cm-1 rose as temperature increased, indicating the increase of crystallinity.

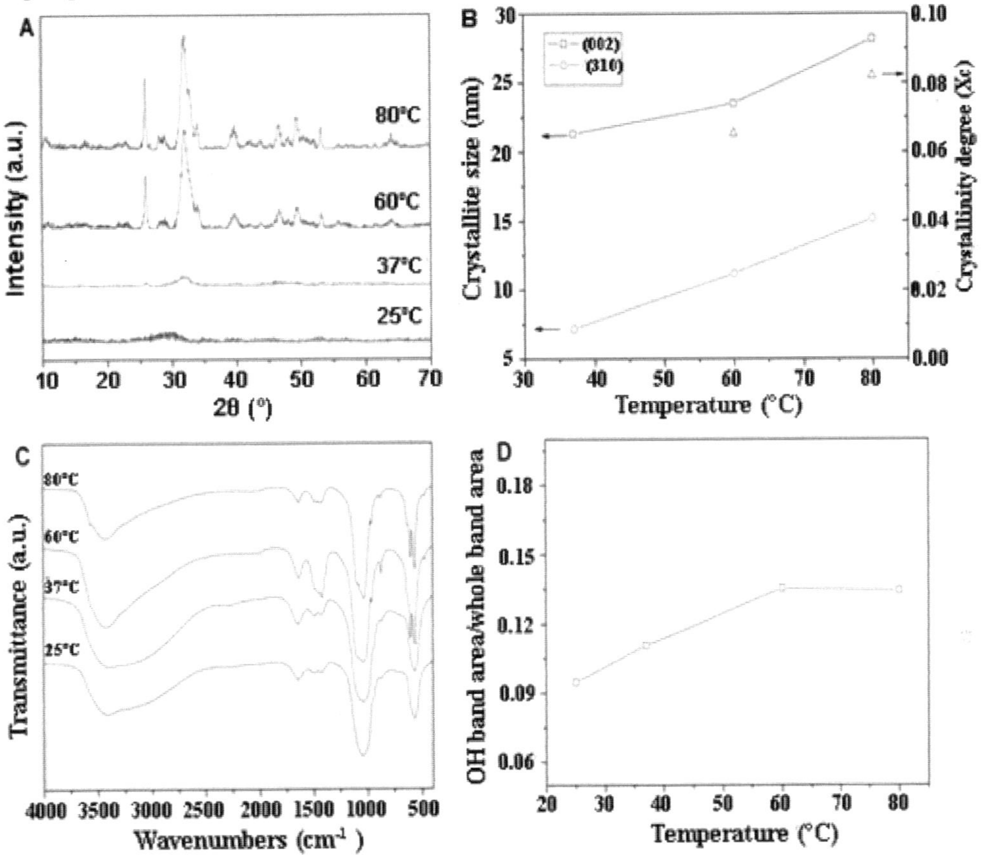

Figure 1. XRD patterns (A), the calculated crystallinity degree and crystallite size from XRD patterns (B),FT-IR spectra (C), the calculated crystallinity degree from FT-IR (D).

Figure 2 shows morphology of nCaP prepared at different reaction temperatures (25°C, 37°C, 60°C and 80°C). The corresponding size results obtained by SEM observation are shown in figure 3.The particles obtained at 25°C had spherical shape with size of 39.44±1.81nm. When the temperature was increased to 37°C, particle's size increased to 69.3±3.82nm. When temperature further increased to 60°C and 80°C, the particles were becoming needle like shape. Moreover, the particle sizes increased a lot, which were 29.58±1.28nm×121±3.25nm (60°C) and 29.07±1.35nm×175.59±5.18nm (80°C). The crystal growth is along the long axis direction, which is consistent with XRD results.

Figure 2. SEM images of the samples.

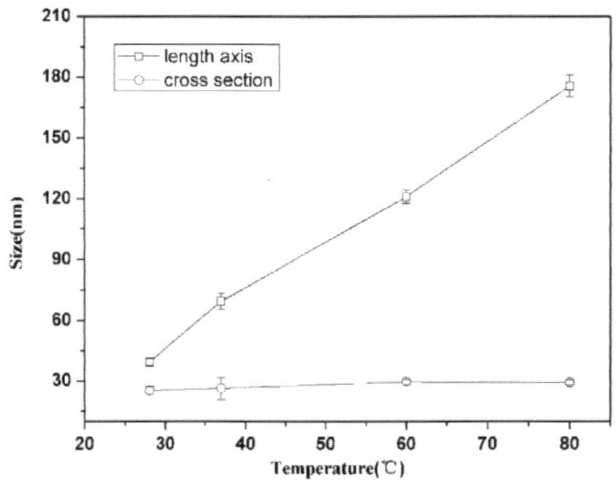

Figure 3. Crystallite sizes measured by SEM.

Table 1. BET areas and calculated diameter of nCaP.

CaP	37ºC	60ºC	80ºC
specific area(m2/g)	110.7241	119.3882	112.6077
diameter(nm)*	17.1592	15.9139	16.8727

*d=6/ρ·S (d-diameter, ρ-density, S-specific area)

Table 1 shows the specific surface area of nCaP. Indicating that nCaP synthesized at different temperatures possessed similar specific surface area (110.7241 m2/g, 119.3882 m2/ and 112.6077 m2/g), corresponding to 17.1592nm, 15.1592 nm and 16.8727 nm in diameter respectively.

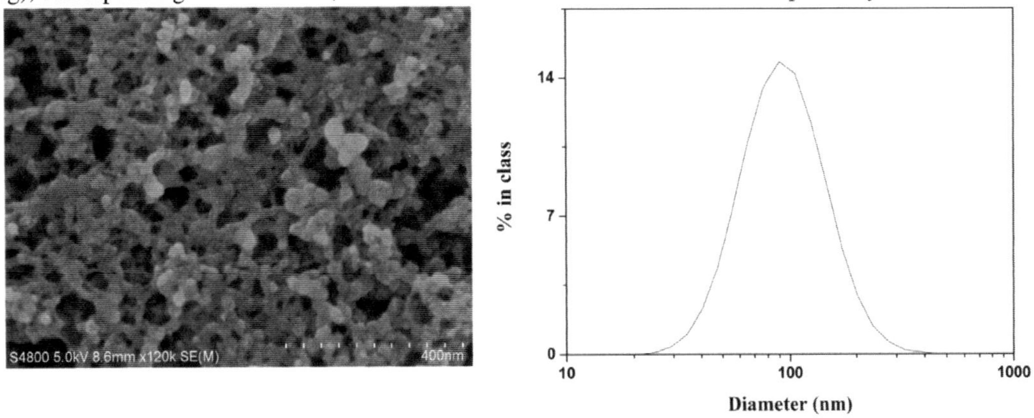

Figure 4. Size distribution curve of PAA modified nCaP.

After ultrasound treatment with addition of PAA, nCaP can be stabilized and dispersed in aqueous solution to form stable suspension. As shown in figure 4, the intensity averaged particle size was 92.9nm with a narrow size distribution (PDI:0.19). The agglomeration of nCaP was declined drastically after surface modification.

3.2. Drug loading property

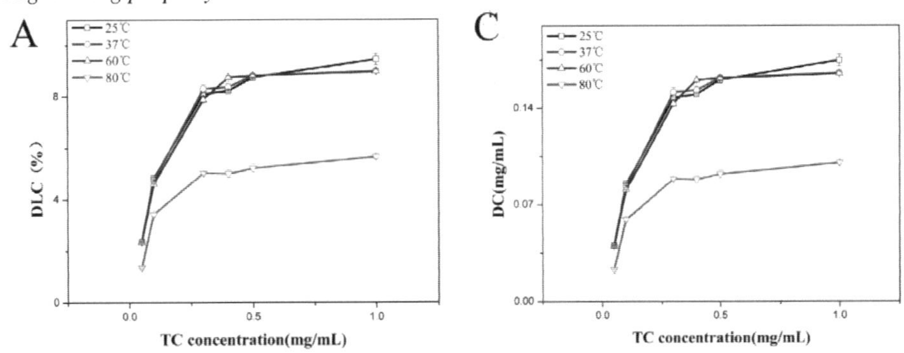

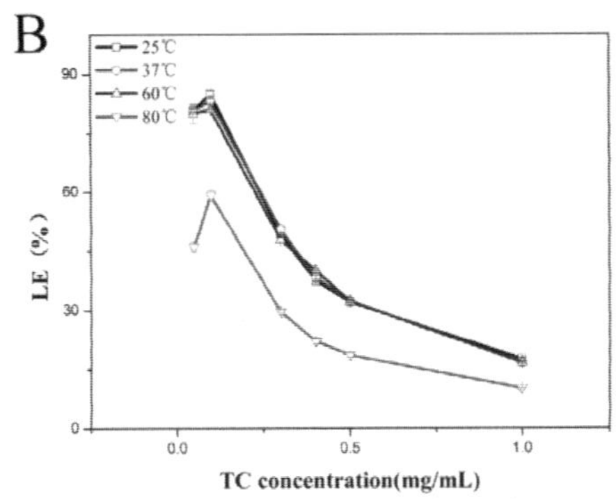

Figure 5. The DLC(A), LE(B) and DC(C) of nCaP to TCH.

Figure 5 interprets the variation of DLC, LE and DC of nCaP to TCH synthesized at different temperatures. The highest DLC values of nCaP are 9.44±0.21%, 8.99±0.1%, 8.97±0.07% and 5.67±0.07% for 25°C, 37°C, 60°C and 80°C corresponding to the highest LE values of 85.04±0.84%,

83.33±0.52%, 81.02±0.23%, 59.15±1.41% and the highest DC values of 0.1745±0.0045mg/mL, 0.1658±0.0020mg/mL, 0.1649±0.0016mg/mL, 0.1005±0.0013mg/mL respectively. There was no significant difference in drug loading property of samples synthesized between 25°C and 60°C. However, the DLC decreased a lot when temperature further increased to 80°C, showing that low crystallinity could improve the loading of TCH on nCaP. Samples have close specific surface area (table 1) and should possess similar adsorption ability to TCH. But results displayed different adsorption ability. This might be due to much more Ca^{2+} binding sites on the surface of nCaP with lower crystallinity, leading to higher adsorption to TCH via interaction between surface Ca^{2+} of nCaP and amide groups of TCH. In this experiment, the size and shape of nCaP particles generate little influence on drug loading.

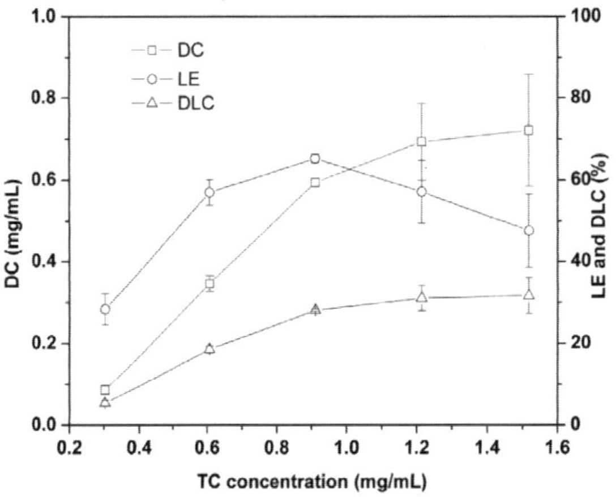

Figure 6. DC, DLC and LE of nCaP (25°C) stabilized by PAA to TCH.

Figure 6 depicts the DC, DLC and LE of nCaP stabilized by PAA. Compared to the results of pure nCaP (25°C,Fig.6), loading of TCH on PAA stabilized nCaP was largely promoted, showing highest DLC and DC with 31.7±4.4 % and 0.72±0.14 mg/mL. To conclude that PAA surface modification can enhance the adsorption ability of nCaP to TCH by inhibiting the agglomeration of nCaP.

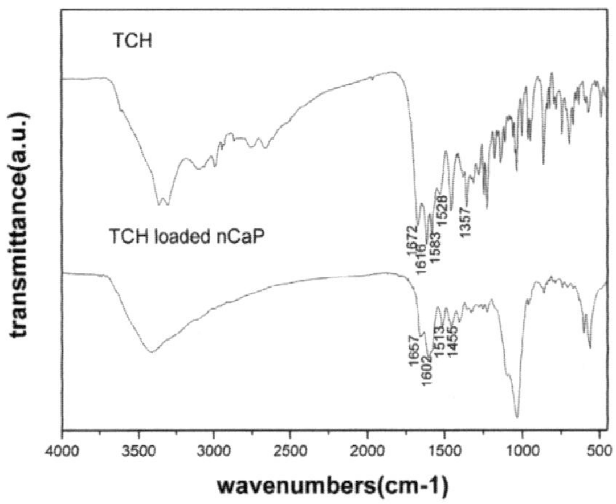

Figure 7. FT-IR spectra of TCH and TCH loaded nCaP.

Figure 7 shows FT-IR spectra of TCH and TCH loaded nCaP. The characteristic vibrations of TCH were observed at $1700cm^{-1}$-$1200cm^{-1}$ and $850cm^{-1}$-$650cm^{-1}$, indicating TCH was successfully loaded on nCaP. Moreover, some peaks of TCH were moved after being adsorbed on nCaP, for example, 1672cm^{-1}, 1616cm^{-1}, 1528cm^{-1} and 1357cm^{-1}. The bands at 1672 cm^{-1} and 1528cm^{-1} were due to C=O stretching, NH_2 bending and C-N stretching, showing TCH was adsorbed on nCaP via the interaction between amide group of TCH and surface Ca^{2+} of nCaP.

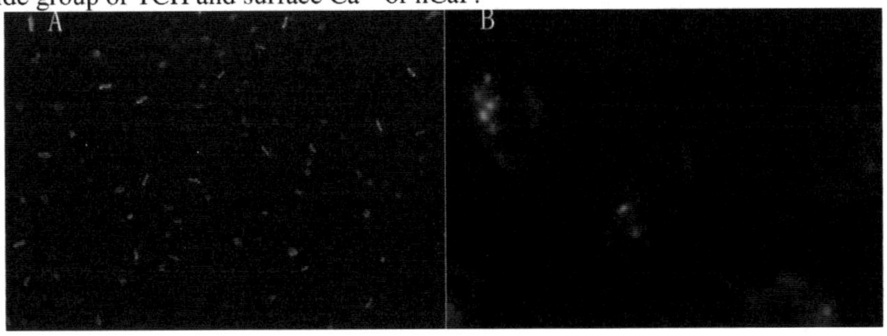

Figure 8. Fluorescence microscope observation of TCH loading nCaP (A) and intracellular TCH loading nCaP (B).

As shown in fluorescence microscope observation (figure 8), the fluorescent signal was observed in HepG2 cancer cells, indicating TCH was transported into cells by nCaP.

3.3. Drug release profiles

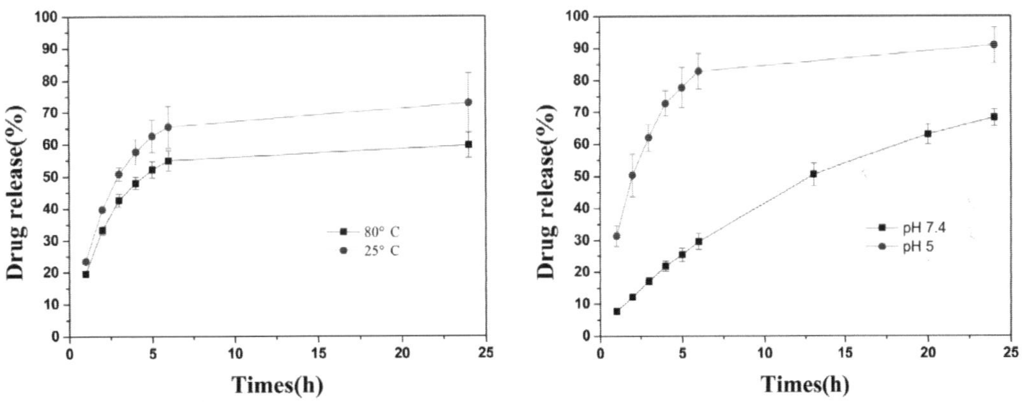

Figure 9. drug release profiles of nCaP with different crystallinity and PAA stabilized nCaP.

Figure 9 shows the drug release profiles of nCaP with different crystallinity and PAA stabilized nCaP. Results showed that TCH was released slowly in PBS solution; however the release of TCH was enhanced much in NaAc-HAc buffer solution (pH 5.0). This indicated that nCaP was pH sensitive drug carrier. Furthermore, after 24h, the rate of released drug from nCaP synthesized at 25°C (72.95±9.35%) was higher than that of nCaP synthesized at 80°C (59.64±4.00%), showing that the drug release was influenced by the crystallinity of nCaP. Higher crystallinity corresponded to lower drug release rate. Additionally, PAA stabilized nCaP showed a little faster drug release profile than pure nCaP, indicating that PAA surface modification increased the drug release from nCaP due to its well dispersing property (lower agglomeration). However, PAA stabilized nCaP could result in higher released drug amount than pure nCaP due to its higher drug loading capacity. Therefore, the released drug concentration could be adjusted in a wide range to satisfy the requirement of drug dose for treating disease.

4. Conclusions

The experimental results show that CaP nanoparticles with different crystallinity could be obtained by adjusting reaction temperature. CaP nanoparticles with lower crystallinity exhibited higher drug loading capacity due to much more Ca^{2+} adsorption sites on the surface. CaP nanoparticles could be stabilized and dispersed in aqueous solution by PAA, leading to enhanced drug loading capacity. TCH loaded CaP nanoparticles could be easily transported into cancer cells, indicating drug could be delivered into cells by CaP nanoparticles efficiently. Experimental data proved that nCaP were pH sensitive drug carrier, suggesting that nCaP could achieve the controlled drug release in intracellular acidic environment. Furthermore, nCaP showed different drug release profiles depending on crystallinity and surface modification. Increasing crystallinity could slow drug release rate, indicating the drug release profile could be adjusted by crystallinity. PAA stabilized nCaP showed faster drug release profile than pure nCaP, and led to higher released drug amount depending on its higher drug loading capacity. nCaP with adjustable drug loading and release properties is promising to become candidate as drug carrier for disease treatment.

Acknowledgment

This work was supported by the National Natural Science Foundation of China (51672206), the National Key Research and Development Program of China (2016YFB1101302), the Science and Technology Partnership Program, Ministry of Science and Technology of China (KY201602002), the Wuhan International Science and Technology Cooperation Project (2016030409020217), the Fundamental Research Funds for the Central Universities (2016-YS-014).

References

[1] Epple M, Ganesan K, Heumann R, Klesing J, Kovtun A, Neumann S and Sokolova V 2010 *J Mater Chem.* Application of Calcium Phosphate Nanoparticles in Biomedicine **20** 18−23

[2] Lin K L, Xia L G, Gan J B, Zhang Z Y, Chen H, Jiang X Q and Chang J 2013 *ACS Appl. Mater. Interfaces* Tailoring the Nanostructured Surfaces of Hydroxyapatite Bioceramics to Promote Protein Adsorption, Osteoblast Growth, and Osteogenic Differentiation **5** 8008−8017

[3] Campoccia D, Montanaro L and Arciola C 2006 *Biomaterials* The significance of infection related to orthopedic devices and issues of antibioticresistance **27** 2331–2339

[4] Gristina A 1987 *Science* Biomaterial-centered infection: Microbial adhesion versus tissue integration **237** 1588–1595

[5] Di Silvio L and Bonfield W 1999 *J Mater SciMater Med* Biodegradable drug delivery system for the treatment of bone infection and repair **10** 653–658

[6] Wenke J, Owens B, Svoboda S and Brooks D 2006 *J Bone Joint Surg Br* Effectiveness of commercially-available antibiotic-impregnated implants **88** 1102–1104

[7] Hanssen A 2005 *Clin Orthop Relat Res* Local antibiotic delivery vehicles in the treatment of musculoskeletal infection 91–96

[8] Hoff S, Fitzgerald R and Kelly P 1981 *J BoneJoint Surg Am* The depot administration of penicillin G and gentamicin in acrylic bone cement **63** 798–804

[9] X Wu *et al.* 2012 *J. Bionic Eng.* Preparation of mesoporous nano-hydroxyapatite using a surfactant template method for protein delivery **9** 224-233

[10] Zakaria S, Zein S, Othman M, Yang F and Jansen J 2013 *Tissue Eng., Part B* Nanophase Hydroxyapatite as a Biomaterial in Advanced Hard Tissue Engineering: A Review **19** 431−441

[11] Poinern G, Brundavanam R, Le X T, Nicholls P, Cake M and Fawcett D 2014 *Sci. Rep.* The Synthesis, Characterization and In Vivo Study of a Bioceramic for Potential Tissue Regeneration Applications **4** 6235

Microstructure and mechanical properties of the as-cast Mg-Zn-Mn-Ca alloys

Junjian Fu[a], Ke Liu[b,*], Wenbo Du[c,*], Zhaohui Wang[d], Shubo Li[e] and Xian Du[f]

Beijing University of Technology, Beijing 100124, China.

E-mail: [a]fujunjian@emails.bjut.edu.cn, [b,*]lk@bjut.edu.cn, [c,*]duwb@bjut.edu.cn [d]wangzhaohui@bjut.edu.cn, [e]lishubo@bjut.edu.cn, [f]duxian@bjut.edu.cn

Abstract. As-cast Mg-Zn-Mn-Ca alloys are the candidates for medical implants. The microstructure and mechanical properties of the as-cast Mg-Zn-Mn-Ca alloys were investigated by using optical microscope (OM), X-ray diffraction (XRD) and scanning electron microscope (SEM). The results suggested that the addition of Zn significantly refined the grain size of the as-cast magnesium alloys, and hence improved the mechanical properties of the alloys. Meanwhile, the secondary phase mainly distributed in the grain boundaries and the solution treatment made parts of secondary phases dissolved into the matrix, resulting in enhancement of the plasticity of the as-cast Mg-Zn-Mn-Ca alloy.

Keywords: Mg-Zn-Mn-Ca alloys, microstructure, mechanical properties

1. Introduction

Medical metal materials, such as traditional metallic biomaterials of stainless steels, titanium and cobalt-chromium-based alloys, are widely used in medicine fields because of their high strength, good toughness, fatigue resistance, corrosion resistance and excellent processing molding ability. However, the disadvantages of the traditional materials are the possible release of toxic metallic ions and particles through corrosion. First, they will lead to inflammatory cascades which result in biocompatibility reduction and tissue loss [1]. Second, the traditional material implants, remaining as permanent fixtures in the case of plates and stents, must be removed by a second surgical procedure after the sufficient heal of tissue, which increases costs and further morbidity to patients. At last, the elastic modulus of current metallic biomaterials did not match with that of natural bone tissue, which causes stress shielding effects [2]. Therefore, a new material that is non-cytotoxic, degrade and biocompatible need to be developed.

Magnesium and its alloys have shown the potential to serve as biodegradable medical metal implant to cure bone fracture and vascular tissue due to their degradability, biocompatibility and appropriate elastic modulus (40-45GPa). The value of the elastic modulus is very close to that of human bone (10-20GPa) [3]. However, the fast degradation rate of magnesium in vivo-environment limited their clinical application. Although the traditional industrial magnesium alloy, such as AZ31 and WE43 containing Al and RE elements showed a relatively high strength and good corrosion resistance, Al and RE are neurotoxin elements. The elements of Al and RE can cause nerve toxicity and restraining growth to human body [4]. It is great important to find a new kind of magnesium alloys for the production of medical metal implants.

Actually, there are only few kinds of elements that can be used as enhanced elements of biological magnesium alloy, including Zn, Ca and some very small amount of RE elements. The Ca is an essential element of the human body, and is also an important component for human bone. It was reported that Ca can improve the mechanical properties and corrosion resistance. The optimum performance in mechanical properties, corrosion resistance and biocompatibility was obtained when the Ca content less than 1% in magnesium alloys [5]. In addition, Zn is also an essential element for human body. It can promote the body growth, enhance the immune function and accelerate the repair of damaged tissue. It was reported that adding Zn element into magnesium alloys improved both the tensile strength and the corrosion resistance successfully [6,7].

In the present investigation, the elements of Zn, Ca and Mn were added into the magnesium, and new kinds of Mg-Zn-Mn-Ca alloys were produced. The microstructure and mechanical properties of the Mg-Zn-Mn-Ca alloys were investigated by using optical microscope (OM), X-ray diffraction (XRD), scanning electron microscope (SEM) in order to find the effect of alloying elements on the mechanical properties.

2. Experiment procedure

2.1. Materials preparation

Magnesium alloys were prepared from pure magnesium ingots, pure zinc chips, Mg-5%Mn and Mg-9.4%Ca master alloy. The materials were melted by electrical resistance furnace under the protection of SF_6 and N_2 mixed gas in the graphite crucible at 720°C, and the ratio of SF_6 and N_2 is 1:10. The addition order of alloying elements was Zn, Ca, Mn. The melting was stirred for 5 min, and then preserved for 20 min at the melting condition for stabilization. At last, the melting was poured into mild steel mold which had been preheated at 200°C. The as-cast alloys with nominal compositions of Mg-2Zn-0.2Mn-0.2Ca (defined as Alloy1) and Mg-4Zn-0.2Mn-0.2Ca (defined as Alloy2) were produced. The chemical compositions of as-cast alloys A and B were examined by XRF, as shown in table 1.

Table 1. Chemical compositions of the Mg-xZn-0.2Mn-0.2Ca alloys (wt.%).

Alloys	Zn	Mn	Ca	Mg
Alloy1	1.96	0.25	0.15	Bal.
Alloy2	4.03	0.23	0.23	Bal.

2.2. Microstructure and mechanical properties

Specimens with the shape of bulk were cut from the achieved alloy ingots then the samples were mechanically wet ground with 250 to 7000 silicon carbide grit papers. Microstructure observation was conducted by using an optical microscope (OM, AXIO2IMAGER A2m) and a scanning electron microscope (SEM, HITACH S3400N) equipped with EDS analysis. Tensile tests were performed by using an universal testing machine at a displacement rate of 1.0 mm/min at ambient temperature. Specimens with a diameter of 5mm and a gauge length of 25 mm were machined for tensile test. For each testing material, three specimens were examined. Vickers micro-hardness test was conducted by using an automatic digital HDX-1000 TMC/LCD micro-hardness tester with a load of 0.1Kg and a dwelling time of 10s.

3. Results and discussion

3.1. Microstructure of the Mg-Zn-Mn-Ca alloys

The optical micrographs of Mg-Zn-Mn-Ca alloys with different Zn contents were shown in figure 1. It showed that the grain size of as-cast Mg-Zn-Mn-Ca alloys decreased as the Zn content increased. The grain size was about 215μm when the content of Zn was 2wt%. When the Zn content increased to 4wt%, the grain size reduced to 140μm. It indicated that the Zn has a great effect on grain refinement.

The distribution and the morphology of the secondary phases were also shown in figure 1. The secondary phase was mainly strip-like distributing at the grain boundary while some granular phases were also found inner grain. It was observed that the amount of the secondary phases increased with the increase of Zn content. The EDS analysis (figure 2) showed that the secondary phases at the grain boundary and inner grain were both composed of Mg, Zn and Ca elements. The tiny peaks of MgZnCa phase were found at the XRD graph (figure 3) but the EDS analysis showed that the atomic ratio of Zn/Ca was not precise 1.25.

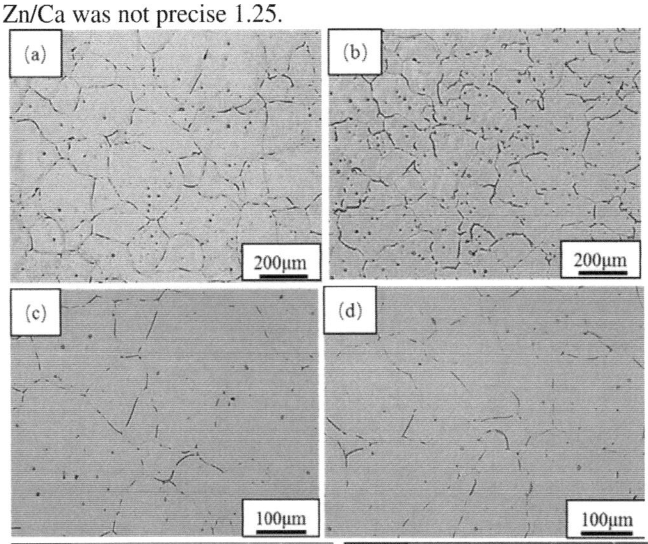

Figure 1. Optical microstructure of the as-cast Mg-Zn-Mn-Ca alloys: (a)(c) Mg-2Zn-0.2Mn-0.2Ca, (b)(d) Mg-4Zn-0.2Mn-0.2Ca.

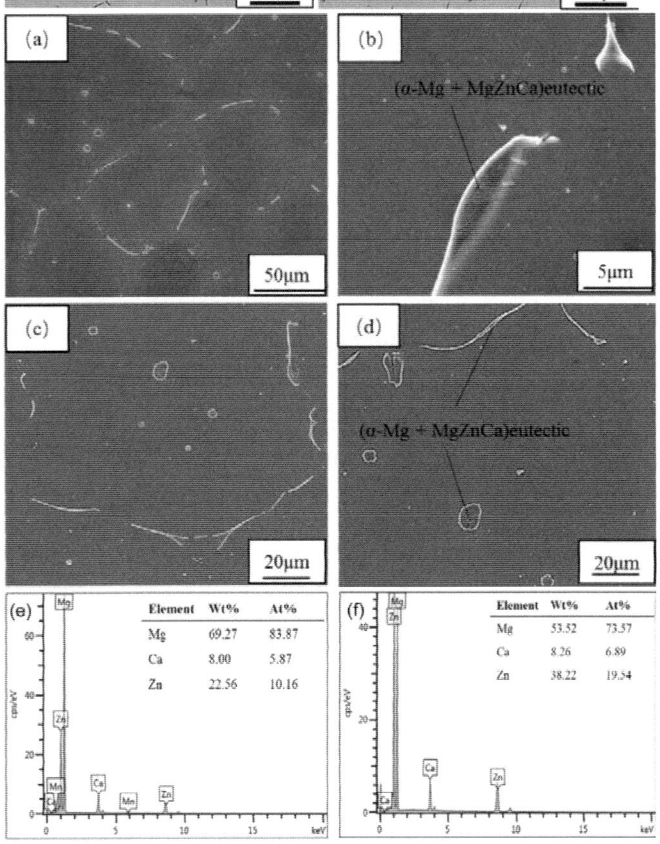

Figure 2. SEM morphology of as-cast Mg-xZn-0.2Mn-0.2Ca alloy:

(a) (b)2wt.%Zn, (c) (d)4wt.%Zn, (e) EDS analysis of the eutectic (MgZnCa+α-Mg) phase of as-cast Mg-2Zn-0.2Mn-0.2Ca alloy, (f)EDS analysis of the eutectic (MgZnCa+α-Mg) phase of as-cast Mg-4Zn-0.2Mn-0.2Ca alloy.

3.2. Mechanical properties of the Mg-Zn-Mn-Ca alloys

Figure 4 was the typical strain-stress curve of the Mg-Zn-Mn-Ca alloys, and the mechanical properties of the Mg-Zn-Mn-Ca alloys with different content of Zn were summarized in table 2. The yield strength (YS) increased slightly with the increment of the Zn content. The ultimate tensile strength (UTS) and elongation also increased with the increment of the Zn content. The value of the ultimate tensile strength increased significantly than that of the yield strength. The yield strength increased by 18.2%, and the ultimate tensile strength increased by 15.4%.

The vickers hardness value of the as-cast Mg-2Zn-0.2Mn-0.2Ca alloys was 49.3Hv. The hardness value of the as-cast Mg-4Zn-0.2Mn-0.2Ca alloys was 53.4Hv. The result showed that Zn can improve the strength of the alloy slightly because of the presence of the secondary phase.

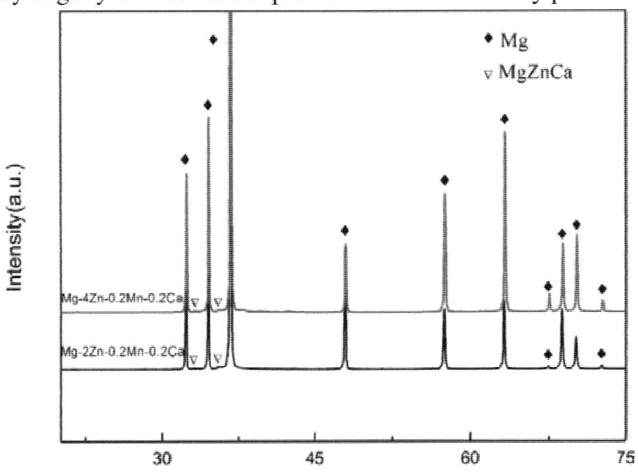

Figure 3. XRD patterns of the as-cast Mg-Zn-Mn-Ca alloys.

Table 2 showed mechanical properties of as-cast Mg-Zn-Mn-Ca alloys. Compared the as-cast magnesium alloys, the enhancement of the UTS was mainly due to the grain refinement, solution strengthening and secondary phase strengthening. As reported, the segregation of alloying element at the front of grain growth formed an intensive constitutional undercooling in a diffusion layer ahead of the advancing solid/liquid interface, which then restricted the grain growth and promoted the nucleation of the primary magnesium grain [8]. Some secondary phases were dispersed in the matrix and reacted with the dislocation during deformation, resulting in improvement of mechanical properties.

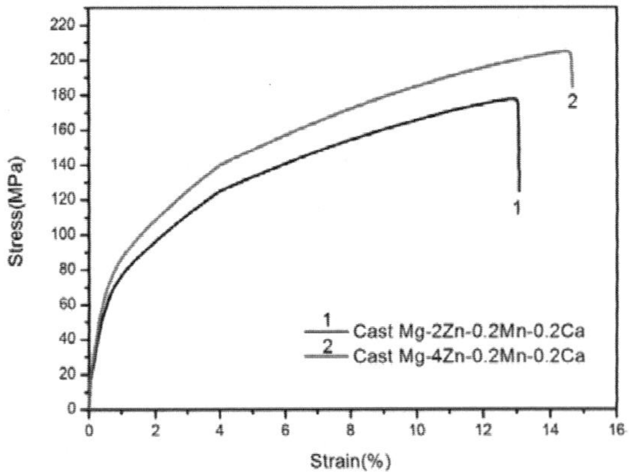

Figure 4. Typical stress-strain curve of as-cast Mg-Zn-Mn-Ca alloys.

Table 2. Mechanical properties of the as-cast Mg-Zn-Mn-Ca alloys with different content of Zn.

Specimen	Yield strength (Mpa)	Tensile strength (Mpa)	Elongation (%)	Hardness (Hv)
as-cast Mg-2Zn-0.2Mn-0.2Ca	58.4	177.6	13.1	49.3
as-cast Mg-4Zn-0.2Mn-0.2Ca	69.0	205.0	14.7	53.4

4. Conclusion

The main secondary phase in Mg-2Zn-0.2Mn-0.2Ca and Mg-4Zn-0.2Mn-0.2Ca alloys is MgZnCa phase. The volume fraction of the secondary phase increased as the Zn increased.

The addition of Zn had a great effect on refining grain, which could improve the mechanical properties of the as-cast alloys.

Acknowledgement

The project was supported by National key research and development program (2016YFB0301101), the Natural Science Foundation of China (51401005), Beijing Natural Science Foundation (2142005), Beijing Municipal Commission of Education (KM201410005014) and RiXin Talents Plan of Beijing University of Technology (2015-RX-L11).

Reference

[1] Niinomi M 2002 *Met Mater Trans A* Recent metallic materials for biomedical applications **33** 77–86
[2] Nagels J, Stokdijk M and Rozing PM 2003 *Shoulder Elbow Surg* Stress shielding and bone resorption in shoulder arthroplasty **12** 35–9
[3] Staigera MP, Pietaka AM, Huadmaia J and Dias G 2006 *Biomaterials* Magnesium and its alloys as orthopedic biomaterials: a review **27** 1728
[4] H D Li, Y M Wang and L Li 2006 *World Elem. Med* **13** 28–31
[5] Li Y C, Hodgson P D and Wen C E 2011 *Mater Sci* The effects of calcium and yttrium additions on the microstructure, mechanical properties and biocompatibility of biodegradable magnesium alloys **46 (2)** 365
[6] Erikson K M *et al.* 2005 *Environ Toxicol Pharmacol* Interactions between excessive manganese exposures and dietary iron-deficiency in neurodegeneration **19 (3)** 415
[7] Song Y W *et al.* 2012 *Corrosion Science* The effect of Zn concentration on the corrosion behavior of Mg-xZn alloys **65** 322
[8] Erlin Zhang and Lei Yang 2008 *Materials Science and Engineering A* Microstructure, mechanical properties and bio-corrosion properties of Mg-Zn-Mn-Ca alloy for biomedical application **497** 111–118

Influence of solid-solution treatment on microstructure, mechanical property and corrosion behavior of biodegradable Mg-Zn-Ca alloy

Xuan Nam Ly, S Yang and Y Qin

School of Materials Science and Engineering, Nanjing University of Science and Technology, Nanjing 210094, China

E-mail: yangsen@njust.edu.cn

Abstract. The influence of solid-solution treatment on microstructure, mechanical property and corrosion behavior of Mg-Zn-Ca alloy was studied in the present investigation by SEM, tensile test, electrochemical and immersion test. The results show that the microstructure of Mg alloys after solid solution treatment significantly changed, a large number of the second phase ($Ca_2Mg_6Zn_3$, Mg_2Ca) dissolved into the α-Mg matrix reaching a supersaturated state, and the grains size was bigger than before solid solution treatment; the mechanical properties were obviously improved. In particular the tensile strength of 0.5wt.% Ca of Mg alloy reached 220MPa and the ductility reached 16.6%. Compared with the as-cast Mg alloys, the corrosion potential after solid-solution treatment slightly shifted negative, but the corrosion current density significantly decreased. After solid solution treatment, the surface corrosion was not serious and the result of weight gain was lower compared with those of the as-cast Mg alloys.

Keywords: Mg-Zn-Ca alloy, solid-solution treatment, microstructure, corrosion behavior

1. Introduction

Magnesium alloys with a low density, the elastic modulus closest to that of the human bone than other traditional metallic alloys. In addition, magnesium alloys have a excellent biocompatibility, degradability, making it a promising candidate in bone implant [1]. However, the as-cast magnesium alloys serving as implant materials are limited due to their quick corrosion rate and poor dispersion strengthening [2]. Heat treatment can improve the performances of magnesium alloys through changes their microstructure, relief of the defect in the process of cast [3-5]. The solid-solution treatment can make the microstructure of magnesium alloys more homogeneous than that of as-cast Mg alloys, resulting in influence on the mechanical property, corrosion performance of the alloy. The effect of solid-solution treatment on corrosion and electrochemical mechanisms of Mg-15Y alloy in 3.5 wt.% NaCl solution was investigated by Zhang [6], the results indicated that the E_{corr} and corrosion rate of as-cast samples were both lower than those of solid solution-treated samples, and both increased with increment of solid solution-treated time and the corrosion resistance of Mg-15Y sample gradually deteriorated with immersion time increasing. Chen [7] was investigated the effect of heat treatment (T5, T6) on the biodegradable property of ZK60 alloy, the results showed that both the mechanical properties and degradation behavior were improved after T5 treatment due to the formation of small and uniformly distributed MgZn phases, but T6 treated samples were severely corroded due to the formation of large amounts of second phases accelerating the corrosion rate owing to the galvanic corrosion.

In the research, Mg-Zn-Ca alloys with different Ca content were prepared, and then studies the influence of solid-solution treatment on mechanical property and corrosion behavior of magnesium alloys.

2. Experimental

The magnesium alloys were prepared from pure magnesium (99.95%), high purity zinc (99.99%) and Mg-30%Ca intermediate alloy (mass fraction) through smelting in a graphite crucible in a resistance furnace protected by flux. The melt was held at 720°C for 10 min and casted at 690°C into a steel mold. The final compositions of Mg alloys are given in table 1.

The samples were cut from the as-cast ingot to solid-solution treatment, which were carried out at 445°C for 12h in the air with protected by carbon powders, and then quenched into water at room temperature.

Table 1. Chemical composition of the as-cast magnesium alloys in wt.%.

Design alloy	Analyzed composition (wt.%)							
	Zn	Ca	Fe	Al	Si	Cu	Mn	Mg
Mg-5.0Zn-0.5Ca	6.52	0.46	0.056	0.03	0.035	0.021	0.015	Bal
Mg-5.0Zn-1.0Ca	5.62	0.98	0.031	0.045	0.118	0.02	0.016	Bal
Mg-5.0Zn-2.0Ca	5.46	1.90	0.026	0.022	0.049	0.028	0.018	Bal

Microstructure observation was conducted on an Olympus GX41 optical microscope (OM) and a XQUANTA-200 scanning electronic microscope (SEM), and the phase constitutions were analyzed by a X-Ray diffraction (XRD, Bruker D8-Advance).The samples for mechanical testing look like dog-bone is shown as figure1, and tensile rate is 0.6 mm/min.

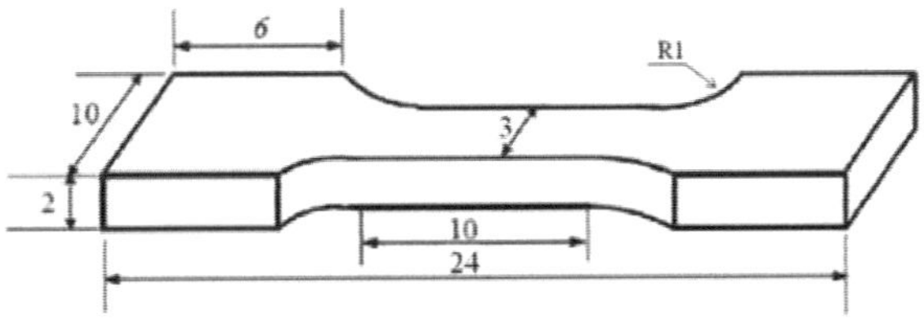

Figure 1. Schematic of tensile specimen.

The electrochemical testing was carried out using specimens with 1cm^2 of working area exposed in Hank's solution at 37°C. The measurement system using a three-electrode configuration of the Zennium electrochemical workstation, with a platinum electrode as the counter, a saturated calomel as a reference and the samples as the working electrode. For all measurement a voltage scan rate was 10 mV/s from -3V_{SCE} to -0.5V_{SCE}.

Table 2. Chemical composition of Hank's solution (g/L).

Compounds	Content
NaCl	8.00
KCl	0.40
$Na_2HPO_4.12H_2O$	0.12
KH_2PO_4	0.06
$CaCl_2$	0.14
$NaHCO_3$	0.35
$MgSO_4.7H_2O$	0.20

The immersion test was performed using samples with only one side of 1cm^2 exposed in Hank's solution at 37 °C, this working surface is ground with SiC papers up to 1200 grit, and then finely polished up to 0.5μm diamond powder liquid and cleaned by alcohol and dried in air. The samples were immersed in 120 ml Hank's solution after weighed, and Hank's solution is refreshed every 24 h. The samples are removed from Hank's solution and dried in dying box and is weighed after immersed for 24 h, 48 h and 72 h, respectively.

3. Results and discussions

3.1. Microstructure

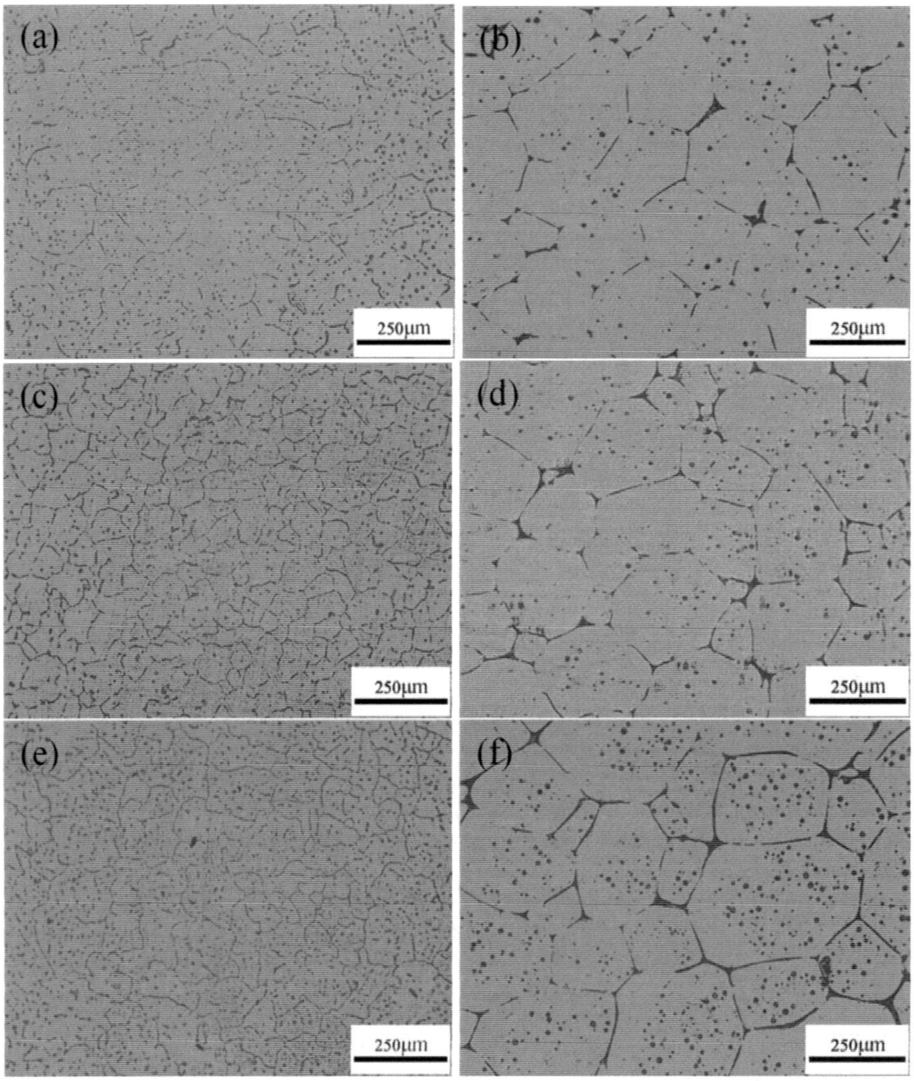

Figure 2. Optical micrograph (OM) of the magnesium alloys.
, OM of Mg-5Zn-0.5Ca alloy before(a) and after(b) solid solution treatment
, OM of Mg-5Zn-1.0Ca alloy before(c) and after(d) solid solution treatment
, OM of Mg-5Zn-2.0Ca alloy before(e) and after(f) solid solution treatment

The optical micrograph of the magnesium alloys are shown in figure 2. It can be seen that the microstructure of Mg alloys consisted of Mg matrix phase and the second phases, which distributed on grain interior and grain boundary. After solid-solution treatment, a lot of intragranular spherical secondary phases and the strip-like second phases at grain boundaries were dissolved into α-Mg matrix phase, the grains of Mg alloys obviously grow up and their size was significantly increased. The average grain size is increased from 104μm to 197μm for Mg-5Zn-0.5Ca alloy, from 86μm to 189μm for Mg-5Zn-1Ca alloy and from 65μm to 178μm for Mg-5Zn-2Ca alloy, respectively.

Figure 3 shows the X-ray diffraction patterns of as-cast Mg alloys. It can be seen that Mg alloys was mainly composed of α-Mg matrix phase and $Ca_2Mg_6Zn_3$ phases, and a new phase of Mg_2Ca was detected with the increase of Ca content. The Mg alloys after solution treatment at 445 °C for 12 h had similar phase constitutions to the corresponding as-cast Mg alloys, but the number of second phases are decreased quickly due to dissolved into the α-Mg matrix reached a supersaturated state.

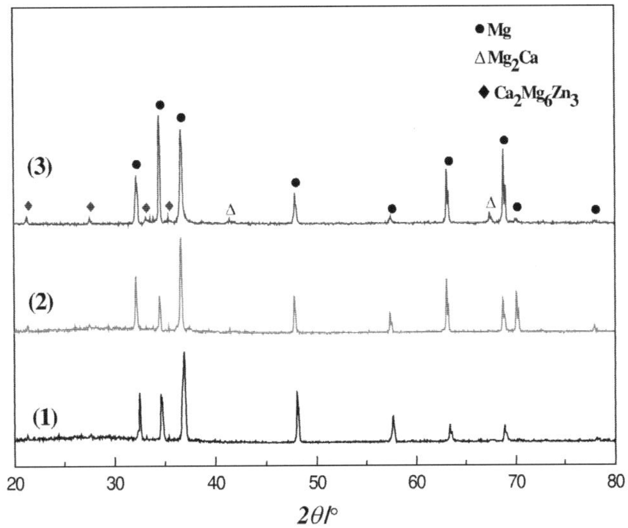

Figure 3. The XRD patterns of Mg alloys after solid solution treatment.
Mg-5Zn-0.5Ca (2) Mg-5Zn-1.0Ca (3) Mg-5Zn-2.0Ca

3.2. Mechanical properties

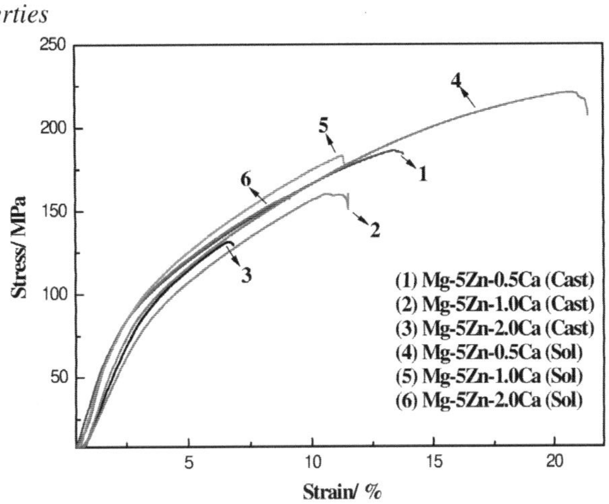

Figure 4. Tensile curves of the magnesium alloys at room temperature.

The stress-strain curves of the as-cast Mg alloys and after solid-solution treatment obtained from the tensile test are shown in figure 4. After solid-solution treatment, the yield strength (σ_{YS}) of magnesium alloys is slightly reduced compare with that of as-cast Mg alloys due to the grain size of Mg alloys is bigger than before; but the ultimate tensile strength (σ_{UTS}) and elongation (ε) of Mg alloys were significantly improved, the results are attributed to the microstructure of Mg alloys is more homogenous than that of as-cast Mg alloys and could be effectively decreased the defects in the cast process. In addition, the solid- solution treatment could improve the mechanical properties of as-cast Mg alloys through the second phases which reinforcement the α-Mg matrix phase during tensile test. It can also be seen from the table 3 that after solid-solution treatment σ_{UTS} of 220.7 MPa and ε of 16.6% are obtained in the Mg alloy of 0.5wt.% Ca, compared with those of the corresponding as-cast Mg alloy improved 18.5% and 66.7%, respectively.

Table 3. Data of Mg alloys for tensile test.

Alloy	Status	σ_{YS} / MPa	σ_{UTS} / MPa	ε/ %
Mg-5Zn-0.5Ca	As-cast	72.0	186.2	9.96
	Solid-Solution	70.1	220.7	16.6
Mg-5Zn-1.0Ca	As-cast	75.8	160.3	7.55
	Solid-Solution	69.3	183.1	8.3
Mg-5Zn-2.0Ca	As-cast	78.9	131.4	4.1
	Solid-Solution	71.2	158.3	6.56

3.3. Electrochemical results

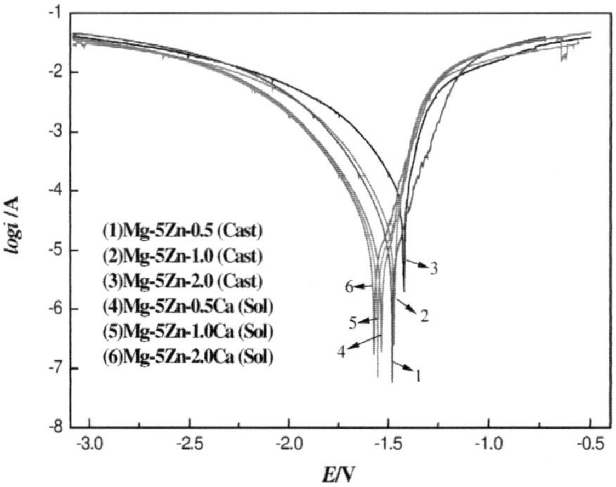

Figure 5. Tafel curves of magnesium alloys in electrochemical test.

The Tafel curves from electrochemical test of the magnesium alloys samples are shown in figure 5, and the corresponding electrochemical parameters are summarized in table 4. It can be seen that the corrosion potential (E_{corr}) of as-cast Mg alloy had a little changed, but significantly increase in the corrosion current density (I_{corr}) can be observed with increasing Ca content. The result shows that the corrosion resistance of as-cast Mg alloys is deteriorated with the increase of Ca. Compared with the as-cast Mg alloys, after solid solution treatment the corrosion potential of Mg alloys slightly shifted negative, but the corrosion current density is significantly decreased (show in table 4). With the addition of Ca, the number of second phases ($Ca_2Mg_6Zn_3$, Mg_2Ca) will increase, which beneficial to the formation of more micro-galvanic coupling between α-Mg matrix phase with the second phases. After solid- solution treatment the microstructure of Mg alloys is more homogenous than before and a large amount of second phases dissolved into the α-Mg matrix phase which can decrease probability and activity of micro-galvanic corrosion, resulting in significantly improved the corrosion resistance of Mg alloy.

Table 4. Data of magnesium alloys for electrochemical test.

Alloy	Ecorr/*V*		Icorr/($\mu A/cm^2$)	
	As-cast	Solid-Solution	As-cast	Solid-Solution
Mg-5Zn-0.5Ca	-1.421	-1.532	48.27	35.15
Mg-5Zn-1.0Ca	-1.467	-1.552	56.25	44.98
Mg-5Zn-2.0Ca	-1.479	-1.570	84.55	55.17

3.4. Electrochemical results

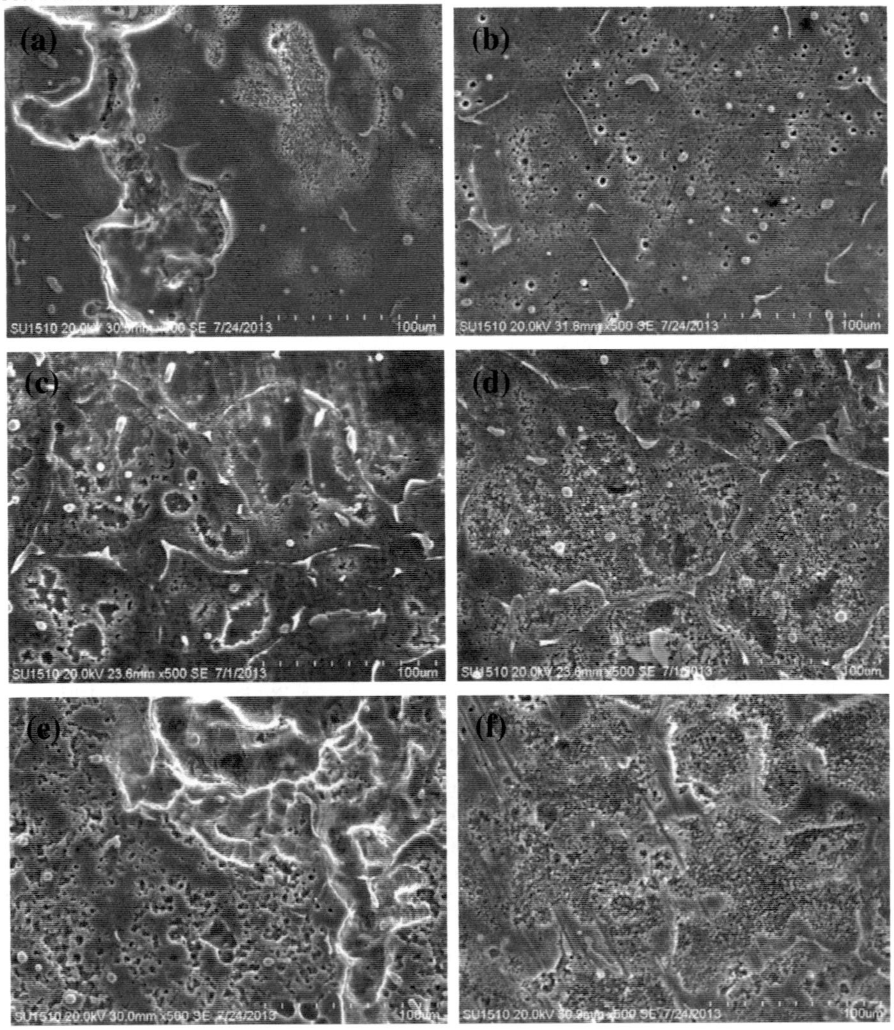

Figure 6. Surface morphologies of Mg alloys in immersion test for 3 days after cleaning the corrosion products. Morpholigies of as-cast(a) and solution treated(b) Mg-5Zn-0.5Ca alloy samples; , Morpholigies of as-cast(c) and solution treated(d) Mg-5Zn-1.0Ca alloy samples ;, Morpholigies of as-cast(e) and solution treated(f) Mg-5Zn-2.0Ca alloy samples

The surface morphologies of Mg alloys from immersion test which were carried out in Hank's solution at 37°C for 3 days after cleaning the corrosion products are shown in figure 6. It can be seen that the surface morphologies of Mg alloys have been appeared a lot of holes, which deduce that the corrosion characteristic of Mg alloys was started from pitting corrosion, which along surface direction and depth direction constantly grew up at the same time. With the increase of Ca content, the surface morphologies of Mg alloys observed more big holes, and the corrosion on the surface is more serious than that of low Ca content of Mg alloy as shown in figure 6 (a), (c), (e). After solid-solution treatment, the grain of as-cast Mg alloys is obviously grown up and a lot of the second phases dissolved into the α-Mg matrix, which beneficial to reduce the formation of micro-galvanic coupling and could be effectively impede the corrosion pits growth up during the immersion process (shown in figure 6(b), (d), (f)). In the word, the corrosion resistance of Mg alloys was improved by suitable solid solution treatment.

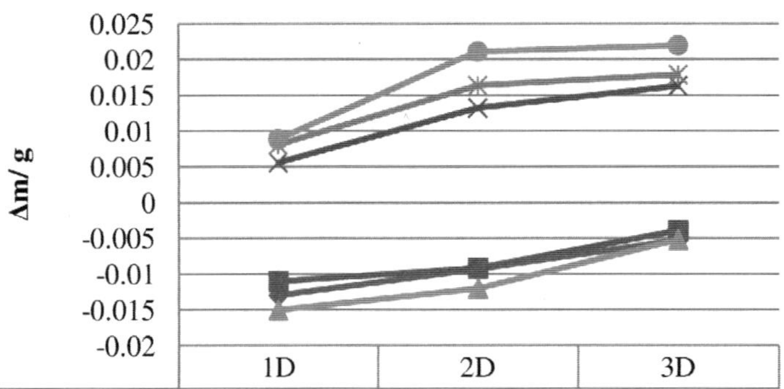

	1D	2D	3D
Mg-5Zn-0.5Ca(Solu)	-0.01301	-0.00932	-0.00532
Mg-5Zn-1.0Ca(Solu)	-0.01104	-0.0092	-0.00395
Mg-5Zn-2.0Ca(Solu)	-0.01503	-0.01203	-0.00525
Mg-5Zn-0.5Ca(Cast)	0.00558	0.01319	0.01627
Mg-5Zn-1.0Ca(Cast)	0.00806	0.01632	0.01782
Mg-5Zn-2.0Ca(Cast)	0.00883	0.02112	0.02197

Figure 7. Weight gain curve of the magnesium alloys.

Figure 7 shows that the weight-gain curves of Mg alloys after immersed in 37°C Hank's solution for different times. The corrosion of Mg alloys in the Hank's solution is combined result of forming the corrosion product film on the surface and a dissolution process [8]. It can be seen from the figure 7 that the weight-gain of Mg alloys increased with increasing immersion time, and the result of weight-gain of Mg alloys after solid-solution treatment is lower than that of as-cast Mg alloys. The FT-IR patterns of the precipitates on the surface of before and after solid-solution treatment of the Mg-5Zn-0.5Ca alloy in Hanks' solution for 3 days are show in figure 8, we can seen that the precipitates layer contains hydroxyl radical (OH^-) and phosphate radical (PO_4^{3-}) suggesting that the dissolved metal ions (Mg and Ca ions) combine with ions in the Hank's solution (OH^- and PO_4^{3-} ions), magnesium hydroxide is formed, Mg and Ca ions combine with the PO_4^{3-} forming Mg- and Ca- containing phosphates and HA, which to improve the biocompatibility of Mg alloy.

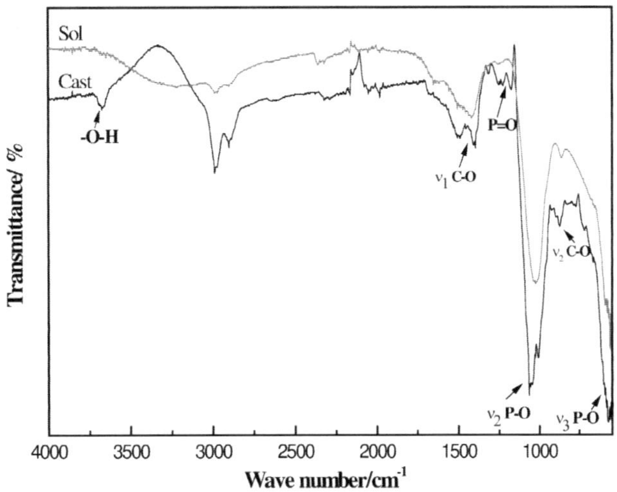

Figure 8. FT-IR pattern of the precipitates on the surface of before and after solid-solution treatment of the Mg-5Zn-0.5Ca alloy in Hanks' solution for 3 days.

4. Conclusions

The as-cast Mg alloys is mainly composed of α-Mg matrix and the $Ca_2Mg_6Zn_3$ phase, and a new phase of Mg_2Ca is appeared with increasing Ca content. After solid-solution treatment at 445°C for 12h, the Mg alloys had similar phase constitutions to the corresponding as-cast Mg alloys, but a large number of second phases dissolved into the α-Mg matrix reached a supersaturated state, the grain of Mg alloys obviously grew up and the average grain size is increased from 104μm to 197μm for Mg-5Zn-0.5Ca alloy, from 86μm to 189μm for Mg-5Zn-1Ca alloy and from 65μm to 178μm for Mg-5Zn-2Ca alloy, respectively.

After solid-solution treatment, the yield strength (σ_{YS}) of Mg alloys is slightly reduced compare with that of as-cast Mg alloys, but the ultimate tensile strength (σ_{UTS}) and elongation (ε) of Mg alloys were significantly improved; σ_{UTS} of 220.7 MPa and ε of 16.6% are obtained in the Mg alloy of 0.5wt.% Ca, compared with those of the corresponding as-cast Mg alloy improved 18.5% and 66.7%, respectively.

The corrosion resistance of Mg alloy is significantly improved by suitable solid-solution treatment. The precipitates layer on the surface of solid-solution treated Mg alloys after immersed in Hank's for 3 days contains Mg- and Ca- containing phosphates and HA, which have a good biocompatibility.

References

[1] Li N and Zheng Y 2013 *J. Mater. Sci. Technol.* **29** 489-502
[2] Staiger M P, Pietak A M, Huadmai J and Dias G 2006 *Biomaterials* **27** 1728-34
[3] Liu C L, Xin Y C, Tang G Y and Paul K.Chu 2007 *Mater. Sci. Eng.* A **456** 350-7
[4] Zhou W, Shen T and Aung N N 2010 *Corr. Sci.* **52** 1035-41
[5] Feliu Jr S, Samaniego A, Barranco V, El-Hadad A A, Llorente I and Adeva P 2014 *Corr. Sci.* **80** 461-72
[6] Zhang X, Zhang K, Li X G, Deng X, Li Y J, Ma M L and Shi Y 2012 *J. Rare Earths* **30** 1158-67
[7] Chen J, Tan L and Yang K, 2016 *Bioactive Mater*
[8] Liu D, Liu Y, Huang Y, Song R and Chen M 2008 *Prog. Natur. Sci. Mater. Inter.* **24** 452-7